While numerous historical works exist on specific East European cities, nothing has hitherto been published from an historical geographical standpoint on the spatial development of trade. *Trade and Urban Development in Poland* adds to this, and uses the experience of Cracow to illumine general patterns of trade and urban growth in central and eastern Europe over several centuries. Dr Carter emphasises the spatial aspects of commodity analysis during the later medieval and early modern periods, and traces the impact of political circumstance on commercial progress and mercantile evolution. He describes the regions and places of especial significance for Cracow's trade development, and examines the principal trading flows and commodity movements within the overall context of European economic and social change. Based upon an intensive analysis of primary sources, *Trade and Urban Development in Poland* breaks new ground in its examination of the impact of commerce on urban growth over the *longue durée*, and will make a major contribution to our understanding of the historical geography of central and eastern Europe.

Cambridge Studies in Historical Geography 20

TRADE AND URBAN DEVELOPMENT IN POLAND

Cambridge Studies in Historical Geography 20

Series editors:
ALAN R. H. BAKER, RICHARD DENNIS, DERYCK HOLDSWORTH

Cambridge Studies in Historical Geography encourages exploration of the philosophies, methodologies and techniques of historical geography and publishes the results of new research within all branches of the subject. It endeavours to secure the marriage of traditional scholarship with innovative approaches to problems and to sources, aiming in this way to provide a focus for the discipline and to contribute towards its development. The series is an international forum for publication in historical geography which also promotes contact with workers in cognate disciplines.

For a full list of titles in the series, please see end of book.

TRADE AND URBAN DEVELOPMENT IN POLAND

An economic geography of Cracow, from its origins to 1795

F. W. CARTER

Head of the Department of Social Sciences,
School of Slavonic and East European Studies,
University of London

CAMBRIDGE UNIVERSITY PRESS
Cambridge, New York, Melbourne, Madrid, Cape Town, Singapore, São Paulo

Cambridge University Press
The Edinburgh Building, Cambridge CB2 2RU, UK

Published in the United States of America by Cambridge University Press, New York

www.cambridge.org
Information on this title: www.cambridge.org/9780521412391

First published 1994
This digitally printed first paperback version 2006

A catalogue record for this publication is available from the British Library

Library of Congress Cataloguing in Publication data

Carter, Francis W.
Trade and urban development in Poland: an economic geography of Cracow,
from its origins to 1795 / F. W. Carter.
p. cm. – (Cambridge studies in historical geography: 20)
Includes bibliographical references.
ISBN 0 521 41239 0
1. Kraków (Poland) – Economic conditions. 2. Kraków (Poland) –
Commerce – History. I. Title. II. Series.
HC340.3.Z7K732 1993
330.9438′6–dc20 92–27971 CIP

ISBN-13 978-0-521-41239-1 hardback
ISBN-10 0-521-41239-0 hardback

ISBN-13 978-0-521-02438-9 paperback
ISBN-10 0-521-02438-2 paperback

For Krysia

Contents

Figures

Tables

Preface

This book has arisen out of my long-term interest in the cities of north- and south-east Europe. The method of approach is not that of the historian, but an historical geographer, who wishes to examine the wider spatial impact of a city through its trade links over an extended time period. In some ways the work acts as a counterbalance to my earlier research on Dubrovnik and its commercial contacts with the Balkans and eastern Mediterranean. A similar approach has been adopted here with Cracow, in a bid to extend our knowledge of trade and urban development in Poland and other parts of east-central Europe over several centuries.

It is not intended here to present English readers with a concise history of Cracow, a task better left to more qualified scholars. Neither is it primarily an attempt to portray the city's political history, economic development or cultural life, although such influences have been taken into consideration where necessary. Aspects of city life (daily problems, survival, etc.), while a growing topic of interest for urban historians, is only utilized here if it was thought to have some impact on the wider spatial consequences for Cracow's trading connections. The book is an effort to try and see one European city against the backcloth of the whole continent, interpreted through commercial activity funnelled through its emporium during the Late Middle Ages and Early Modern Period up to the end of the eighteenth century. It is hoped that such a study will provide insights into the changing spatial patterns of commodity analysis through time over parts of the European continent.

The first two chapters provide background information on the scope of the problem, resource material availability, and previous research on Cracow's historical geography. The city's early development up to the time it received its municipal charter in 1257 is discussed in the following chapter, while the next two chapters appraise the political situation and development of the city's European trade network during the Late Middle Ages up to the year 1500. This approach is adopted for the ensuing two chapters for the Early Modern Period as far as the Final Partition of Poland in 1795. These latter four chapters provide the bulk of the study, with greatest emphasis placed on the various commodities traded, their transit nature, and reasons for the demand and supply of individual goods during specific time periods. These factors are placed against a continental dimension to illuminate the changing vicissitudes related to political and economic events. The final chapter discusses the city's emergent spatial trade patterns on a European scale, in an effort to highlight those regions and places having a particular significance for Cracow. In sum, they reveal the results of an immense number of commercial judgements over time and space, which brought wealth to this city. In more contemporary terms, some remnants of this former trading glory may still be seen in Cracow's present urban landscape.

Inevitably a book of this size, the product of more than a decade's labour, will have its imperfections. The broad sweep of such an historical topic as the trade and development of Poland could fill many books; here it is hoped that at least some parts of that large canvas have been dealt with in sufficient detail to inspire further curiosity amongst its readers to pursue the topic elsewhere.

It is hoped, too, that the complexities of place-name evidence will not deter the English reader; wherever possible I have tried to utilize known English equivalents, usually only applicable to the largest urban centres (e.g. Warsaw, Prague). Elsewhere I have adopted contemporary place-name usage wherever possible. The minefield of place-name evidence in this part of the world, where some towns and cities can have three or more possible forms depending on the historical time period, is well known. I have tried to standardize them as much as possible, but some may have slipped through the net. What follows is my sole responsibility, but the outcome could never have been achieved without the help of numerous friends, colleagues, teachers and others too numerous to mention. This is the result of their efforts as well as my own.

Acknowledgements

I should like to take this opportunity of expressing my sincere thanks to all those colleagues, both in Britain and Poland, who have given their invaluable help and assistance in the evolution of this work. Financial support was made available through the receipt of several University of London, Hayter, Travel Grants enabling me to make many visits to Cracow, and a stipend from the Institute of Geography at the Jagiellonian University during my stay in that city.

In Poland, I was indeed fortunate in having had the advice and encouragement of Professor M. Klimaszewski, and Drs Bromek, Kortus and Groch of the Institute of Geography in Cracow, who freely placed their goodwill, knowledge and help at my disposal. A special appreciation is owed to the late Professor W. M. Bartel, of the History of Law Department at the Jagiellonian University, to whom I am indebted for his constant encouragement and enthusiasm for my research topic, and to Professor J. M. Małecki, of the Department of Economic History at the same university, for his careful reading of the text and helpful suggestions. Moreover, Professors Podraza, Bieniarzówna, and Kieniewicz (Department of History, Jagiellonian University) and the late Dr K. Pieradzka, provided useful comments and fortitude during my stay in Cracow. In addition, I wish to thank officials at the various archival, library and cartographic services, which I consulted, who spared time and effort to assist me.

In England, a debt of gratitude is owed to my colleagues in the Department of Geography at University College London, including Prof. W. R. Mead, Drs R. A. French, H. C. Prince, and R. J. Dennis for their valuable criticism and academic expertise along with Dr P. Nunn, Mr W. J. Campbell and Mr P. R. Schooling for their help over technical problems in graphic presentation and spatial analysis. The many illustrations, maps and plans were drawn and photographed by the Cartographic Unit at University College London; for applying design expertise and helpful counsel the assistance of Mr K. Wass, Mr A. Newman, Miss A. Mason, and Miss L.

Saunders are gratefully acknowledged. Mr C. Cromarty reduced and printed the illustrations and Miss A. Swindells typed the final manuscript. I would also like to thank Cambridge University Press, with its long legacy of making Poland and its history accessible to English readers, and especially Mr R. Fisher for his organizational proficiency. To each of them I am greatly indebted for their various skills. Finally, I wish to express my appreciation and deep gratitude to my wife, Krystyna, for her constant help and encouragement. To them, and all those others who gave generously of their time and energy, I offer my deepfelt thanks.

1
Introduction: Cracow in context

Cracow, ancient capital of Poland, today represents a lively industrial, scientific and cultural centre. It is a city of priceless relics and monuments to Poland's national culture which miraculously have survived the disasters of war; the oldest documentary sources date from the mid-tenth century, and describe the city as an important commercial centre on the Vistula river, at the junction of several great international trade routes leading from Southern Europe to the Baltic Sea, and from the countries of the Near East, via Silesia to Western Europe.

Whilst it would be wrong to overemphasize the role of geographical conditions in the development of this city, it should be remembered that they played a significant part in Cracow's destiny through their influence on human activity both within Poland and against the wider background of the European continent. The city was surrounded by hills, of which on one limestone outcrop named Wawel its first settlement was located, seventy-five feet above the large river Vistula. From here the outlying countryside could be observed, rich in natural resources and with its fertile soil suitable for grain production; likewise fruit and vegetable growing was also pursued whilst the large forests were full of game. Timber was plentiful, as was clay and building stone. From the city's early days iron-, lead-, silver- and even gold-ore were exploited locally, together with rock salt from nearby Wieliczka, a commodity always high in demand.

All these initial advantages contributed to a rapid growth of Cracow in the Early Middle Ages, when it became one of the residential seats of the Polish kings under the Piast dynasty. One of these, King Kazimierz I Odnowiciel (Casimir the Restorer) 1034–58, transferred his royal residence from Gniezno in Greater Poland (Wielkopolska) to Cracow, transforming a place noted for its castle and cathedral both of which had been built on Wawel Hill in the eleventh century, into the capital of Poland, a title it was to retain until the early seventeenth century.

Cracow within Poland

Amongst the various epithets lauded on this city one of the most famous is 'Cracovia, totius Poloniae urbs celeberrima' (Cracow, Poland's most celebrated city). Certainly during the twelfth century Cracow was the capital of a Poland which gained itself a powerful base in central Europe, in spite of strong competition from the grand dukes of Kiev and the Bohemian kings. The country's boundaries were successfully established between the Oder and Vistula rivers, in spite of internal fragmentation which divided the country into separate dukedoms on the death of King Boleslaw III Krzywousty (Wrymouthed) in 1138. Unfortunately, his efforts to maintain Cracow as the leading seat of Polish authority failed and the country fell into a period of disunity and political strife. Added to this misfortune was the terror of a Tartar attack on the city in 1241 which destroyed most of the settlement's wooden structures, leaving only stone edifices to survive. Nevertheless, with the unification of Poland in 1320 the country entered a period of considerable prosperity, which naturally enhanced Cracow's position.

The emergence of urban settlements in the Polish lands is veiled in the mists of time but by the ninth century there was clear evidence that trading-handicraft quarters in them had been developed. In the southern part of the country in Little Poland (Małopolska) geographical factors played an important role in the growth of the urban network, none more so than around Cracow where three physical regions (plateaux of Little Poland, the sub-Carpathian basins and foothills of the Carpathian mountains) merged. Other towns, perhaps less fortunate geographically than Cracow, were also emerging in Little Poland, including Przemyśl, Sandomierz, Opole, Wiślica and father north-east Lublin.[1]

In other parts of the country too, rudimentary types of towns were developing as evidenced by archaeological research in the deeper cultural layers of places like Gniezno, Poznań, and Wrocław where finds have indicated signs of non-agricultural craft occupations and trade exchange.[2] They were also often associated with a strongly fortified site nearby. An urban settlement pattern was clearly emerging during the tenth and eleventh centuries with towns having their own hinterland areas; Cracow, Sandomierz and probably Wiślica in Little Poland, Wrocław, Opole, Głogów and Legnica in Silesia, Gniezno, Poznań and Kalisz in Greater Poland, Kruszwica and Włocławek in Kujavia, Łęczyca and Sieradz in central Poland, Płock in Mazovia, and Gdańsk, Wolin and Kołobrzeg in western Pomerania (Fig. 1).[3]

By the reign of Bolesław Krzywousty (1102–38) signs of an urban hierarchy had appeared. At the top, larger urban settlements performing roles as regional capitals and important economic centres included places like Cracow, Wrocław, Wolin, Gniezno and Kołobrzeg with population

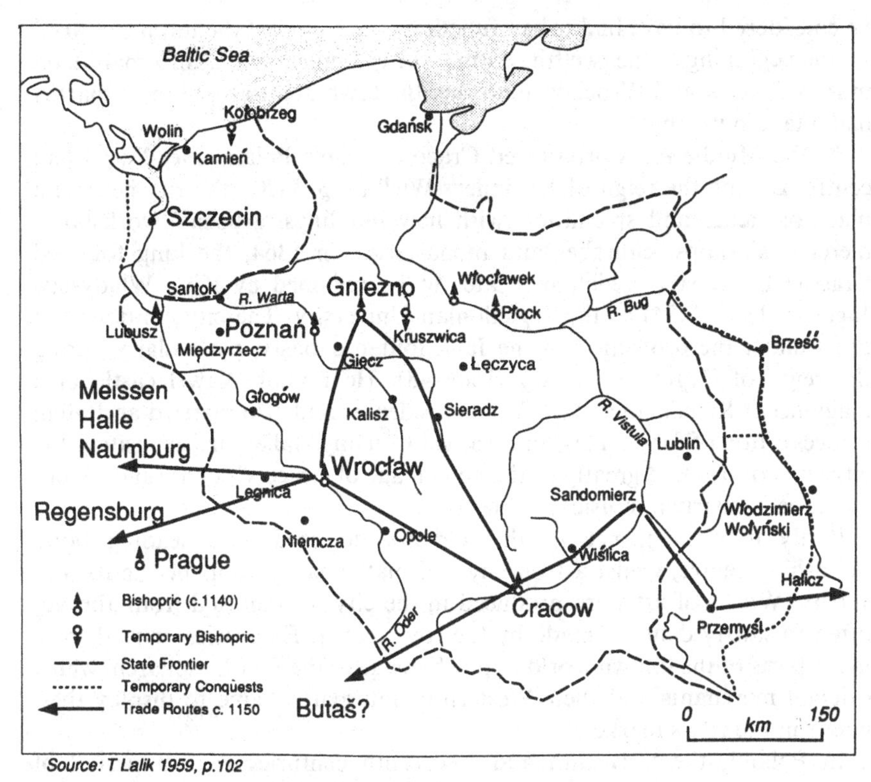

Source: T Lalik 1959, p.102

1. Poland: Towns from 10th–12th centuries.

totals of between 4,000 and 5,000 inhabitants. Below them were settlements with between 1,000 and 3,000 people including Przemyśl, Opole, Kruszwica, Legnica, Głogów, Poznań, Włocławek, Sandomierz, Płock and Szczecin. A further twenty settlements contained up to a thousand inhabitants (e.g., Lublin, Sieradz, Łęczyca, Kalisz), and there were over 250 small rural market centres.[4]

In the past much debate amongst Polish academics centred around the pre-urban/urban phase of the country's settlements. Some historians, and particularly legal historians, tended to favour the view that a place only became a town on obtaining its 'town rights', whereby rural centres were upgraded according to when they adopted the code of either Magdeburg or Lübeck Law. This restricted many of them to a date of origin from around the mid-thirteenth century, coinciding with a phase of German immigration into Poland at that time. More recent archaeological research has tended to refute this view, with proof that economic significance within the settlement hierarchy could also be utilized as a criterion for urban status.[5] Thus, Cracow, which did not receive its code of Magdeburg Law until 1257, could be considered to have had urban functions since at least the ninth century.[6] By the beginning of the twelfth century, for example, there is information on inns in Cracow and Wrocław and often the town abattoir and meat market had a tavern nearby.[7]

As the Middle Ages progressed Cracow became Poland's leading urban centre. During the reign of Kazimierz Wielki (1333–70), the city became a place of riches and splendour, with new buildings, artisans' workshops, merchants' shops, churches and monasteries; in 1364, the king founded Cracow University, which was later to be renamed by King Władysław Jagiełło (1386–1434) as the Jagiellonian University. The city continued to grow under the protection of the Jagiellonian dynasty, particularly during the reign of Zygmunt I Stary (1506–48). He rebuilt Wawel castle in a magnificent Renaissance style for his residence, and also married an Italian princess, Bona Sforza. This, in turn, led to many Italian artists visiting the city and contributed greatly to the golden age of Cracow's cultural development, particularly in music and literature.

Partly in consequence of this Cracow emerged as a leading book publishing centre, whilst astronomy and mathematics flourished in its university. Works of art were produced in the city or imported from abroad, often to satisfy demand made by the royal court. Emissaries arrived from many parts of the known world to pay homage to the Polish monarch, whilst oriental merchants and their Western counterparts, came to display their goods in the city's market.

In Poland, the thirteenth and fourteenth centuries were a period of economic prosperity. This was evidenced by the reform of the country's constitution, increase in the urban network, development of monumental architecture and colonization of the forested areas, which all contributed to a denser population distribution. Estimates for this have been based on the analysis of the 'Peter's pence' income registers between 1318 and 1359, a very active period of tax collection for the Holy See.[8] In a country thought to have around 940,000 inhabitants during the reign of Kazimierz Wielki, evidence suggests that on average small Silesian parishes had about 400 inhabitants, whilst in Little Poland it was between 700 and 800, with small towns having around 500 citizens.

The average density for Little Poland and Mazovia, both noted forested areas, ranged from 6.6 to 6.8 inhabitants per square kilometre; the most crowded part was in the Löess region containing Cracow, Wiślica and Sandomierz, with nearly 40 percent of Little Poland's population living in the environs of the capital. This suggests the attraction of residing close to such an important military and commercial centre, and may have been one of the reasons for the transfer of the Polish capital from Greater Poland to the Upper Vistula valley in an earlier century. The 'Peters' pence' registers allow estimates to be made of the 56 urban settlements in the Cracow diocese for 1340; the capital clearly stood out from the rest with 12,000 inhabitants (13,000–14,000 if its close suburb of Kazimierz is added), with the largest of the other urban centres only having between 4,000 and 10,000 residents. Despite the fairly reliable nature of these figures, efforts to estimate later years are full of imprecision. The ravages of fire and plague took their toll, but it is thought that by 1500 there were around 14,000 inhabitants in the city rising to 18,000 by the mid-sixteenth century.[9]

After 1609, Cracow's importance diminished when Warsaw was declared the new capital of Poland by King Zygmunt III Waza (1587–1632), head of a new dynastic line of the Vasa (Swedish) family. The seventeenth century was an unfortunate era for Poland, characterized by wars which were disastrous for the country, and the continued development of Cracow in particular. The mid-century was an especially bad period for the city with the Swedish occupation, which not only led to local devastation, but also the loss of many art treasures ruined or stolen by the invading armies. Cracow's population declined to about 10,000 inhabitants during the eighteenth century, which like the rest of the Polish state entered a period of gradual decay, particularly after the First Partition in 1772.[10] Such a division of the country had unfortunate implications for Cracow, only to be followed in 1795 by the Final Partition of Poland; then the city was captured and occupied by Austrian troops and eventually became part of the Austro-Hungarian Monarchy.[11]

Cracow within Europe

According to Bairoch, the years between 1000 and 1340 saw the establishment of Europe's urban pattern which was to last fundamentally until the nineteenth century.[12] He has calculated that of the 182 cities with 20,000 plus inhabitants in 1800, three-quarters of them in 1300 already had 5,000 or more residents and a further eighth, between 2,000 and 5,000. Thus Cracow, with its estimated 12,000 inhabitants in 1340, clearly represented an urban centre within the European spectrum by the beginning of the fourteenth century. Furthermore, Bairoch maintains that places in excess of 8,000–12000 inhabitants which Cracow probably just had, at that time, could be classified as large cities.

Although one may argue that Cracow lay on the margins of this division from a population size viewpoint, other criteria such as function, places it firmly in the larger city category. In its early years it took on the mantle of a religious centre for, as Davies has stated, 'Kraków is without parallel in Europe as a city which, from its origin, has been dominated by the Royal Court and the Cathedral'.[13] Certainly, the impact of Christianity, which arrived at this Slavic stronghold by way of Vienna, came when Prince Mieszko I (960–92) accepted Catholic baptism in 965. The Christian faith spread throughout the region, enabling his successor King Bolesław Chrobry, 992–1025, (Boleslaus the Brave) to establish a bishopric in Cracow at the beginning of the eleventh century, and establish unity amongst the Slav tribes. By the sixteenth century the Jewish ghetto was also well established on the southern outskirts of the city in the suburb of Kazimierz (founded 1335), where the Synagogue remained outside the city walls.

Cracow also had a commercial function, not only at local or regional/national level, but as a city where international trade played a decisive part in its economy. As it is hoped to show in later chapters of this book, Cracow was an intermediary city, where merchandise flowed from Western to Eastern Europe and vice-versa, as well as beyond the confines of the continent to the Near and Middle East. Within Europe it lay on the international trade route from Kiev to Prague, Vienna and Cologne, and, whilst not a maritime city, was a member of the Hanseatic League with its links between the Baltic and North Sea. Predominantly, however, as a commercial city of the continental interior, it was on a vital crossroads not only between the North and South of Europe, but also the continent's eastern and western margins. By the seventeenth century the rise of oceanic commerce and the decline of overland trade routes to Asia took their toll on the city's commercial significance, but it still retained some importance at the national and regional level.

As the nation's capital, Cracow also had a significant administrative function, particularly as the seat of the Royal Court. It was to Cracow that various dignitaries and emissaries from foreign states came, enhancing its reputation and bringing new ideas and innovations from other parts of Europe. The court also generated its fair share of administrative duties associated with the running of the state, promulgation of its laws and controlling finances. The organisation of regional and local administration have to be added to this list, as well as a military presence to ensure the safety of the monarch. With the transference of these capital-city functions to Warsaw at the beginning of the seventeenth century, Cracow lost the political impetus derived from its former status. Thus some of the wider European implications associated with this role were to be lost forever.

Another function suggesting Cracow's urban prominence within the European scene was that played by its university. Thirteenth- and fourteenth-century Europe experienced a blossoming of centres for higher learning of which Cracow was no exception, joining some seventy-odd other European cities with one or more universities by the end of the fifteenth century.[14] During the fifteenth century over 18,000 students were admitted to Cracow University from all strata of society and various European countries.[15] Graduates and teachers from Cracow travelled on to Hungary, Bohemia, Moravia, Germany, Austria and Italy, including the University of Bologna, where, between 1448 and 1480, seven of Cracow's former students filled the Chair of Mathematics. Perhaps Cracow University's most famous son entered the Faculty of Philosophy in 1491 for a four-year grounding in contemporary astronomy and mathematics – Mikołaj Kopernik (1473–1543) – better known as Nicolaus Copernicus propounder of the heliocentric theory of the universe. His residence in Cracow also coincided with a period of active research in geography at the university, which analysed earlier geographical literature, devised new teaching methods, followed an interest in the voyages of discovery, and studied the geography/ethnography of Eastern Europe (e.g. Maciej Miechowita, 1457–1523).[16] Finally, it should not be forgotten that Cracow possessed other levels of education with the oldest cathedral school having been founded on Wawel hill in 1110, and the first secular secondary school established in 1588.

It may also be argued that Cracow possessed a cultural function as the leading city in Poland. Evidence of Romanesque architecture is best seen in surviving remnants of the second Wawel cathedral (*c*. 1075–1142), where St Leonard's Crypt exhibits certain Rhineland influences, whilst the Benedictine abbey of Tyniec (later eleventh century) near Cracow was equally inspired by models from distant European art centres, as was St Andrew's Church (post-1086) in the city. The Cistercian Order in Little Poland from the thirteenth century, had direct contacts with Burgundy and artistic links with Italy and Germany, as reflected in the Early Gothic of Mogiła Abbey (*c*. 1250), outside Cracow, whilst more general Gothic architecture is found in the cathedral with a mixture of Cistercian and Northern French traditions. A combination of Late Gothic and Renaissance styles are found in Cracow University, whilst part of the city's fortifications against possible Turkish attack are evident in the monumental Barbican building (1498–9). True Renaissance style was imported from Florence for the Wawel by 1500, with the work of Fiorentino and Berecci, and the later Lombard version seen in the rebuilt Cloth Hall (Sukiennice) begun in 1556. With the coming of the Counter Reformation after 1574, Cracow's architects and stonemasons first adopted forms from Italian Mannerism, to be followed in the early seventeenth century by the more austere style of the Early Baroque, so beloved of the Vasa dynasty. St Anne's Church in Cracow contains High Baroque styles from later in that century, while some palatial town houses exhibit evidence of Baroque Classicism.[17]

Besides architecture, other evidence suggests that Cracow occupies an outstanding position in the history of Polish art. Painting, sculpture and other artistic work reached its zenith in the fifteenth century, when a large number of foreign artists, particularly from Nuremburg, lived in the city. Much of the initial credit for this must be given to Wit Stwosz (1438–1533), of unknown nationality who lived part of his life in Cracow, and whose greatest contribution is the altar piece in St Mary's Church, carved between 1477–1489. Cracow's own school of painting was under the early influence of

Italian, Flemish, German/Bohemian, Russian and Byzantine styles, but by the end of the fifteenth century more original works free of these traits emerged. Even the unfavourable economic, political and cultural situation of Poland after the beginning of the eighteenth century did not deter Cracow's art movement, particularly after a School of Art was established at Cracow University.

Cracow was a noted centre for printing; the first work, a wall calendar, dates from 1473, whilst two years later the first book printed in Poland (*Explanatio in psalterium*, by Jan de Turrecrematus) was issued. In 1513 the first book in Polish called *Soul's Paradise* by Biernat was printed in the city, and thirty years later literary works by Rej (*Krótka rozprawa między trzema osobami Panem, Wójtem a Plebanem* – A short dialogue between Lord, Bailiff and Parish Priest), Modrzewski (*O karze za mężobójstwo* – Penalty for Genocide), and Orzechowski (*Fidelius subditus*) appeared. The university library already contained 20,000 volumes by 1630, and thirty years later the first periodical in Poland was published – *Merkuriusz Polski*.

Finally this leads us to the question of whether Cracow was a significant European city?[18] Certainly the above functions of the city's past suggest it ought to have been, together with contemporary UNESCO recognition that its historical heritage conservation is of the highest priority. Yet an important city is not always well-planned, beautiful, or a provider of gracious living; in the final analysis a city is only as great as its human associations allow, and it is highly probable that most European states have had only one great city at any one time period. Obviously the necessary requirements for such a city must include some guarantee of at least a tolerable life, and should never be a dull place in which to reside. It has to have some initial attractions, possibly as a significant place of early worship, coupled with strategic considerations and a role as a centre of trade. In the final consideration one is searching, by delving into a city's past, for some elusive complex ingredient which may help to explain the basis for that continental approval. Here attempts are made in a case study of Cracow, to try and discover at least some of the strands which contributed to the development of that complex urban process, eventually leading historically to European, and more recently, world recognition as a city of considerable value and significance. This study, covering a period of several centuries, concentrates on its commercial links with the rest of the continent and beyond, in the hope that at least a small part of that elusive ingredient may be better understood.

Focus of the study

This book studies the spatial extent of Cracow's trading influence over time and the underlying causes of its changing commercial fortunes, not only within the boundaries of Poland, but also in a wider context of the European continent and beyond. A perusal of the relevant literature suggests that few studies have been made from a spatial viewpoint on the commercial links of cities over a long time span, or on the ramifications such activity had on the development of a city like Cracow. There is, however, plenty of more general evidence that the extent of the market and the scale of production tended to reinforce each other, whilst greater efficiency allowed a city to sell over a larger area; in turn this provided sufficient demand to permit a profitable division of labour.[19]

Nevertheless, the general idea that there was a gradual development from local to long-distance commercial activity is a concept which appears too precise and definitive though logically sound and supported by historical evidence. Important cities also sprang up on the basis of long-distance trade and, right from their inception, their economy was dependent to a great degree on specialization, contacts with distant markets, and large-scale financial complexity. Merchant initiative, for example, could generate certain types of craft production, rather than having to rely on artisans, with limited trading capabilities, selling their own goods which were surplus to local needs. Furthermore, close ties with the surrounding local region could, in some cases, also act as a brake on development: more often than not a lot of energy was spent by the urban oligarchy trying to forge advantageous links with the local landed aristocracy and the purchase of land, rather than encouraging an exchange of food for goods in the local market.

Pre-classical economies gave deep insights into the role of cities and their hierarchies; classical economics has been less successful in this field. Space was conceived by followers of the later doctrine as a production factor and an obstacle to the mobility of economic goods; the aim of the classical economists was to explain the opposition which started between cities and the rural world, but not the role of geographical location. Modern theory puts more emphasis on communication costs and optimal relation networks, which created rather a synthetic approach to urban forms, functions and networks.[20] Certainly exchanges offered by cities to their regions were not just economic; from early times, cities have been seen as centres of civilization, producers not only of goods, but also of cultural models. For example, the diffusion of Christianity throughout the Roman Empire in the fourth century AD took place largely through the conversion of the urban middle and lower classes; similarly the Roman Catholic church became much more strongly urban orientated during the Middle Ages, whilst monastic buildings originally located in peripheral rural areas were gradually engulfed by urban expansion.

Nevertheless, Cracow was a centre from at least as early as the tenth century, not only of local/regional trade but also long distance commerce. The existence of trade with east-central Europe in the three centuries between the earlier Scandinavian expeditions and the rise of the Hanse,

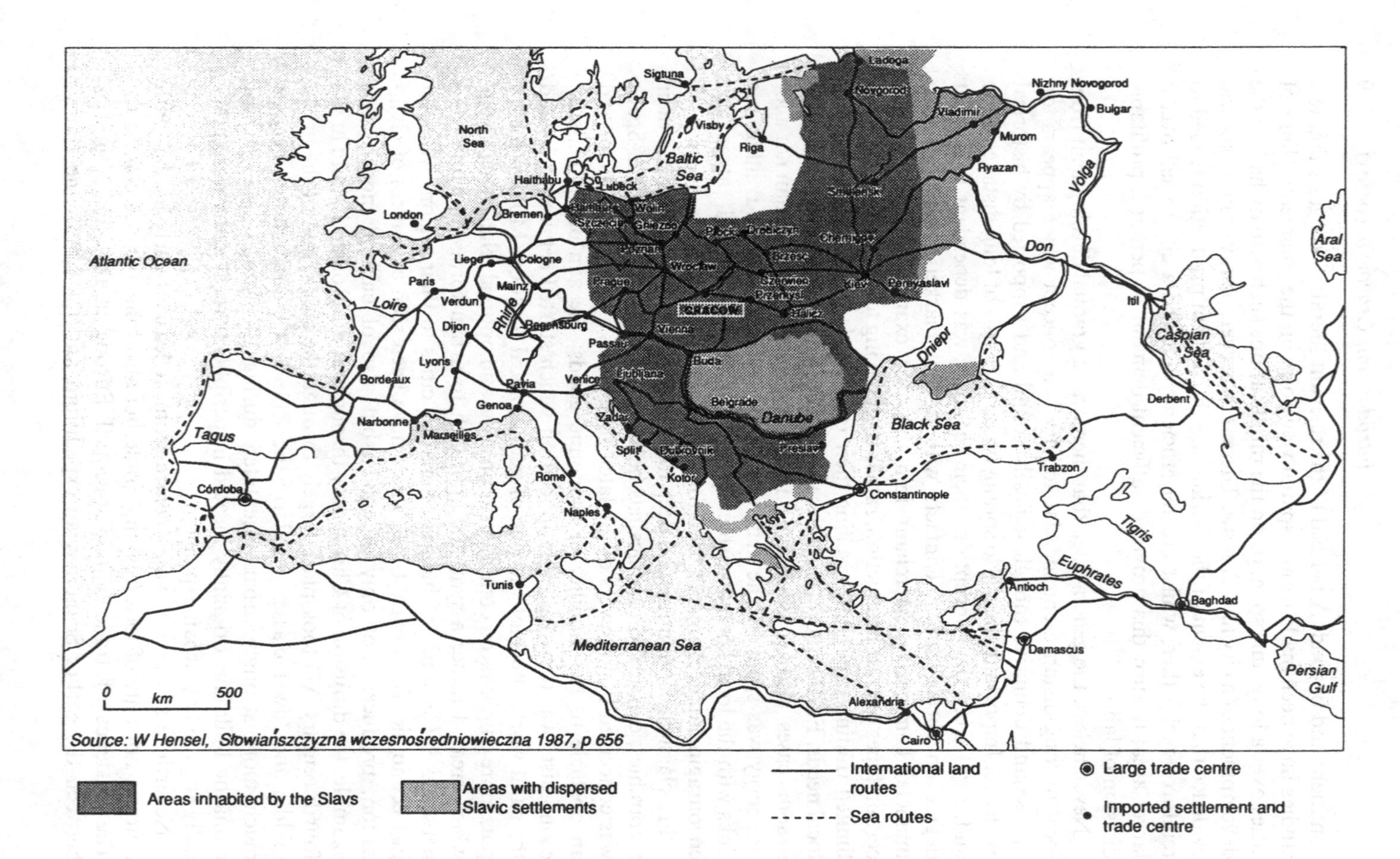

2. International trade routes across Europe c.800–1100 AD.

suggests that merchants arrived in the region to exchange goods from distant countries, with Slavonic merchants from Bohemia and Russia wishing to barter with them for slaves, horses and wine. This evidence corresponds with that given by a later traveller, Ibrahim Ibn Jakub, about the Russian and Slavic merchants who came to Cracow and Prague in the mid-tenth century. The road from Mainz ran through Bohemia and Poland and, in 965, Prague was visited by the above Spanish traveller, who left a description of Bohemia and the neighbouring lands. Whilst Prague was described as a rich city by reason of its trade, he also referred to Russians and Slavs coming to Prague from Cracow with their wares, whilst Moslems, Jews and Turks (the latter from Chezaria and the Crimea) arrived with various types of furs.[21] It seems fairly safe to assume, therefore, that in the tenth and early eleventh centuries, an important trade route ran from Mainz to Kiev via Regensburg, Prague, *Cracow*, and Przemyśl. This was a period when long-distance trade was beginning to intensify interconnections between the east, north, south and west of the European continent, and was dependent on commercial links by its cities with distant parts of the known world. It is important to note that Cracow was on that commercial network, albeit perhaps not as significant as Prague, Venice etc., but at least playing a secondary trading role in this early period of commercial development (Fig. 2).

Even the broad features of Cracow's contemporary commercial pattern will be unfamiliar to many people; moreover, as is commonly the case with the geography of a complex economic urban unit, the present makes little sense until it is related to the evolutionary process which has produced it. This fact alone provides some justification for this study. The problem therefore arises as to how far it is necessary to revert into the past to provide an adequate explanation of the present? Through its wide commercial activity, Cracow managed to develop and prosper as a commercial emporium from the Later Middle Ages until the late eighteenth century. It gradually built up great wealth, amassed mainly from its trading activities and was known from the Appenine Peninsula to the Black Sea in the south, and the Baltic and North Seas in the north.

While certain precise dates have been chosen for the major chapter headings, explanatory forays into the years before it became a great trading emporium have been included, for one must remember that commercial prosperity is an on-going process, with a need for some explanations and reflections prior to the main period of study. Therefore, some outline of Cracow's early development has been included, to examine initial trading links and evidence of commercial contacts up to the mid-thirteenth century. In 1257, Cracow received its Municipal Charter under Magdeburg Law, which signalled its acceptance as an important urban centre of that time, and seems a significant, logical date for the real commencement of its commercial development.

By the High Middle Ages hundreds of cities were located throughout much of Central Europe, serving as centres of commercial, cultural and religious activity.[22] All these cities had certain factors in common; they all conformed to a basic physical pattern with each city containing a high density of dwellings and public buildings which were usually confined by a high wall and/or a complex system of fortifications. Each had distinct forms of economic activity such as regular markets and a particular form of economic organization, such as the guild system. These cities also had a well-defined class pattern which included the citizen or 'burger' group with a superior status clearly separating it from the ordinary urban/artisan class, and more pointedly from the mass of rural inhabitants. Furthermore, each city had a governmental system based on rule by an oligarchic elected city council, as opposed to the alternative form of hereditary noble aristocracy. Thus cities like Cracow had not only the general criteria of size and function, but also possessed particular privileges based on self-protection and self-administration, which set them apart from the ordinary villages of the surrounding area or region. These privileges were granted and confirmed by feudal rulers, often in the form of charters and, like Cracow's Municipal Charter, were always looked upon as the most significant and treasured documents in the possession of the urban community.

Cracow's political situation between 1257 and 1500 is discussed so that a better understanding may be had of the various implications this had on trading links. The city's importance as a royal capital was extremely significant; from its early origins, Cracow had been dominated by its royal court and cathedral. During the thirteenth century fears had waned of spreading German colonization, but were followed by the Tartar raids, a terrible scourge which began to threaten from the eastern steppes and which nearly wiped out a large proportion of southern Poland's population; as a result of this menace Cracow took over the former role of Poznań as the defensive capital and bastion of Western freedom from the nearby frontier of danger. Fortunately, the threat of Tartar raids gradually diminished, partly due to their inability to adapt to the forest environment of this part of Europe and to their mode of life and military tactics. Political relations with other Polish towns were also relevant for Cracow's growing emporium along with political links afforded by neighbouring states in Hungary, Silesia, Bohemia and Prussia. Merchants from these states came at the invitation and under the protection of the Polish king, attracted by the defensive nature of the city dominated as it was by the royal castle.

The year 1500 has been taken as an important benchmark in this study. This, from a Polish viewpoint, may not seem a logical break, for the city's development of its so-called 'Golden Era' of commercial prosperity extended at least until the mid-sixteenth century. However, as Bushkovitch has clearly pointed out, 'the question of the rise and decline of the Polish cities cannot be considered only within the boundaries of Polish history'.[23] One must take into account consideration of economic development in Central Europe as a whole and, further, the relations of Central Europe with England and Holland. Perhaps in this case study, where Cracow is being analyzed against the background of the whole European continent (and beyond), then dates more applicable to general commercial movements at a continental level are necessary. The last years of the fifteenth and the first of the sixteenth centuries were to prove a turning point in the commercial life of Europe. During the sixteenth and seventeenth centuries the political, commercial and cultural relations of Europe became, for the first time in history, world-wide in extent. The Early Modern Period was to see not only a new sea route with the East brought into use, thanks to the discoveries of the Portuguese navigators, but also the discoveries of Columbus and the Cabots which opened up parts of the American continent to European colonization, exploitation and trade.

In the Middle Ages, European commerce developed, in the Baltic and North Seas, new regions of activity which were bound up by land and sea routes with the centres of the Mediterranean culture. With the discovery of new ocean routes both to the East and the New World, there was in effect a re-orientation of European commerce. These two events changed the geographical 'values' of European lands. The Baltic and Mediterranean lost their former centrality and supremacy in European trade, but nevertheless continued to play an active part in European commerce.

Cracow, situated on the borders between east and west, also had to take account of events in eastern Europe. Whilst the Fall of Constantinople had been a significant event for Europe in 1453, the year 1526 was a momentous one for Christendom. The Turkish wars with Hungary had been going on intermittently for years, but the success of Suleiman the Magnificent (1520–66) in the great battle of Mohács in 1526 saw the total defeat of the Hungarians, a factor not unnoticed by Cracow's merchants. This event, together with the fall of Moldavia in 1504, and further Ottoman successes in and around the Crimea, meant declining access for Cracow's trade interests in the Black Sea region at the beginning of the Early Modern Period.

Finally the year 1795 was taken as an obvious concluding point for this study. In spite of Tadeusz Kościuszko's historical oath for freedom on 24 March 1794 in Cracow's Market Square, as Commander-in-Chief of the Polish Insurrection, his brave demands were to no avail and within a year Poland ceased to be part of the map of Europe. This was a significant date not only for Poland, but the whole of Europe; it signalled the emergence of three imperial powers, Russia, Prussia and Austria over a large territory of the European continent, which for so long had been the backcloth for Cracow's commercial activity. After the Partition of Poland in 1795, Cracow experienced the vagaries of foreign domination following the Prussian and

Austrian occupations. This partition, along with the transfer of the Polish capital to Warsaw in 1609, and the numerous and disturbing wars in Poland during the seventeenth century, perhaps contributed most to the decline of Cracow during the Early Modern Period. The year 1795 not only meant 'Finis Polonia', but sounded the death knell on an amazing period of commercial development for this southern Polish city.

Methodology

The methodology presented in this book is based on historical geography and, more specifically, with urban historical geography. At no time was there any ambition or consideration of writing a history of Cracow, a job better left to Polish historians; neither is it meant to be a study of the city's economic history, or its contemporary geography. That, too, has to be left to more competent specialists in the various academic fields of interest. This book is, however, an attempt to examine the geographical links of the city's commercial contacts over several centuries and try and interpret them in a spatial manner. Readers interested in the philosophical underpinnings and methodological debate on trends in recent urban historical geography, for example in Germany and Britain, may look elsewhere.[24] Here, emphasis is on the empirical approach, hopefully providing new material for a better understanding of one city's former role in European trade. Through such a study geography allows one to marry a sense of space to the historian's feel for time; moreover, it has been long accepted by historians themselves, that the unity of geography and history provides benefits for both subjects.

Certainly, we must accept that both history and geography are integral sciences, standing alongside the more precise systematic sciences. Nevertheless, without history and geography it is not possible to achieve a full understanding of the world, either now or in the past. Even so, as Andrew Clark stated over three decades ago, 'There is, surely, no need to chew once more the old cud of the relationships of geography and history.'[25] Unfortunately, there is no one definition of historical geography, as each researcher tends, like the urban historian, to put forward his/her own notions and stress. Obviously, one is dealing with space and its intricate associations between the physical environment and human response. In early works, much historical geography dealing with pre-modern times was concerned largely with the rural landscape, but some more recent studies have shown greater interest in the city and the encounter between the physical background and human activity within it. Certainly in a study of Cracow's trade one cannot forget the role played by individual merchants, nor the impact of politics and culture, through significant historical events which have helped to shape the city's fortunes.[26] Moreover, perhaps the most crucial yardstick should be not just yet more facts about a place, but how that place fits into a wider concept of generality, and this is attempted in the latter stages of this book.

Geography's connections with economic history are perhaps not as apparent. Geographical change is certainly a possible theme for economic history. Phenomena associated together in space may have a greater or lesser impact on each other, and their interaction may extend through space for longer or shorter distances. Such factors are part of the very heart of a geographer's interpretive analytical methods. Obviously there is intense interest in location, distribution and circulation, the results of which may be quantified and maps produced through statistical–cartographical techniques. Such results are of interest to economic historians, but many of the geographical data, like that used by economic historians, are not quantifiable; if they are, then on occasions they are not able to be manipulated for any clear interpretive purpose. Nevertheless, in the study of economic growth (or lack of it), the two disciplines may find mutual interests, particularly from the general approach of geographical description and interpretation of economic phenomena, and the need to relate economic factors with other relevant information in an areal manner. It is most probable that historical geography will have much to contribute for the economic historian. Unfortunately, a study of the past by geographers is often dominated by the need to help explain the present, which is not fundamentally an historical approach. Even so, much has been done in terms of cross-sectoral methods for a reconstruction of past geographies, either singly or in the 'sequent-occupance' or 'successive stage; ideas, as evidenced by the works of Darby and Whittlesey.[27]

Perhaps Clark has made the clearest call for links between these two disciplines, when he states, 'There must be more emphasis on the geographical structure of change; on the changing patterns of phenomena and relationships through area.'[28] It is hoped that this study of Cracow's changing commercial links with the European continent may add a little to Clark's clarion call. Not that the work of economic historians should be overlooked. Here the research of many such specialists, particularly from the countries of central and south-east Europe has been consulted, including the sterling pre-Second World War work done on commodity prices and wages by Polish academics. This enabled a comparison to be made of commodity price changes in Cracow over time for many of the goods analyzed in the text.

The painstaking collection of data on the city's commercial links has been broken down according to commodity analysis. It soon became clear that this analysis had to be presented cartographically against the backcloth of the whole European continent. Although small by comparison with some other continents, Europe is a complex area. Commercial connections had to be interpreted not only from the standpoint of Cracow itself, but against the ever changing cultural geographies of the European environment, especially as the social forces gathered strength, forcing upon the continent fresh and effective methods towards different goals. Stress on the time–place continuum in Europe has many faces, and like Proteus, have to be broadly portrayed. The problem of sharp cultural divisions, so evident in the contemporary scene, have persisted in the area since ancient times. Even the question of the continent's territorial extent has led to different perceptions of its boundaries and hence to much confusion. It should be remembered that Europe does reach eastward towards the Urals, and is not just the peninsular part of the continent, from which the homeland of western civilization emerged. In this context it is most important to understand this fact when dealing with a city like Cracow, which for long in its history may be judged to have been located on the margins between the eastern and western parts of the continent, not only physically, but also from cultural, economic and other viewpoints.

Historical geography as it is understood today, namely the study of 'changing geography or of geographical change', developed in England at about the turn of the twentieth century. Since the Second World War, one of the most important shifts in the subject of European scholars has been the switch in focus from early history to the historical geography of modernization. Certainly as this study of Cracow's trade relations closes at the end of the eighteenth century it falls firmly into the first camp. This does not mean it is of no value to those pursuing the latter course, because nineteenth and twentieth century Cracow cannot be properly understood without some knowledge of its previous background; there is some concern about the city's uniqueness, and obviously there will be some value judgements and selection procedures regarding the material utilized, but this should not deter from the overall attempt to give some wider understanding to Cracow and those concrete empirical facts which may lead to the establishment of broader generalities for the urban historical geography of the continent as a whole.

Conclusion

This book attempts to provide a better understanding of commercial development within Europe over a considerable time span, based on the trading links established by one of its cities, namely Cracow; it is hoped to discover how different, if at all, it was from cities elsewhere on the continent, and to what degree its trading links extended spatially over the continent at certain periods of its economic development. The mapping of these trade links will be the first stage in this analysis, to portray the changing structure of the city's commerce over time, and more specifically the growth/decline and location of its markets particularly between the end of the fourteenth and late eighteenth centuries.

Once completed, this material can then be subjected to further analysis so that the various patterns of trade, and the city's sphere of commercial influence may be defined within the broader pattern of European trade exchange. The year 1500 has been chosen as a nominal date forming a benchmark between the Late Middle Ages and Early Modern Period against the wider backcloth of the European continent.

2
Source materials and published literature

Previously Cracow's trade relations with the rest of the European continent from its beginnings to the end of the eighteenth century have not greatly interested many scholars, particularly of non-Polish descent. One has only to peruse the geographical literature on the city both in Poland and elsewhere to realize how little has been produced on any aspect of Cracow's historical geography. Perhaps in Poland this is related to the lack of scholarly interest in urban historical geography, whilst non-Polish urban historical geographers have been deterred by the enormity of the task, involving linguistic problems and the interpretation of documentary evidence from archival sources. In this case, a study of Cracow's trade relations with the rest of Europe over a long time period has involved the author in the collection of data from over a dozen countries, in order to attempt a spatial analysis of the city's commercial influence on a temporal basis.

Literature on the historical geography of Cracow

Before one can appraise the limited literature on Cracow's historical geography, it is necessary to define what is understood by the limits of the sub-discipline. Certainly, historical geographers are concerned with understanding the past, and it is this that separates them from the interests represented by other branches of the subject. Traditional historical research has tended to rely heavily on empirical analysis, and there is little evidence to suggest among recently published papers that this is changing. In 1984, Norton pointed out that 'Most historical geography is short on theory',[1] a fact which perhaps irritates those scholars keen on methodological or theoretical concepts in the subject; there has been some adaption to the new technology such as computer cartography, and computerized methods of record linkage and data management,[2] but overall the task still remains of trying to forge a mass of interesting and usually disparate information from empirical sources, into some form of coherent entirety. In this sense, the study of Cracow's trading relations up to 1795 falls firmly into this latter category, although some of the new technology has been utilized for illustrating price changes and map presentation.

Today historical geography is usually considered to be a study of the reconstruction of past geographies and of geographical change through time: this contrasts with earlier ideas when the sub-discipline was seen as the history of geography, as a study of the history of geographical discoveries, the mapping of the earth's surface, or an analysis of historical changes in state/nation boundaries.[3] At present, more emphasis is placed on the geographical synthesis concerned with the changing character of regions, places (including cities like Cracow), and landscapes. Such studies can have a practical purpose for they allow a deeper understanding of how economic and cultural activities developed, flourished and decayed in the past, in which lessons may be learnt for planning the geography of the future. In turn, these studies can reveal how human beings have changed the face of an area, be it local, regional, or on a continental scale, and tie in with contemporary debates about conservation (in Cracow's case historical urban conservation) and management of world resources.

In Poland, the role of historical geography as a sub-discipline of geography, has experienced increasing popularity in recent years. According to Krawczyk,

> Many misconceptions and divergent opinions about historical geography as a discipline of science have been formed. Although this notion is used by numerous research workers, both historians and geographers, the methodologically comprehensive definition of historical geography is still lacking in our science. The reasons for such a situation are various; among them one of the most important is the fact that neither historical geography, virtually never an academic discipline in Poland, was practised in a methodological way, nor research centres permanently specializing in this type of studies were organized. Hence the divergent definitions of historical geography, its aims and scope of investigation.[4]

This is seen in attempts to define the subject by certain Polish scholars.[5] Krawczyk continues that 'an immediate and at the same time attainable aim of historical geography should be, on the one hand, the reconstruction of a possibly more detailed picture of environment in particular time sections, and on the other hand discovery and examination of mutual relationships between man and the environment, with an attempt to define their direction, intensity and consequences'.[6] The present trend of historical geography in Poland seems to be moving towards studies of past landscapes within the confines of Poland's territory (historical synergetics) and historical human geography using existing sources for a better understanding of human life in the past (food, disease, climatic influences etc.) and its relationship with the natural environment.

This study of Cracow's commercial links and its role as a European emporium adopts a somewhat different approach. Firstly, the study is not confined within the territorial limits of one country (i.e., Poland) but explores the trading contacts of one city, Cracow, with the whole of the European continent and in some cases beyond. Secondly, whilst past landscapes and environmental factors are important they only play a secondary role in the final spatial analysis of the city's trading patterns over time. Use of the existing documentary, archival and secondary published sources play a significant part in the study but their interpretation relates more to an explanation of how economic activities develop, flourish and decay and their spatial impact through a study of long-term trends, often scorned by some scholars, particularly historians.[7]

In 1979, Edward D. Wynot published a critical appraisal of urban history in Poland.[8] According to him, 'Polish historians, however, only recently have become interested in the systematic examination of the origin, rise and development of cities.'[9] With reference to the city under study here he states, 'Cracow [Kraków] the only other city that rivals Warsaw in mystical symbolic importance for the Poles and their history, is decently covered by scholarly studies. Most are devoted to either general topics unrestricted to chronological divisions, or more narrow subjects dating from prior to the nineteenth century' but admits that 'industrial and economic development receive comparatively little attention'.[10] It is precisely in this field of economic development, that the present study of Cracow's commercial trading contacts with the rest of Europe from an historical geography viewpoint, can add some new interpretations into the city's growth from its earliest origins to the end of the eighteenth century when Poland as a political entity disappeared from the map of Europe.

Given this general background it is not surprising to find that there are few publications on the historical geography of Cracow. Only two have a marginal connection with the present study. One was written some fifty years ago by the English historical geographer, Arthur Davies in a study of several Polish towns and had some historical dimension with reference to Cracow, but was not based on deep research and was really a descriptive account of the city's historical background.[11] The other study was part of a larger more contemporary analysis of the city by the Italian geographer, Piero Innocenti, who explained the historical past of Cracow in an introductory account for post-Second World War development.[12] Amongst Polish geographers the only work undertaken from an historical perspective is by Karol Bromek and Zygmunt Górka, who examine specific aspects of the city's growth, namely population and urban development of the city centre.[13] All these studies are only concerned with the city itself, whereas this present study tries to relate Cracow to wider European commercial developments as a whole and place them within a spatial/temporal framework.

Related studies

Works having a similar approach to this study, appear to fall into two major groups; firstly, there are those which attempt a parallel type of approach to that found here, namely a spatial analysis of trade relations undertaken by geographers on other cities like Cracow, and secondly studies carried out by scholars from other disciplines, who have looked at commercial links between Cracow and other places and regions in Europe.

The first group contains a small number of studies which have tried to measure in some way the impact of trade for a particular place over time. The nearest example is the author's own study of Dubrovnik (Ragusa)[14] which revealed the commercial contacts this Dalmatian city created both with the Apennine and Balkan peninsulas from the thirteenth to nineteenth centuries, in order to find out its commercial sphere of influence through trading links. What this study showed for south and south-east Europe, it is hoped the present research on Cracow will reveal for the more northerly parts of the continent.

Studies on the urban historical geography of specific towns and cities are more forthcoming. One of the most significant changes in the approach of historical geographers in Europe since the Second World War has been the shift of focus from earlier history to the historical geography of modernization. This has produced wide study of the intertwined processes of industrialization and urbanization, and the extensive social and economic changes involved therein. One may mention here the work of James Bater on St Petersburg and the present author's study of Prague, both the products of a new relationship of geography and history in the field of Russian and Central European studies,[15] while Bastié's work on Paris may also be added to this category.[16]

The second group may be divided into two sections, namely studies by Polish scholars relevant to this work, and those by their non-Polish counterparts, all from disciplines other than geography. Here it should be stressed that such works selected as source material for this study were of paramount importance in support of the main theme. For geographers there seems to have been little in this selection that was controversial. Certainly detailed, meticulous scholarship in an empirical framework must be respected, even if it has implicit value judgements and unidentified selection procedures as to the matter to be examined. But the critical criterion must be contribution to the establishment of generality, not the multiplication of facts about places. One must clearly accept this view, in an attempt to perhaps placate the historians who may read this work.

Of the Polish scholars, the work of three historians were extremely valuable. Firstly, one may note the detailed research of Stanisław Kutrzeba, with his useful analysis of Cracow's trading relations with parts of the

European continent, carried out in publications emanating from the early part of this century.[17] Secondly, from the post-Second World War period, there was the study of Cracow's regional market by Jan Małecki[18] and the publication of Mariusz Kulczykowski.[19] Valuable though they were, these studies mainly concentrated on limited time periods: for example Kutrzeba's work is predominantly concerned with the medieval period, Małecki's on the sixteenth century, and Kulczykowski has worked on the second half of the eighteenth century. Such detailed analysis of the late seventeenth and early eighteenth centuries is still a task awaiting some able Polish scholars. Other works found to be supportive for the author's geographical theme were the publications of Krystyna Pieradzka, Marian Małowist, Jerzy Topolski, Zenon Guldon and Lech Stępkowski.[20] The partially published series on the history of Cracow, is also of interest.

Although a rich literature already exists on the various aspects of Cracow's history, some of the earlier elaborations do not meet the requirements and standards of contemporary historical knowledge. It is largely for this reason that a team of Cracow historians was asked to undertake this collective work under the editorship of Janina Bieniarzówna, Jan Małecki and the late Józef Mitkowski. Of the planned five-volume series only two are relevant to the present study; volume one will contain the city's medieval history up to 1500, and the second has already been published and covers the era from the sixteenth to eighteenth centuries.[21] The whole series will eventually include the city's history from its earliest beginnings to the post-Second World War period.

Finally, mention should be made of the useful analysis of the road network within the Cracow region up to 1500 by Bożena Wyrozumska[22] and detailed results from road inspection for specific years, e.g. 1570 and more generally for 1765[23] which enabled this author to map the major routeways so vital for commercial success.

Amongst the non-Polish scholars whose work has a bearing on the present theme, particular mention should be made of an early work by the German historian Johannes Müller, who looked at the impact of trade routes on Nuremburg's commerce in the Middle Ages and refers to the importance of Cracow in this pattern.[24] More recent post-Second World War studies are particularly illuminating for Cracow's trade links to the south of the Carpathian Mountains through studies by the Czech historians, Rudolf Fišer, František Hejl and Ondrej Halaga,[25] and the Romanian scholar Mihail Dan.[26]

Source materials

The observation by N. J. G. Pounds about the greater abundance of documentary evidence after about 1500 throughout much of Europe[27] may be valid, but it is often difficult to discover change during a period in certain areas and in some fields of human activity, like agriculture or urban development. For trade, the availability of preserved registers from any period before or after 1500 can prove an invaluable source for commercial history, providing a base for statistical analysis of goods exchanged. Interest for historians, economic historians and historical geographers has been aroused by the growing number of commercial customs registers which have become available for research purposes, some in published form.[28] Although many suffer from the obvious weaknesses of fragmentation or gaps in annual series, they can provide a rich source of detail when trying to construct a past period of commercial exchange.

Nevertheless, certain reservations have to be made about their use. Customs registers are usually part of a closed communication system between an official and his immediate superior in which we are mainly late onlookers; such documents may also suffer from unequal precision, accuracy and care in recording detail. Thus, the view that 'one just went to the records, read them, thought about them, read some more, and the records would do the rest'[29] may be one viewpoint on archival sources, but this is becoming increasingly insufficient from a methodological position in a world of ever demanding scholars. More recently a call has been made to historical geographers by the late J. B. Harley, who maintains that 'Amidst a search for new philosophies, methodologies and techniques to vitalize the practice of historical geography, the independent study of historical evidence has been neglected'[30] and that there has been 'a lack of conviction in studying the historical record for its own sake'.[31] In this study, a primary concern is with verifying evidence of trade flows between Cracow and the rest of Europe over a lengthy time period.

In the light of such comments one must begin to tread warily amongst the minefields of East European commercial data sources, both at the primary and secondary levels; in studying the spatial ramifications of trade exchange, documentary sources usually give time and origin/destination, but often neglect to give details on merchants, carriers, carters, etc., unless for some specific reason, e.g., law suit, last will and testament, or particular financial benefactor. While it may be true that 'Much historical geography has been based on a false consciousness which personified places which were nonetheless described without reference to the peoples who inhabited them',[32] studies of trade exchange are often limited by the data sources which have survived. Historical geography can only enrich these studies by looking at human pursuits in history and apply spatial methods to the data sources.

With all these reservations about the use and abuse of data sources, one may now summarize those utilized in this work for a study of Cracow's trade links over time. The major period reviewed in this study is from the mid/late fourteenth to the end of the eighteenth centuries, largely because extant documentary sources are available in the Cracow state archives; in general Polish archives contain only fragmentary material on commerce except for two rich sources, namely the customs ship mooring registers for Gdańsk and Elbląg (sixteenth to eighteenth centuries)[33] with gaps, and the incomplete registers for Cracow covering a two-hundred-year period from 1589 to 1792.[34] As Małecki has stated 'In spite of the great value offered by Cracow's customs registers they remain largely unknown as a source of research. In the field of commercial history this material, consisting of several thousand hand-written pages, has undoubted value.'[35] Other primary documentary sources consulted consist of extant material from several town archives of commercial centres along the major trade routes between Cracow and its various trading places. They were all used to supplement the above series and add further information on the course of trade.

It is therefore appropriate to discuss these numerous sources according to their geographical distribution, namely first and most important the Cracow state archival material, followed by data from other Polish towns and, finally, evidence collected from non-Polish documentary sources. There is also some appreciation of cartographical and monetary sources.

(i) Archival material

(a) Cracow state archives

The present work is based primarily on Polish archival material, much of it in the state archives, Cracow, together with data from town archives elsewhere in Poland and other parts of eastern Europe. The Cracow archives yielded the largest quantity of source materials, especially the customs registers and other books concerned with commodity trade exchange. The city's enormous collection provided information of basic importance on the subject of Cracow's commercial links with other countries. Using the trade documents as a basis, it was possible to collect subsidiary materials concerning the leading Cracow merchants; it was they who played the most significant part in forming Cracow's prosperity, became members of the city administration and developed commercial links within an area ranging from the Baltic to Black Seas, and other regions of the known world of that time. Such data was found to exist in private correspondence, wills and testaments, and city council reports, as well as within the customs registers themselves, which all helped to throw light on the organization of Cracow's trade with the rest of Europe and beyond.

With the granting of a municipal charter to Cracow in 1257 under Magdeburg Law, the stage was set for the town's development. The earliest documentary source dates from just forty-three years later, namely 1300, when the first court jury book was written and included information on the selling of houses, legal contracts, wills and testaments etc. During the nineteenth century attempts were made to put all the documentary sources into some proper classified order; this basic organizational procedure still remains evident in the present archival structure.[36]

Cracow's archival sources were utilized in the present study according to the two major periods under analysis, namely the Later Middle Ages from 1390 to 1500 and Early Modern Period from 1500 to 1795. Documents from the Later Middle Ages bear some resemblance in form to those of other towns such as Cologne, Basle and Genoa. Source materials may be divided into four basic categories:

(i) Varied documents.
(ii) Resolutions of the Town Council.
(iii) Municipal books, especially accounts.
(iv) Lists of town taxes.

The varied documents were not found to be of great use, as they were not all of the same type, with a mixture of public and private acts. One may find by chance, however, commercial statements made by certain witnesses acting as a third person for some foreign king, or data from the Polish king's income sources (such as that for Kazimierz Wielki dated 1358 on income from customs duty, weights, etc.)[37] and important for information from the first half of the fourteenth century onwards. Town Council Resolutions were largely concerned with matters fiscal and date from the fifteenth century; unfortunately it was not compulsory to write all the financial resolutions in the town books, and details depended very much on the writer, who on occasions may have added some marginalia pertinent to commercial transactions etc. The municipal books referred largely to accounts and the law courts; the latter were mainly concerned with property changes and did contain some occasional information on trade and commercial transactions. Documents pertaining to the post-1400 period have appeared in published form,[38] notably containing Cracow's account books from the end of the fourteenth century (*c.* 1390) and for certain periods in the ensuing century. The probable reason for their disjointed runs relates to the lack of significance accorded to accounts books as soon as the year had ended; in fact it is surprising that so many of these books have survived. It was a different matter with some of the other municipal books on court decisions or public council resolutions, which were carefully preserved, although some may have been destroyed by fire in the nineteenth century. Finally, the lists of town taxes show greater care was taken with details on income and expenditure by the town council for administrative purposes. Nevertheless, it appears that Cracow's economy lacked any form of planning, as there is no evidence of an annual budget, and no information is given on the relationship between income and expenditure; everything seemed to be organized according to earlier traditions. The first of these town lists was probably from the end of the fifteenth century but, as no date was noted, this remains uncertain.

From within this fundamental framework certain documentary series provide information on the city's commercial links, though often in an incidental way. Cracow's income was derived from varied sources. Firstly, it came from property ownership as seen in the series 'Census civitatis' begun in 1358, 'Pecunia pastoralis' (from 1367) and 'Advocatia' (commenced 1430, 1460–72) and also from certain town enterprises such as quarrying (begun 1375), brickworks (commenced 1390), limeworks (1375 onwards) and brewing (1400–56). Income was also obtained from leases e.g., fishing rights (from 1364) and customs duty from herrings (1390–7). A second income source was through municipal monopolies. These included weights and metal melting rights, production of larger weights for use in the lead and silver trade, and provision of smaller weights for lighter commodities such as wax and certain fats. There was a separate town law on the monopoly of melting these goods. The manufacture of silver weights and the melting down of silver was also a municipal monopoly as well as the right to cut cloth, first introduced in 1358.

A third form of municipal income involved administrative charges for civil justice (documents available for 1391–2, 1414, 1431), and money received from various town fines, whilst service provision was also credited to the town accounts for supplying piped water etc. (first documented 1399), brewing facilities (1453 onwards) and maintenance of bridges. Finally, a fourth source of town income came from taxes; these involved taxes on individual factors such as roads (1367 onwards), for the city's safety at night ('Exactio vigiliae nocturnae' began in 1366) and from customs duty. This latter category often provided interesting data on trade links, together with specific information on certain commodities. For example, the 'Census quartarum', starting in 1323, provided the first evidence of a customs duty on cloth, and occasionally added details on cloth prices for lengths from different European textile centres.[39] Salt ('Forum salis', 1405 onwards) and wine ('Weygelt', from 1493) also had their own separate customs books.

Municipal expenditure was recorded under six major headings. Royal expenditure came first, with various details on the cost of gifts (1391–1487), requisition of waggons for royal use (1431–87) and the administrative costs of custom duty practices (1390–3). General administration came second with the upkeep of the Town Hall ('Pretorii Necessaria') and its services, together with council costs and ambassadorial visits; also included in this category was expenditure for banquets and other receptions. The third category concerned municipal supervision and safety in Cracow; this included the supervision of commerce ('affusores'), and prevention of fire. Also details on funds for the police ('circulatores') and defence were noted as well as maintenance of bulwarks, walls, towers, men-at-arms and watchmen, together with judicial administration. Cleansing facilities and communications came fourth, and included costs on street cleaning. The fifth category was concerned with municipal services including the upkeep of wells, maintenance of the town aqueduct, and public clocks, whilst the final, sixth, category specifically referred to municipal charities which included alms to the poor and to the Church.[40]

It should also be mentioned that on occasions in this work reference was made to the archival section in Cracow on Kazimierz, a town with its own separate identity at this time. As a town in its own right, Kazimierz had independent status from Cracow after 1335, when it was founded by the king, Kazimierz Wielki; it remained so until 1802 when it was joined again to Cracow. Its documentary sources were organized in a form similar to that of its larger neighbour, but the total archive only contained just over 200 pergamon documents and nearly 900 books, much smaller than Cracow archives with about 3,600 books to its credit.[41] Again, some of the documents in the Kazimierz section of Cracow's archives proved useful for information on trade between this part of the city and the rest of the European commercial world. However, in this present study such commercial links have generally been included under the name of Cracow, because from a spatial analysis viewpoint there seemed little gain in differentiating between the two separate parts of the present-day urban centre.

The period after 1500 was one of much greater documentary evidence than earlier times. An invaluable source for research on commercial links may be found in preserved customs registers. These allow for more detailed attempts to be made at statistical analysis with regard to commodity exchange. Cracow archives contain various examples of such books which can be basically divided into three major groups. The first group, under the title *Regestra thelonei civitatis Cracoviensis* (coded 'sygn. 2115–250') contain 136 manuscript books recording the amount of customs duty received by the city during the period 1589 to 1792, although unfortunately with considerable gaps, for the years 1689–1700, 1713–36, and 1773–91. The second group, known as *Regestra novi thelonei civitatis Cracoviensis* (coded 'sygn. 2251–2270') contains twenty books and covers the period from 1659 to 1679, whilst the third and smallest group consists of registers on bridge tolls entitled *Regestra pontalium Cracoviensis* (coded 'sygn. 2271–2273') and these books record details for the years 1615–28.

These three groups of customs duty registers are varied in nature, resulting from different payment methods utilized by merchants and their employees according to a variety of customs tariffs. For example, three books note enlistment pay for customs related to the royal treasury and include data on goods sent to, and received from abroad. Another example is the excise register on customs collection by the Swedish government during their occupation of Cracow, 1655–66 (coded 'sygn. 2170') which is full of interesting

Table 1. *Selected examples of commodity movement over the Carpathian Mountains (to/from Cracow), January 1593*

Date	Information	Customs duty
4	Wojciech from Mikulów brought in hops; utilizing 2 horses.	
	Ezechiel Gilar from Opava declared 7 vats of scythes and 4 cetnars (400 kgs.) of iron for making cutlery.	1 grosza (gr.)
8	Urban from Zabrze delivered 46 pails of Moravian wine; had more but had sold it in Oswięczim along with containers, utilizing 12 horses.	29 gr.
10	Ezechiel from Opava declared 10 cetnars (1,000 kgs.) of cutlery steel, and 300 scythes in a box, for Bartel Sambek.	
11	Marek, a Jew from Prague, declared 1 ½ ells of white axamite cloth; several ells of other cloth.	2 złoty (zł.) 2 gr and 2 halerz (hl.)
	David, a Jew from Opava, quilts and Jewish caps; utilizing 4 horses.	15 zł. 9 gr. 2 hl.
	Jakob, a Jew from Prague, declared stockings, gloves and other goods.	17 gr.
	David Vloch, a Jew from Prague, declared stockings, gloves and other small goods.	12 gr.
	Issac, a Jew from Prague, declared 11 white quilts and half a roll of mohair cloth.	5 gr.
	Volf (a Jew) brought these goods in a cart with 3 horses.	1 gr.
12	Ezechiel Gilar from Opava paid debts for 7 vats of scythes and 4 cetnars of cutlery iron brought into Cracow on 4 January.	2 zł. 28 gr.
	Kuba, from Bielsko, brought in rye seeds which he had purchased for himself; using 4 horses.	2 gr.
13	Ezechiel Gilar from Opava declared 20 chamois skins; using 4 horses.	3 gr.
14	Stanisław from Ivanovice declared a quantity of salt; using 6 horses.	3 gr.
15	Jura from Opava brought in transit goods to Cracow, for Bartel Sembek – vats and steel; using 11 horses.	5 gr. 6 hl.
17	Kaspar Frilich from Bielsko declared 56 rolls of Bielsko linen; using 4 horses.	1 zł. 14 gr.
18	Severin from Bielsko declared 7 rolls of Bielsko linen.	1 zł. 5 gr.
19	Bartoš from Lipnik declared 19 oxen.	6 gr. 9 hl.
	Lurin from Lipnik brought in vegetables for himself; using 3 horses.	1 gr. 6 hl.

Table 1. (*cont.*)

Date	Information	Customs duty
	Kaspar Frilich from Bielsko declared 12 ćwiercień (3 bushels) of rye, 3 ćwiercien of cereals, and 2 ćwiercien of beans.	5 gr.
	Piotr from Lipnik declared 8 ćwiercien (2 bushels) of malt barley, 2 ćwiercien of cereals, beans, 1 stone of pepper, 2 rolls of paper, 3 korzec (3 bushels) of groats.	8 gr.
23	Andris from Opava declared 2 barrels of Opavian beer for the Royal Court; using 8 horses.	4 gr.
24	Bartosz, a knight from Bartošvice, declared 5 barrels of plums and 4 bags totalling 140 stones (17½ cwts) of wax; using 8 horses.	13 gr.
25	Hanzel Prister from Frystat declared 4 warps of Silesian cloth and 2 rolls of linen for himself.	5 gr. 9 hl.
27	Aron, a Jew from Prague, declared 3 otter and 3 fox skins; left for Prague having bought 4 horses and a cart in Poland.	4 gr. 9 hl.
	Marcian from Bielsko, declared 4½ rolls of Bielsko linen.	3 gr.
	Ludwik from Opava declared 10 cetnars (1,000 kgs.) of processed wool.	6 gr.
28	Daniel from Frystat declared 13½ stones of fleece wool, and 13 skins.	4 gr. 9 hl.
	Isak, a Jew from Prague, declared 20 mink skins, one wolf skin, 28 tanned weasel skins and 2 fur coats made from wolf and mink skins; he also had axamite cloth in lieu of payment, which he brought to Cracow for sale, together with his grey squirrel coat. Goods transferred to the carter; duty paid on goods by Aron, a Jew, the previous day.	1 zł. 13 gr.
29	Irzy from Bielsko declared 32 rolls of Bielsko linen cloth; using 4 horses.	24 gr.
	Jurek from Rożnów, declared 1½ barrels of Opavian beer; using 4 horses.	6 gr.
30	Matis Singler from Jičin declared 8 barrels of Eger wine and 2 small barrels of the same wine for Sigmont Alance a citizen of Cracow; using 33 horses.	16 gr.
	Matis from Brno declared 2 firkins of pepper from Kischburg.	3 gr.

Source: Wojewódzkie Archiwum Państwowe w Krakowie: *Register Thelonei Civitatis Cracoviensis* (Księgi celne, Rkp. 2117).

material on excise obtained by the Swedes from all merchandise, whether large or small, brought by them to Cracow. A similar example is the excise register ordered by the Polish government (sejm) for 1658–9 (coded 'sygn. 2174'); although it proffers little actual information, it does enable some analysis to be made on commercial profit margins obtained by various Cracow merchants.

The oldest extant customs duty book preserved in Cracow archives dates from 12 June 1589, and is based on a royal privilege granted by King Zygmunt III Waza (1587–1632). It contains information on the value of merchandise, its size, number of oxen per waggon, use of bridges, and (if applicable) a tariff on horses and carts. Different goods were levied at varied rates as, for example, cloth and wine. More precisely the registers contain the date, name of merchant or carter, name and quantity of merchandise, town of origin/destination and value of goods, number of horses and carts, and value of customs duty obtained (Table 1). However, much depended on the precision of the customs official as to how well or badly facts were recorded, so that on occasions complete information is lacking.

In 1659 a new customs duty system was introduced under King Jan Kazimierz (1648–68) with a basic rate of one hundredth the value of the merchandise. However, for some years the registers contained duties recorded both at the new and old rates (e.g., 1679). Later books in the eighteenth century show changes in customs registration, and the type of goods recorded (e.g., 1746–64) whilst a separate payment was made on wine (known as 'burkowa'), and there was a fixed rate for horses and carts. Later customs books on wine, referred to as 'beczki', exist for the years 1702–92 (coded 'sygn. 2241–2250'), but contain many gaps in the series. There is also an interesting catalogue (coded 'sygn. 3177–3263') to the Cracow Congress of Merchants; these contain only Christian merchants and was reformed from an earlier organization in 1722; their aim was to keep a tight control on royal trading privileges, particularly in relation to foreign and Jewish merchants who traded in Cracow.[42] Unfortunately their deliberations are of much less value as source material because their registers only record certain goods (usually intoxicating beverages), or tend to present the sum total for merchandise traded.

One of the main weaknesses of the Cracow customs books as a source of commercial activity is the differing precision, accuracy and attention to detail in which they were written. Perhaps the most reliable and of greatest value are those for the periods 1600–50 and 1750–1800. Even so they do not give the full picture of Cracow's trade, and the latter period is very much dependent on the bridge toll books, for it was obligatory for carters to pay customs duty before entering the city. Also the customs books did not record merchandise which belonged to the nobility, the Church, or peasants if it was their own produce brought to the city for sale; this was especially so in the case of peasants supplying goods to the local market via the Vistula river, such as wood. Shopkeepers who purchased goods 'for personal use' were also exempt, as obviously were articles smuggled into the city.

In spite of all these disadvantages, the Cracow customs duty books are of great research value, not only in the field of commercial history, but also in helping to study the past historical geography of Cracow's commerce and its many and varied trading relations with other places both locally, nationally and on an international scale.[43] Together with documents referring to Kazimierz suburb,[44] they provide a rich source of economic material and include details on the prices of various foodstuffs, handicraft workers' wages, and other material channelled through the city's administrative machine, e.g., *Libri taxorum* begun in 1554, and 'Foralia' (coded 'sygn. 1511–1522') which notes cereal trade prices after 1772. Similarly, documents on shopkeepers' associations operating from the Cloth Hall (Sukiennice) give details of goods sold, based on municipal protocols and acts dating back to 1431 (coded 'sygn. 1523–1533'). There is also considerable material on house rents in the city as well as that collected from shop owners, and covers the period 1545–1774 (coded 'sygn. 1964–1975') although it is very fragmentary in nature. Customs duty was also levied according to weight ('Wielka i mała waga') and operated from 1628–1771 (coded 'sygn. 2274–2313'). Specific customs duties were placed on certain beverages such as wine, honey drinks and vodka which all came under the collective title of 'ducillaria', together with wine corkage dues, for the years 1503–92 (coded 'sygn. 2314–1352'). Purchase and supply of barrels containing such beverages were controlled through the city council and entered into special registers known as 'szrotlin' and 'nigra signa'; these beverage tax records exist for the period 1548–1635 (coded 'sygn. 2353–2367') and occasionally give data on place of origin. City income was also obtained from brewing, which during the years 1557–84 was called 'ternarii' (coded 'sygn. 2368–2379') and from 1588–1613 known as 'groszowe' (coded 'sygn. 2380–2390'). Documents relating to guild activities in Cracow are recorded from the fifteenth to mid-nineteenth century (coded 'sygn. 3010–3176') and give details of each organization and the importance of handicraftsmen in the city. Finally, on occasions use was made of private correspondence, particularly with reference to merchant activity. Private letters originating from Cracow citizens and sent to provincial areas of Poland are held in alphabetical order according to family name (coded 'sygn. 3264–3298') and provide interesting material for studying family relations and trading links with the more distant parts of Poland from the sixteenth to seventeenth–eighteenth centuries. Perhaps most remarkable is the book containing postal rates of a Cracow merchant, one Montelupi, who recorded these matters from 1620 to 1626 and wrote them in Italian; it is part of the section on private correspondence for the sixteenth century (coded 'sygn. 3246–3283').

This list of documentary sources to be found in Cracow archives by no means exhausts the rich material found there,[45] but does include the major series utilized for this study. Extant documents on Church affairs (coded 'sygn. 3299–3397') preserved from the fourteenth century onwards, on hospitals (coded 'sygn. 3398–3539'), the university (coded 'sygn. 3558–3660') and other diverse materials under the section group 'Varia' (coded 'sygn. 3561–3568') were found to be of less interest for studying the city's commercial activity.

(b) Other Polish archives

Archival material on customs duty in various other parts of Poland is much less forthcoming than that from Cracow. The former rich sources of archival books on customs duty in Warsaw were destroyed by the Nazis in 1944, when they were stored in the archival treasury (Archiwum Skarbowe) and Krasiński Library. Fortunately, some of this material had been worked on previously by Polish scholars such as S. Kutrzeba and F. Duda (Włocławek customs books) and R. Rybarski, particularly for sixteenth-century Poland; unfortunately the rich and valuable customs registers for other parts of Poland for the period 1718–90 held in the Krasiński Library had not been subjected to academic research before their destruction.[46]

It is not surprising, therefore, to find that most Polish archives and libraries contain only fragmentary material on customs books and matters pertaining to commercial transactions. The two exceptions are the Cracow archives, previously described, and the ship mooring books from Gdańsk and Elbląg ports on the Baltic coast. These registers record shipping movements from the first half of the seventeenth and second half of the eighteenth centuries for Gdańsk, and from the end of the sixteenth and whole of the seventeenth century for Elbląg,[47] although they suffer important gaps in their chronology. Gdańsk archives also contain small amounts of material on goods transported along the Vistula river, e.g., for 1784, which have been analysed by S. Hoszowski.[48]

The remaining Polish archives are poorly represented by custom register material. In Poznań archives there are some books from the Augustów and Kalisz chambers of commerce, the former covering the months of April–June for 1607, and the latter from 1647 to 1655.[49] Other towns with some archival material on customs duty that have survived include Chrzanów, Olkusz, Krzepice, Siewierz, Koziegłowy and Oświęcim.[50]

(c) Non-Polish archival sources

In a study which contains wider geographical implications than just Cracow itself or its national setting, there was a need to consult archival material from countries outside Poland. For the former Hungarian territory, specific town archives were used from places now situated in former Czechoslovakia.

Commercial centres in Slovakia, then under Hungarian rule, were extremely useful in analysing Cracow's trans-Carpathian trade. Use was made of the so-called 'thirtieth' (i.e. 3.33 per cent of a good's value) which was a customs duty levied on merchandise by Hungarian royalty, and had been in existence since the thirteenth century.[51] Although some of the archival material has been destroyed, that surviving in towns such as Levoča, Prešov, Bardejov, Trenčín etc. gives some evidence of past trade links with Cracow. Documents formerly belonging to noble families in Upper Hungary (Slovakia) are now in Bratislava archives and give details of trade movement through the Carpathian valleys into Poland, e.g., Orava, and originally stored in Orava castle; documents on the 'thirtieth' duty for such strategic commercial centres as Tvrdošín, and Žilina are now also located in Bratislava.[52] Unfortunately, some of these series contain considerable gaps which deter statistical analysis, but they do give evidence on the import/export of goods and occasionally give specific destinations, particularly for the sixteenth century. They are in chronological form according to the Christian calendar, and contain a merchant's name, his destination and type of goods applicable for customs duty. Much of the archival material was written in Latin (e.g. 1529–33) but contained phrases and terminology used in the Czech/Slovak languages; the register for 1537 was entirely in Czech. This set of registers was intensively used by J. Janáček in his study of the Czech cloth trade in the sixteenth century.[53] Another useful source for Moravian commercial links may be found in the 'thirtieth' customs books from Trenčín especially for the years 1530–6.[54]

Customs registers from other former Czechoslovakian towns also proved of interest, although many suffered from fragmentation, and short runs. Prague archives have only one customs register surviving from the sixteenth century, dated 1597 which referred to the 'ungelt' (Archiv. hl. m. Prahy coded No. 2054) and gives some idea of Prague's trade pattern in that year. Customs registers from other provincial towns in present-day former Czechoslovakia were utilized; merchants' books from Jihlava (Městský archiv. nos. 1888, 1890) contained some details on links with Cracow. Bardejov archives contain some useful sources, especially for the period around the mid-fifteenth century, and these have previously been used by such scholars as Fejérpataky, Iványi and, more recently, Deák.[55] Marečková has used seventeenth-century material from this archive especially in her study of the town's merchants,[56] but these too suffer from incompleteness over long time runs. Similarly, the archives in Levoča contain a useful series *Regestrum tricesimale super inductionum et eductionum mericum*, starting in 1553 which give a merchant's name, and/or that of his carter, the date and the amount of the 'thirtieth' duty paid, on all goods brought into, or taken out of Bardejov. These have been carefully analysed by Pavel Horváth in his study of trade relations with Poland in the second half of the sixteenth century.[57] Košice was another important trade centre and its town archives have documents referring to customs duty and import/export trade but, as Granasztói states, 'Unhappily, the town archives do not have sources with long time series.'[58]

In present-day Romania, archival sources from some Transylvanian towns proved useful, such as Braşov (Kronstadt), Sibiu (Hermannstadt) and Cluj (Klausenberg). Customs duty in Braşov was a 'twentieth' (5.0%) of the total merchandise value and called 'vigesima', whilst in Cluj it was the 'thirtieth', as in Slovakian towns, and called 'tricesima'; these two customs duty levies therefore obtained 5% and 3.33% respectively, of a good's total value. In Sibiu, a mixture of the two systems operated, the 'vigesima' in the sixteenth and early seventeenth centuries, and the 'tricesima' for most of the seventeenth and all of the eighteenth centuries. For Braşov, only a few customs registers have survived (for the years 1503, 1529/30, 1542–1550) and have been summarized in published form.[59] Only one customs tome (in Cluj) exists from the sixteenth century (1599), but is complete from the early seventeenth century up to 1637.[60] The Sibiu customs books have long runs for the seventeenth and eighteenth centuries, (with gaps) and have been summarized by Panova.[61] The Cluj registers have been used for links with Cracow[62] by Mihail Dan whilst he and S. Golderberg have utilized the Braşov books for commodity analysis, including the cloth trade.[63] Customs data from the Turnu-Roşu trading post near Sibiu has also been studied for Transylvanian commercial contacts by Demény, including for Poland in the seventeenth century.[64] Published documentary sources on trade links between parts of the Ottoman Empire and Poland were used from works by Abrahmowicz, Guboglu, Bogdan, and Hurmuzaki.[65]

Finally, various published material was scanned for archival information from a variety of authors and regions. Polish–Lithuanian commercial links have been analysed by Wawrzyńczyk, the Budapest customs registers by Perenyi, German contacts, especially for cattle, by Klier, and an appreciation of the Sund registers analysed by Nina E. Bang.[66] All these, in turn, helped to throw some light for this study on the international dimension of Cracow's trade during the various periods under review.

(ii) Cartographic sources

Cartographic evidence is particularly critical for the study of commercial trade links over a past period, especially against a backcloth of the whole European continent. Whilst Cracow forms the focal point of this study, maps of the city itself were of less importance than their regional, national and international counterparts for an examination of its spatial impact on trade. Nevertheless, some maps of the city were consulted, mainly to judge the growth of the urban area over time and to find specific place names related to the city's internal structure. From the outset it was decided, for the sake of clarity and reasonable uniformity, to utilize the contemporary version of place names within the present boundaries of European states, rather than a Polish version throughout the text, e.g., Košice not Koszyce. It was also decided not to use former German place names if a contemporary Polish version existed, e.g. Gdańsk not Danzig, Wrocław not Breslau. The exception to this general rule was only applied when a familiar English form was available, thus Cracow not Kraków, Warsaw not Warszawa, or when the use of the Polish form seemed better suited to the historical context, e.g. Lwów not L'vov, Królewiec not Könisberg for places outside present Polish territory.

For the city of Cracow itself, both contemporary and historical maps were utilized. For the former, the *Plan Krakowa* (Pań. Przed. Wyd. Kartog. Warszawa, 1975) proved invaluable, whilst various city maps from earlier times were used for the latter. The recently published *Katalog dawnych map wielkoskalowych Krakowa XVI–XIX wieku* (Pań. Wyd. Naukowe, Warszawa-Kraków, 1981) helped illuminate areas within the city on a large scale for the sixteenth to nineteenth centuries, whilst other individual city maps were useful from the seventeenth century, e.g. *Delin: Obsidionis Cracovia* of 1657 (Bibl. Jagiellońska) and 1655 (*Pufendorff Samuel De rebus a Carolo*, Norimbergae, 1696). Swedish engineers produced an interesting map in 1702 entitled *Geometrisches Plan von Schloss Wawel und die Stadt beidem: Cracau Mitt dehro Herumbligenden.* Two maps from the end of the period under review here come from two Polish authors, namely K. Bąkowski ('Kraków i najbliższa okolica w XVIII wieku') from his book *Historia Krakowa* p. 179 which he published towards the end of the nineteenth century, and the detailed city map of 1785, *Śródmieście Krakowa* (według planu Kołłątajowskiego) at a scale of 1:3286, and reproduced in *Rocznik Krakowski*, Vol. IX, by S. Tomkowicz during the early years of this century. Various graphic illustrations of the city also exist and, though skilfully and artistically produced, were not found of great significance for this study.

Much greater importance was attached to various maps which helped trace trade routes and commercial roads over the period under review. Contemporary maps included *Okolice Krakowa, mapa turystyczna* (Pań. Przed. Wyd. Kartog. Warszawa, 1974) and *Mapa województwa krakowskiego* (ibid.) at scales of 1:200,000 and 1:500,000 respectively. In this historical context maps at a regional level were extremely helpful in constructing initial trade routes; for example, Bożena Wyrozumska's commercial road map of the Cracow region in the sixteenth century (Załącznik do *Lustracja dróg województwa krakowskiego z roku 1570* Kraków, 1971), that of W. Śląski for the Baltic coastal region (*Szlaki handlowe Pomorza w XII–XIII w.* pub. Warszawa, 1969), and that of S. Weyman for the Poznań and Kalisz region ('Poznański i Kaliski węzeł dróg w XVI wieku',

Przegląd Zachodni, Poznań, 1953). Similarly, S. Domanowszky's map of the Spiš region of Slovakia gives early trade routes through that area ('*A Szepesség régi kereskedelmi útvonalai*', Budapest, 1913).

At a national level, Stefan Weymann's map of trade routes through Poland during the Piast period clearly shows the significance of Cracow as a commercial centre, (*Mapa dróg handlowych Polski Piastowskiej*, Poznań, 1938), as does Marian Małowist's work *Carte des routes commerciales en Europe Centrale au début du XV siècle*, Brussels, 1931), and Teresa Wąsowiczówna's outline (*'Some important commercial roads in medieval Poland, approx. 1200 AD'. Archaeologia Polonia* Vol. II, Warszawa, 1959). The extent of former Polish territory is seen in A. Sujkowski's publication (*Szlaki drogowe na ziemiach dawnej Polski od czasów najstarożytniejszych do XV wieku*, in *Geografija ziem dawnej Polski*, (Wyd. M. Arcta), Warszawa-Lwów-Lublin-Łódź-Wilno, 1921, p. 144–5); similarly, A. Wawrzyńczyk has published a map of trade routes in the sixteenth century covering Poland/Lithuania ('*Mapa dróg handlowych w XVI* w.', Warszawa, 1956). This may be seen as complementary to a reproduced version which appeared in Kosman's *Historia Białorusi* (Warszawa, 1979, p. 120). Other cartographic sources of interest were found firstly in *Études Historiques Hongroises*, Vol. I, Budapest, 1975, which includes some evidence of trade routes in Hungary for 1541–1568 by I. Sinkovics, and secondly, one for Silesia in the sixteenth century ('*Mapa gospodarcza Śląska dla XVI wieku*', in *Dzieje Śląska w wypisach*, (S. Michałkiewicz and J. Sydor, editors), Warszawa, 1964, p. 41).

On an international scale, works vary from those showing trade routes between two states such as Dziubiński's map of Polish–Ottoman trade routes in the sixteenth century (*Przegląd Historyczny* Vol. CVI (2), and an interesting trans-Carpathian trade-route map drawn by C. Dobrzański, (*Mapa dróg handlowych krakowsko-węgierskich wieku XIII–XV*) at a scale of 1:500,000, in Cracow, 1910. A more extensive area is covered by O. R. Halaga for the trade routes between Hungary/Poland–Prussia ('Cesty uhorsko-pol'sko-pruského obchodu') in *Slovanské Štúdie*, Vol. XVI, Bratislava, 1975, p. 188. At a continental scale two useful maps were those by H. Obuchowska-Pysiowa ('Ważniejsze ósrodki i drogi handlowe w Europie w XVII w.') at a scale of 1:12 million, in her work *Handel wiślany w pierwszej połowie XVIII w.* (Warszawa, 1954) and A. V. Florovskij (*České země v síti Evropských obchodních cest do konce 18 století*, Praha, 1954), in his book on Czech–Russian trade relations.

Specialist commodity trade-route maps were also utilized wherever possible. Examples of this cartographic category include that for cattle by Jan Baszanowski ('Zasięg regionów hodowlanych i rynków zbytu wołów', in *Z dziejów handlu Polskiego w XVI–XVIII wieku: handel wołami, Gdańsk*, 1977, together with *Szlaki wołowe w granicach Rzeczypospolitej*; for minerals by Danuta Molenda (in *Górnictwo kruszcowe*, Warszawa, 1963); more specifically for copper (T. Dziekoński *Metalurgia miedzi* ..., Warszawa, 1963, p. 368); and for salt by J. Wyrozumski (*Saliny Polski do schyłku XIV wieku* in *Państwowa gospodarka solna w Polsce do schyłku XIV wieku* (P.W.N.), Kraków, 1968) and a plan for Wieliczka (*Plan miasta Wieliczki Marcina Germana z 1638 r.*) reproduced by A. Keckowa in *Zupy krakowskie*, Warszawa, 1969.

Finally, on a more general note great use was made of the English edition of the *Historical Atlas of Poland*, based on the fourth edition published in Wrocław in 1981. This atlas was invaluable as it illustrates the various territorial changes over the whole historical period under review, and permits a clear understanding of the evolution of Poland's frontiers ranging from the early Piast dynasty to the country's partitions and beyond. Commentary on each map is given, together with a separate place-name index. Further information on specific areas and regions of Poland in times past is well represented in *Katalog dawnych map Rzeczypospolitej Polskiej w kolekcji Emeryka Hutten Czapskiego i w innych zbiorach* (P.A.N.) Wrocław-Warszawa-Kraków-Gdańsk, 1978; this work contains reproductions of early maps of Poland which were originally collected by E. H. Czapski, and range from the end of the fifteenth to mid-nineteenth centuries. They were particularly useful for tracing early place names of urban and rural centres trading with Cracow, which today no longer exist, or have changed their original names. Other maps consulted for this work include Meig's *Karte des Königreich Galizien und Lodomeirien*, dated 1779–82 and produced in Vienna, and W. Semkowicz's map for the same period of the Cracow region (*Mapa województwa krakowskiego w dobie Sejmu Czteroletniego 1788–1792*). Lastly, a general appreciation of Polish cartography has recently been published by Bogusław Krassowski,[67] which gives an overview of major contributions on that theme since Poland first appeared on a map; this early evidence was compiled if inaccurately by that great Alexandrian scholar, Claudius Ptolemaeus (Ptolemy) between 90 and 188 AD.

(iii) Monetary matters and prices

Throughout this study there has been reference to monetary matters and commodity prices. Although these are only of secondary importance to the major thrust of the work, some explanation of their sources is necessary. An attempt has been made to chart the progress of price fluctuations in particular commodities which passed through Cracow, and where reasonable data runs exist, then time series graphs have been produced. In order to have some impression of these financial changes it is first imperative that some knowledge of the Polish monetary system is outlined.

The monetary situation in Poland over the years covered by this study may be divided into five separate periods. The first period covers the fourteenth and fifteenth centuries and opens with a confused monetary situation around 1300; a lot of thin metal coins (bracteates) of varying value circulated in the country, together with much foreign currency brought in by merchants. One of these currencies was the 'Prague Grossus' which the Polish king, Casimir the Great (Kazimierz Wielki, 1333–70), adapted using its design to form the silver 'Cracow Grossus' (Grosz). Poland's first gold coin, the ducat, was introduced about 1320 by his predecessor, Ladislas the Short (Władysław Łokietek, 1306–33). No other Polish gold coinage was minted during this period, and many important financial transactions were concluded in Hungarian ducats. The major Polish monetary units were the silver 'half-grossus' and the 'denarius' both introduced by King Ladislas II (Władysław Jagiełło, 1386–1434) and continued by his successors up to King Alexander (1501–6). During this period other trading towns like Cracow, e.g., Toruń and Gdańsk, began minting their own currency, as did the provinces of Silesia and Pomerania. The second monetary period in Poland spanned the sixteenth century; the 'grossus' (grosz) had lost value through inflation and become a rather cumbersome coin for the growing number of merchants, who were rapidly expanding their business transactions. A new monetary unit was established, the 'złoty' which was equivalent to 30 grossi, but there were many varieties of coinage circulating throughout the country, largely from various Polish provinces, together with a surfeit of gold Hungarian ducats. This period also saw the introduction of the 'taler' a thick silver coin much used in foreign trade, and introduced by King Sigismund I (Zygmunt I Stary, 1506–48). The third monetary period in Poland covers the first half of the seventeenth century, an era of peace and prosperity in Poland under the reigns of Sigismund III and Ladislas IV (Zygmunt III Waza, 1587–1632; Władysław IV, 1632–1648) respectively. City mints like those of Cracow, Poznań, Gdańsk, Toruń and Elbląg, were very active, but the value of coins fell. In 1586, at the beginning of Sigismund III's reign, a gold ducat was worth 56 grossi; when he died in 1632 it was worth 165 grossi (i.e. trebled); similarly the 'taler' rose during his reign from 35 to 90 grossi *vis-à-vis* the gold ducat. Numerous mints were producing small change at lowering values, in order to make quick profits, and this led to price rises for both the 'taler' and 'ducat', although the quality and weight remained unchanged.

Poland's fourth monetary period stretched from the mid-seventeenth to mid-eighteenth centuries, with the earlier part of this era marred by wars and invasions. In turn, these factors affected the monetary situation, which not only deteriorated markedly, but was also plagued by the circulation of false coinage, purposely introduced into the Polish market. This included the 'szelągi' (shelags) i.e. shilling introduced by King Jan Kazimierz (1648–68) and worth about a third of a 'grossus', and the 'tynf' which also first appeared during his reign, and had a nominal value of 30 grossi, but with only 13 grossi worth of silver content. Poland also suffered from a large influx of foreign currency at this time, particularly from north-west Europe. The early seventeenth century saw a continuation of this trend, with monetary activity centred on Leipzig; from here the two Polish kings belonging to the Saxon dynasty (August II Mocny, 1697–1733 and August III 1733–63) flooded Poland with coinage, much of it forged, which in turn contributed to the downfall of the state's money mart. The final period spanned the second half of the eighteenth century and saw monetary changes introduced by King Stanislas Augustus Poniatowski (1764–95) to counteract previous mistakes made by the Saxon dynasty. New minting regulations introduced in 1766 and 1787 led to much tighter control on coinage production, now centred in Warsaw; Poland's first paper money was printed in 1794, a year before the end of King Poniatowski's reign and the disappearance of the Polish State from the map of Europe.[68]

Monetary fluctuations and a history of price development in Poland will help understand the price commodity analysis which accompanies certain sections of this study. The first serious attempt to analyse the role of money in Poland, particularly in relation to the price revolution of the sixteenth and seventeenth centuries, was by A. Szelągowski in 1902.[69] However, this work was based on rather fragmentary material, and it was not until the inter-war period that a historical study of the prices for some commercial centres in Poland was undertaken; this was part of a larger programme conducted by an International Committee for the History of Prices.[70] Five major Polish cities (Gdańsk, Warsaw, Lublin, Lwów and Cracow) were studied, but projected work on Poznań, Toruń and Wilno was interrupted by the Second World War. The aim of the project in Poland was to analyse prices and earnings for chosen cities from the fourteenth to early twentieth centuries. Fortunately, works on Cracow by Pelc and Tomaszewski, covering the years 1369–1600 and 1601–1795 respectively, were published before the Second World War, and thus invaluable for a study of Cracow's trade relations over these two periods.[71]

The number of commodities and other details published in the Cracow volumes vary between 60 and 100 topics, divided into production/consumption categories, and occupational salary differentials. Even so, the information has its weaknesses, which sometimes curtail the possibility of assessing data on long-term price movements. Amongst the most serious limitations are quality variation of merchandise, seasonal price fluctuations, the abundance of price types (market, retail, wholesale etc.), the small number of quoted prices for some years, whole years missing, and long gaps in price data for some commodities.

In spite of these shortcomings, prices can provide a useful barometer on economic conditions. In medieval times, money seemed to be constantly in short supply due to insufficient silver, the melting and re-issue of old coinage and subsequent currency debasement, and sharp price rises. The latter was common throughout Europe until the mid-fourteenth century, when plague and high morality reduced demand and prices stabilized or fell. This trend was reversed during the late fourteenth century, becoming the 'price revolution' of the sixteenth century, in which, according to John Nef, silver mined in central Europe played a considerable part.[72] The devaluation process in Europe was well under way by the mid-fifteenth century and continued with fluctuations till well after the mid-eighteenth century.[73] These general trends may also be detected in Cracow's own price movements, or more generally in the country as a whole. Prices in Poland experienced an overall downward tendency in the fifteenth century with many oscillations, but the first half of the sixteenth century began to show a slight increase; the second half of that century felt a strong general price increase amongst all goods, especially foodstuffs. While there was a threefold escalation of prices and wages during the sixteenth century, living costs grew fourfold; therefore purchasing power dropped by a quarter.[74]

Throughout Europe the seventeenth century was one of crisis which Trevor-Roper believes was closely associated to relations between society and the state.[75] Its implications were widespread, among them continued intensity of price rises. In Poland devaluation of the grossus (grosz), begun during the reign of King Stefan Batory (1576–86), culminated in the disastrous monetary crisis of 1617–23, and followed in the 1650s and 1660s with a new one. The former was part of a more general European process, but the latter was national, connected with events related to the mid-century wars – the Cossack uprisings and the Swedish invasion.[76] Up to mid-century the greatest impact of price rises was felt by grain/livestock products, the least amongst imported goods (e.g., spices, wines). In Cracow, nominal prices remained fairly stable up to the second decade of the seventeenth century, but the commencement of the Thirty Years War (1618) seems to have caused a recession in some goods.

If the seventeenth century had been dogged by difficulties and misfortunes, its economic stagnation was gradually replaced by a livelier eighteenth century, which took off in different parts of Europe between 1720 and 1750. Poland was slower to respond to this new development, beginning its upsurge of nominal prices in the late 1730s; whilst the whole century was characterized by violent price oscillations, the general trend was upward towards the end of the century. In Cracow, prices suffered wide fluctuations, although the overall trend was downward, reaching a trough in the 1740s, to be followed by hesitant increases over the next thirty years. A low occurred around 1775–9 connected with the troubles associated with the Partition, but the years 1780–95 saw revival, part of the impetus created from greater regional exchange of goods between Little Poland and other parts of the country.[77]

Conclusion

It is essential that any study in depth should be supported by a rich body of sources and ancillary literature. In Cracow there is obviously a vast range of material, covering many facets of the city's development. Primary sources in Cracow's archives and other centres connected to it by commercial links have been supplemented with secondary published material from several countries and languages. Some attempt has also been made to review previous inquiries into the city's commercial activities in the historical perspective, in order to place the focus of this work in its wider context. Before embarking on a general temporal–spatial analysis of Cracow's trading connections between 1257 and 1795, a brief synopsis of the city's early urban growth up to the time of its Municipal Charter and the political situation from 1257 to 1500 is necessary in order to set the scene for later chapters.

3

Cracow's early development

Fernand Braudel has stated that, 'Every town grows in a given place, is wedded to it and does not leave it, except for rare exceptions. The site is favourable, to a greater or lesser degree; its original advantages and drawbacks stay with it for ever.'[1] North of the Danube river many settlements were emerging between the seventh and tenth centuries as tribal centres, fortified religious sites and princely seats; they provided a haven of refuge, commonly sited in an advantageous defensive position. Evidence suggests that they were built surrounded by wooden stockades, strengthened by earth and masonry, within which lay permanently inhabited huts for craftsmen, workshops and often a stone-built church.[2] By the late ninth century, the most important settlements dominated large areas of country and were often bishops' seats and princely castles. Gradually a separation took place between the princely/ecclesiastical functions of the settlement and the trading/artisans' quarters, a phenomenon already well defined by the eleventh century. By then, craftsmen, retainers, traders and artisans usually lived in a separate enclosure beyond the castle walls. These elementary Slav towns north of the Danube appear, until about the twelfth century, to have been largely based on local native effort with some external influences from the Church and small groups of German merchants and traders. As Rörig maintains, 'The medieval town should not be thought of as a small, self-contained unit, but only within the framework of the organic inter-relationship of towns. Long-distance trade provided the basis on which urban life on the grand scale was founded and prospered. Each town was dependent on another for its functioning.'[3]

Cracow provides a useful example of one such early Slavonic town. Archaeological evidence has shown that it was a site of early settlement[4] occupied from about the sixth century AD by the Slavonic (Vislane) tribe. Sometime during the seventh/eighth century one of the limestone outcrops known as Wawel which formed a bluff above the Vistula river was fortified; it was ringed by strong embankments made from wood, stone and earth, and formed part of a chain of castle settlements in the ninth century, through an area which became known as Małopolska (Little Poland). Further fortifications were built at the foot of the hill's northern side; within the precincts of Wawel and its suburb (Okół), a settlement of wooden buildings was established containing workshops for blacksmiths, forgers, carpenters, potters, bootmakers and craftsmen working in bone and other materials. The surrounding countryside was rich in natural resources; the fertile alluvial soil produced grain, fruit and vegetables, the forests were abundant with game and timber, whilst local supplies of clay and building stone were readily available. Iron, lead, silver and even gold ores were early exploited, together with rock-salt deposits at nearby Wieliczka. Material wealth was also accompanied by spiritual fervour. Christianity led to the foundation of a cathedral and churches on the Wawel during the eleventh century,[5] and all this suggests that during the early medieval period, Cracow was establishing itself as a centre of both political and ecclesiastical importance. During the eleventh century Cracow became one of the residential cities of the Polish kings, its position further enhanced in the ensuing century through its role as capital of the Polish state. All these factors contributed to the rapid growth of the city, which cannot be fully interpreted without some appreciation of its geographical situation. Again quoting the historian Braudel, 'We have no need to be apprehensive about calling on geography for its contribution. It probably has too much to say, but what it says is clear; it deals in known facts. There is no difficulty in summarising them.'[6]

Advantages of geographical position

The site of Cracow (50.04 N., 19.58 E.), is on the same parallel of latitude as Land's End (50.03 N.) and on the same meridian as Tromsö (19.00 E.) and Corfu (19.56 E.). The situation of the settlement is favourable both towards its immediate hinterland and the more extensive overland routes; its geographical position is one of the keys to the city's success. It lies on a contact point between three physical regions: the Cracow Upland, the Carpathian foothills and the Sandomierz basin, where the two former regions are separated from each other by the latter lying to the east and the Oświęcim basin to the west.[7] The narrow area between the two basins is known as the Cracow Gateway (Fig. 3). The Sandomierz basin lies within a triangle formed by the Lublin Plateau, the Carpathians and the Uplands of Little Poland, drained by the Vistula and its tributary, the San. Natural routeways lead out of the basin in easterly and westerly directions – to the upper Dniester valley in the Ukraine, by way of the Przemyśl Gate, and to the shallow Oświęcim basin on the upper Vistula above Cracow, respectively. From the latter a very low plateau-watershed gives easy access to the Racibórz basin in the upper Odra valley and thence north-westwards to the

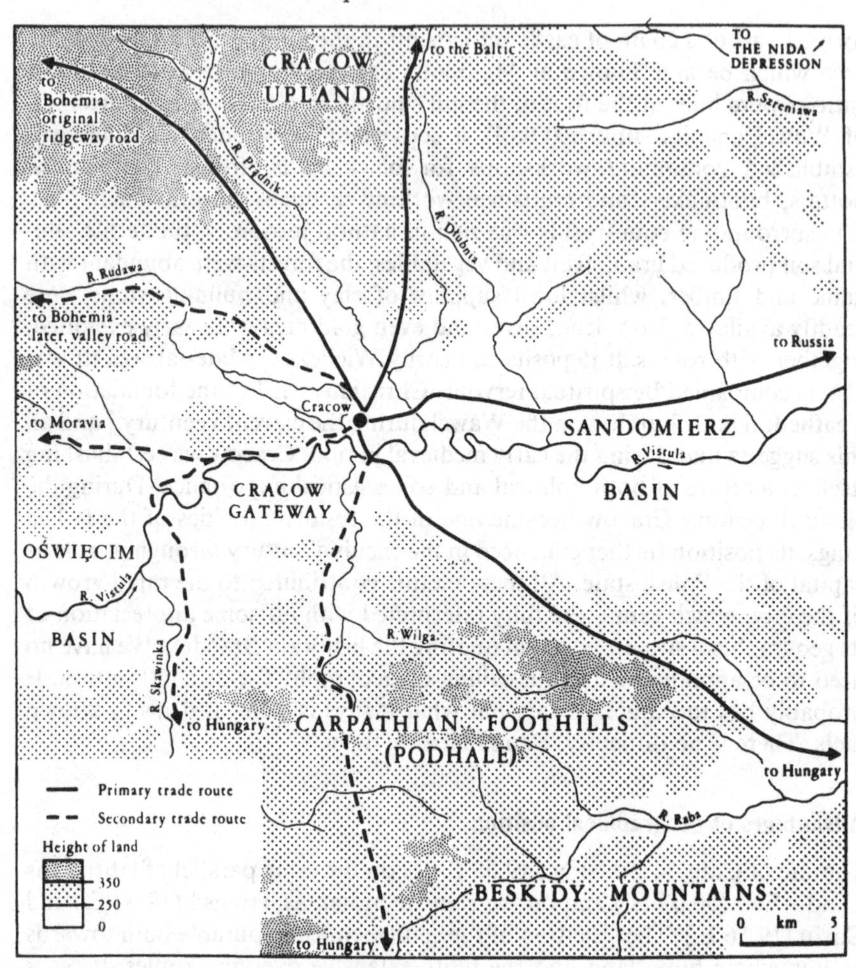

3. The geographical situation of Cracow.

Silesian Plain. From the Oświęcim basin there is also unrestricted passage south-westwards via the Moravian Gate to the Danube.

To the south of Cracow lie the Carpathian foothills and the Beskidy mountains. This well-dissected area contains a number of basins such as those of Nowy Sącz and Krosno, and also the large east/west longitudinal depression of Podhale. From here several mountain passes give access through the Carpathians to the valleys and basins of Slovakia. Directly to the north of Cracow lie the uplands of Little Poland with well-developed karstic features and an undulating terrain. This limestone plateau is drained mainly to the north-east, toward the more gentler relief of the Nida depressions; from here routes lead to the plains of central Poland and on to the Baltic coast.

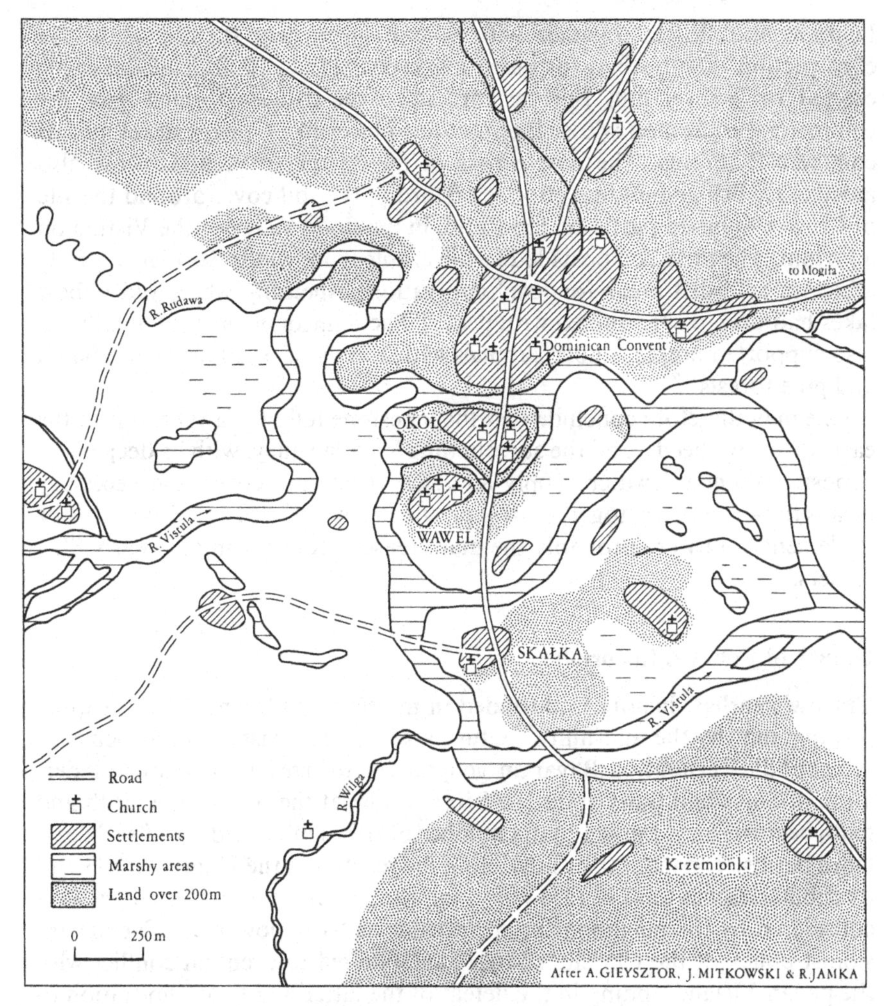

4. The site and situation of Cracow between the 10th and 12th centuries.

Cracow's exceptionally advantageous position was further enhanced by its site location (Fig. 4). The first point to notice is that Cracow lies in the Vistula valley on the southern part of the Silesian–Cracow monocline;[8] about ten kilometres to the south of the city one finds the rich soils of the Carpathian flysch zone, whilst the Cracow area is characterized by horsts and grabens formed from limestone outcrops. These outcrops, like the Wawel, Skałka, Krzemionki and Tyniec, provided excellent defensive sites above the Vistula river for early settlers. Secondly, the hydrographic network[9] provided adequate water supply at a point where an arm of the Vistula river flowed, whilst several smaller tributaries, like the Rudawa, Prądnik,

Dłubnia and Wilga, together with several lakes, lent themselves to the construction of suburban fortifications, especially at the foot of the castle erected on Wawel hill. Furthermore the river courses formed axes for commercial routes, on whose intersections places like Cracow developed as commercial centres. The narrowness of the Vistula river at Cracow also aided early bridge building efforts.[10] Thirdly, the soil cover around the site of Cracow is based partly on löess and, in the flood valley of the Vistula on alluvium. The löess soil provided fertile arable land, whilst the loamy, sandy alluvium afforded areas for local cultivation, especially where old oxbow lakes had been partly infilled with peat.[11] The immediate vicinity of Cracow also supported ample timber supplies with stands of oak/hornbeam, beech and pine forests.[12]

One may therefore conclude that the limestone hill of Wawel supplied the early defensive needs, and the surrounding Vistula valley, with its deeper soil deposits, abundant water supply and rich vegetation cover, the necessary means for sustenance; together they provided a satisfactory environment for settlement, and were probably the main reasons for enticing the early Slav settlers.[13]

Growth of Political Importance

Cracow's earliest history is shrouded in mystery and legend, but it is quite possible that by the mid-ninth century a settlement existed on, or beneath, Wawel hill. Certainly a Bavarian geographer referred to a political federation of Slavs and Balts about AD 850,[14] whilst at the turn of the ninth and tenth centuries it is known that a number of tribes inhabited the Vistula and Odra basins. The main tribe inhabiting the reaches of the Upper Vistula and its tributaries was the Vislane (Wiślanie) who, by the mid-ninth century were considered by the neighbouring Moravians as 'very powerful'. The expansionist policy of the Christian Moravian state led to eventual conflict with the pagan Vislane, ending in the defeat of the latter and their annexation to the Great Moravian Empire between AD 875–879.[15]

It is difficult to know what the various tribes inhabiting Southern Poland at this time were like. According to Goudy some of the early ninth- and tenth-century texts mentioned the Vislane, who were supposed to be a Polish tribe settled along the Upper Vistula river.[16] The Great Moravian Empire was severely weakened by the newly arrived tribe of Hungarians from the east during the second half of the ninth century; the empire suffered internal strife leading to defeat and eventual dissolution by the end of the century,[17] leaving the Vislane free from external rule. Thus in the tenth century the Vislane state emerged and embraced the entire upper reaches of the Vistula river together with part of its central basin,[18] and centred on Cracow.

This independence was apparently short-lived. The natural successor to the Great Moravian Empire was its near neighbour, Bohemia. Geographically Bohemia was more favoured than Moravia, for it had natural borders protected by mountain ranges and forests. It also possessed a strong ruler in Boleslav I who succeeded to the Bohemian throne in AD 929. He was a man obsessed with political expansion; he looked eastwards, involving himself in treaties with several tribal rulers who recognized him as their leader. He obtained the territories of the Slenzani, Dadosezani, Bobrane, Opolini and Golensizi tribes in the Upper Odra valley, and the Vislane with their main centre, Cracow. His son, Boleslav II (967–99) held on to the eastern territories[19] and according to twelfth-century Czech sources, the province of Cracow belonged to Bohemia up to AD 999. This, however, conflicts with Polish historical sources,[20] although it is quite likely that Cracow remained part of Bohemia at least during the AD 970s and 980s.

To the north, a second, and politically more important, state was founded based around the middle Warta basin; it was centred on Poznań and Gniezno, in the region known as 'Polska', Greater Poland (Wielkopolska) by the thirteenth century. Mieszko I (963–92) became Poland's first historical ruler and founder of the Piast dynasty;[21] his marriage to Dobrava (Dubravka) daughter of Boleslav I in AD 965 led to closer ties between the two Slavonic countries, and one of the results of Mieszko's marriage to a Christian princess was his conversion to the new faith in AD 966.[22] There followed a general spread of Christianity throughout his kingdom for he saw to it that his subjects adopted his religion. This movement extended to the south in the region of the Upper Vistula basin, an area first referred to in written records of the fifteenth century as Little Poland (Małopolska),[23] and prior to that as simply the 'lands of Cracow and Sandomierz'. For Cracow, the spread and general adoption of Christianity was significant; in AD 969 the first bishopric was established there, a year after its counterpart in Greater Poland, Poznań.

Between AD 967 and 990 Mieszko I consolidated his territory and seized Moravia to the south; this fact helps to refute Czech claims of retaining Cracow up to the end of the tenth century. Perhaps even more significant was Mieszko's linking of Poland with the Church of Rome; in AD 990 he formally submitted to the supremacy of the Holy See. Admittedly, there were ulterior motives for this move, for Mieszko wished to avoid the authority of the German Church, and it helped in his conflict with the common foe, the Union of the Veleti, a pagan state that lay to the west of the Odra river. Indirectly, this was of significance for Cracow;[24] diplomatic advances and agreement with the Holy See proved invaluable when Mieszko I decided to dissolve his treaty with Bohemia. The war that followed (989–92) and his eventual success meant that he was able to incorporate both Silesia and Cracow into his dominions.[25] On his death in 992, Cracow found itself part of a new state covering about a quarter of a million square kilometres

5. Poland during the late 10th and early 11th centuries.

(100,000 square miles), a country at that time larger and more powerful than the Scandinavian countries, Hungary or Bohemia.

Mieszko I's successor, Bolesław Chobry (the Brave) (992–1025), quickly reunited all the Polish provinces under his rule after their division in 992 amongst Mieszko's sons. Moreover, according to Slocombe, 'In or about the year 999 he recaptured Cracow.'[26] However, it is not certain whether this had not already been achieved by his father.[27] What is certain, is that Bolesław I continued his father's expansionist policy and enlarged the state beyond the ethnic boundaries of the Polish tribes (Fig. 5). In AD 1000 the Holy Roman Emperor, Otto III and the Papal legates visited the new Polish state and founded a church province, centred on Gniezno; besides the archbishopric of Gniezno, the bishopric of Wrocław, Kołobrzeg and Cracow (for the Cracow province) were attached, and owed allegiance to it.

By the beginning of the eleventh century, therefore, Cracow found itself situated in a state which had become one of the greatest powers in Europe, as much by reason of territorial size as of superior Polish military organization and political alliances. Geographically, Poland now embraced almost the entire basins of the Odra and Vistula rivers; this was an area based on natural geographical foundations characterized by the amalgamation of two river basins which, together with their tributaries, formed a distinct physiographical entity.[28]

At the death of Bolesław I, Poland had attained a degree of influence and a surface area which would not be experienced again for several centuries. During his reign commercial prosperity was as marked as the military progress of the nation. New cities were built; trade flourished in centres like Cracow, while foreign merchants were encouraged to enter the country. Nevertheless, the political condition of Poland was that of a military autocracy, with the cities, towns and rural settlements firmly held in the king's grip. After his death in 1025 the new state, under his son Mieszko II (1025–34), succumbed to internal strife and external attack.[29] In 1034, a very dark period began, the meagre contemporary records suggesting the threat of foreign incursions. In 1037, the Bohemian ruler Břetislav I (1034–55) raided Poland with such thoroughness that the country did not fully recover for over a century. He ransacked several parts of the kingdom, including the largest urban centres, among them Cracow, Wrocław, Gniezno and Giecz, which were plundered and destroyed; thousands of people were slain and great treasures taken.[30] For two decades the embryonic Polish state remained in disarray, but thanks to the efforts of its new leader, Casimir I Odnowiciel (the Restorer)[31] (1034–58), former losses were regained and religious life in the country reorganized. The foundation of a monastery at Tyniec near Cracow, the greatest Benedictine centre in Poland, dates from his reign; his contacts with the ecclesiastical world of the Lower Rhine and Flanders led to recruitment of men from there to the higher echelons of the Polish Church, including Aaron, Archbishop of Poland, to Cracow in 1046 (or 1049).[32]

Casimir's reign also had one other significant act of special importance for Cracow. Lack of precise information clouded the issue, but at some time during this period the city became the capital of Poland. Frančić believes it was made the country's leading centre in 1037–8;[33] Garbacik asserts that from about 1058, Cracow was generally considered to be the common capital of Poland.[34] Certainly Casimir's successor, Bolesław II Śmiały (the Bold) (1058–79), inherited a well-consolidated country, in which Cracow acquired the status of a ducal seat. Kętrzyński maintains that 'the alleged removal of the capital from Gniezno in the north to Cracow in the south', meant that 'the centre of gravity of Polish policy was said to have been shifted from the North and West to the South and the East'.[35] Whatever the circumstances, it proved of lasting significance, for Cracow remained the Polish capital for five and a half centuries, until 1609.[36]

Cracow was also to witness during Bolesław II's reign a contest for supremacy between church and state, a contemporary phenomenon in many European countries. In 1079 the Bishop of Cracow, Stanisław Szczepanowski, a prominent Polish churchman and leader of the country's clerical faction, was accused of treason in circumstances not unlike those of Thomas à Becket in England. This accusation was symptomatic of the struggle between foreign learning and native conservatism, between western culture and eastern custom, and the assertion of authority between the King and the Pope. The actual details are entangled in a maze of popular folklore and poor knowledge of Poland's political situation. What facts are known reveal that a number of Polish nobles led by the Bishop of Cracow were involved in an anti-royalist plot, strengthened by foreign intervention originating in the German lands. Bishop Stanisław excommunicated the Polish king, only to be sentenced to death in return by the royal court. According to popular tradition Stanisław, like Thomas à Becket, was killed in the city's cathedral, by the king's henchmen, but an alternative version states he was found guilty and executed in April 1079.[37]

Popular revulsion at this deed resulted in stronger support for Rome, further discontent led to open rebellion by the nobility, and Bolesław II eventually had to leave Poland, shortly before his death in obscure circumstances (1081–2) in Hungary.[38] His brother, Władysław Herman (1079–1102), a man strongly influenced by the Polish nobility, was elected by common consent to occupy the throne. Perhaps in an effort to atone for his brother Bolesław's mistakes, Władysław was responsible for building a new cathedral on Wawel hill (1096–8, finally completed in 1142)[39] and restoring several churches (e.g. St Andrew's in the city).[40] During his reign the Bohemians and Hungarians profited from the continued internal strife, making several almost totally successful incursions; the city of Cracow was seized by the invaders, along with Little Poland, which remained under Czech rule until about 1085.[41]

Joint rule by Władysław Herman's sons after his death in 1102 was characterized by dissension and turmoil. One of them Bolesław III, commonly called Krzywousty (Wrymouthed), ruled alone from 1107 to 1138; for Cracow this was a period of relative calm, interrupted only by a disastrous fire in 1125. His bitter experience of disputes over the succession with his brother Zbigniew led Bolesław III to produce an elaborate plan for his heirs; his eldest son Władysław II (1138–46) was given the country's 'senior' province with its capital in Cracow and stretching from western Little Poland through central Poland to the Baltic shore of eastern Pomerania and Gdańsk.[42] This naturally consolidated the importance and role of Cracow. In 1180, however, during the reign of Casimir II Sprawiedliwy (the Just) (1177–1194), a congress was called at Łęczyca, in which the concept of a 'senior' province was abolished in preference to creating for the king a

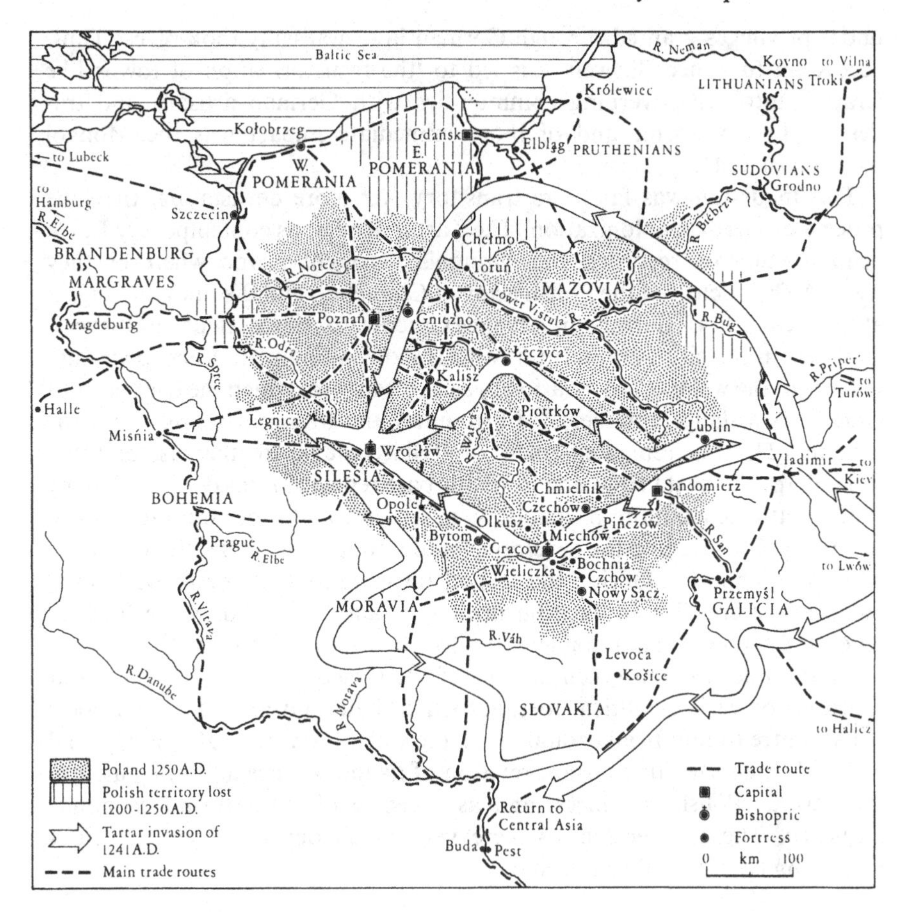

6. Poland during the first half of the 13th century.

new hereditary province of Cracow; in this way Cracow maintained its position as the chief city of Poland and symbol of unity.[43]

The first half of the thirteenth century saw political changes which were to have profound effects both on Poland and her capital city. Internal dissent and the political disintegration of the monarchy towards the end of the twelfth century led ambitious neighbouring states to cast envious eyes on Polish territory. The earthquake in Cracow in 1200, probably the first recorded,[44] heralded a new era for Poland with threats from German colonization in the north and the first Mongol invasion from the east (Fig. 6). Although German political expansion gave rise to much Polish anxiety,[45] Cracow and Little Poland seemed relatively untroubled by the activities of the Teutonic Knights to the north.[46] Some German influence came to Little Poland through Silesia; there the immigration of Germans, under privileges granted through German law, gradually took place. Thus, according to Bruce Boswell, this led to 'the transformation of towns like Cracow into self-governing communities with German artisans and traders'.[47] Less welcome and of more immediate impact was the Mongol invasion of 1241.

This incursion was far more transitory, yet more devastating, than the process of German colonization. The Tartar armies, as contemporary European documents recorded the Mongols, wrought havoc wherever they went.[48] They defeated the Polish army at Chmielnik and the Silesian knights at Legnica in 1241; during this advance Cracow was devastated[49] and several churches in the city were burnt down. Cracow, along with other towns and villages in the wake of the invasion, was laid waste (of which the trumpeter of Cracow is a daily reminder).[50] Fortunately the Mongols retired from the area in 1242, but remained rulers of Red Ruthenia, to the east of Little Poland, for the next fifty years; from this base repeated raids destroyed many Polish settlements; more particularly, in Cracow and other trading centres, the Mongols prevented the development of strong commercial ties with their eastern neighbours. During the second invasion in 1259, Cracow, Lublin, Sandomierz and Bytom were burnt down[51] and in 1287 only the fortified cities of Cracow and Sandomierz resisted the Mongol invaders.[52]

By the mid-thirteenth century therefore, Cracow had experienced four centuries of varied political fortunes which had converted it from a local tribal centre to unofficial capital of a large state; both internal struggles and external threats led to constant revision of its importance and role amongst the various Polish provinces, and as a centre of cultural and economic influence. These latter factors were very much dependent on the city's increasing commercial significance.

Early evidence of commercial contacts

Trade is central to the development of urban life. In Western Europe the general decline of trade and town life, which had begun in the third century AD reached its nadir by the eighth. The invasions of Europe in the ninth century merely added to the economic and social decline. Of course trade had not ceased; neither had coin currency or mercantile exchange, but it had ceased to be significant. Another key factor, closely connected with the collapse of trade and urban centres, was the breakdown of communications and decline in the power of government. By the mid-eleventh century in Western Europe these conditions already belonged to the past, for a new age of expansion had begun; towns and trade began to revive, new lands were brought under cultivation, and the frontiers of Europe were expanding. Perhaps most important of all, the number of inhabitants were increasing, a fact reflected dramatically in the rise of new urban centres.

The revival of city life and recovery of trade found commerce and industry as natural partners. A merchant's success often depended on knowing where to find goods that could be sold far afield, and traders would gather wherever production was concentrated. The great stimulus to urban concentration of industry was the prospect of a ready market, whilst a haven of refuge in those troubled times proved appealing to merchant and artisan alike. A city provided the merchant with a base and source of possible associates, whilst a company of merchants helped diminish risks and provide a source of more capital for larger scale enterprises. Perhaps the greatest need for a merchant was a sense of security, not only in his home town, but on his commercial travels, in places where he came to buy and sell.

Małowist has observed, however that 'Comparing the phases of development of western and eastern Europe in the Middle Ages one can easily observe important differences. Rapid development in the economies of most of the eastern countries began late in the twelfth and thirteenth centuries, when the west had already reached the height of its medieval economic development.'[53] It appears that the period from the tenth to twelfth centuries in eastern Europe was one of rather slow economic development; Bohemia and south-west Russia were the most economically advanced at this time, whilst Poland's economy was still rather at a low level of production. As expected, extensive and primitive agriculture was predominant, whilst commerce was characterised by luxury products with a limited market demand. The large majority of towns in eastern Europe were small agglomerations near the castles of princes and their courts, inhabited by people employed in handicrafts or trading and closely integrated into the needs of the local ruling groups.[54]

Cracow, blessed with a good natural position on the crossroads of two major routeways, and a safe haven of refuge, went through its embryonic stages of commercial development between the ninth and thirteenth centuries. It satisfied Sjoberg's prerequisites for the emergence of a pre-industrial city, namely 'a situation conducive to repeated contacts among peoples of divergent cultures'[55] which in turn permitted a constant accretion of social and technological skills in the area, and the emergence, through its merchant class, 'of a well-developed social organization, particularly in the political and economic spheres'.[56] The oldest accepted written record of Cracow dates from 965 AD; in that year a Spanish Jew Ibrāhīm ibn Ya'qūb visited Prague and left a description of Bohemia and its neighbouring lands.[57] He came from distant Cordoba, and travelled through central Europe; of Cracow he wrote: 'the distance between Prague and Cracow is equivalent to three weeks travel', and 'to Prague come Russians and Slavs from the city of Cracow with their wares',[58] implying that already by the mid-tenth century Cracow was known as a trading centre on the route from Prague to the east. Brutzkus has stated: 'There can be no doubt that even at this early date . . . this important entropôt on the frontiers of Poland, Russia and Hungary . . . was part of a trade route used in the tenth and early eleventh centuries which ran from Mainz to Kiev through Regensburg, Prague, Cracow and Przemyśl.'[59] However, the earliest Polish chronicles suggest that except for merchants going to Russia, Poland was a little-known country.[60]

It is quite possible that the various small settlements that were established around the foot of Wawel hill were beginning to play an increasingly important role in long-distance commerce; their convenient location along a trade and water route encouraged commercial contact. Two main inland trade routes passed through Cracow; one of them ran from Košice through Levoča (in Slovakia), to Nowy Sącz, Czchów, Bochnia and Wieliczka to Cracow,[61] and thence northwards through Miechów, Piotrków, Łęczyca to Toruń and beyond along the lower Vistula to the Baltic Sea. From Gdańsk[62] there was contact with western Europe and the Arab world,[63] lively connections having existed since the ninth century. The second route linked Bavaria with Ruthenia through Wrocław, Opole, Bytom, Olkusz to Cracow thence eastwards to Lwów and the Crimea.[64] Thus even in these early days Cracow was assuming the role of a commercial intermediary between east and west; these two trade routes were to play a vital role in the later development of the city.

Although as Lalik states, 'Up to the end of the twelfth century, we know very little about the merchants of Cracow',[65] its commercial importance is referred to in documentary sources. The Arab/Sicilian geographer and cartographer Muhammed al-Idrīsī (1100–66) refers to Cracow in his *Nuzhat al-muštāk fi'htirāk al-āfāk*, as 'a large town in Poland, lying to the north of the Tatra mountains, having a much frequented market. Routes connect it with other important towns such as Miśnia (Meissen), Gniezno, Halle etc.'[66] The eleventh-century Jewish author Yehuda ha-Cohen wrote about the trade route from south-east Europe going via Przemyśl to Cracow and the west; a second route went through Slovakia.[67] A document from the mid-twelfth century, known as the Nienburg fragment, refers to 'Cracowa urbs, ecclesia et mercatus',[68] whilst Mitkowski argues that 'Cracow at the turn of the tenth and eleventh centuries was a centre for international trade, a centre of political power and the seat of a bishopric . . . and a chronicle dated from the beginning of the twelfth century rightly calls Cracow not only a town (civitas) but also the Polish capital'.[69] Moreover, in the twelfth century al-Idrīsī wrote that Cracow contained many markets, and that within its walls lived many qualified artisans and scholars who spoke and wrote in fluent Latin.[70] Thus the urban character of early medieval Cracow presents itself as a centre of craftsmenship, international and local trade, and of scholarship.

Recent archaeological finds have confirmed the existence of products of

Cracow's craftsmen from the early Middle Ages.[71] Traces of open hearths have been located near the churches of St Mary (Mariacki) and St Adalbert (Św. Wojciech) together with numerous specimens of locally produced pottery and metal objects.[72] Some forms of artistic work in Cracow date back to the earliest period of Christianity in the city, as witnessed in local seals and money. Already during the reign of Władysław Herman (1079–1102) coins were being minted here, whilst the oldest known inventories for Wawal Cathedral (1101 and 1110), record utensils and implements made by Cracow craftsmen;[73] chalices manufactured in Cracow began to be utilised in churches both locally and farther afield.[74] Workshops where non-ferrous metals were worked have been located on the site of the somewhat later St Adalbert Church; this is suggested by crucibles found in the cultural layers beneath the early medieval occupation level with traces of silver and bronze.[75] The development of craft work may also have been connected with the growth of agricultural production around the city, notably the working of iron. Archaeological material from finds around the Okół settlement supports the existence of metalworking, where there appears to have been a quarter for smiths and founders.[76] Early medieval pottery, based on local clay supplies, appears also to have been an important craft industry in the city.[77]

Much of the commerce entering Cracow during the early Middle Ages was transit trade. Ibrāhīm ibn Ya'qūb refers to merchants from Russia carrying goods to Prague which included slaves, flour, lead and furs.[78] The export of slaves from Slav regions to Spain, Arabia and towards Baghdad was a common phenomenon during the tenth century,[79] whilst valuable furs, forest and agricultural products, and metals from the Urals, along with oriental goods, were sent en route to the Baltic coast.[80] It seems that during the tenth century Bohemia and Kievan Russia were more advanced on the road to economic development and feudalisation than Poland and the other western Slav countries.[81] Commercial relations between the Baltic coast via Russia with Byzantium and the Arabs, has been proved from money circulation found along the international trade routes.[82] It was not until the second half of the eleventh century that there seems to have been a certain amount of economic progress in central and southern Poland; this idea is supported by the appearance of regular issues of local coinage (e.g. in Cracow under Władysław Herman), which in time supplanted the foreign currencies in internal circulation.

One local product from the Cracow region, which had a wider significance was salt. As the economies of these early medieval states were partly dependent on livestock for their existence, there was a dual demand for salt, for the people themselves, and for their cattle. Fortunately for Cracow rich salt deposits, dating from the Miocene period are found in the Carpathian region, where the salt formations were strongly folded at the time of the

Carpathian orogenic movements.[83] The salt deposits lie immediately under the shallow quaternary clays in the Wieliczka–Bochnia region, some ten to thirty kilometres south-east of the city, and meant they were fairly easily exploited by the local population.[84] Although archaeological evidence suggests that salt was obtained here during the prehistoric period,[85] more intensive development did not begin until the mid-tenth century; previously salt had only been obtained from surface springs, but discovery of salt rock deposits, improved mining methods and better implements, e.g., metal pans, led to greater exploitation.[86] Documentary evidence first mentions salt from Wieliczka in 1123[87] and from Bochnia in 1251.[88]

Longer distance trade in salt from the Cracow region in the early medieval period arose due to the dearth of known deposits, or insufficient production from domestic sources, in neighbouring states. The most salt deficient area was Silesia; although rich in many other minerals it lacked this vital commodity.[89] Little information also exists for salt mining in Greater Poland during the thirteenth century,[90] but the Baltic coast was supplied by deposits at Kołobrzeg;[91] similarly both the Mazovia and Sandomierz regions lacked this mineral. Further afield, shortages were experienced in Bohemia–Moravia;[92] Slovakia and Hungary were dependent on salty streams at this time[93] so that despite the scanty documentary evidence it is reasonable to assume that exports of this commodity from the Cracow region were found on sale in these areas, along with those from Halle in Saxony, Kołobrzeg and Russia.[94]

The quickening pulse of economic life in Cracow during the twelfth and thirteenth centuries attracted people from the surrounding countryside. Besides the merchants specialising in commerce with more distant places, and the town tradesmen (butchers, bakers, artisans) without whom city life is impossible,[95] there were local traders from the immediate vicinity who found in Cracow a ready market for their produce. Mitkowski believes that 'towards the end of the twelfth century, and during the first decades of the thirteenth century Cracow may have had some 5000 inhabitants'.[96] Gieysztor, again quotes al-Idrīsī, who, about 1154, wrote, 'Her [Poland's] towns flourish and the population is numerous ... Among her towns is Cracow [Kraku]. It is a beautiful and large town with a great many houses, inhabitants, markets, vineyards and gardens.'[97]

Many of the local products were based on the exploitation of woods and waters[98] (animal skins, honey, wax, fish) and were supplemented by the more permanent agricultural production of grains and livestock.[99] Produce was delivered to the Cracow market in carts or waggons pulled by horses or oxen,[100] and a document dated 1256 specifically refers to them as a method of connecting Cracow with the numerous surrounding villages.[101] Wąsowiczówna's work on the medieval Polish system of roads emphasizes the importance of local and regional routes in the development of early medieval town centres and their significance for the growth of local markets.[102]

Finally, mention should be made of the impact this economic development had on the cultural life of Cracow and the early impetus it gave to the city as a centre of scholarship. Commercial success led to more regular contact with distant lands and ideas and helped to promote a more common level of culture. During the twelfth century some of Cracow's libraries were founded; in the Cathedral, by the Benedictine monks in Tyniec, the Cistercians at Mogiła, and in the city's Dominican Convent (founded 1222).[103] Unfortunately, as a result of domestic feuds and foreign wars the foundation of libraries did not immediately lead to the development of vernacular literature; not until the thirteenth century was an important chronicle (Chronica Polonorum) written by the Bishop of Cracow, Wincent Kadłubek (1150–1223).[104] Even so, according to two inventories taken at the beginning of the twelfth century, the library of the Cracow chapter possessed the basic Church and secular literature.[105] Polish pupils found their way to the leading schools of western Europe,[106] such as Lyons and Paris. Court annals continued to be written, perhaps the most famous by a foreign Benedictine monk in the court of Bolesław Krzywousty (Wrymouthed) between 1116 and 1119, which extolled the concept of state unity and importance of the ducal dynasty.[107] Few early paintings have survived, but notable amongst them is the mid-eleventh century *Sacramentarium* in Tyniec, originally from Cologne.[108] Illuminated manuscripts and paintings began to be imported into Cracow from both the east and western Europe, whilst its merchants brought pictures of the Madonna from abroad, richly decorated with gold and jewels; mosaics were introduced from Byzantium in the twelfth century.[109] Monumental sculpture also flourished, but on a more modest scale than in Lombardy or the Low Countries.[110] All this cultural and scholastic activity reflected a curiosity of the world beyond the city's limits, and helped sharpen the awareness of its citizens to problems of general interest to their society. This was certainly reflected in the thinking of the new spatial plan of the city, which emerged during the sixth decade of the thirteenth century.

The Municipal Charter of 1257

The thirteenth century was a period of flourishing development in the history of European towns and cities. Their rise and the expansion of commerce had some consequences of general importance. After the invasions of the ninth and tenth centuries a commercial class had emerged and the older town centres had little understanding of its needs. This new section of society had to deal with the financial demands of everyday life, such as questions of credit and debt, price regulation, conditions of sale, etc. Such matters could only be competently dealt with by the citizens themselves, and a right to some form of self-government was therefore vital for their future prospects. In some countries the struggle was a long one, the citizens fighting for the rights to choose their own laws and govern their own cities. The more enlightened European rulers realized the necessity to accept these forces of change and to grant those privileges demanded by the rising merchant class.

The Polish rulers, wishing to take advantage of increasing economic activity and the development of trade relations with neighbouring countries, encouraged the founding of towns, giving them a legal status based on western and central European examples.[111] Henryk Brodaty (the Bearded) (1232–8) was particularly sensitive to the significance of contemporary social and economic changes, and knew how to take advantage of them to strengthen his authority. The legal aspects of urban life were set out in special charters. These charters established relations between the feudal lords and the colonists who settled in the towns and cities. The idea was based on examples taken from German towns, whose citizens had regulated their own legal status earlier, in order to create favourable conditions for economic development. The most important changes resulting from this new 'German law' (ius Theutonicum), were the replacement of dues to the feudal lord in kind, by money, and the granting of quite extensive judicial and administrative self-government to a city's inhabitants. In most cases the charters were based on the Magdeburg law, or its Polish variants, namely the Środa (ius Novi Fori Sredense) and Chełmno (ius Culmense) laws.[112]

The Magdeburg municipal law was introduced to Cracow by way of Wrocław. There is some conjecture that the first charter for the city already existed during the reign of Leszek Biały (the White) (1194–1227); this idea is based on the building of St Mary Magdalene Church and a better organized market square during the years 1211–17 under the influence of the more regular Silesian town plans.[113] Perhaps more realistically, Bromek has suggested that the whole process of the city's foundation charter was a long-drawn-out operation which initially began about 1220 and lasted over a one-hundred-year period.[114] What is certain is that the acceptance of German law by Polish towns was called 'locatio civitas', and was widely adopted during the first half of the thirteenth century; much of the stimulus for its acceptance came from Henryk Brodaty, the shrewd ruler of Silesia, who saw such charters as a means of obtaining fiscal rewards and invigorating the local economy by attracting both foreign merchant capital and craftsmen and miners into the area.

At this time Cracow controlled several craft and market villages, each under the influence of various lay and ecclesiastical lords. In this situation it was necessary to strengthen the city's single municipal authority under a leader (sołtys), in order to free the local inhabitants from direct dependence on the feudal lords. Documentary evidence from 1228 and 1230 refer to such authority being vested in one Piotr, sołtys of Cracow ('scultetus Cracoviensis') and his officials (ministri).[115] By suggesting this idea, Henryk Brodaty sought to ingratiate himself with Cracow's leading citizens, whom he hoped would later use their influence to help him become master of the state capital. Nevertheless, the sołtys remained an official of the city's ducal ruler; the latter's power, though still effective, was now restricted only to matters concerning the provision of market facilities.[116] Unfortunately the work of organizing Cracow, begun by Henryk Brodaty and continued by his son, Henryk Pobożny (the Pious) (1238–41), was interrupted by the first Mongol invasion of 1241 when Cracow was completely destroyed.[117]

Three years after the Mongol invasion, Magdeburg law was applied to the merchant craft-settlement of Okół, which lay to the north of Wawel hill on a dry Pleistocene alluvial fan. Here a prototype attempt at planning a regular shaped market square was tried out, which acted as a forerunner for the much greater venture which followed in 1257.[118] Meanwhile documentary evidence from 1250 confirmed that the city was still under the influence of its own sołtys, one Salomon ('Salomon scultetus Crfacoviensis'),[119] and enjoyed a certain measure of self-government.[120]

The mid-thirteenth century saw the granting of municipal charters to a number of urban centres in Poland; Poznań (1253)[121] soon followed by Kalisz (1253–60)[122] and others including Cracow in 1257.[123] This was part of a process in which according to Ostrowski, 'Separate settlements built by various nobles and clergy often grew up around the largest burghs, for the feudal lords liked to settle near the seats of their dukes. As a result of this process, the large early medieval towns, such as Poznań and Cracow, were each more a growing group of settlements than a compact municipal organism; to use the modern term, they were primitive "conurbations".'[124] The actual confirmation of the legal charter of municipal status for Cracow was signed by Bolesław Wstydliwy (the Shy) (1243–79) together with his mother Grzymisława, and his wife Kunegunda-Kinga at the Kopernia assembly (near Pińczów) on the 5 June 1257.[125] Financial donations were made by these two women. Further contributions came from three wealthy Silesian immigrants (Getko i.e. Gideon, Jakub and Ditmar) who became town founders (locatores) and city fathers (advocati). Their donation to Bolesław secured for themselves the office of wójt (advocatus),[126] a sort of mayoral position in the city. By virtue of the charter these burgesses were granted considerable benefits, such as complete administrative autonomy in municipal matters, independent jurisdiction under Magdeburg law – with the provision that final appeals could be made to Magdeburg city itself – and the right of legislation;[127] the sphere of action of the municipal administration was, however, limited to the territory and inhabitants of the towns.

Bolesław's municipal charter stressed the need for a complete overhaul and reorganization of the city which, in future, was to be centred around a

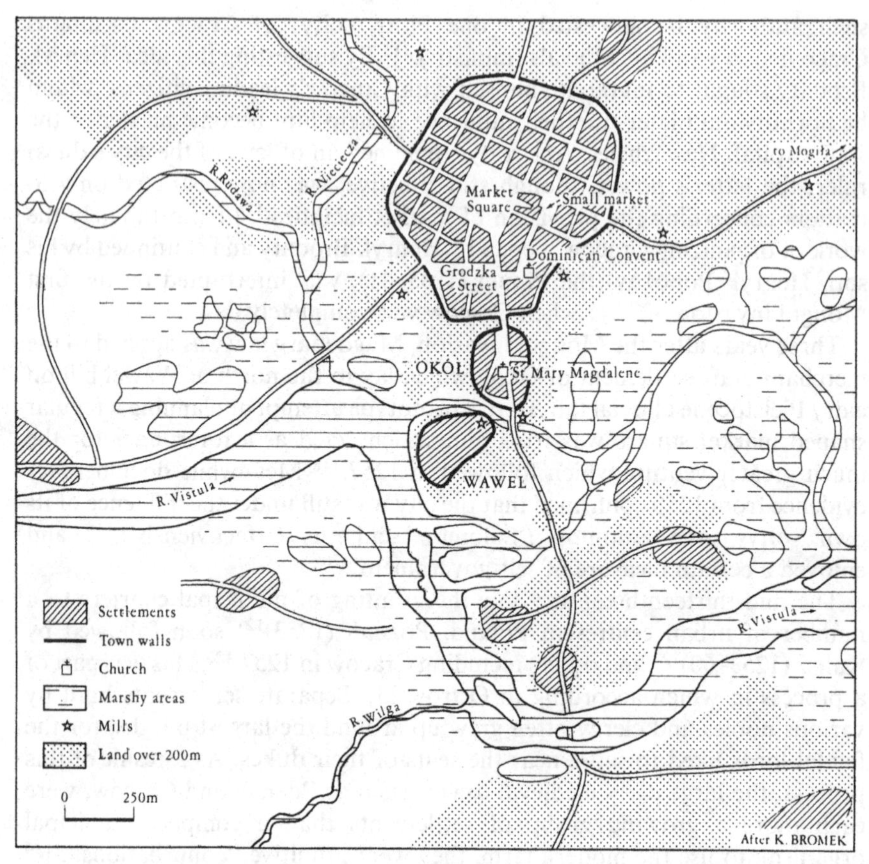

7. Cracow at the time of its municipal charter, AD 1257.

concisely planned area completely surrounded by walls. The hub of this newly planned area was to be the 'circulus' or market (rynek) which was to be located next to the smaller market-place (mały rynek) first laid out in the prototype plan of 1244. Around the new larger market square a network of streets was designed, which intersected each other at right angles giving a gridiron effect (Fig. 7).[128] The one exception to this rule, partly dictated by the terrain and the location of the ruler's residence on Wawel hill was Grodzka Street[129] which left the market place at an acute angle. The total planned area covered over 30 hectares and was based on similar designs for the market squares of Wrocław (1242), Poznań (1253) and certain Silesian towns.[130] The new urban plan was able to solve the previous problem of road congestion in and around the market area, which had arisen over the years as a result of the irregular street pattern therein. Finally, the everyday running and upkeep of the newly planned city was in the hands of the city council, administered by the hereditary wójt's office; this was a privilege gained in exchange for the profits from annual city rents and land holdings, which became the main source of Cracow's ducal and later monarchical income.

The year 1257 provides a convenient halting place in the historical development of Cracow. It marks a watershed between its earlier phase of growth and the period following the granting of the municipal charter; several substantial changes were made in the ensuing years, not only in real size, for Cracow was steadily expanding, but also in the form and style of urban administration and local government. By the end of the century it had absorbed the area around Okół,[131] and during the first decades of the fourteenth its walls had been extended to include the settlement on Wawel hill.[132] In future it was the duty of Cracow's council to 'preserve the laws and uphold the honour and benefits of the city'.[133] It was now responsible for settling market disputes, for supporting crafts and commerce, and especially for controlling the application of weights and measures in an honest way. The council was empowered to call meetings of the whole population, when promulgating city laws, whilst the municipal bench administered justice and the wójt was responsible for the city's defence in case of emergency. The council was elected on an annual basis, the members retaining the right, at the end of its term, to nominate their successors.[134] The city therefore faced a bright future, in which its first municipal authorities enjoyed a great measure of freedom in carrying out their duties, and Cracow during the latter part of the thirteenth century could be said to have been truly autonomous. Unfortunately this idyllic situation was not to last, for the turn of the century saw increasing power struggles and greater tension, as the various provinces were encouraged to coalesce into the unity of a single state. This, in turn, meant fighting for the control of Cracow, which had already become Poland's symbol of natural unity.

Conclusion

Thus Cracow, blessed with an advantageous natural position in southern Poland, and a haven of refuge in troubled times, went through its embryonic stages of commercial and cultural development between the ninth and thirteenth centuries. During the reigns of the first dukes and kings, especially Mieszko I, Bolesław I (the Brave) and Bolesław II (the Bold), Cracow gained a powerful position within the newly formed Polish state. The latter was achieved in spite of strong competition from the grand dukes of Kiev and the kings of Bohemia, with its boundaries established in the Odra and Vistula river basins. The oldest documentary evidence from the mid-tenth century describes Cracow as an important commercial centre on the Vistula river at the junction of international routes leading from the south to the Baltic Sea and from the countries of the Near East via Silesia to Western

Europe. All these factors contributed to the rapid growth of the city; in the eleventh century it became one of the residential seats of the Polish monarchy, symbolized by the castle and cathedral on Wawel hill, whilst in the twelfth century Cracow was established as the state capital. During the thirteenth century the increasing size and prosperity of Cracow demanded the reform and modernization of its earlier type of organization and changes in the routine of everyday life. This was accomplished through the adoption of a municipal charter in 1257, which enabled the city to reorganise its administrative functions and spatial pattern to meet the rigours and demands of the fourteenth century, when Cracow entered a new phase of her development.

4

The political situation of Cracow, 1257–1500

By the mid-thirteenth century, Cracow found itself in a changing political world; the two great empires, bequeathed to Europe by the early and central Middle Ages for later times, were in disarray. The Eastern Empire, based on Constantinople, was only a shadow of its former self, whilst the Holy Roman Empire lapsed for a time after the death of Frederick II (Hofenstaufen) in 1250. In western Europe, particularly in France, England and Spain, the twelfth and thirteenth centuries had witnessed a growth of stronger royal administration and a general acceptance by the people that hereditary monarchy was an unavoidable part of political life in their lands. This belief was by no means universally accepted throughout Europe; in Scandinavia the monarchies were not powerful, whilst in eastern and central Europe, the princes were encountering resistance among their more influential subjects. Moreover, during the latter part of the Middle Ages the Poles, Hungarians and Czechs were able to develop as sovereign states, unhindered by very powerful neighbours to the east or west. Freedom from such distractions meant the rulers of Poland, Hungary and Bohemia were to become increasingly engaged in a struggle for supremacy among themselves; the prize for such efforts was a chance for unity under a powerful dynasty which alone might save it from the ambitions of German or Russian expansion.[1]

In Poland, lack of natural protection had always laid the country open to aggressive neighbours, and forced its medieval rulers to watch carefully the vicissitudes of political units along its eastern and western borders. The ambitious plans of the Bradenburg Margraves in the west, with their covetous eyes on the Polish lands, and the incursions and raids of the heathen Lithuanians and Tartars in the east, were always a threat during periods of the thirteenth and fourteenth centuries. To the south, the great wall of the Carpathian Mountains afforded some protection although the numerous passes and cols could allow possible military, cultural and economic influences to threaten Polish territory; throughout the fourteenth century there were continual conflicts arising with the Czechs and their German overlords.

Conversely, Poland's Hungarian neighbours were always regarded as possible allies against trouble from either Bohemia or the east. Northward, the Baltic Sea had always proved a successful obstacle to serious invasion, but the infiltration of Teutonic Knights into Prussia was eventually to lead to open conflict in the fourteenth century.[2]

In Little Poland the struggle for Cracow recommenced, after the retreat of the Tartars in 1242, only to end with the accession of Bolesław V, Wstydliwy (the Shy), 1243–79; the one significant feature of his comparatively long reign was the increasing power of the clergy both in political and religious affairs. Territorial disintegration had characterized the period prior to his rule, when the parcelling and portioning of the country meant that almost every province of Poland had become divided and further subdivided among the many petty princes and princelings; Cracow alone remained one and indivisible, although its power had been severely shaken by the incessant struggles for its throne, which Bolesław's reign only merely interrupted.[3] Moreover, continual civil wars and internecine struggles in Poland during the first half of the thirteenth century were further complicated by the Tartar invasion; all these events led to insufficient production within the country and the need to invite German settlers into Poland with the offer of protection, free land and other forms of preferential treatment. The first colonists came from Germany to Silesia and, subsequent to the Tartar invasions, gradually migrated through Poland.[4] One offshoot of this movement was the settlement of a population almost entirely German in origin; all matters concerning their own welfare were solved according to their own German judicature. An example of this is well illustrated in Cracow. The municipal charter of 1257 was established 'zu germanische Recht', as was the planned town,[5] for, according to Pounds, 'It is likely that the Germans who created it could be persuaded to come only by the grant of a formal charter of privileges – German law, as it was termed; the Polish custom was too indefinite and uncertain to attract them.'[6] Nevertheless, the Wawel with its royal castle and settlement beneath it, remained a strong symbol of Slavic presence in the city.

Importance as a royal capital

By the mid-eleventh century, Cracow was generally considered to be the common capital of the country, a position it was to hold for the next five centuries.[7] Nevertheless, it was not without competitors, for in the eleventh century Kruszwica, the old Polish capital, and Gniezno, the metropolitan see, had actually achieved some importance and had become prominent in the political life of the country; but in the ensuing period of partition, when political localism and regional particularism had become more rife, both cities were ruined. Later the centre of national life was transferred from

Greater Poland to Little Poland, where Cracow, from the advantage of her position, soon became the seat of the monarchy and one of the most influential cities in Europe.

The municipal charter of 1257 further strengthened Cracow's position; anyone was admitted to the freedom of the city who could either produce his baptismal certificate (*litterae genealogiae*) or prove through witnesses, that he or she was of legitimate descent and a Roman Catholic. The newly received citizen had to take an oath of allegiance to the city, and likewise those who later migrated elsewhere were obliged to resign their freedom of Cracow. As the residence of both king and bishop, Cracow also contained within its walls the whole nobility that held court and state offices; they mostly had their own residences in the city. Nobility and clergy were exempt from municipal laws, and also excluded from city rates.[8]

Perhaps even more important in the long run was the psychological impact Cracow had on the country. The city was perceived by the country's inhabitants as their capital and by the end of the thirteenth century it had become the focus of sentiment, in spite of political disintegration, for Polish unity.[9] Not only was Cracow the capital of Little Poland, it was rapidly emerging as the arch-rival to Gniezno for the title of ecclesiastical centre for the whole of Poland.[10] When Władysław Łokietek (the Short),[11] 1306–33, became king, as Knoll has pointed out, 'Cracow was known in Poland as "the seat of the kingdom", and "the royal city and dwelling place", where, "the Polish crown . . . had been since antiquity".'[12]

The physical appearance of the Wawel hill was also changing. Between 1257 and 1300 the whole complex was rebuilt with a complete new system of defensive walls.[13] In 1265 a stone castle was begun to replace the old wooden structure,[14] which from its commanding view of the surrounding countryside, provided an excellent base for future Polish kings. As Knoll reminds us, 'What may be overlooked is the fact that Cracow is a city of great military significance, since its acropolis, Wawel Hill, is an excellent strongpoint which in 1288 was already well fortified.'[15] In spite of setbacks like the burning down of the cathedral on Wawel in 1306,[16] Cracow and its citizens did obtain some privileges from Władysław Łokietek's rule such as the promise that the castle wall would never be joined up to the city walls,[17] staple rights, major renovations to the cathedral, and extensions to the castle in 1320.[18]

Although the years of Władysław Łokietek's rule were turbulent ones,[19] punctured with seemingly endless wars, some of them proved significant. During the years 1313–14 he conquered the province of Greater Poland, and was able to form a nucleus from which to restore the Polish kingdom and revive the concept of 'regni Poloniae'; his coronation in Wawel cathedral in 1320 marked the unification of the formerly disjointed Polish lands under a single rule. From then on, until the late eighteenth century,

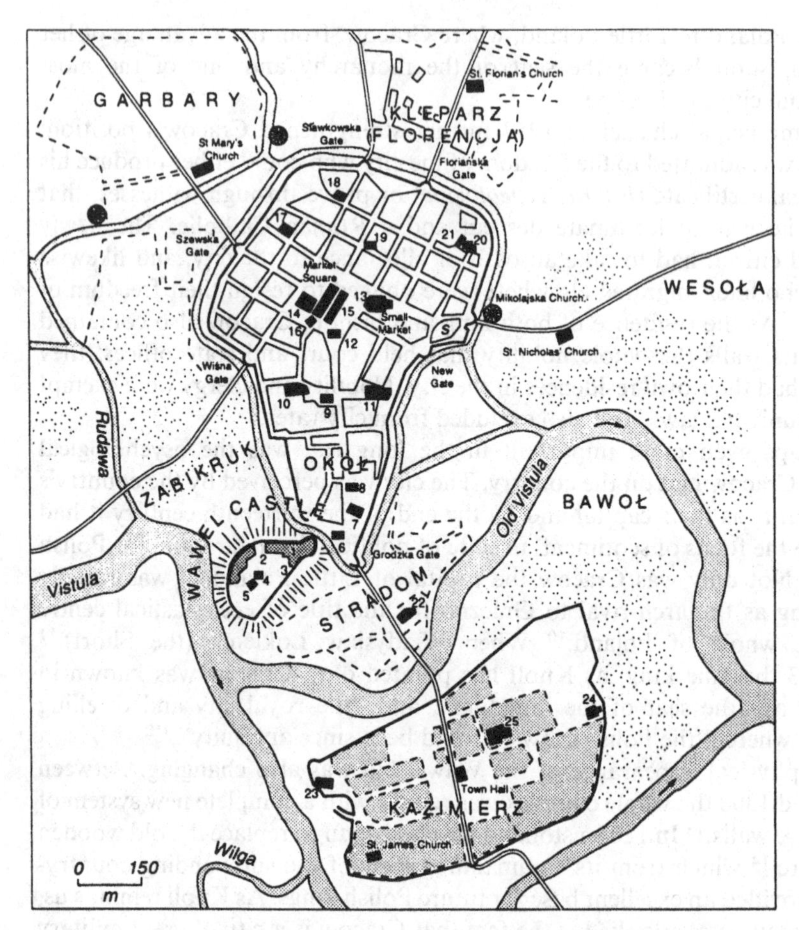

Wooded area

Mills

Walls and gates

S Stronghold

1 Royal Castle
2 Cathedral
3 Rotunda of St. Felix and Adauctus' Church
4 St. Michael's Church
5 St. George's Church
6 St. Idzi's Church
7 St. Martin's Church
8 St. Andrews Monastery
9 All Saints Church
10 Church and monastery of the Franciscans
11 Church and monastery of the Dominicans
12 St. Adalbert's Church
13 Church of the Virgin Mary
14 Town Hall
15 Cloth Hall and merchants stalls
16 Stuba Communis
17 St. Stephen's Church
18 St. Mark's Church
19 St. John's Church
20 Church of the Holy Cross
21 Hospital of the Holy Spirit
22 St. Hedwigs Church
23 Church and monastery "na Skalce"
24 St. Lawence's Church
25 Church of Corpus Christi (being built)

8. Cracow in the mid-14th century.

Cracow was the place for the coronation and interment of all Polish monarchs and up to the seventeenth century the permanent residence of the royal court.

Łokietek's successor, Kazimierz Wielki, the only Polish king to be called 'the Great', 1333–70, ushered in an age of considerable prosperity for Cracow. The king favoured the development of the city, seeing in its power and wealth a sound basis for his rule. He was particularly active in the building of churches, castles and fortresses throughout the country, replacing the wooden structures with more solid materials. Much of this work is evident in Cracow, and of whom it has been said he 'found Poland dressed in timber and left her dressed in brick'.[20] Moreover, he granted charters for the building of new town settlements around Cracow (Fig. 8), establishing to the south Casimiria (now Kazimierz) in 1335,[21] and Florencja (now Kleparz) in 1366. The former site incorporated the small villages of Stradom and Rybaki; five years later (1340) Bawół was added, together with a canal which gave the new town a fortified position being completely surrounded by watercourses connected to the R. Vistula. To the north of Cracow the latter settlement was founded around the church of St Florian.[22] Both these new towns were to become future rivals of Cracow in the wake of commercial development during the later Middle Ages. Kazimierz Wielki's interest in education was also reflected in the growing number of schools founded in Cracow for its citizens during the second half of the fourteenth century. His concern for higher education culminated in the foundation of Cracow University in 1364, the second, after Prague, to be established in Central Europe;[23] it was based on the Italian model, where law was the principal subject and enabled the training of state officials. Finally, mention should be made of Kazimierz Wielki's attempts to give Cracow greater financial significance through the establishment of its own currency. In 1338 the large Cracow grosz (grossi Cracovienses) was first minted, with the further hope of creating a unified Polish monetary system; continued debasement by the royal treasury and greater stability enjoyed by the Prague groschen (grossus Pragensis), unfortunately failed to achieve this aim.[24]

Further proof of Cracow's increasing political significance came in 1364. The great Congress of Cracow on the instigation of the Polish ruler, led to the meeting of all the important rulers and princes of central Europe.[25] Initially the impetus came from the King of Cyprus, whose island was threatened with invasion by the Turks; he consulted Kazimierz Wielki on matters of defence against the Ottomans which eventually led to this congress which Knoll believes, 'If one were to single out a specific event which symbolized the emergence of Poland as an important central European power, it would be difficult to make a better choice then the great Congress of Cracow in 1364.'[26]

The death of Kazimierz Wielki in 1370 and the succession of Ludwik

Węgierski (of Hungary), in 1370–82, did not diminish the significance of the royal capital. From the outset Ludwik extended great favour to the city, giving it free access to all the main trade routes, and confirmed staple rights for the city's merchants. Such privileges were not given without reason, for the king wished to solicit the goodwill of Cracow's citizens in the hope of securing the Polish throne for his younger daughter Jadwiga. His plans were realized and her accession in 1384 was followed two years later by an important historical and political event. In 1386 she married the Lithuanian Prince Władysław Jagiełło (1386–1434) and thus added a new large province to the Polish kingdom and to the cause of Christianity. Wawel castle became the seat of the new Jagiellonian dynasty which was to last for over two hundred years. This event, as Frančić confirms, increased the city's ranking as an important political centre.[27]

Within the city other changes were taking place. Between 1380 and 1400 the Drapers' Hall (Sukiennice) was rebuilt in stone in the centre of the main square; this replaced the original building which had consisted of four rows of booths severed by a narrow gangway and covered by a timber roof.[28] This was also a period of great local activity; in 1388 Cracow city bought up the village of Grzegorzowice (today Grzegórzki) with its quarry and a year later the small settlement of Dąbie. About 1394 the Church of St Barbara began to be built, and in the next year first mention was made of parchment manufacture in the city.[29] In 1399 an aqueduct was constructed (called 'rurmus'), tapping water from the Rudawa river by means of 'a special contraption and conducting it throughout the town', in which 'separate pipelines even supplied water to the palace'.[30]

The fifteenth century was to see further consolidation of Cracow's position as the leading city of the Polish state. Externally, Cracow's age of splendour coincided closely with the era of the Jagiellonian dynasty. The Union of Poland and Lithuania had strengthened the country politically, militarily and economically; the historic victory of the new state over the Order of Teutonic Knights at the famous battle of Tannenberg (or Grünwald / Zielone Pole) in Prussia (1410),[31] marked the beginning of Poland's rise as a great power. The following year Cracow's Wawel castle provided prison facilities for 51 knights captured in the battle; confiscated banners decorated Wawel cathedral, whilst other trophies and *objets d'art* taken as spoils of war were distributed among the treasuries of several city churches.[32] On Władysław Jagiełło's death (1434) he was succeeded by Władysław Warneńczyk (1434–44) whose rule had no direct influence on the city. This could not be said of his successor, Kazimierz IV Jagiellończyk (the Jagiellonian), 1447–92, particularly after 1454 when he married Elizabeth of Hapsburg.[33] She was a person of refined taste and learning, whose main interest was promoting closer cultural ties between Cracow and Nuremburg as well as other centres of German art. In 1465 the first book printed in Cracow was by the German printer Günther Zayner,[34] followed in 1491 by the first Slavonic incunabula from the Cracow press of Sweipolt Fiol, who combined the printer's trade with that of embroidery.[35]

Internally, the city continued to grow and prosper. Intensive building activity characterised fifteenth-century Cracow, including such notable buildings as the Mariacka church and Barbakan Gate.[36] The new settlements of Kazimierz and Kleparz continued to develop, (Fig. 8) whilst the area around Stradom,[37] known as 'pons regalis' developed handicraft trades geared to the needs of Wawel castle. Other settlements outside the city walls developed specialist functions such as Garbary (leather and tanning).[38] For Cracow the second half of the fifteenth century was a period of flourishing intellectual life, contacts with foreign countries, and flowering of artistic skills, in painting, sculpture and craftsmenship. This era is best illustrated by the work of Wit Stwosz,[39] one of Europe's greatest late-Gothic sculptors, whose work can still be seen today in St Mary's Church (Mariacki) – the high altar, and in Wawel Cathedral in the form of Kazimierz Jagiellończyk's tombstone. Underlying much of this success, however, was the need for continued favourable political relations with the city's close and more distant neighbours.

Relations with other Polish towns

The growth of town and country in Poland led to an increasing stratification of the population. The urban population varied and was related to actual town size. At the beginning of the fifteenth century large centres,[40] such as Cracow with about 14,000 inhabitants and the smaller towns like Poznań (4,000), Bochnia and Sandomierz (2,000 each), were inhabited by three distinct groups of burghers, excluding the gentry and clergy, who themselves often constituted a considerable percentage of the population. At the top of the pyramid stood a small group of merchant patricians who, in the course of the fourteenth century, assumed complete control over the political and economic life of the main urban centres. It is not surprising, therefore, that Cracow's relations with other Polish towns often revolved around the role played by commercial politics and the presence of trade wars between different urban market settlements.

Part of the problem was the development of urban particularism and protection of each town's own commercial importance.[41] Moreover, the concept of stapling rights encouraged inter-urban rivalry; many towns obliged merchants passing through them to unpack and offer their wares for sale to the burgesses before going on, and was a serious hindrance to inter-local transport. In 1306 Władysław Łokietek granted Cracow such exclusive and unlimited staple rights for the rich copper trade;[42] this placed the Cracow merchant in the important position of mediator, and considerably increased his significance. The city's richer merchants became members of the patrician class, contributing their skills to the banking business and even lending money to monarchs. Unfortunately, Cracow was not alone in imposing stapling rights, for other Polish towns obtained such privileges during the thirteenth–fifteenth centuries, such as Sandomierz (1286), Kazimierz (1335), Bochnia (1369), Lublin (1392), Poznań (1394) and Kleparz towards the end of the fifteenth century.[43] Disagreements over stapling rights often led to trade wars and boycotts between towns, in Cracow's case especially with the nearby, but independent settlements of Kazimierz and Kleparz.[44] Sometimes the merchants evoked the monarch's displeasure, as in Cracow after a dangerous insurrection in 1311.[45] Privileges granted to the city were somewhat curtailed, and the king's favour turned to the town of Sandec (Sącz)[46] which began to rival Cracow's commercial role as a trade mediator for goods en route to Hungary and the East from the Baltic and Western Europe. High customs duties were imposed upon the capital, hitherto exempt, the situation remaining sensitive for several years, although Sandec did not obtain stapling rights till 1554.[47] Political conflict over commercial rights continued up to the mid-fourteenth century when Kazimierz Wielki confirmed Cracow's monopolistic position over other Polish towns forbidding non-Cracow merchants from utilizing trade routes to the East, much to the annoyance of such towns as Lublin, Sandomierz,[48] and especially Lwów in 1403.[49]

Another factor evident during the thirteenth–fifteenth centuries was the increasing immigration of people to Cracow from other parts of Polish territory. Analysis of trade guild membership reflects growing polonization of the city; in 1403 only 13 per cent of the members were Poles, but by 1450 this had increased to 29 per cent and in 1500 reached 41 per cent.[50] Mitkowski has stated that 'the Polish element in Cracow was both numerous and strong during the period from 1257 to the end of the XIVth century. This is the era of the greatest concentration of German colonization in Cracow and the era of their greatest economic and political domination.'[51] Certainly increased contact and intermixing with the exclusively rural Polish population and from other urban areas could have explained this phenomenon; moreover stronger ties by the newly founded university in the city with other parts of the country, and contacts by the Polish clergy, encouraged students from other areas to study in Cracow. The acceptance of numerous Poles to the freedom of the city also swelled the number of inhabitants, many coming from the nearby urban centres of Kazimierz and Kleparz.[52] According to Ladenberger, Poles formed about half of Cracow's population in the fourteenth century, whilst they completely dominated the towns of Kazimierz and Kleparz (Table 2).

The third largest group after Poles and Germans were the Jews who constituted a separate national, religious, cultural and legal group in Polish

Table 2. *Ethnic structure of Cracow, Kazimierz and Kleparz population during the fourteenth century*

Ethnic Group	Cracow	Kazimierz	Kleparz
Poles	*Circa* 5,000	*Circa* 1,500	*Circa* 1,000
Germans	*Circa* 3,500	–	–
Jews	*Circa* 800	–	–
Hungarians (and later, Italians)	*Circa* 200	–	–
Others	*Circa* 500	–	–
Total	10,000	1,500	1,000
Court, soldiery, clergy		*Circa* 2,500	
Grand total		*Circa* 15,000	

Source: T. Ladenberger, *Zaludnienie Polski na początku panowania Kazimierza Wielkiego*, Lwów, 1930, p. 63.

towns.[53] Liberties originally granted them in 1264 by Bołesław Wstydliwy were extended by Kazimierz Wielki to the whole Polish Kingdom (Statute of Wiślica, 1346) and once more confirmed in 1453 by Kazimierz IV (Jagiellończyk).[54] Jews in the surrounding villages and smaller towns had played an important role in the Cracow region since the end of the thirteenth century. They devoted themselves to the traditional activity of money lending, which promoted the growth of a commodity/money economy; one of their number, Lewko, became banker to Kazimierz Wielki and to his successor.[55] Kazimierz Wielki granted privileges to the Jews of Little Poland in 1367.[56] During the last decade of the fifteenth century, however, anti-Semitism in Cracow led to many Jews resettling in the nearby town of Kazimierz.[57] In 1495 the king, Jan Olbracht expelled all Jews from Cracow resettling them in the Jewish quarter of that town.

Cracow's relations with other urban centres in Poland were closely tied with her development as an important trade centre during the period from her municipal charter to the end of the fifteenth century. Harmonious relations appeared to be most noticeable the further towns were from Cracow, such as Warsaw, Jarosław or Wilno (after 1386)[58] where little competition arose over trade and transit commerce.[59] Closer to home questions of customs tariffs[60] and control of trade routes[61] often led to domestic disputes, although in general Cracow had an important influence on the overall development of other towns in Little Poland at this time.[62] Perhaps of equal significance at this time was Cracow's connections with her various foreign neighbours, not least the lands of the Magyars to the south (Fig. 9).

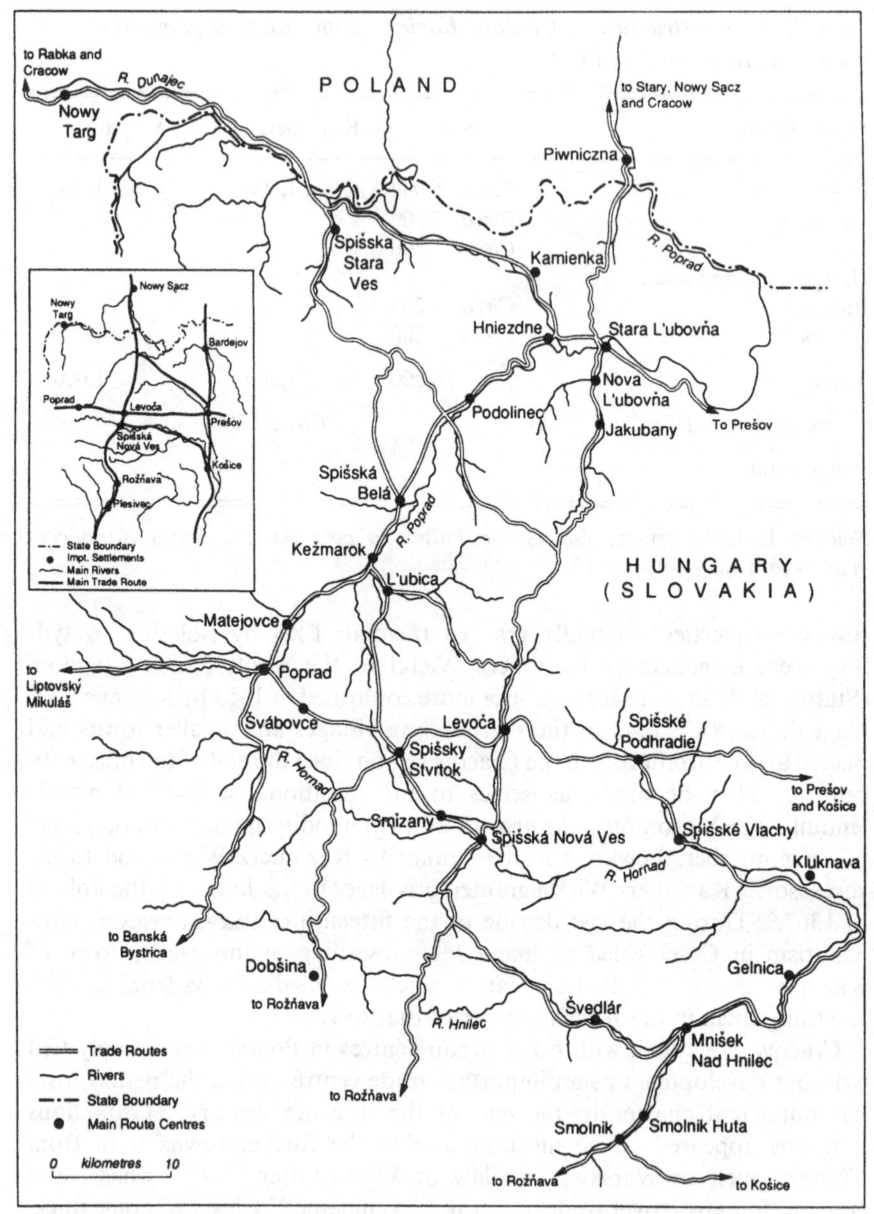

9. Cracow's land routes with northern Hungary in the Later Middle Ages.

Relations with Hungary

For the medieval monarchs of Poland, the Magyars and their Angevin rulers provided a traditional ally against the Czechs of Bohemia and the omnipresent threats from the east. Closer ties between Poland and Hungary originally emerged during the reign of Władysław Łokietek (1306–33) when fears of an expansionist policy by the Bohemian rulers brought them closer together. Geographical proximity may also have contributed to greater Polish–Hungarian friendship. Both states were neighbours, profiting from close cultural and economic contact, yet far enough apart to prevent those sorts of unconstrained connections which might have led to inspiring incursions into each other's territories; the anfractuious and roughly broken terrain of the Carpathian Mountains helped dispel such ideas. Perhaps of equal, if not greater, importance in the political relations of these two countries was the close family ties between the ruling houses of Poland and Hungary.

It was against this favourable background that Cracow developed her ties with Hungary (Fig. 9) and, more particularly, those nodal commercial centres on the main trade routes passing through Poland to Budapest and the river Danube. Internal struggles in Hungary during the thirteenth century had led to support and intervention by the Austrian Hapsburgs; unfortunately for them they backed the losing side and surrendered the strategically significant city of Pressburg (Bratislava) to the Hungarians.[63] Moreover, commercial contacts between Hungary and Western Europe began to decline as a result, so the Magyar merchants began to look for alternative routes to Germany and the Baltic via Poland,[64] particularly through Cracow.[65] The first evidence of an Hungarian merchant in Cracow dates from 1301.[66]

Political relations between Poland and Hungary took a favourable turn during the reigns of Kazimierz Wielki and the Magyar king Karol Robert (1308–42).[67] He was concerned about securing the throne of Poland for his son, Ludwik, in the advent of Kazimierz Wielki dying without issue. A document signed in 1339 was granted in principle[68] but the agreement was incomplete and required further amplification and more precise definition. The thought of living in Cracow obviously appealed to Ludwik, for it was at that time 'not only a rich town but also one of considerable political importance'[69] and in this document it was agreed that he and his brothers would respect and preserve the laws and privileges of Cracow, if Kazimierz died childless, and the crown would go to the house of Angevin. Further discussion on the succession problem was again confirmed in 1351[70] whilst Cracow's commercial interests were additionally enhanced by these closer political ties. In 1358, Kazimierz Wielki, with Hungarian approval, decreed that all merchants from and to Hungary en route to Prussia, Silesia or

Bohemia, must travel via Cracow; this in turn meant Cracow now had control over trade between the important Prussian emporium of Toruń with Hungary.[71] A decade later Ludwik I (King of Hungary since 1342) gave all Polish merchants access to Hungary free of all taxes and customs tariffs as far as the city of Kassa (Košice).[72] This in turn greatly profited Cracow, since Košice had strong commercial ties with the Polish capital and had imposed stapling rights on all merchants in 1361.[73]

On the death of Kazimierz Wielki in 1370, Ludwik I (Węgierski) succeeded to the Polish throne and thus created a Polish–Hungarian alliance; this was the outcome which Kazimierz Wielki's brother-in-law, King Karol Robert planned and hoped for in 1339 – an eventual Angevin succession in Poland. Kazimierz Wielki's nephew managed to keep Poland as a unified country, although Ludwik always regarded Poland as his secondary kingdom. The Polish aristocracy, a deciding factor in the succession of 1370 and of Ludwik's heir in 1382, obtained from the king the grant of special privileges at Košice in 1374; which formed the basis of a Polish 'Magna Carta'. For Cracow, his reign was a period of excellent commercial prosperity.[74] The city also benefited from a quarrel which prompted Ludwik to close the trade route from Toruń to Hungary via Sandomierz;[75] furthermore Cracow's greatest competitor for the Hungarian trade, Nowy Sącz, was forbidden by Ludwik to utilize the newly opened river route via the Dunajec and Vistula to Prussia in 1380, forcing Sącz merchants now to bring their goods through Cracow.[76] With the death of Ludwik in 1382, Cracow's most successful period of commercial politics was over.

Ludwik's eventual successor to the Hungarian throne in 1387, Sigismund of Luxemburg, was much less disposed to Poland and its new ruler Jadwiga, or her husband Władysław Jagiełło. Sigismund's fear of their claim to an hereditary right for the Hungarian throne eventually led to a Polish invasion of Northern Hungary in 1396, disrupting trade with Košice.[77] Moreover, Sigismund's internal policy of downgrading Košice at the expense of developing other Hungarian towns eventually led to the economic decline of most of them and was reflected in Cracow's trading contacts there. The damage done to Polish–Hungarian relations could not be quickly restored after Sigismund's death in 1437, although his successor, Elizabeth, and the Polish king, Władysław Warneńczyk, did improve the situation. Władysław obtained sixteen mainly German inhabited mountain cities in the Spisz (south of the High-Tatra Mountains) in 1412, thus giving Poland a share in the mining activities of upper Hungary, the richest in Europe at that time.[78] Privileges given by Ludwik were confirmed in 1440,[79] giving freer access to Hungarian towns and reversing some of Sigismund's policies. Władysław Warneńczyk's short reign (to 1444) unfortunately only led to a continuation of Sigismund's ideas and were further disturbed in 1473 when the Polish king, Kazimierz Jagiellończyk, confirmed Cracow's stapling rights,[80] thus encouraging Hungarian merchants to by-pass the city. Even an agreement signed in Cracow in 1498 between Poland, Lithuania and Hungary guaranteeing merchants from each country free trade in the others[81] was not a success due to the unstable political conditions of the time. The century therefore ended on a rather sour note for Cracow and its merchants in their relations with Hungary. Unfortunately, possible connections with Bohemia over the period 1257–1500 were not to provide a much happier story.

Relations with Bohemia

For the first three decades after obtaining its municipal charter, Cracow found the kingdom of Bohemia of little interest either politically or economically. Internal problems and a disastrous foreign policy culminated in defeat for the Czechs in Marchfeld (Lower Austria) in 1278 and loss of their king, Přemysl Otakar.[82] Failure to expand southwards beyond their natural territorial boundaries therefore led Přemusl Otakar's successor, Václav II (1278–1305) to cast his eyes over the possibility of eastward expansion for the kingdom of Bohemia. Poland at this time was troubled with internal strife and badly needed a strong ruler. Václav's eastward ambitions were carefully prepared; prudent diplomacy won him the support of various Silesian lords[83] providing Václav with a footing in Poland. In 1291 he gained the favour of the inhabitants of Little Poland with promises to guarantee all the privileges and prerogatives of the knights, clergy and urban centres of that territory. In return the population of the Dukedom of Cracow elected him their lord.[84] Václav was not without opposition, for an anti-Czech coalition led by Przemysł II of Greater Poland styled themselves as 'heirs to Cracow' with the intention of regaining territory lost to Václav together with the right to rule over Cracow. Their failure, combined with the assassination of Przemysł II (1279–96), paved the way for Václav to be crowned Polish King (Wacław Czeski, 1300–5) in September 1300.[85] He now ruled over about two-thirds of all Polish territory, which he considered an extension of Bohemia, and in turn was part of the Holy Roman Empire. His early death and assassination of his son, Wacław III in 1306 left vacant both the Polish and Bohemian thrones.

The moment had arrived for one of the minor Polish princes, Władysław Łokietek to establish his claims to the Polish throne. In 1306 his army laid siege to the Czech garrison in Cracow[86] and, with support cleverly obtained from Hungary, occupied Little Poland; moreover Łokietek inherited two significant factors from Václav's reign. Firstly, fiscal reforms initiated by Václav in Bohemia (1300), based on the Prague groschen (grossus Pragensis), provided a stable coinage throughout much of Central Europe. During his reign there is evidence that 'silver groschen were brought to Cracow'.[87] Secondly, because Václav ruled Poland largely 'in absentia' he was dependent on an efficient system of representatives; he therefore introduced into Poland his 'capitanei' (in Polish starosta) responsible for numerous administrative tasks including tax collection, garrison control and judiciary organisation.[88] The starosta was, according to Knoll, 'in effect, a territorial prince',[89] who was to achieve immense royal importance during the later reign of Kazimierz Wielki.

As far as Cracow was concerned one of the more negative aspects of Václav's reign was the founding and granting of a municipal charter to Nowy Sącz, in 1292.[90] This new settlement lay on the main commercial artery from Cracow to Hungary (Fig. 9) and controlled not only the land route to Cracow and beyond, but Sącz merchants could also utilize the natural river route down the Dunajec and Vistula right through to Toruń, thus by-passing Cracow and avoiding any staple rights or compulsory customs tariffs the city imposed. Václav's reasons for establishing the settlement were not only financial gain from trade tariffs imposed on Hungarian trade, but also a means of defending the Cracow region from possible attack by his adversaries in Hungary.[91] The seeds of discontent and dispute sown by Václav in founding Nowy Sącz were to linger on long after he and Bohemian rule had departed from Little Poland.[92]

With the demise of Václav, Bohemian influence in Cracow, as in much of the rest of Poland declined. Internal troubles in Bohemia continued throughout much of the first half of the fourteenth century only to be relieved by the accession to the Czech throne of a strong ruler Charles (Karel) IV, who managed to consolidate his country, kept the nationalist issue in cold store and created the 'Golden Age' of Bohemian history. Contact with Cracow was minimal, although in the last year of his reign (1378) he granted a special privilege whereby Cracow merchants were allowed to trade in Prague, the Bohemian capital.[93] Nevertheless, as Kutrzeba has stated, 'Cracow's relations with Prague and Bohemia were of little importance as they had nothing to offer each other.'[94]

The Hussite Wars of the early fifteenth century according to Bradley 'were less glorious and more damaging than is usually admitted. They lasted for almost twenty years; and the following half century was far from peaceful. Throughout these long years the Czech kingdom lived in cultural and political isolation.'[95] The calmer waters of the second half of the fifteenth century saw greater contact between Cracow and Bohemia; Czech merchants increasingly came to the city after mid-century.[96] Greater contact between Poland and Bohemia came after 1471 when the Czech nobility thought that only a foreign monarch could effectively play the role of arbiter and conciliator in their disunited kingdom. The Czech Diet turned to Władysław (Vladislav) brother of the Polish king Kazimierz Jagiellończyk, and elected him to the throne. Unfortunately, the choice was not a happy one, for Władysław failed to help in solving Czech problems and quickly complicated those of Poland. Struggles with Hungary by Władysław did little to help Cracow's trading prospects in the Magyar lands, although by the end of the century closer trading contacts between Prague and the city were evident especially amongst Jewish merchants.[97] Another ingredient in this complicated story of political relations was the role played by Silesia.

Relations with Silesia

Even before the mid-thirteenth century attempts were made by the Silesian Dukes, Henry I (the Bearded) (1201–38) and his son Henry II (the Pious), (1238–41), to unite all the lands of southern Poland, from Lubusz and Wrocław, to Sandomierz and Lublin, only to be frustrated in their plans by the Mongol invasion of 1241.[98] Later Silesian efforts at Polish occupation were thwarted during the second half of the thirteenth century, by Bohemian expansionist plans under King Václav (Władysław) who managed to make several of the Silesian duchies his vassals.[99] More significant for Cracow was the development and growth of Wrocław as the main trading centre for Silesia and a future competitor in trade with the East.[100] Wrocław obtained its municipal charter in 1242 and had it reconfirmed in 1261,[101] four years after Cracow. Perhaps even more consequential was the granting of staple rights to Wrocław in 1274,[102] three decades before Cracow. In practical terms it meant that much of Germany's trade with Poland went through Wrocław, which at this time had far more contacts with Toruń and northern Poland than with Cracow.[103]

More direct involvement of Silesian interests in Cracow came during the late thirteenth century. The death of the Polish king Leszek Czarny (the Black), 1279–88, who died without issue, meant the greater share of his inheritance, including the city of Cracow, went to Duke Henry IV Probus of Wrocław.[104] The smaller share of the inheritance went to Władysław Łokietek, half-brother of Leszek Czarny, who by devious means managed to make himself master of Cracow. Henry IV, on hearing of this event sent an army from Silesia to lay siege to Cracow. Władysław Łokietek beat a hasty retreat,[105] not to become master of Cracow again until the death of King Václav in 1306. Henry IV continued his ambitious plans to achieve the Polish crown; he had managed to obtain and unite Silesia with Little Poland but his tragic death in 1290 led to the disappearance of effective Silesian control in Cracow.[106] It was now the Bohemian, King Václav II, who gained supremacy over the Duchy of Cracow in 1291–2.

The early years of the fourteenth century saw the Polish king, Władysław Łokietek, successfully ruling his new kingdom, if with some difficulty. Meanwhile, in 1310, Jan of Luxemburg arrived in Prague with a small imperial army to claim the Bohemian throne, which he was to rule until

10. Cracow, Poland and surrounding states in the 14th century.

his death in 1346 at the battle of Crécy. Against all natural expectations the reign of King Jan was a success; although he abdicated all internal power to the Czech nobles, he was very successful in acquiring foreign territory and keeping it.[107] When the Bohemian crown reduced the Silesian dukes to vassalage between 1327 and 1329, the Polish king, who was elsewhere engaged in battle against the Teutonic Knights, was powerless to resist. Władysław Łokietek had taken some precautions, however, leaving a garrison of Hungarian troops in Cracow (July 1327) to defend the city against King Jan's possible invasion.[108] Fortunately for Cracow such a catastrophe was averted as the Bohemian king moved northward in an alliance with the Teutonic Order; their combined armies terribly devastated Greater Poland,[109] especially in the war of 1328–32 against Władysław Łokietek. Nevertheless, Cracow remained perilously close to the Polish/Silesian border (Fig. 10).

As Knoll perceptively notes, 'originally part of the Kingdom of Poland, Silesia had followed an independent path of political and even cultural development', of which a striking characteristic had been the 'penetration into the region of German and Bohemian influences; which had 'brought significant changes in the region'.[110] Further evidence of the de-polonization of Silesia came in 1335, when King Kazimierz Wielki renounced all claims to Silesia, formerly Poland's most densely populated and wealthiest region; in return the Bohemian king renounced all pretences to the Polish crown.[111] Kazimierz Wielki's chance for revenge was to come later, and directly benefited Cracow.

During the fourth decade of the fourteenth century relations between Wrocław and Cracow deteriorated, mainly over problems of customs duty and trade connections with the route eastwards to Ruthenia, and the towns of Włodzimierz and Lwów. The only route went via Cracow but, according to its stapling rights of 1306, copper alone was subject to such restrictions;[112] these did not affect Wrocław merchants travelling to the East. The new customs duties imposed by Kazimierz Wielki in 1348, however, did impinge on their commerce;[113] in revenge Polish traders were retained in Wrocław,[114] an act which prompted Kazimierz Wielki to occupy Ruthenia in 1351–2 for the Polish crown and close all routes through it for non-Polish merchants.[115] Although this new situation affected Prussian traders, especially from Toruń, it was felt most acutely by Wrocław merchants, who could no longer take their caravans by any other route than via Cracow.[116] Consequently a trade war emerged between Cracow and Wrocław which was to last for over a decade.[117]

Even after Kazimierz Wielki's death (1370), renewed confirmation of Cracow's stapling rights by his successors in 1372 and 1387[118] did little to ease the situation for Silesian (and Prussian) merchants.[119] Although alternative routes to Ruthenia were found around Poland via Lithuania,[120] pressure to utilize the shorter trip through Cracow finally materialized in 1402, when all routes through Poland were reopened.[121] In spite of further confirmation of trade access to the East for Wrocław traders in 1417[122] (subject to customs duty), the conflict with Cracow continued; a trade war between the two cities existed from 1440 to 1490 aggravated by the fact that Cracow's stapling rights were reconfirmed as compulsory for Silesian traders in 1457 and again in 1473.[123] As Myśliński has stated, 'in spite of all documentary evidence it seems that the kings of Poland from Kazimierz Wielki to Jagiellończyk wanted profits from the Ruthenian trade to be kept for their own Polish towns'.[124] The last decade of the fifteenth century saw little improvement in the situation. In 1490 Silesia closed its borders to Polish merchants; this affected particularly traders from Cracow who were cut off from links with Germany, especially Nuremberg and Leipzig.[125] Polish retaliation led in 1499 to the prohibition of certain goods, especially cereals and oxen, being exported to Silesia, with disastrous results for transit towns like Wrocław.[126] For Cracow, therefore, the century ended on a difficult note in her relations with Silesia and particularly Wrocław; a similar pattern was to emerge with regard to her northern neighbours, Prussia and its main city, Toruń.

Relations with Prussia

During the thirteenth century the expansionist plans of the March of Brandenburg and the Teutonic Order of Knights into Polish territory were to provide forerunners for the later Prussian state. Conterminous with these events was the large migration of German colonists into the Polish lands, attracted by the rapid economic growth and ample opportunity for settlement.[127] The last quarter of the thirteenth century was to witness a consolidation of these German colonists into many parts of Poland, where efforts were made to build a united state. Cracow itself felt repercussions of the phenomenon in 1311 when a rebellion of German burghers, under their bailiff, wójt Albert, occurred against the Polish ruler, and was only put down with some difficulty.[128] Germanization of Western Pomerania led to the dissolution of its political ties with other parts of Poland, whilst German influence, though much weaker, was also experienced in Silesia.[129]

In 1308, King Władysław Łokietek sought the help of the Teutonic Order to regain Eastern (Gdańsk) Pomerania seized the previous year by the Brandenburg Margraves. Unfortunately, he was opposed by important internal pressures, above all by the Bishop of Cracow (Muskata),[130] and failed to obtain the necessary military personnel to combat the March. A treaty was therefore concluded with the Teutonic Knights, who promised to relieve Eastern Pomerania from Brandenburg occupation and return it to Poland. Instead, after defeating the occupants, they established themselves in Eastern Pomerania, founding their capital at Malbork on the lower Vistula.[131] Lawsuits instituted against the Order by Władysław Łokietek over Eastern Pomerania brought little success, further complicated in 1331 by their occupation of Kujavia (Fig. 10) a province to the south of Toruń.[132] For Cracow all this activity meant that trade outlets to the Baltic coast via the lower Vistula were now effectively controlled by the Teutonic Knights.[133]

The reign of King Kazimierz Wielki was to see little fundamental change in the situation. In 1339 he renewed his predecessor's legal action against the Order to regain Eastern Pomerania, but to no avail; further pressure through family ties in Western Pomerania and the hope of a united coalition against the Knights failed[134] and in the Treaty of Kalisz (1343), Kazimierz Wielki relinquished all claims to Eastern Pomerania in favour of the Teutonic Order.[135] This state of affairs remained unaltered until the first decade of the fifteenth century, but the union of Poland and Lithuania under the Jagiellon dynasty in 1382 did mean the former had to defend the latter against possible incursions from the Teutonic Order.

Within Prussia the main inland trade centre was Toruń (crossroads of six main routes through Poland) and its coastal counterpart was Elbąg.[136] Their contacts with Cracow were long standing, particularly Toruń for the copper trade. During the rule of the Order, Cracow's links with them were intermittent depending on the prevailing political relations between Poland and the Teutonic Knights at any period of time.[137] Kazimierz Wielki's closure of trade routes through Poland to Ruthenia and the East (1352) was acutely felt by Toruń merchants. In retaliation, trade routes through Prussia were closed to Polish merchants in 1360,[138] which particularly affected Cracow's commercial links with Flanders via the Baltic ports. Pressure was placed on Kazimierz Wielki by Polish traders, routes were once more open, and no trade wars took place for the remaining years of his reign. His successors were not so fortunate; the Prussians fully realized the importance of stapling rights and their significance for Polish trade. Toruń received her staple rights in 1382[139] and, less than a decade later, the Teutonic Knights applied the principle to their whole territory.[140] This intensified trade rivalry between Cracow and Toruń, the law of 1389 being finally rescinded in 1403,[141] allowing Cracow merchants freer use of the lower Vistula river.

In 1410 the Poles defeated the Knights in the battle of Tannenburg (Grünwald) and from then on the acquisitions of the Order were gradually eroded, a process to which restive subjects and disobedient Knights actively contributed.[142] However, Eastern Pomerania, although its return to Poland was discussed at this time, still remained in the possession of the Teutonic Order. Continued conflict in two later wars (1421–2 and 1431–5) ended commercial relations between the Baltic coast and the Polish lands, although some guarantees of access to Polish shipping along the lower Vistula were made.[143] The inter-war years saw Cracow traders once more traversing the Prussian lands en route to Bruges, but a further disagreement in 1440 effectively ended use of the Vistula by Polish ships.[144] Toruń, one of Cracow's main commercial rivals, suffered heavily during the first half of the fifteenth century and, in spite of rigorously applying her stapling rights,[145] began to suffer from increased competition; the rapidly developing Prussian port of Gdańsk (Danzig) was growing at her expense.[146]

Within Prussia dislike of the arrogant Knights led to the formation of the so-called 'Prussian Union' which looked southward to Poland for help.[147] It readily came from King Kazimierz Jagiellończyk, who began the Thirteen Years War by invading Prussia in 1454. Support for Poland came from the cities of Gdańsk, Toruń and Elbąg, the Teutonic Knights eventually making peace at Toruń.[148] Poland recovered Eastern Pomerania (renamed Royal Prussia) and won back access to the Baltic Sea that it had lost in 1308. For Cracow the political scene remained rather uneventful, although old disputes between the various Prussian towns over trade remained.[149] Perhaps more significant was the overall decline of commerce ploughing along the lower Vistula and adjacent lands, merchants changing to safer, and more certain alternative routes.[150] After the Peace of Toruń (1466), Polish rule stretched from the Baltic to the Black Sea for a period of 32 years; Poland became a trading partner with access to the most profitable trading routes crossing the European continent,[151] and it was open to Cracow to exploit these opportunities to the east and south towards the Pontus Euxinus.

Relations with the East

In the thirteenth century peoples from central Asia, known as the 'Golden Horde'[152] swept through the Caspian Gate, conquered Kiev in 1240 and most of the Russian principalities, before defeating the Polish army at Chmielnik and the Silesian knights of the Teutonic Order at Legnica (1241) (Fig. 6). Cracow was devastated in 1241,[153] and again on further raids in 1259[154] and 1287.[155] Although these Mongol invaders had short sojourns on Polish territory, they still remained masters of Red Ruthenia (Red Russia), directly to the east of Little Poland, for another 50 years after 1242; from this base their repeated raids laid waste many of the towns and villages in the Polish lands. It also prevented Cracow and other trading centres from developing strong commercial ties with their eastern neighbours.

In the second half of the fourteenth century, and in the course of the fifteenth century, the disastrous economic and demographic consequences of the Tartar/Mongol invasions were overcome, although they did not cease.[156] During the fourteenth century the powerful rule of the khans of the Golden Horde were on the wane, when serious dynastic rivalries in the khanate occurred.[157] Cracow's trade links with the Black Sea were mainly through intermediary towns like Włodzimierz and Lwów[158] for, as Knoll points out, 'Polish–Ruthenian relations before 1340 are imperfectly reflected in those documents which have survived', but 'It is probable that Łokietek paid little attention to his eastern neighbour because of his involvement in the west.'[159] His successor, King Kazimierz Wielki adopted a more positive eastern policy and between 1340 and 1352 occupied Red Ruthenia with the cities of Przemyśl, Lwów, and Halicz; by 1366 he had extended his rule over the Ruthenian principalities of Chełm, Bełz, Włodzimierz and Podolia.[160] Already by 1344 new trade routes through the area were created by Kazimierz Wielki, to the great benefit of Cracow and other cities involved in east–west transit trade,[161] and reconfirmed in 1371.[162] Moreover, in 1380, Cracow received a privilege from the prince of Łucho-Włodzimierz, which lowered customs duties for her merchants by half.[163]

One of the political reasons for favouring a dynastic alliance between Poland and Lithuania in 1382 was the chance for the former to regain Red Ruthenia and Podolia, which had become estranged from Poland during the reign of Ludwik Węgierski. In August 1385 one of the most important treaties in Polish history was signed at Krewo in Lithuania;[164] here the grand duke of Lithuania, Iogaillas, (Władysław II Jagiełło) promised to unite his Lithuanian and Ruthenian lands to the Polish crown. In reality it meant that by this union a state was created about four times as large as the original Polish realm. It included the old Russian capital of Kiev, now a border town against attacks from the Golden Horde; to the south-east there stretched over 48,000 square miles (150,000 km^2), of thinly settled steppe land, belonging to Lithuania in name only. More significant for Poland, these lands provided a bulwark against further Mongol/Tartar incursions including the great principality of Moscow, then under their control.[165] In 1387 Poland, with the help of the Lithuanians, was able to renew the incorporation of Red Ruthenia and Podolia and to assert its authority over the principality of Moldavia (1387–1497), which had previously been under Hungarian rule and of minor significance. All this political activity was of profound importance for Cracow. The city was now capital of a state which had common access to two seas – the Baltic and the Black Sea. It also meant accessibility to the direct trade route to the Orient, Cracow now becoming a partner in the rich eastern commerce that flowed through the cities of Caffa (Theodosia, in the Crimea) and Akkerman (now Belgorod-Dnestrovsky).[166]

Profits from this lucrative trade inevitably led to conflict between Cracow and Lwów, particularly over the sensitive issue of stapling rights which occurred in 1403.[167] Much of the competition was for the growing markets of Moldavia and the right to trade in 'ad partes Valachiae'.[168] Unfortunately, whilst both cities were locked in their personal trade wars, something far more sinister was emerging – the growth of the Ottoman Empire. King

Władysław III, with the help of the Bishop of Cracow (Zbigniew Oleśnicki),[169] organized a cruasade to try and save Byzantium, already threatened by Ottoman incursions. Sadly all was to no avail; the Polish king was killed at the Battle of Varna (1444) and the Ottoman troops successfully conquered Byzantium. Proposed plans for the Ottoman occupation of Hungary were then considered but delayed for several decades, due to their defeat by the Hungarian military leader Jan Hunyadi in 1456.[170] Nevertheless, the Turks now controlled access to the Black Sea ports; feverish diplomatic activity on the part of Polish ambassadors,[171] together with the privilege of Mahomet II in 1456,[172] still enabled Cracow merchants and other Polish traders to keep in contact with the eastern trade. Meanwhile, King Kazimierz IV Jagiellończyk, successor to Władysław III, signed a treaty with the principality of Moscow in 1449 establishing the Polish/Lithuanian boundary about 90 miles (150 kms) west of Moscow; this secured the eastern boundary for 37 years.[173]

The Turkish capture of the Genoese Black Sea colonies of Caffa (1475), Kilia and Akkerman (1484) was a severe blow to Oriental trade conducted by Lwów and Cracow.[174] This meant that Poland was now cut off entirely from sources to eastern goods, further complicated by the appearance of yet another enemy – the Tartars from the Crimea. These people were ruled by the khans of the Girey dynasty (e.g., Khan Mengli Gherai) and became vassals of the Ottoman Empire. They pursued a pastoral economy, augmented by the spoils of raiding and plunder. The Crimean Tartars now became a hostile force invading the borderlands of Poland and Lithuania, for their favourite targets were the settled agriculturalists to the north and north-west, a rich source of booty, livestock and, most of all, slaves.[175] Raids into Red Ruthenia and Little Poland intensified after 1492, pushing further into Polish territory with Ottoman help; they reached the banks of the Vistula and Cracow in 1498–9.[176] Kazimierz IV Jagiellończyk saw this danger for Moldavia, coming to the assistance of its ruler, Prince Stephen, who became a vassal of Poland in 1485.[177] In spite of Polish tributes paid to the Turks in 1490[178] and some Polish military successes against the Tartars, defeat for the Moldavians in the forests of Suceava in 1497 and Polish recognition of Ottoman conquest, effectively spelt the end of Cracow's trading interests in the area; Poland's entente with the Turks further alienated the Moldavian rulers against possible contacts with Polish merchants and their goods.

The other great external threat to the lands of Kazimierz IV (which covered about 336,000 square miles (870,000 km^2) in 1491–1 was the ally of the Ottoman Empire, namely the tsar of Moscow. Tsar Ivan III (the Great) carefully manipulated the disturbed political situation at the end of the fifteenth century;[179] he allied himself with the Crimean Tartars, and twice attacked Lithuania (1486–94 and 1500–3).[180] His efforts were rewarded with the capture of all the principalities on the Oka and Desna rivers as far as the middle Dnieper River, further closing possible avenues of commercial contact for Polish merchants in this area. Thus by the end of the fifteenth century, Cracow, along with other Polish trading towns like Lwów, saw a decided shrinking of their potential market area in the East, which, along with overall changing trading conditions, was to lead to reorientation of their commercial attitudes in the ensuing centuries.

Relations with the West

While much of Cracow's commercial activity was closely associated with the political events of central and eastern Europe, it should not be forgotten that her trade links reached the more western margins of the European continent. The thirteenth century, which had been an era of trouble and confirmed weakness in Polish history was, beyond doubt, a period of prosperity for medieval culture and trade in the West. Medieval Poland's strongest connections with this area came as a result of her existence as an ecclesiastical province of the Holy See, and recognized by Christianity's highest authority. The bishops of Cracow continued to have close relations with Rome, and the Duchy of Cracow, along with other Polish provinces, was under the patrimony of the Church and paid the Peter's pence.[181]

The reign of Kazimierz Wielki saw an extension of Poland's political relations with the West. Polish embassies succeeded each other at the Papal court in Avignon (1305–76) without a break; if trade 'followed the flag', then the foundation of the Lyons fairs continued to attract merchants, particularly the Genoese, to the valley of the Rhône, the main overland trade route between Italy and north-west Europe. King Kazimierz Wielki sent reinforcements to the Pope during the Italian wars of the fourteenth century and ties between the Polish capital and Avignon were considerable. During the Later Middle Ages however, the River Rhine became the most important routeway in west-central Europe, with navigational facilities from Zurich to the North Sea. The Rhineland routes handled the largest volume of traffic, but increasing emphasis at this time was being placed, particularly by Venice, on the trans-Alpine passes (e.g., Brenner) which led to Augsburg, Regensburg, Nuremburg and Prague, linking Cracow with western Europe.[182]

The other major political entity for Cracow at this time was the Low Countries, especially Flanders, the route focus centre of north-western Europe. Political development in western Europe in the Later Middle Ages was frequently connected with the series of conflicts between France and England, often termed the 'Hundred Years War'. Flanders, with its wealthy wool manufacturing towns, was closely connected with England through its trade, and supported the English cause. Bruges, with whom Cracow had known commercial links, reached the peak of its prosperity between 1300 and 1350, and then declined, disadvantaged by the eastward shift of European trade routes towards Brabant and the Lower Rhineland. Its successor was Antwerp, which enjoyed unrivalled prosperity as part of Brabant from 1406 till the end of the sixteenth century, and again was known to have commercial ties with Cracow.[183]

King Kazimierz Wielki's political interests also stretched northward; he allied himself with Valdemar Atterdag of Denmark, and had himself named, together with the Scandinavian kings, protector of the archbishops of Riga. From early times the Baltic had attracted merchants as a rich source of supplies of food products, especially grain, as well as the raw materials vital for the construction of shipping fleets. By the mid-fourteenth century the former organization of the Wendish cities in the southern Baltic, began to develop into a larger and more powerful unit known as the Hanseatic League, of which Cracow was part, if only as a peripheral member. The League carried goods from such ports as Gdańsk (Danzig), Królowiec (Königsburg), Riga and Revel, to Lübeck and other North Sea destinations, enabling Cracow to be linked via the Vistula river with places such as Bruges, Antwerp and London. Its political organization held together a confederation of some sixty–seventy trading towns, of which Lübeck was the undisputed leader; even in 1500, the Hansa as a power bloc, still seemed a formidable political force to many in northern Europe.[184]

Conclusion

In studying Cracow's political relations during the Later Middle Ages, three characteristics emerge – in order of importance – firstly, the close ties of Poland with both Hungary and Lithuania at various periods of time, secondly, the rise and fall of the Teutonic Order along the Baltic coast and lower Vistula valley, and thirdly, the growth of Turkish power in Europe. These three factors must be thoroughly understood before an appreciation of the city's commercial relations can be attempted. The link between politics and commerce was strong. The quest for profits dominated the minds of Cracow merchants. Their achievements depended, however, very largely on the vagaries of political events. It was the continental trade of Europe at this time which formed a genuine link between different regions and which helped to give a sense of economic coherence to politically divided and sometimes warring areas. The foresight of Cracow's merchant class in utilizing the rights of staple and developing transit trade between Hungary and Western Europe is to be commended, together with the skilful diplomacy used in contacts with the Infidel. The fear of Teutonic hegemony in Poland is understandable seeing that the Order had similar commercial interests to her own. The overall considerate rule of the Polish monarchs is to be applauded for they allowed Cracow to develop her commerce during a period of intense unrest and mistrust amongst the powers of central and south-east Europe.

5

European trade through Cracow, 1257–1500

The main basis for Cracow's prosperity was trade. It was upon trade, particularly transit trade and industry that many of Cracow's citizens had to depend for their means of livelihood. The period from the granting of its municipal charter in 1257 to 1500 was to witness probably the greatest era of commercial success in its history, when the city was one of Europe's more significant emporia.[1] Moreover, the years around 1500 may still be regarded as a convenient and satisfactory point at which to draw a line between the Later Middle Ages and the Early Modern Period. The sixteenth century awoke to new political and economic situations throughout the European continent and beyond;[2] the growth of the Spanish Empire, discovery of America, new routes to the East, growth in the money supply, eclipse of the Mediterranean ports at the expense of the Atlantic seaboard,[3] decline of the Hanseatic League and, perhaps most important for Cracow, the increasing pressure of the Ottoman Turks[4] now being felt at the eastern extremities of Europe. Thus, in many ways other than the purely geographical, the sixteenth century was to see a slow process of change; adoption of radically new techniques, different business methods and independent solutions in areas experiencing a similar husbandry, all contributed to a wind of economic change blowing through much of the known world. Prior to these events, one still found cities like Cracow tied to the circumstances of medieval commerce. It is therefore necessary here to discover more systematically what importance Cracow had for the men who made their living by trade, the commodities they handled, the routes they followed, the markets they served, and the way they organised and conducted their businesses.

Trading organization

Medieval commercial enterprises displayed a wide variety of sizes and patterns.[5] At one extreme, the seller might be a small shopkeeper or a craftsman who sold his own products.[6] At the other, he was a member of a firm engaged in large-scale long-distance commerce and banking, including high financial dealings with popes or kings.[7] In between one found every conceivable kind and scale of business. The economic climate in the fourteenth and fifteenth century was, however, dictated, according to Thrupp, by 'War financed by heavier taxation, contractions of long-distance trade, the inroads of plague, inter-town competition for markets that were for long periods more or less static, deliberate holding down of production by privileged guilds, and continued shortages of money supply', which 'cloud the late medieval scene until the 1470s.'[8]

It is therefore useful to differentiate between three kinds of commerce associated with the Late Middle Ages.[9] There were the local markets in most towns where the products of the immediate region were bought and sold.[10] Secondly, there were the larger urban centres which not only transacted local business, but also dealt in commodities which had emanated from the more distant parts of Europe.[11] Finally, there was a limited traffic in the more glamorous, tramontane products of the East.[12] Whilst the first category may be termed provincial traffic, the second could be designated continental commerce and the third inter-continental trade; for a considerable part of the fourteenth and fifteenth centuries, Cracow could be said to have been a participant in all three categories.

Cracow was firstly a local trade centre. One may guess that the greatest volume of commerce was to be found in the numerous exchanges between the city and the many small villages and even hamlets in the Cracow region; the city was the focus for the numerous regional products which were bought, sold and bartered. At this level there was little need for the professional merchant, for it was the agricultural products of the surrounding area, or goods made by the small handicraftsmen, that reached Cracow's market place; small quantities were the general rule where producers and consumers could deal with each other without the intervention of a trader. It was therefore mainly local products being exchanged by local people: animals, locally woven cheap cloth, and jumble ('kramme'), the goods and services of small craftsmen, smiths and carpenters above all.[13] Trade exchange was mainly centred on stalls put up in the market place, and held on a weekly or bi-weekly basis. Less frequently fairs were held at designated times throughout the year,[14] which attracted buyers from other larger centres intent on acquiring the specialities of the region, a common feature throughout much of medieval Europe.[15]

Cracow also participated in the continental trade of Europe. In its way this meant that the city, through its transit trade connections had a genuine link with the different regions of the continent which provided a sense of economic coherence to often politically divided and sometimes warring areas. Moreover, up to and during the course of the fifteenth century the urban centres of Europe were the sole nuclei of commerce and industry, to such an extent that little of it was allowed to escape into the surrounding countryside. A sharp division of labour existed between the latter practising only agriculture, whilst the former controlled trade and the manual arts.[16] Critical in this scenario were the merchants. From the thirteenth century onwards, they lived under the protection of their individual town, reciprocal guarantees often being given between urban centres. The merchants, in turn, traded in a wide spectrum of merchandise, of which some of the main commodities included the basic necessities such as grain, salt, fish, timber and metals, less essential products such as quality cloth and wine, and sheer luxuries.

The merchants of Cracow were no exception to this overall picture. The organization of trade had two separate forms, namely the merchant companies (either single or in groups) and the mechant unions or associations. There were many types of commercial companies which varied according to the prevailing economic conditions, imposition of customs duties, and economic relations. Admittedly, during the Middle Ages their form was more simplified than today, but even then one cannot compare those of Cracow or even in Poland as a whole, with the Italians and the commercial empires of Venice and Genoa.[17] Even the commercial centres in Germany pale beside such institutions as the Casa di San Giorgio,[18] whilst the companies in Cracow, although similar in form to the German towns, were economically weaker.

The simplest form of organization was the single company, in which one man provided the capital, ideas, energy, risks and reaped the profits for himself. He was often a man who had built up a business in the city which eventually obtained materials from some distance, produced goods in quantities beyond local needs, and exported them. One of the largest such merchants in Cracow during the first half of the fifteenth century was Sweidniczer[19] whose company had connections as far away as Flanders[20] and Hungary,[21] and dealt in a variety of goods on a large scale. Only at the end of the fifteenth century were firms of a similar size and importance to be found in Cracow again. More common in the city was the specialist merchant who only traded in one commodity, and had connections in one area, e.g. Flanders or Silesia. Sometimes there were merchants who specialized in one transit trade route such as Hungary to Flanders, or Lwów[22] to Wrocław.[23] Help in such a company was often forthcoming from assistants, usually relatives, who were needed when goods had to be sent beyond the bounds of the city, to organize their actual transport, etc. Real transportation was carried out by a 'Furman' or waggon driver, who was not just a carter but was responsible for merchandise from collection to delivery according to specific instructions. They are rarely mentioned specifically in Cracow's documents, but two examples are particularly noteworthy.[24] Furmen were not tied to one merchant, but free agents who worked on whatever commissions they wished, stating their own individual terms, etc. Another person who provided services for a company was the 'lieger' or commercial agent who was responsible for selling a Cracow merchant's goods in some distant place like Flanders or Košice. He may have been a Cracow citizen or a foreign merchant, selling goods supplied from Cracow. There were also commission agents who collected goods from Cracow companies, delivered the merchandise at their own expense to some other place, sold them and, with the money, bought other goods to take back to Cracow. Their reward was usually a quarter of the profit earned on any transaction.[25]

Individual companies, however, often lacked sufficient capital to participate in long-distance trade, especially to places like Flanders or the Black Sea coast, where large sums of money were involved. This was especially true in cases where it was only profitable to transport large quantities of goods for sale, and entailed hiring boats to Flanders or caravans to the East. Moreover, in an era when routes were not all that safe, risks were high and individual merchants, except perhaps the richest, were hesitant to participate. The answer was found in commercial partnerships,[26] which provided a larger capital sum and reduced financial loss through risk. Partnerships were found in Cracow from the end of the fourteenth century, particularly for merchants involved in the Flanders trade and also commerce with Moldavia.[27] Unlike foreign partnerships mentioned in Cracow documents, e.g. from Nuremburg[28] or Wrocław,[29] the city's own companies were not complicated and usually had only two or three partners; lead-mining ventures were the only exception.[30] Some of Cracow's partnerships were formed for only one venture, such as sending goods to Flanders in exchange for other merchandise; furthermore, merchants could belong to more than one partnership, but always for a set time period such as one, two or perhaps three journeys. Capital was supplied jointly, often with one merchant making the journey himself;[31] sometimes the city's merchants joined partnerships from other places such as Toruń or Gdańsk, but this was later forbidden. Contribution to the partnership was not always equal in amount or type; some merchants gave money, others did the work, and profits gained were divided amongst the partners accordingly.[32] On occasions, such agreements led to misunderstandings and disputes, but sparcity of information in Cracow's law-case documents suggests such conflicts were settled out of court by the merchants themselves.

Continued commercial growth led to the second form of trade organization, namely the merchant union/association/corporation. The great disadvantage of the individual companies and partnerships was their inability to obtain and defend the privileges of greater size. The object of these merchant associations was, while allowing traders to carry on their normal dealings, to create favourable conditions for the protection of commerce as a whole.[33] In Cracow the necessity for such an organization did not seem to arise until the beginning of the fifteenth century. Prior to this the functions of such an organization were performed by the city council. Privileges were given to its merchants in the city's name[34] and acts were passed to guarantee commercial freedom for its merchants both within the city and outside its walls. The acts not only stressed the need to protect their merchants from abuse, but reflected certain aims for improving their situation in the commercial world. The town council consisted mainly of handicraftsmen and councillors, the latter were mainly merchants. The composition of the city council in Cracow was rather different from other cities, for whereas the latter usually had a predominance of handicraftsmen, already during Kazimierz Wielki's reign it was suggested in Cracow that it should consist half of handicraftsmen and half councillors.[35] Although this idea was never applied, it set the pattern for later times, and therefore the city council had a more aristocratic character than other medieval centres, composed mainly of a patriciate, which was dominated by a large number of merchants.

In Cracow, as in other Polish urban centres, guilds for handicraftsmen had existed long before the fourteenth century,[36] but the city was the first in Poland to have a merchants' guild. In 1405, Polish merchants were forbidden by the Hungarian king, Zygmunt Luksemburg (1387–1437) from trading in his country. Places most involved in the Hungarian trade (i.e. Cracow, Sącz and Dobczyce) agreed to a meeting of their merchants in Cracow; it resolved to form a community of common traders[37] and become members of a west European organization known as the 'Kumpanie der Kaufmann'; in this way the merchants could continue their participation in the Hungarian trade. The oldest extant city council book (dated 1410) records the names of merchants elected as senior members of this merchants' corporation.[38] No further reference is made to the corporation for fifty years[39] and then silence until the eighteenth century. As a result Cracow remained remote from other Polish trading towns like Lwów, Poznań, Gniezno, Lublin and Kalisz which through mutual jealousies, distance and self-preservation, would have found it difficult anyway to work together on a larger scale. Cracow's strength lay in the privileges guaranteed by the town council and the power of the Polish king but, beyond the country's boundaries, e.g., in Flanders, his influence was less; for this reason Cracow was to accept a rather loose relationship with the Hanseatic League.

Within Cracow one of the strongest privileges obtained from the city council was the law on stapling rights. The staple rights of many European towns consisted in the privilege of forcing all passing merchants/caravans/ships etc. to stop, unload, and offer their goods for sale, and to reload and proceed on their way only if there was no demand for them. Conditions varied from town to town, some goods being entirely exempt, whilst other merchandise was strictly controlled; in some places goods had to remain exposed for as much as two weeks.[40] The staple right (Stapelrecht/prawo składu) was rigorously enforced in parts of Germany, particularly along the Rhine Valley; in Poland the idea was also very appealing due to the transitory nature of much of the trade passing through the country. The concept of stapling came to Poland from western Europe via Silesia and western Pomerania *circa* 1250;[41] its application spread eastwards with certain Polish towns utilizing the idea quite early (e.g., Nysa, *circa* 1250), and others much later (e.g., Lwów, 1379, Torún 1382). Cracow obtained her first staple rights in 1306.[42] Stapling in Cracow was only applicable to foreign merchants, referred to in documents as 'Hospites' (Latin) 'dy geste' (German), 'obcymi' (Polish) and were deliberately designed to protect Cracow's traders. Due to its fortunate geographical location, Cracow was able to make great use of this idea and was 'in the best position for this type of trade in the whole of Poland'.[43] Although the 1306 document refers only to copper, the reconfirmation of the city's stapling rights granted by Kazimierz Wielki in 1354,[44] included all commodities and was basically aimed at protecting Cracow from the commercial exploits of other non-Polish towns, notably Torún[45] and Gdańsk,[46] and illegal dealing of foreign merchants in the city e.g., from Nuremburg.[47] Jews were treated separately and subject to special laws; one of the oldest dates from 1485 which explicitly forbad them from taking part in any commerce,[48] but they were allowed to sell goods during the time of fairs, and on market days i.e. on Tuesdays and Fridays.

Many other ordinances and laws were promulgated by Cracow's city council regarding the security of fair play and equal competitive conditions between merchants. There were strict directives on production control in the city,[49] and rigid rules on the problem of weights and measures,[50] money and exchange values, especially as transit trade to Flanders and Holland was common. The use of bills of exchange, which began in Italy *circa* 1300 and became the general practice in long-distance trade by the mid-fourteenth century,[51] did not seem to be of much importance in Cracow. Evidence for them from the city's archives is rare before 1500, with only two known examples from 1429 and 1492.[52] Up to the end of the fourteenth century the main form of insurance made use of a merchant's real estate but, as the fifteenth century progressed, greater use was made of credit connections in Cracow. *Circa* 1400, therefore, real estate credit was replaced by a merchant's personal credit; merchants and commission agents gave information to a registrar notary who entered it into a special book.[53] Difficulties arose, however, if goods were lost by a Cracow merchant whilst he was abroad, in which case only the king could make protestations on their behalf.[54]

Finally it should be remembered that Cracow also participated in intercontinental trade. As Hay has noted 'No trade had been more stimulating in earlier centuries than that between Europe and the East.'[55] Compared with the local and even continental exchange of merchandise, the quantities involved here were relatively small, but their economic significance was as high as their intrinsic value, namely the importation of precious commodities. Although such products from Asia Minor, the East Indies and the Far East were largely in the hands of Venice and Genoa during the thirteenth and early part of the fourteenth centuries, for Cracow it was Caffa in the Crimea that was important, as it tapped the resources of southern Russia and formed one outlet for the overland routes to India and China. For Cracow, the key to success lay in the exchange of such precious oriental goods, for the specialities of the various European regions, e.g., amber, furs and manufactured products en route to the East, which changed hands in her market place. Vital in all this trade, for the city at local and continental level, was connection to an efficient system of land and river routes.

Land and river routes

Commercial distribution was an expensive operation for, apart from toll stations, the main element of cost was undoubtedly transport. In the whole period prior to the railway age, transport was wasteful of time, equipment and manpower. Between the collapse of the Roman Empire and the eighteenth century efficient road making seems to have been a lost art. In general a road was a strip of land on which people had the right to travel, rather than an improved surface, whilst over moors, through passes or forests, and across swamps it was often little more than a path. In easier, flatter and drier country it was often wide enough to allow a horseman to ride more quickly on an unbroken surface, whilst cattle could more easily be herded along them finding food en route; wheeled vehicles, however, found the going difficult in wet weather. On such roads the normal speed was the walking pace of a man or animal, in which twenty miles (32 kms), was a good day's journey, and some daily stages were much shorter.[56] The shortcomings of the roads were in part counterbalanced by the widespread use of rivers,[57] but even these suffered from some natural defects. Spring floods, summer low water and winter ice impeded navigation, and there were many perpetual shallows, rocks and rapids. Most waterways were, however, capable of use; downstream traffic moved easily, but upstream movement was slow, since poles, oars or simple sails were the only power available for combating the current. Moreover the disadvantages of river transport were in some ways even greater than that of the roads; imposition of tolls by some feudal lord on river traffic passing through a section of his territory provided easy prey, whereas carts and pack animals could more easily avoid such fiscal dues, plague, banditry, warfare, or such physical obstacles as floods and landslides, simply by taking an alternative route.

This was the background which Cracow, like other medieval commercial centres, had to accept and make allowances for if its commerce, particularly

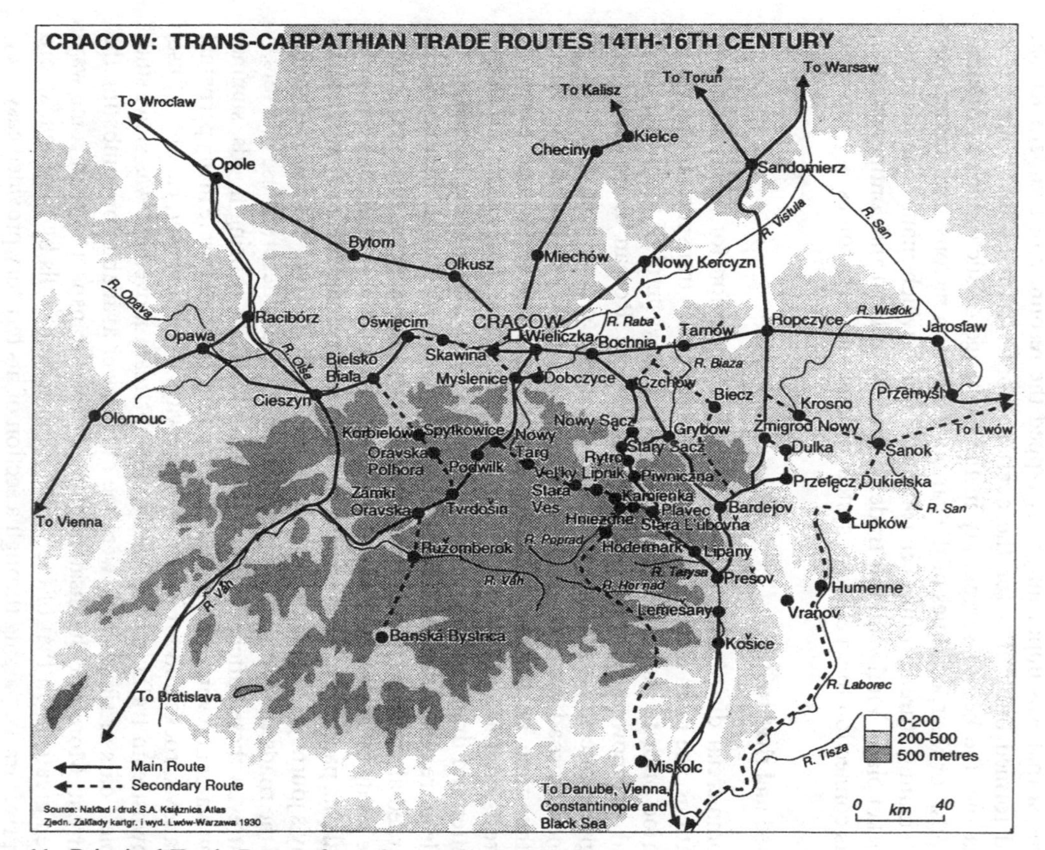

11. Principal Trade Routes from Cracow in the Middle Ages.

transit trade, was to be a financial success. At the local level within the Cracow region, goods including fresh meat, fish, wheat flour and wine were mainly carried from village to village by oxen-pulled carts (e.g. 1261).[58] Peasant carting was, however, in the nature of things seasonal and could not supply the needs of trade all the year round. However, horses pulling carts are also noted as carrying merchandise between the villages of the Cracow region (e.g. 1256)[59] but these were usually articles of greater value, or larger sized loads.[60] Horse transport appears to have been utilized by local knights for conveying valued articles like gold or cloth, whilst the village peasantry relied mainly on oxen for conveyance.[61] On more long distance and continental trade in the Later Middle Ages, merchants rarely travelled with their own goods; these were entrusted to regular carters (furmen); it therefore became more necessary for the merchant to get his instructions to his agents quickly, to receive their replies, and to appraise market conditions in which he wished to do business. It is not surprising, therefore, that, in comparison with local trade, regular traffic along the main lines of communication was in the hands of men who specialized in the business of transport and acted as messengers and common carriers. For this the merchants relied upon couriers who travelled much more quickly than the carters. A courier on horseback could cover between 60 and 100 kms in a day, but if he travelled on foot he could not expect to do more than 30 to 40 kms.[62]

Four principal communications lines radiated from Cracow, to the north, south, east and west (Fig. 11). The northern route to the Baltic coast was known as the Toruń or Prussian (Toruńska/Pruska) road and was already in use at the beginning of the fourteenth century.[63] The central part of this route left Cracow going via Miechów, Kurzelów, Przedbórz, Pilica, Piotrków, Łęczyca, Brześć in Kujavia to Toruń.[64] From there it followed the banks of the Vistula river to Gdańsk. The route also had a small branch which by-passed Toruń (Brześć, Inowrocław, Bydgoszcz, Tuchola, Chojnice and Skarszewo) allowing access to the Baltic coast for Cracow merchants, at times when the Toruń staple laws were seen to be too prohibitive. The tendency to introduce the principles of staple trade actually provoked a sharp conflict with the Prussian towns towards the end of the fourteenth century leading to a dislocation of communications with Gdańsk (Danzig) and Toruń.[65] For Cracow, however, also significant was the Great (Wielka) road,[66] which partly utilized the Prussian route as far as Łęczyca, then branched off towards Konin for Poznań,[67] and on to Skwierzyna, Santok and the Baltic ports of Szczecin, Stralsund and Greifswald,[68] and called the Flanders (Flandryska) route after about 1400.

The route from Prussia connected at Cracow with the road southwards to Hungary. It too had been in use since the thirteenth century. It linked Cracow with Bochnia and Lipnica and then went on to Czchów,[69] Nowy Sącz,[70] to Stary Sącz and Rytro.[71] It then continued through the

Carpathian Mountains via Stara L'ubovna and Prešov[72] and on to Košice.[73] This route together with the previous one, was the main artery for trade links between Košice and Gdańsk, which formed the basis for exchange of merchandise between Hungary and Flanders/England, during the latter part of the Middle Ages. Although the principal route, it was not the only one of importance. A shorter section went from Košice via Bardejov and Grybów[74] to join the main route at Czchów; its use was restricted, however, probably because it had fewer toll-stations. There was also an alternative route connecting the Baltic coast with Hungary which by-passed Cracow and was the source of much friction in the world of commercial politics. This Mazovian road (Mazowiecka) was already in use prior to 1320[75] running from Toruń via Łęczyca, Wąchock, Opatów to Sandomierz,[76] with tributary roads along certain sections.[77] Only in the second half of the fourteenth century was the route extended southwards from Sandomierz to Hungary, via Żmigród Nowy and the Dukielska Pass to Bardejov.[78]

Routes following a westward direction from Cracow were numerous, but the Wrocław (Wrocławska) road going to the north-west was most important for the city and much more significant than those leading on to Prague and Vienna. The road left Cracow for Olkusz,[79] on to Będzin, Bytom, Toszek, Opole and Brzeg to Wrocław.[80] A shorter version of this route went from Cracow to Oświęcim and joined the former road at Brzeg.[81] An alternative route came in to use during the mid-fourteenth century leaving Cracow for Lelów, Krzepice and Oleśnica[82] but was regarded as often unsafe by merchants sending goods to Wrocław.[83] In the reverse direction Wrocław traders were always looking for alternative routes through Little Poland to by-pass the Cracow staple laws. As trade competition intensified in the fifteenth century, three routes were found to avoid the Polish capital. The most important was from Wrocław eastwards to Wieluń, Opoczno, Radom, across the Vistula at Kazimierz Dolny and on to Lublin,[84] then via Hrubieszów to Lwów.[85] Two less significant alternative routes from Wrocław existed, one via Lelów and Działoszyce to Tarnów,[86] to the east of Cracow, and the other via Opatów and Sandomierz to Lwów.[87]

Eastward routes from Cracow were mainly concerned with connecting the city to the important emporium of Lwów.[88] The principal road to the east followed the banks of the Vistula out of the city to Opatowiec, Wiślica, Pacanów and Połaniec to Sandomierz.[89] Here there was a bifurcation, one branch going north-east to Lublin and on to Włodzimierz en route for Kiev, and south-east to Lwów. During the first half of the fourteenth century, this important latter road went up the San river valley from Gorzyce, Krzeszów to Kopki where it left the valley and cut across country to Lubaczów, via Gródek to Lwów.[90] During Kazimierz Wielki's reign, however, an alternative section from Kopki via Jarosław[91] and Przemyśl[92] developed, which rejoined the old route at Gródek.[93] This latter diversion was more attractive to some merchants, because these two larger urban centres offered more opportunity for trade, especially Przemyśl. Although this remained the principal artery from Cracow to Lwów, the route was long and uncomfortable, particularly for heavy loads. It is not surprising, therefore, that a second, shorter version, known as the 'Ropczyca' road was established. This route left Cracow for Bochnia, and on to Tarnów over the Biała river, continuing to Ropczyce[94] and finally arriving at Jarosław via Rzeszów to join the main Sandomierz-Lwów road.[95] At Ropczyce a tributary branch went south to Krosno, across to Sanok and on to Lwów,[96] but this was only utilized during periods of difficulty through trade wars or adverse weather conditions.

These four axis routes therefore connected Cracow with the more distant trade centres in Europe and beyond. Although merchandise arrived in the city from far-away places, constraints sometimes prohibited Cracow's merchants, or their representatives, from travelling to their markets of origin. For example, the staple laws in Lwów were compulsory for all merchants travelling on the Tartar (Tatarska) road to southern Russia, the Crimea and Sea of Azov.[97] If caught attempting to by-pass Lwów, traders had their goods confiscated and were forbidden to utilize the city's warehouses.[98] Conversely, Cracow's merchants were allowed to travel via Lwów to Moldavia, with prior permission, but strictly forbidden to trade amongst themselves in Lwów, having to buy and sell from local traders.[99] These facts should not suggest, however, that Cracow's merchants were limited to the confines of Lwów, Wrocław, Košice or Gdańsk. Evidence has shown that her merchants travelled to Italy, especially Venice, taking a route via Vienna, Neunkirchen, the Semmering Pass, Bruck, Leoben, St Veit, Villach, Pontebe Chiuse, Cremona, St Daniel, Portogruaro and Treviso,[100] as a result of agreements made between Kazimierz Wielki and Rudolf IV, Prince of Austria. Venetian merchants were allowed to trade freely in Cracow and vice versa.[101] An alternative, longer route to Italy was also utilized via Nuremburg, Augsburg, Innsbruck, Bruneck, Capo di Ponte, Conegliano to Treviso and Mestre, utilized from about 1350 when Nuremburg merchants arrived in Cracow and gave permission to travel via their city.[102] Perhaps Cracow's strongest trade links during the Later Middle Ages, however, were with Flanders and the Low Countries, a position enhanced by membership of the Hanseatic League.

River routes were also a possible means of sending certain merchandise through Cracow. The main alternative to wheeled traffic was the barge and the boat; moreover, in the Middle Ages as in modern times, carriage by water was much cheaper than by land. This was one reason why river traffic was able to bear the heavy tolls, whilst traffic in heavy goods, such as timber and salt, over long distances was only possible where cheaper waterways were available. As Postan states, 'The classical country of river navigation was east of the Elbe and more especially east of the Oder – in Lithuania, Poland, Galicia. Among the Western Slavs there were whole societies – villages and regions – which lived on and by their broad sluggish rivers.'[103] It appears that during the Early Middle Ages the Vistula river was little used as a trade route due to the weak development of Polish commerce up to the thirteenth century.[104] With the growth of towns in this century and expansion of trade in the fourteenth, especially transit trade from Hungary to the Baltic coast, river routes gained in importance. Wood seems to have been the first article in such trade, being floated down the larger rivers of the Carpathian foreland, e.g., Dunajec, and along the Vistula to Gdańsk.[105] Soon, however, goods from Hungary, particularly copper and iron, were being exported from Nowy Sącz down the Dunajec and thus avoiding Cracow's staple laws; this in turn led to a trade war between the two places in 1329, and was resolved when Sącz merchants agreed to send all their goods through Cracow, and not use the river route except for transporting salt.[106] No mention was made of wood as it was not an article subject to staple dues. It was half a century before the agreement was again put to the test[107] but once more Cracow asserted her authority; in spite of occasional freedom for the Sącz merchants the agreement remained intact well into the next century.[108]

During the fifteenth century Cracow merchants continued to use the Vistula river for transporting some of the heavier loads to the coast. The imposition of stapling laws through Prussian territory in 1389 was imposed on all transit trade[109] and, although it added costs to Cracow's goods, particularly copper, it did mean that there was also a guarantee of safe passage along the lower Vistula.[110] In 1403 the staple laws were lifted for Cracow merchants[111] to the great benefit of such traders as Jerzy Morsztyn who imported in his own ships large loads of cloth from Flanders to the city for resale.[112] Intermittent wars interspersed with guarantees of free transit along the Vistula for Polish boats[113] continued throughout much of the century, to be more clearly defined in 1505 by King Aleksander (1501–6) who declared the Vistula free of stapling obligations.[114] For Cracow's merchants, therefore, there was a continual need to adapt to the changing conditions imposed on both river and land routes throughout the Later Middle Ages, and reflected in the various routes taken in order to ensure that her goods reach their destination.[115]

Hanseatic League

Cracow's involvement in long-distance transit trade at a continental and inter-continental level attracted the attention of the Hanseatic League. Initially the League emerged during the thirteenth century from the union of

German merchants abroad which developed into a loose confederation of German cities,[116] organized under the leadership of Lübeck.[117] Separate German cities, which had already evolved their own spheres of trading activity, felt the imperative need to safeguard their trade routes by concerted action. The *raison-d'être* of the League was the absence of any ruler capable of protecting the commercial cities' interests, but it was an ad hoc political entity and never created a firm constitution of its own.

The League came into being about 1260–5, with its centre at Lübeck, and at one time or another its membership included nearly a hundred cities.[118] A condition of membership prescribed that cities should be situated on the seacoast, at the estuaries of rivers or on the banks of navigable streams and, although the rule was not stringently applied, it does suggest the importance of water transport in the Hanseatic world. The League's commercial territory operated over a wide area, between Russia and the Low Countries in one direction and between northern and central Germany and the Baltic ports in the other. Its merchants searched for merchandise as far away as Portugal[119] and Andalusia (salt),[120] and Novgorod (Russian products),[121] for 'It is well known that the Hanseatic trading system, thus extending from the Gulf of Finland to the Straits of Gibraltar, was essentially based abroad.'[122] One of the main supports for the League's trade came from the Baltic,[123] where purchased commodities were sent to Flanders (especially Bruges),[124] by its merchants in Revel,[125] Riga,[126] Danzig (Gdańsk),[127] Stettin (Szczecin),[128] Rostock,[129] and most of all from Lübeck.[130]

The Baltic cities were located in advantageous geographical situations where the numerous inlets provided port facilities, together with deep estuaries which were connected to the hinterland by long, slowly moving rivers. The hinterland was rich in all the resources of forest, field and mine, a natural wealth which only needed the demand of trade to begin their active exploitation.[131] The gentle currents of even the smallest rivers allowed the light-draught craft of the Middle Ages to help open up the country to trade,[132] while the longer rivers made connection over passes or lower watersheds with the headwaters of the South European rivers and brought the commerce of the south and Far East into the Baltic sphere of influence. According to Semple, 'This is the explanation of the fact that even such remote cities as Breslau (Wrocław) and Cracow, the emporiums of the upper Oder and Vistula, were eventually drawn into the commercial league of the coastal towns,'[133] giving access to 'The Russian trade-routes which were perhaps the most vital arteries of the Baltic commerce of the Middle Ages.'[134]

Given this background of the League, how and when did Cracow become involved in its organization? The first evidence of connections between Poland and the Hanse dates from the beginning of the fourteenth century; at a meeting of the latter organization a letter was sent to Poland inviting it to

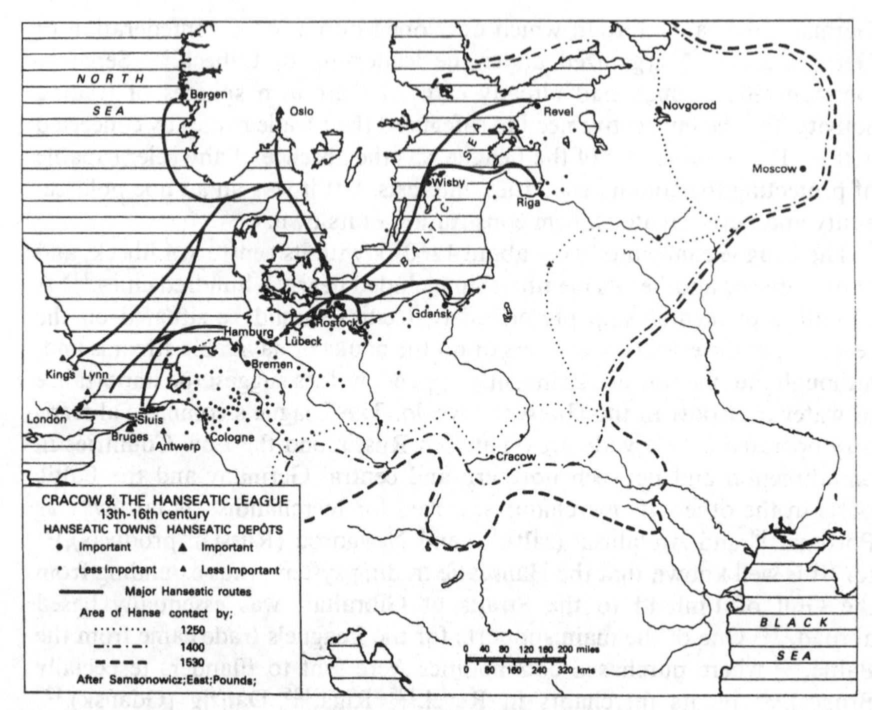

12. Cracow and the Hanseatic League, 13th to 16th centuries.

send a representative.[135] At this time the only Polish city to have connections with Flanders was Cracow, but there is no proof of this first probable link. During the second half of the fourteenth century, when the League was experiencing a particular period of strong economic growth, a letter (dated 1368) was sent from a Polish city to a meeting of the Hanse being held in Stralsund,[136] but again one may only speculate that it came from Cracow. The first positive information on Cracow's link with the League dates from 1376,[137] whilst ten years later, at a meeting of the Hanse in Marienburg (Malbork), there was a request to call upon the Prussian towns to help Cracow merchants reach Flanders.[138] Thereafter, Cracow is frequently mentioned in Hanse correspondence, and may have joined the League sometime after the end of Kazimierz Wielki's reign (in 1370). By 1401, Cracow is clearly counted as amongst the cities located within the League[139] (Fig. 12). Two and a half decades later a letter firmly reiterates Cracow's membership, and suggests that it had been a partner for some time.[140] Further correspondence from 1438 confirms Cracow's membership of the League.[141]

The initial impetus for Cracow to join the Hanseatic League probably came from those German merchants who had originally emigrated to the city. This, combined with the strong commercial connections between

Cracow and Bruges, at that time dominated by Hansard merchants, must have contributed to the decision for affiliation. The benefits of trading through the Bruges counting house, then under the League's authority, the defence of commerce in the Flemish lands[142] and other towns allied to the Hanse, and advantages of safe conduct for Cracow merchants displaying the Hanseatic flag, were privileges that must have seemed attractive to the city's traders. Conversely, certain aspects of the League's activities must have appeared rather alien to the city. Geographically, Cracow was situated at one of the farthest extremities of the League's influence, its nearest neighbours being Wrocław in Silesia to the west and the Prussian town of Toruń[143] to the north; there were no settlements immediately to the east or south that were members. Moreover, Cracow was also alien to the political activity of the League, centred on the north German towns[144] and the Baltic,[145] and a factor which eventually led to a weakening of the association. Although Cracow was in the first rank of importance for Poland and the Poles, it could not have been compared at this time with the large commercial cities such as Lübeck, Bremen,[146] Antwerp[147] or Cologne[148] which each had its own political policy. Cracow, like other Polish towns, only had its own industry and commercial importance to rely on, and could therefore not afford to meddle in the larger political intrigues of Europe. Perhaps this helps to explain Cracow's rather passive attitude towards many aspects of the League's organization.

Membership of the Hanse provided certain rights and demanded some basic obligations. For example, all member cities were allowed to attend the sessional meetings (Hansetage) of the League held on average once every eighteen months; the first extant document inviting Cracow dates from 1429 (although there may have been earlier ones), concerning a meeting in Lübeck towards the end of the year.[149] Similar invitations, usually sent from Lübeck via Frankfurt on Oder and Wrocław or through Gdańsk, are found up to 1487,[150] when meetings were held only once every six or seven years. Unfortunately, Cracow was conspicuous by her absence from such meetings[151] sometimes with justifiable reasons, such as late arrival of invitations,[152] but other times without any justification.[153] Continued absence eventually led to Cracow being deputized for by some other Prussian town.[154] Copies of the minutes of all these meetings (i.e. voting decisions) were sent to Cracow,[155] but none seem to have been preserved in the city's archives.

Cracow's obligations to the League also seem to have been defective. Each member city was supposed to contribute financial help, but Cracow never did as witnessed, for example, in 1494 when a tax was placed on each city; Cracow was specifically mentioned as not having made a tax return[156] but the only sanction against towns which defied decisions (Rezesse) of the assembly, expulsion, was in practice largely unenforceable; in the case of the larger centres they found it easy to secure readmission, or distant cities like Cracow which the League was reluctant to expel anyway. Only twice in 1407 and 1430 is it recorded that Cracow upheld her obligations, on both occasions supplying armed personnel for protecting the League's interests.[157] Protection of merchants' interests appeared of minor significance for Cracow, having little recourse to demand the Hansards' help. Only in the rare case of accidents[158] or complaints to Bruges about short measure on Flemish and English cloth,[159] is it possible to see how Cracow called on the League for help or intervention.

As the fifteenth century progressed and Cracow's participation in the Flanders trade began to wane, so did the importance of the Hanseatic League for the city's merchants. For a time formal relations between the two partners remained, and invitations to the Hanstage assembly still arrived in Cracow as a right of membership, but links weakened; by the sixteenth century there are no traces of information on the partnership, which it may be assumed ceased to exist. The Hanse itself was being eclipsed at this time for, as Małowist notes, 'Each sector of this organization acted politically in defence of its own interests. The Swedish, Danish, Polish and Russian merchants, like the English and the Dutch in the west, tended to free themselves from the commercial domination of the Hanse.'[160] Antwerp and Bergen became the chief markets of Northern Europe, whilst the Dutch emerged as the new carriers of merchandise (as witnessed by the Sundtolls),[161] creating for themselves an impregnable position in northern commerce. As Postan has observed, 'They now traded as far east as Breslau (Wrocław) and Cracow, and the towns of Prussia themselves began to look with dismay at the Dutch entering Poland.'[162] For Cracow and its merchants, the years after 1500 were to provide a new challenge to the city and its commercial organization.

Transport costs

The question now arises as to how much it cost to take merchandise by road from and to Cracow? Postan maintains that 'the final test of a transport system is not its density on the map but its effects on costs; and the costs were doubtless higher than the plethora of routes and quasi-routes might suggest to the uninstructed. What the average costs in fact were, no historian could now so much as guess, and it is doubtful whether the guess would be worth making even if it could be made. The most salient feature of trading costs in the Middle Ages was their infinite variety – a variety which would distort and falsify any attempt to strike an average for the system as a whole.'[163] Certainly inland transportation for Cracow's trading empire had many problems. The shortcomings of the medieval road system were in part counterbalanced by the use of rivers, in spite of their many defects. Safety was more important than speed and the human enemies might be worse than the physical. To these dangers of travel were added the financial burdens. Medieval princes and land owners collected tolls from travellers and traders but spent little of the revenue on road improvement. A merchant wishing to send goods from Cracow to Košice had to pay tolls on the Polish side at Żmigród, Sanok, Grybów and Czchów[164] (Fig. 13), and on the Hungarian side at Makowice, and go through stapling rights at Bardejov,[165] Prešov, Sabinov, Levoča and then again at Košice.[166] Furthermore, the condition of this route in the fifteenth century meant that it could not be used throughout the year, lying as it did through the heavily forested Carpathians. Much of the area surrounding the route, according to the sixteenth-century chronicler Cromeri, was wild; in autumn the rivers overflowed and in spring the slushy, thawing roads made communications difficult.[167] The question of safety in the Carpathian foothill region also proved hazardous. Merchants sending merchandise through the region suffered from brigandage, the smaller traders who accompanied their goods being exposed to physical attack, which in some cases led to their death.[168] Such factors help explain why Postan believes that 'on routes which were so heavily taxed, or were so badly served by transport, or were so profoundly disturbed by war and piracy as to be unsuited to active trade, traders were not in fact active. They frequented instead those routes on which transport was relatively free and cheap.'[169]

Evidence suggests, however, that the Cracow–Bardejov route was far from unpopular. Above all it was necessary for the Cracow merchant to carefully choose his carter, and give him the correct amount of money to dispose of whilst en route. Rewards were very varied, depending on distance, freight and its quantity, and the need to bribe in difficult situations. During the first half of the century carters on the Cracow–Bardejov route were paid from 1 to 3 grzywnas (Latin: 'marca').[170] Articles also differed in conveyance costs; for example, the transporting of 25 cetnars of salt from Bochnia to Bardejov cost 5 florins.[171] Couriers were usually paid from ½ to 2 grzywnas for carrying messages from Cracow to Hungary at this time.[172] Another item which one may add to total transport costs concerned gifts to buy privilege, or influence some important Cracow city councillor, a common and prized gift, for example, from Hungary being wine.[173]

Customs duty also provided some financial burden on the total outlay of a journey. This fiscal imposition was a tiresome strain on transit trade, which the merchant who paid it regarded simply as an exaction, or an unjust levy on his goods. Of all the obstacles placed in the path of commerce none was more annoying or more general. Throughout the Middle Ages customs duty and tolls continued to be a hindrance on all the highways of traffic. At the end of the fifteenth century, there were still 64 of them on the Rhine, 35 on the Elbe, and 77 on the Danube in its course through Lower Austria alone.[174] Likewise in Poland it was customary in the Middle Ages for

CCCX. 1432. mense Martio.

Statuta institorum. De portorio Cracoviae, in Czchów Alba ecclesia et in Będzin solvendo. — De nundinis Cracoviae celebrandis.

Das Gesetcze ist gegeben den Cromern durch dy Herren Iungk vnd Aldt czw des rotes willen czw halden. Geschen yn der ersten fastwochen, yn dem 1432 Iore des hern.

§ 1. Keynen keller yn der wochen zal man awffslissen noch vorkowffen dorynne, beawssen dem dinstage vnd freitage, aws genommen vm iormargkt bey V margk busse.

§ 2. Nymandt zal wider yn kellern noch in hewseren wider am margtage noch sust bey awssen dem yormargkte keynem gaste noch landcromer vorkowffen, awsgenommen vnseren cromern, dy czw kromesteen, vnder der nochgeschriben satczunge bey der obgenanten busse:

Vnder einem gantczen parchen,
Vnder eime stucke blo leymet, Swarcz ader roth,
Vnder eime stucke heidnesche leymet,
Vnder eime gantczen stucke Schwebische, krosnessche ader czanser leimet yn der grösse noch alder gewonheit,
Vnder eime gantczen stucke kamchen, attlas, zammet vnd gulden bortchen,
Vnder eime gantczen Harres, forstat vnd pokoczin,
Vnder eime stuck dyrdumdey vnd schloer vnd drommel zeiden ader wollen,
Vnder eime stuck zeiden schnure vnd bortchen,
Vnder eime thwzin mötczen vnd frawen hewblen,
Vnder eime halben steyn weidgarn,
Vnder eime phunde geczwirnte vnd offene Zeide,
Vnder eime halben steine palmet zeide,
Vnder czween vnczen vnczengolt,
Vnder eime stein bomwolle,
Vnder eime stein bernesche zeffe,
Vnder eime stein wachs,
Vnder eime stein pfeffer vnd kommel,
Vnder eime phunde zaffran,
Vnder einer margk nelken, muscaten, muscatblut, czenemey vnde ingber,
Vnder eime stein alawne vnd weynstein,

Vnder eime stein czin vnd hutczoker,
Vnder einer rezen papir,
Vnder eime steyn reisz, mandeln, feigen vnd rozinken beawssen der fasten vnd dem aduent.

Sneider.

§ 3. Sunderlich ist vorboten den Sneidern, das keyner do heime parchen noch leymet schneiden zal, noch bomwolle, noch weitgarn vorkewffen, noch keyn ander ding, das dy cromer angehoret, bey der obgenanten busse.

§ 4. Auch zal nymandt drumchen, lenczeln vorkawffen wider vnder dem kawffhawsze noch anders wo vnder eime stucke, bey der obgeschribene busse.

Woge.

§ 5. Nymand zal wider yn kellern noch yn gewelben ader hewseren noch anders wo beawssen den cromen keyne der obgenanten war anders wo aws wegen, wenne yn der stad wogen bey der obgenanten busse.

Alzo zal man den Krokeschen Czol ynnemen von den gesten, dy ire gutter her brengen, sunder der landman gibet halb alzo vyl.

§ 6. Von itczlichem czihendem pherde I gr.
Von eime czihenden ochsen eynen halben gr. von einem ledigen I gr.
Von einem itczlichen tuche II gr. an Scharlach,
Von einem centner hummerey, als weidgarn, czichwergk, weisz czwirn vnd sulcherley dingk VI gr.
Von alawn, kömmel, weynsteyn, galmey, czinemey, galiczenstein, zeife, von itczlicher der dinge centner III gr.
Von itczlichem centner Cromerey VI gr.
Von kopper vnd bey (s) von dreyen centnern II gr.
XVI parchen vor ein centner VI gr.
Von hundert scheffel hoppe I firdung,
Von einem centner Bomwolle III gr.
Von einem centner wachs I gr.
Von einem centner zeyde VI gr.
Vom centner sweisleder VI gr.
XVI harres vor eyn centner VI gr.
XXIIII thwzin vor eyn centner VI gr.
Von weiser leymet vnd landleymet, Schumbach, czwelich, von Acht stucke Golcz vor eyn centner VI gr.
Qweckzilber vom centner III gr.
Czigenhor vom centner III gr.
Von einer osterkuffen weyns I firdung,
Von einem tawsent wergk VI gr.
Von einem centner lorber VI gr.
Von einer thonnen herings III gr.
Von einem centner motczen VI gr.
Von felwergk, lampfel, gropfel vom M VI gr.

Von grötczen, hezins, margbalge, Harembalge, lassitczen vnd allerley sulcher dinger vom M VI gr.
Mandeln, reis, feigen, ol, honigk, eyzen, schoffwolle, pech, vnslet, Schmer gibet nicht czol.

Alzo fleget man czw nemen den czol czw der weisen Kyrchen von allerley ware ydoch der landtman gibet is halp.

§ 7. Von allerley gutten kurczin gewand gibt der gast VI gr.
Von einem langen eypreschen, mechleschem vnd von iderm langen gewand geben dy geste VIII gr.
Halbe von derellmunde geben IIII gr.
Kwrtcze engleschen III gr.
Hundert elen Kyrzey geben II gr.
Eyn harres gibet II gr.
Von schweren dornischen tuchen geben dy geste VI gr.
Von einem polneschen geben dy geste II gr.
Von einem harres II gr.
Von hundert leders III fr.
Von czygen vnd schoffelen noch margczal von der margk I gr.
Von schoffen, Schweinen vnd kleinem fye czw IIII heller,
Von grobem fye czw halben gr.
Von hundert fochsbalge vnd margbalge VI gr.
Von M hezins vnd eichornes X gr.
Von M gruczens V gr.
Vom centner wachs III gr.
Von hundert elen leimet I gr.
Von bomwolle, czigenhor vnd aller ander kromery auff dy margk czw rechnen, von einer margk I gr.
Vom centner wolle IIII gr.
Von flader noch der margkwert I gr.
Vom ferde ym wagen czihende II gr.
Von brenden ader swerten das czwenczigste ader III gr.
Von der thonnen herings VI gr.
Vom fas fisschen vnd öls VIII gr.
Von hundert scheffeln hoppens I frdg.
Von eynem eyzen wagen VIII gr. der ym land gesessen ist gibet VI gr.
Von kopper wagen das pherd gibet II gr.
Von einem nest fladeren becher dy gest geben IIII gr.
Alle ledige pherd geben czw II gr.
Von eyner kuffe weyns VI gr.
Kopperwasser, czin, vlader, zensen, zicheln vnd alle gemachte schwere messer, kessel, phannen vnd alle gegerbte fel geben von der margk I gr.
Pelcze, körschen, von der margk I gr.
Hozen, mötczen, görtelwergk von der margk I gr.
Mandeln, reys, bley, kopper, ole ist frey.
Von einem stein vnslet vnd schmer II heller.

§ 8. Wer disen czol helt, der zal auch halden dy lampe vnser frawen czw lobe vnd czw eren.

Den Czol czw Banden von dy aws der Slezie kommen, zal man alzo ynnemen gewonlich.

§ 9. Von einem itczlichem ledigen pherdt I gr. lat.
Von iderem schonen tuch I gr. lat.
Von XI polneschen tuchern IIII gr.
Von wachs, bley, kopper, czin, leder, der koffman gibet vom pherde V gr. lat.
Von cromerey als von III margken II gr. lat.
Von eyner thonnen heringe III gr. lat.

Alzo zal man den dy Iormargkt czw Krokaw anheben vnd nicht lenger haldten.

§ 10. Den Iormargkt noch ostern hebet man an dem tage des heiligen Crewtczes vnd wert gantczer czehen tage vnd II halbe tage vnd nich lenger.

§ 11. Der Iormargkt awff Viti.
Hebet an am tage Sancti Barnabe vnd wert funff tage vnd czwene halbe tage.

§ 12. Vnd dize czwene yormargkte sint frey den gesten, ydoch den quartczol ist man der Stad alleczeit pflychtig.

Behem f. 240 v.

13. Document: Customs Tariff List at Czchów, March, 1432.
Source: *Kodeks dyplomatyczny miasta Krakowa*, F. Piekosiński (Ed.), Kraków, 1882. Document 310, dated March, 1432. Customs tariff list at Czchów, en route from Cracow to Nowy Sącz.

customs duty to be paid at stations lying within the country not just near borders, and found most frequently along the main trade routes. Higher duty had to be paid in the larger towns or key crossroad junctions such as Cracow, Czchów, Stary Sącz and Rytro,[175] Grybów, Biecz, Żmigród and

Sanok,[176] and much lower in centres of less commercial importance, like Myślenice, Dobczyce and Skawina.[177] The significance for trade of these centres was reflected in the total annual sum collected from customs duty; for example, in 1390 this amounted to 1,089 grzywnas for Cracow, 890 grzywnas for Czchów and only 87 grzywnas for Skawina.[178] Admittedly preferential treatment existed in the form of tax reductions for certain traders; in Cracow, Polish merchants only paid half that levied on foreign merchants.[179] In Hungary, customs duty was on the whole lower than in Poland, where a system of special rates operated, known as the 'harminczad', or 'thirtieth' (tricesima, Dreisgstamt, tryatok), because duty was collected at $\frac{1}{30}$th of the nominal value of the merchandise.[180] Furthermore, customs duty tended to vary according to the reign of a king; when Ludwik Węgierski was in power many Hungarian towns were freed from duty obligations,[181] whilst during Władysław Łokietek's reign Cracow merchants were released from paying customs levies in Nowy Sącz.[182]

Other incidental factors could also suddenly influence overall transport costs for the Cracow merchant. During the time of actual transportation of goods, the merchant always had to be prepared for the unexpected. It often happened that carters were detained at certain places en route and waggons held up due to the confiscation of horses and goods on charges of debt by their owner.[183] Similarly events also happened on the open road.[184] On occasions the carters did not supply goods to their correct destination, but sold them freely during the journey, exposing the owner in Cracow to financial loss.[185] Inevitably, this led to conflict arising between the owner and carter over the accounts or even worse failure on the latter's part to properly reckon up the bills.

Finally, commercial contracts with Flanders involved Cracow merchants in some maritime transportation. Goods were sent either down the Vistula to Gdańsk, or overland to Toruń, then from there to Gdańsk by barge along the Lower Vistula. There is a dearth of documentary information on transport costs for such journeys to the Baltic coast prior to 1500.[186] In comparison to overland movement, however, Postan states that 'Sea transport was even cheaper, so long as it followed the well established and regular sea lanes',[187] concluding that, 'Transport and expenses of handling were obviously not the main constituent of costs.'[188] Unfortunately, transport costs were often hidden in documentary evidence under the term 'freight costs' which, according to Vogel, totalled just over a quarter of the outlay for a voyage.[189] Within the global figure for freight expenses it has been estimated that about 5 per cent was for actual transport, a further 10 per cent for customs duty, port taxes and warehousing costs, 3 per cent went on payment for the crew, and 9 per cent on victuals and beverages.[190] How typical these costs were is open to question, for one may cite a much higher percentage for transporting some commodities, like salt.[191]

One may conclude, therefore, that during the Middle Ages the proportion of trading costs to total costs was probably less than at present. In short, for the Cracow merchant, medieval communications, like other trading activities, suffered much more from instability and uncertainty, political in origin, than from high costs generated by an inefficient transport service. In practice, those merchants in Cracow engaged in the main branches of medieval trade could therefore, at most times, find routes which were reasonably cheap, or at least 'not so costly as to justify excessive "traders margins" or greatly to restrict the demand for commodities and their supply'.[192]

The time factor

How slow was commercial traffic for the Cracow merchant? It is generally accepted that there was little improvement in the speed of transport up to the nineteenth century and we are told that 'Napoleon moves at the same slow rate as Julius Caesar.'[193] Nevertheless, the speed of travel varied greatly; a courier on horseback could cover up to 100 kilometres in a day, but if he travelled on foot he could not expect to do more than 30 to 40. Laden carts and waggons moved even more slowly. Yet, for the Cracow trader, the longer the journey the greater the prospect of profit in an age when prices were chiefly dependent on the rarity of the imported goods and where this rarity was increased by distance. It is easy, therefore, to understand that the desire for gain was strong enough to counterbalance the hardships and risks taken with the transport system. Up to the eighteenth century sea journeys were interminable and overland transport almost paralysed.[194] Moreover, the merchant and his carters were subject to a limited range of choice. Perhaps they could choose one itinerary over another to avoid a toll or customs post, but even then they might have to turn back in case of difficulty. They followed one road in winter and another in spring, according to frost or pot-holes, but they could never abandon roads already laid out, which meant drawing on services provided by others. Perhaps the most notable characteristic of the medieval roads was the multiplicity of routes which linked the same urban centres; plurality of roads enabled a greater number of minor places to be included in the network. Furthermore, trade wars between different cities imposed many unnecessary detours into the trade-route pattern.

As part of the continental trade network, Cracow was subject to the same general conditions as other cities. Ideally, one would like to be able to estimate with some exactitude the isochronal pattern for European commerce during the Later Middle Ages. Unfortunately, the scantiness of known information is such as to deter any such attempt, but those isolated details that exist may give some indication of the time factor in continental trade.

Merchants must have been accustomed to two forms of transportation, one carrying information through couriers, and the other actual conveyance of merchandise. When news and mail had to be carried quickly, couriers might cover 64–80 kilometres in a day and some cities and large business firms ran regular fast services. A journey of 1,600 kilometres from Calais to Rome took a relay of express riders 27 days to complete in 1200. By the early fifteenth century a regular mail service carried letters from Bruges to Venice in about 25 days; the Romans had been able to do the same journey in about the same time a thousand years before.[195] Admittedly emergency journeys could be achieved by messengers travelling over 17 kilometres an hour, but these were exceptional for couriers normally did not ride by night.[196] Moreover, even with urgent dispatches there was a margin between prompt and late delivery as evidenced by late-fifteenth-century information, where the difference for Venice to Augsburg was 16 days (maximum 21 days: minimum 5) or Venice–Vienna, 24 days (32:8).[197] In summary, therefore, other calculations would give the same conclusions, namely 'that with horse, coaches, ships and runners, it was the general rule to cover at the most 100 kilometres in 24 hours'.[198]

Conveyance of merchandise was obviously much slower, where one calculated in weeks rather than days. In the carriage of goods the packhorse or the two-wheeled waggon, moving at the easy pace of a horse or oxen, set the speed and the day's journey. A drover who steered a small herd of cattle and a large flock of sheep along 208 kilometres of fairly level English road in thirteen days probably made good time (i.e. 16 kms a day). There was a reliable overland route from Cologne, utilizing a combination of river, cart and packhorse transportation which led southward to the Alpine passes and Italy; by 1500 this route was so well organized that merchandise could traverse it in about two months (and a courier, two weeks).[199] More normally a convoy of horses and mules might cover between 24 and 32 kilometres a day, whilst packhorses took seventeen days to carry loads of wood from La Rochelle to Nîmes (600 kms).[200] Local traffic could have been faster due to the absence of tolls etc. In the thirteenth century peasant carts in Cheshire, England, transported stone from a quarry to the Cistercian Abbey, a distance of 13 kilometres; carters made two complete journeys a day, a distance of about 50 kilometres,[201] mainly during the winter months.

Water transport increased in popularity during the Middle Ages. Surprisingly, even on a favourable network of canals and rivers the speed was little better than overland. Goods came down the Rhône valley at about 33–35 kilometres per day, and it was reckoned normal for barges to take about 18 days to cover the distance between Milan and Venice, a journey a fast boat could do in 3 days.[202] Maritime shipping naturally assumed a greater importance than river trade but slowness was also a predominant feature. Up to the fifteenth century in northern Europe, before the mariner's compass became more general, ships were forced to coast along the shores, ever wary of the threat from piracy. At sea, while ships could make 120–30 kilometres a day, they were usually only able to steer straight for their destination for very short periods. From Lübeck to Bergen might take between 9 and 20 days; a voyage from Venice to London varied between 9 and 50 days.[203] Longer voyages, like those ventured by the Hanse, were always undertaken in convoy for security reasons, and usually made slow progress.

Given these varying times for European commerce in the Later Middle Ages, how did Cracow's experience fit into the pattern? With reference to local trade, perhaps Braudel encapsulates the whole situation when he writes, 'The spectacle remained timeless and unchanging. I still saw convoys of narrow four-wheeled peasant wagons on the Cracow roads in November 1957 going to town loaded with people and pine branches, their needles trailing like hair in the dust of the road behind them. This sight, probably now living out its last days, was just as much a feature of the fifteenth-century scene.'[204] Overland travel may be estimated from a document dated 1402.[205] Goods sent from Cracow to Toruń took 9 days (16/III–25/III), a distance of some 370 kilometres. The land route from Szczecin to Florence, for example, took 5 weeks,[206] whilst a journey down the Vistula could last from 8 to 10 weeks.[207] Again quoting the 1402 document, the goods sent from Cracow via Toruń arrived at Gdańsk by the beginning of April, where they were immediately placed on board ship, which set sail for Flanders. The merchandise arrived in Flanders during the first days of May, i.e. a total journey of some 7 weeks. Variations existed according to the time of year. Samsonowicz has analyzed ships leaving Gdańsk between 1420 and 1460 by month[208] and has noted that the winter months saw only 11 per cent of the 82 ships leave port compared with 42 per cent in the spring, 33 per cent during the summer months and 14 per cent in the autumn. Therefore, though fragmentary in nature, these snippets of information tend to suggest that Cracow's merchants were no more, or less, favoured by the element of time in their commercial dealings, than other parts of Europe given the differences of climate, and other physical controls combined with the human barriers existing throughout much of Europe at this time.

Expensive raw materials

If Cracow was as important commercially as all this evidence suggests, upon what regions and commodities was the commercial pre-eminence of Cracow based? There was an important commerce in raw materials, which can be subdivided into precious commodities – various metals, spices and salt, and into cheaper goods such as skins, wax, livestock products, wood and cereals; and also into manufactured goods concerned mainly with the importation of textiles from western Europe for the Central European and Eastern markets. The primary products came to Cracow principally from the south and east beyond the Carpathian Mountains, such as Hungary (Slovakia) or, as in the case of salt, from sources in the immediate vicinity of the city.

(i) Metals

Minerals seemed to have the greatest significance for Cracow's entrepreneurs in the lands of northern Hungary/southern Poland during the Later Middle Ages. Metals were essential to medieval civilization, but they were used in small amounts and were relatively valuable. Iron was most widely used for it was indispensable for tools, weapons and armour; similarly lead was much in demand as a roofing material and for making water-ducts; whilst, when alloyed with tin, it produced pewter. Copper was used to manufacture bronze, and the precious metals like gold and silver, were in demand for currency as well as for making utensils for both church and lay use and as a form of decoration. Pounds underscores the situation when he states, 'Metals were never produced in sufficient quantities, owing in part to the technical difficulties of mining and smelting, but also to their highly localised occurrence'[209] whilst it should be remembered there was a constant drain on particular European metals such as lead, tin and other precious metals, to the Middle East markets.

In medieval industry mining showed a great diversity of organization. Where minerals were found near the surface, or in outcrops, little capital was needed and the mining unit was small. Deeper deposits required larger outlays for the mine and for the smelting and refining equipment. In some cases Cracow merchants either provided the capital or bought a lease on the right of the mine,[210] granted as a concession in return for some loan, or handed over as collateral. Mining, therefore, became a field for investment by Cracow merchants bringing their capital to the mountains and backwoods of the Carpathian region. Geographical literature has criticized the lack of theoretical underpinnings to the study of mining activity[211] or even the presence of descriptive papers[212] on the contemporary scene, which is even more apparent for the medieval period with its paucity of extant documentary material.[213] Nevertheless, for Cracow at least, some assessment can be made of its mining activity during the Later Middle Ages.

The three most important minerals, according to Cracow documents, were silver, lead and copper, not necessarily in that order, together with less significant ones like iron, zinc, mercury, gold and sulphur. Cracow's interest in mineral exploitation and its transit trade continued throughout the period under review, if anything intensifying towards the end of the fifteenth century. A generalized picture of mineral trade movements can be constructed (Fig. 14), but this does not give complete accuracy since documentary

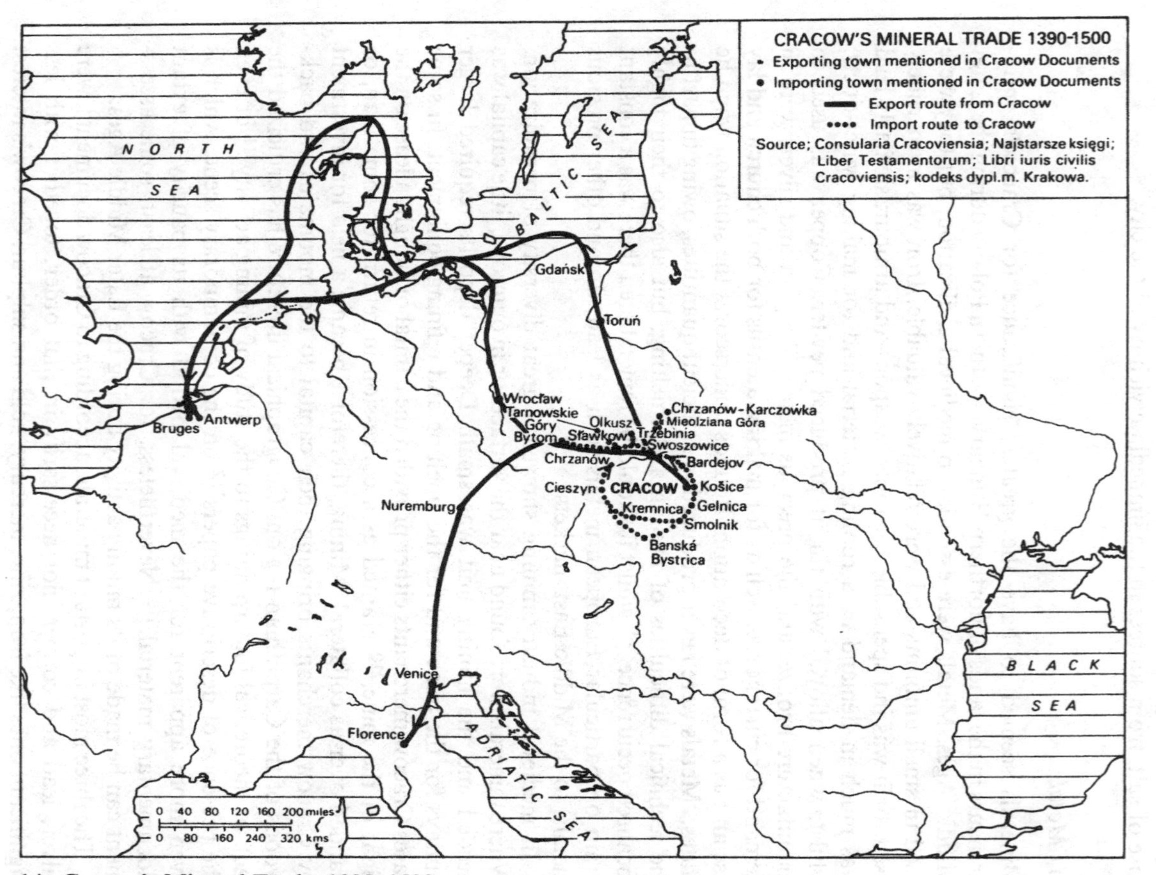

CRACOW'S MINERAL TRADE 1390-1500
Exporting town mentioned in Cracow Documents
Importing town mentioned in Cracow Documents
Export route from Cracow
Import route to Cracow
Source; Consularia Cracoviensia; Najstarsze księgi; Liber Testamentorum; Libri iuris civilis Cracoviensis; kodeks dypl.m. Krakowa.
NORTH SEA
BALTIC SEA
BLACK SEA
ADRIATIC SEA
Gdańsk
Toruń
Bruges
Antwerp
Wrocław
Tarnowskie Góry
Bytom
Olkusz
Sławków
Trzebinia
Chrzanów-Karczówka
Mieolziana Góra
Swoszowice
Chrzanów
Bardejov
Cieszyn
CRACOW
Košice
Kremnica
Gelnica
Smolnik
Banská Bystrica
Nuremburg
Venice
Florence
0 40 80 120 160 200 miles
0 80 160 240 320 kms

14. Cracow's Mineral Trade, 1390–1500.

sources fail to record all the transactions which took place. Such references as occur are often fortuitous, occurring in wills, lawsuits, or accidents rather than as a result of meticulous systematic recording of merchants' agreements. It should also be remembered that many transactions, particularly concerned with precious metals like gold and silver, were never committed to paper. Sometimes archival sources mention the actual mine where the mineral was bought, but in other cases reference is only made to the market town in which it was purchased.

Silver had an international market in the Middle Ages, for credit was then little developed and supplies of silver played a greater part in determining the wealth and political power of sovereign states. As Małowist has stated, 'We already know that very important export articles from Poland to Hungary were lead and salt, and from Hungary to Poland copper, iron, silver and certainly gold.'[214] Conversely, with silver Kutrzeba thought that it did not play an important role in Cracow's commerce,[215] but its significance may have been veiled somewhat by secrecy. Postan clearly maintains that 'Relatively less prominent in the annals of northern trade and in the records of its shipping were the exports and imports of metal. The mining of precious metals, especially of silver, was a great industry and its products played a part in the economic development of Europe so critical that they should not perhaps be treated as mere items in a list of commodities.'[216] Two source areas of silver for Cracow's merchants were Upper Silesia and Upper Hungary. In Upper Silesia the favourable geological conditions[217] led to exploitation of the argentiferous ores around Bytom, Sławków and Olkusz already in the twelfth and thirteenth centuries;[218] during the following century documentary evidence reveals that Bytom silver was sent direct to Cracow.[219] The mines of Upper Hungary, as the Slovak area was commonly called, were opened up for silver mining by Germans in the thirteenth century; silver was being sent to Cracow under strict control by the early years of the fifteenth century.[220]

The real breakthrough in silver mining came in 1451 with a new invention for separating silver from rich argentiferous copper ores with the help of lead.[221] Prior to this event the production of silver in central Europe had reached a low point during the Hussite wars in the twenties and thirties of the fifteenth century; moreover, when the richest and shallowest deposits had been exploited, deeper deposits lay below the water level. Further exploitation was beyond the capabilities of the small producers and individual mine owners, their small capital accumulation being insufficient for continued success; hence many mines were abandoned between 1350 and 1450. Improved technical methods led to the revival of former drowned or buried mines, accompanied by the discovery of ore seams in many old mining centres and in some new places e.g. the region of Tarnowskie Góry.[222] The enormous rise in the value of central European silver for minting coins led to continued exploration for lead ores, since they frequently contained silver. The Silesian mines had small lead deposits with an admixture of silver, but the high silica content made separation a difficult technological task; fortunately Cracow could still be sure of supplies from the Slovak mines. A Cracow citizen, Jan Thurzo, together with the legendary Jacob Fugger[223] became proprietor of all privately owned mines at Staré Hory, Pieský and Špania Dolina between 1494 and 1496,[224] the new company obtained a royal privilege in 1496 which enabled it to export silver from its mines at free market prices in contradiction of the usual obligation to sell all gold and silver at a fixed price for minting purposes.[225] By the end of the century, therefore, Slovak silver was being sent to Cracow, where part was retained for processing at the king's mint[226] and the rest sent via Toruń and Gdańsk to western Europe.[227] Moreover, the overvaluation of silver in the Burgundian coinage from 1465 onwards had turned the Low Countries into a particularly profitable outlet for this metal,[228] with Antwerp the centre for such trade.[229]

Lead, like silver, had an international market despite the problem of weight in transportation. This metal was used either in the pure form for domestic utensils, church ornaments and church roofing, or mixed with other metals to make armour, trappings for horses, chains for livestock and prisoners, candlesticks, arms, etc., and was exported in large quantities by Cracow merchants to Hungary and western Europe. In contrast to the search for silver, Cracow's citizens rarely went outside the local region to obtain lead, firstly because the transport costs would have seriously affected the profit margin and secondly, because there appeared to be enough lead extracted close by to satisfy Cracow's needs.

The most important local lead mine was at Olkusz;[230] here lead deposits were found in the Triassic limestone and dolomite sediments which were first seriously exploited towards the end of the fourteenth century. Royal privileges were granted (the earliest in 1374) for exploration,[231] although mining at Olkusz had been known since about 1250.[232] Lead mining was important at nearby Sławków, first referred to in documents in 1412.[233] Cracow merchants also obtained lead from Trzebinia, Chęciny and Chrzanów,[234] where they owned mines[235] and sent ore in pieces (called 'szrotowane'), to Cracow for export. As Małowist points out, 'The riches of the earth in these regions (Silesia, Upper Hungary and S. Poland) were less than in Bohemia, but they were considerable. There it was ... lead ... in Little Poland.'[236]

The earlier impetus for lead mining had consisted of exploiting rich ore deposits which extended above the level of the water table; once this was reached mining stopped. The mining of lead, of which Cracow held a near monopoly, increased steadily in the later fourteenth century, and reached its medieval peak about 1425. Output then sank for a couple of decades, but

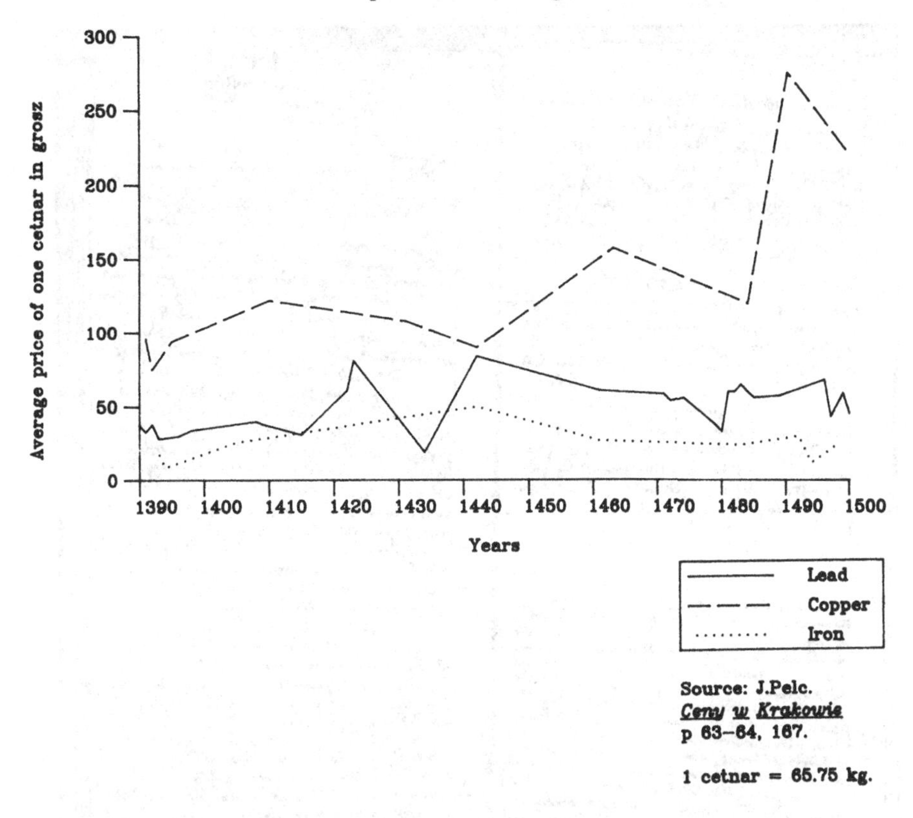

15. Price of Certain Metals in Cracow, 1390, 1390–1500

basic changes in mining technology and water removal after 1450 led to half a century of rich exploitation. This, in turn made fortunes for some mining firms in Cracow and elsewhere, at first managed by city burghers and later by the state and feudal lords. During the first half of the fifteenth century, much of Cracow's lead trade was in the hands of Jan Sweidniczer,[237] whilst Jerzy Morsztyn[238] became a prominent dealer at the end of the century. Documentary evidence, however, does not support Pounds' claim that 'It is not without significance that the price of lead increased steadily in the market of Cracow – not far from mining regions of Bohemia and Hungary – throughout the fifteenth century, at a time when the prices of most other commodities were tending to fall.'[239] Evidence suggests (Fig. 15), the highest price was reached in 1423, followed by the depression of the 1430s and an uneven price decline after a secondary peak in 1442.[240]

The chief market for Cracow's lead was Hungary. The lack of known lead bearing deposits in Upper Hungary meant that already in the thirteenth century it was being exported from Cracow to towns in Slovakia. At the

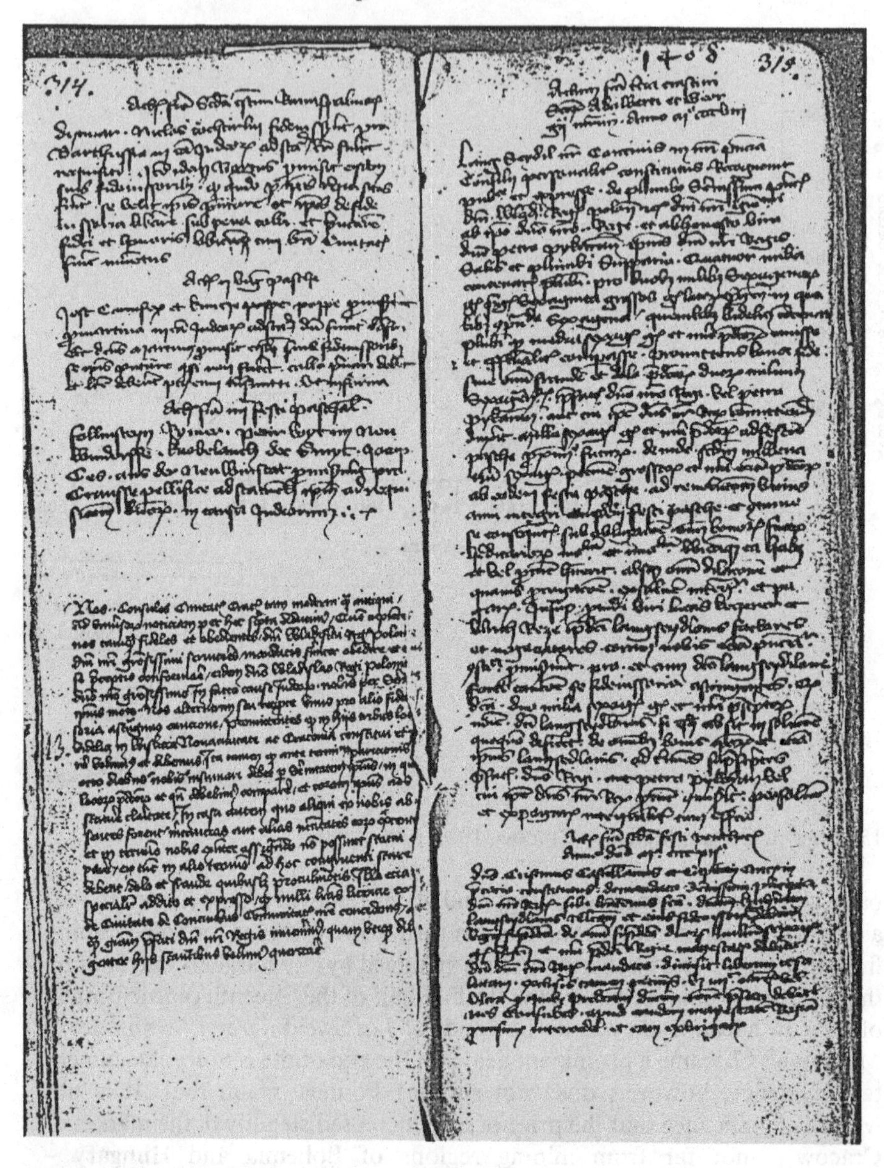

16. Document: Lead from Cracow to Košice, 1408.

Wojewódzkie Archiwum Państwowe w Krakowie: *Consularia Cracoviensia*, Vol. I, p. 315; M/s 427. In 1408, a Cracow merchant, Lang. Seidel sent 4,000 cetnars (258.85 tons) of lead from Prague, via Cracow to Košice in Slovakia (N. Hungary).

beginning of the following century it was one of the city's important export commodities to Hungary;[241] whilst in 1393 lead is specifically mentioned in the customs tariffs of Stary Sącz and Czchów as en route for Hungary via Košice.[242] Further documentary evidence from 1405 refers to a Nuremburg merchant in Hungary with a monopoly for buying Polish lead from Cracow's merchants;[243] reaction came from the mine owners in Olkusz who boycotted the Hungarian market, preferring to send their lead ore to Flanders via Toruń, and Bohemia via Wrocław thus eclipsing Cracow's middleman role. Hungary, however, was too large a market for Cracow merchants to lose, and a satisfactory outcome of the conflict was soon found, as evidenced by a document from 1408 (Fig. 16). This records the dispatch of 260 tons of lead to Košice, to be paid for in Prague currency.[244] Even during the depressed years of the 1430s documents still record lead sales to Hungary[245] but it was the discovery of the cupellation process by Lazarus Ercker[246] at mid-century that was to transform commerce in lead for Cracow's entrepreneurs.

Smelting of the complex silver–copper Slovak deposits required the addition of adequate quantities of lead to the ore during the assaying process. The use of reaction smelting was only possible at this time on ores with a low or non-existent silica content since lead silicates and sulphates coagulated with the other ores producing impurities.[247] Fortunately, the lead ores from the Upper Silesian–Cracow region contained very little of the critical silican dioxides;[248] by importing these lead ores through Cracow's merchants it meant the Slovak deposits could be smelted on site, since the mountains provided adequate supplies of wood for fuel. Naturally, Hungarian demands for Cracow's lead exports increased during the second half of the fifteenth century, providing her merchants with an open market, even in face of competition from Hungarian traders;[249] some technical problems still existed, like the use of the restrictive short-shaft furnace for smelting, and the organization of working methods, often calculated on a short-term basis and low capacity.[250] Nevertheless, the foundations were laid for the halcyon days of the next century, when not only the Fuggers and Thurzos were to reap rich rewards,[251] but also such Cracow merchants as the Boners, Morsztyns and Salomons, together with other less-known entrepreneurs from the city.[252] In comparison, Cracow's lead exports to other areas including Flanders via Gdańsk,[253] paled into insignificance in face of the Hungarian market with its profusion of precious metals – with the exception of lead (Fig. 17).[254]

Copper was apparently of greater importance than either silver or lead, especially in the second half of the fifteenth century, and there is a surprisingly large amount of information on it in Cracow's archives. The reason for this is obvious. Copper from Hungary travelled westwards and northwards across Europe to supply the brass and bronze industries of Germany and the

17. Distribution of North Hungarian Mining in the Later Middle Ages.

Low Countries, much of it being sent by a combination of land, river and sea transport to Antwerp, which, in the sixteenth century, was the chief copper market in Europe. Cracow's merchants participated in this trade, in which their city was one of the significant transit links between mine and market.

The Slovak ore mines were opened up by Germans in the thirteenth century[255] and late in that century copper was exported to Cracow as transit trade for western Europe. By 1306 it was the main Hungarian commodity in the city; the stapling law on copper acquired in that year meant Cracow merchants had a monopoly on its further movement.[256] The tariff lists for 1310 verify that copper was the most important article sent to the city from Hungary.[257] The copper was varied both in type and price, depending not only on its quality but also mine of origin; the most common places mentioned in Cracow documents were Gelnica, Smolnik and, later, Kremnica and Banská Bystrica.[258] Košice remained the commercial centre of Hungary's copper trade up to the end of the fifteenth century,[259] partly due to its proximity to the rich mines of the Gömör district, and partly to its commanding position on the routeway through the Carpathian via Sącz to Cracow. For their part, Cracow merchants such as Jan Bartfal and Mikołaj

Crenmark,[260] around 1400, were heavily dependent on supplies from Košice, as local supplies of copper (at Kielce) were minimal.[261] Their other great worry must have been competition from Nuremburg merchants, equally searching for quality copper ores to send westward, especially to Venice,[262] and from Gdańsk to Flanders.[263] In spite of the occupation of Pomerania by the Teutonic Knights in 1309, Cracow's merchants continued to send copper via Toruń and the Vistula river to the Baltic coast, and there freely negotiated with foreign merchants for transport to Flanders.[264]

The fifteenth century saw an intensification of commercial competition between Cracow and Toruń for mastery of the Hungarian copper trade. Cracow always seemed to have the upper hand with the threat, as in 1442, of route closure for all Prussian merchants travelling to Hungary.[265] The issue was always a sensitive one between the two cities, but with the eclipse of Toruń after the peace of 1466, Cracow's monopoly position with Hungarian copper supplies was assured; but this ideal situation was soon to come under threat. Gradually, during the fifteenth century, the copper trade was being organized through foreign capital, especially merchant companies from Nuremburg.[266] Vital in the new situation was the part played by Gdańsk, with its new warehousing facilities; it, too, wanted its own stapling laws (granted in 1443) which meant that thereafter Cracow must sell its copper to Gdańsk merchants, not foreigners, at the port.[267]

Gradually the transit trade in Hungarian copper to the Low Countries declined in the second half of the fifteenth century, but its export to Gdańsk through Cracow and along the Vistula route continued to prosper, thanks to the commercial activities of such rich Cracow merchants as Jan Sweidniczer; links with his son Jerzy in Toruń led to active participation in Polish–Prussian trade between 1460 and 1477,[268] particularly in copper. Moreover, there were still opportunities for exchanging Polish lead (from Olkusz) for Hungarian copper,[269] especially after 1465 when Jan Thurzo became resident in Cracow.[270] He benefited from the numerous commercial privileges offered inhabitants by the city whilst his son, Jerzy, remained in Slovakia (at Banská Bystrica) to control their mines.[271] Partnership with the rich Augsburg family of Fuggers late in the century,[272] and increased investment in Hungarian mining had its effects on Cracow. By 1490 copper prices in the city (Fig. 15), reached a new peak and Cracow became the main storage centre for Thurzo's ore. In 1494 he built a large copper-smelting works at Mogiła[273] on the outskirts of the city, connected a year later by special route via Cieszyn to the mines in Banská Bystrica.[274] The refined ore (together with separated gold and silver) was then sent via the Vistula to the Baltic coast; the basis was therefore laid in Cracow for renewed prosperity from copper after 1500, and from which Thurzo was to make his fortune.

Of the other minerals sent to Cracow, mention should be made of iron. This was the least tractable of the metals known and used during the Middle

Ages because of numerous smelting difficulties and the great variety of types each requiring different treatment. Much less information is forthcoming in Cracow documents on iron, than either lead or copper; local supplies of ore were not utilized until the sixteenth century.[275] The earliest documentary reference to iron in Cracow dates from 1222 and may have been imported from Italy.[276] During the fourteenth century Cracow merchants began to import iron ore from Upper Hungary, coming from the mines around the Spiš district of Slovakia. No mention was made of it in the 1310 tariff lists, but around mid-century it was bartered for salt from Bochnia.[277] It is first definitely recorded as imported from Hungary in 1380[278] and is further mentioned in the customs tariffs of 1393.[279] As the fifteenth century progressed, iron is mentioned more frequently; in 1405, a Cracow merchant sent it to the city from Hungary, together with a load of copper,[280] whilst the tariff lists for 1432 mention Hungarian iron-ore imports.[281] Usually it arrived in Cracow in smaller quantities[282] than either lead or copper, whilst there was never any mention in documents of merchants who specialized in the ferrous ore trade. Iron prices seem to have reached a peak during the depressed mining times of the 1440s (Fig. 15); there is no information on the purchase of iron or steel goods in Hungary for the Cracow market. Surplus ore requirements over local needs were re-exported from Cracow to Prussia and perhaps further to Flanders.[283]

During the second half of the fifteenth century the mining of sulphur at Swoszowice, near Cracow, became of increasing significance.[284] It had numerous uses; as a bleach for cloth, as a base for colours, soap and explosives, and also utilized for medicinal purposes for treating animals and as an antidote for syphilis.[285] Cracow citizens owned some of the sulphur mines;[286] demand for the chemical as a disinfectant was particularly high after outbreaks of plague in the city and elsewhere. Another chemical traded through Cracow was kermes, a cherry-red mineral (antimony oxysulphide) which was popular as a dyestuff. Quantities are recorded in documents being sent from Cracow to Italy early in the fifteenth century,[287] usually transported via Nuremburg or Flanders, to Venice and Florence. Other examples occur later, when demand for the dye coincided with the period of increased textile production in Italy.[288] Finally, mention should be made of calamine (hydrous zinc silicate) first mentioned as imported to Cracow from Slovakia in 1432.[289] Along with tin it was used in Cracow for bell-making,[290] and also transported to Gdańsk for sale in western Europe in the manufacture of canons and other monumental and ornamental 'dinanderie'.[291]

(ii) Spices

The medieval spice trade was essentially the movement of various types of aromatic seasoning in Asiatic origin from certain Levantian emporia to the ports of Italy; from these places they were distributed, usually overland, to much of western and central Europe. Amongst the goods thought of as precious and expensive merchandise[292] in European trade, the bulk would have been spices, of which there were more than a hundred different varieties.[293] In spite of Gibbon's remark that, 'the objects of orient traffic were splendid but trifling'[294] the demand for spices was noticeable. As Salzman states, 'Medieval man had neither a delicate palate nor the means of gratifying it with choice meats; their meat was coarse and they liked it highly flavoured.'[295] Consequently the spice trade grew in importance throughout medieval times, but it was a trade which had many intermediate stages and innumerable middlemen; Cracow's Genoese merchants residing in the city were amongst those who took on this role.

In 1261 the Genoese became the predominant commercial power in the Black Sea.[296] The chief base for their trade was the Crimean peninsula, which besides many local advantages was ideally located in relation to the caravan routes from Persia, the Far East, and the plains to the north of the Euxine coast. This advantage, too, was the more valuable in the late thirteenth and fourteenth centuries, since the Tartars, whose territories included the south Russian plain, the Crimea and Persia, maintained with considerable success the security of the roads. The greatest trading colony of the Genoese was the city of Caffa (Theodosia), whilst they had trading quarters, along with the Venetians, at the Tartar town of Tana (Azov) at the mouth of the river Don. Tana was very important as a terminal of trade caravans which brought, amongst other goods, spices from India by way of Kabul, Ourengj and Astrakhan, or across Persia to Asterbad, thence to Astrakhan by boat. Apart from their direct trade with Genoa, the Genoese at Caffa traded with Constantinople, the lower Danube, Caucasian coast, and northward via Suceava, Lwów and Cracow with the Baltic coast[297] and Flanders (Fig. 18).[298]

Although there is no positive information that the Geneose living in Cracow and Lwów went to Caffa in the fourteenth century, according to Ptaśnik, 'one may assume so';[299] moreover, he maintains with some certainty that Genoese merchants in Cracow were only involved in the Bruges–Caffa route and had no direct trade contacts with Italy.[300] They were further aided by the commercial policy of Kazimierz Wielki after his accession to the Ruthenian territories;[301] soon after 1350 the king became 'determined to take advantage of the fact that the overland routes to the Black Sea lay under his control and to make Cracow a great emporium acting as an intermediary between Western Europe, Hungary and the East'.[302] Competitors for this lucrative eastern trade soon emerged including the Order of the Mother of God (Zakon najśw. Mariji Panny), who used the land route to transport goods such as spices to Lwów, Cracow and, further, to the main markets in Bruges.[303] Pach has recently provided further documentary evidence of

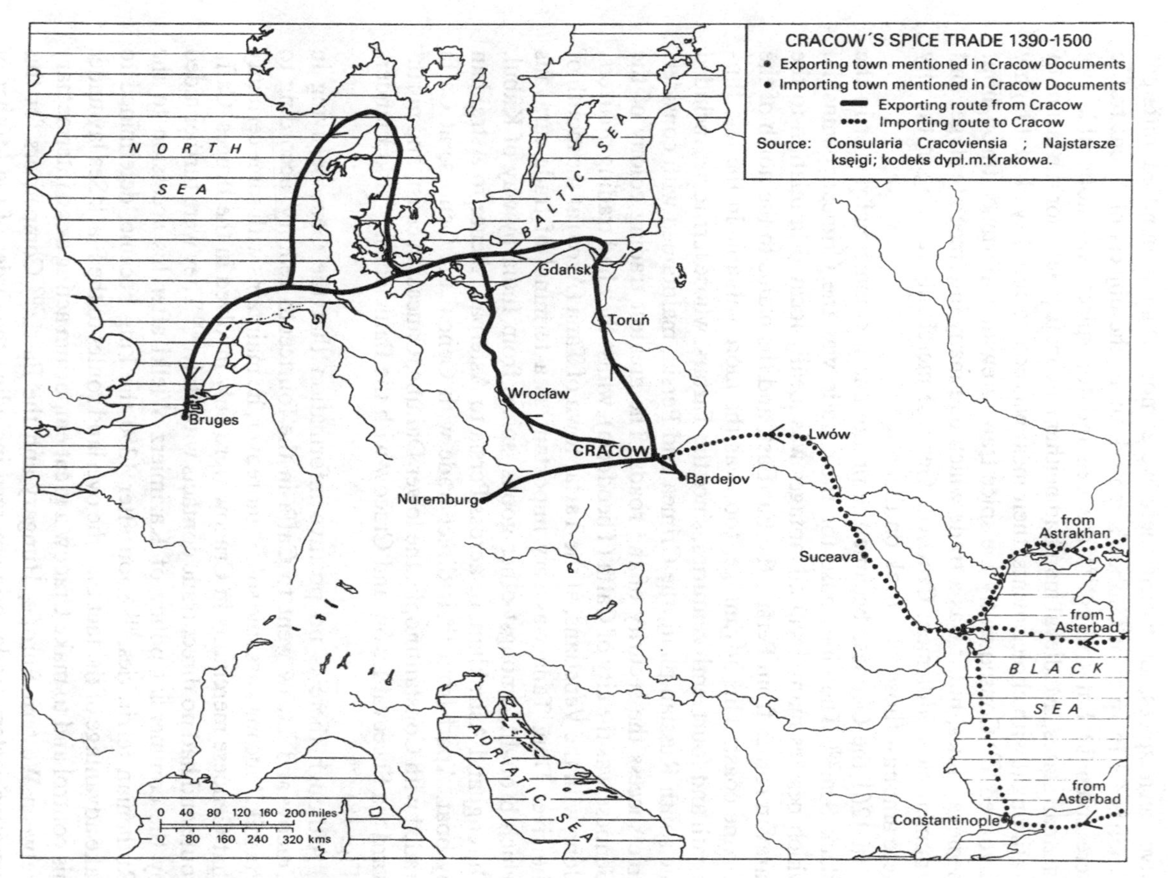
CRACOW'S SPICE TRADE 1390-1500
Exporting town mentioned in Cracow Documents
Importing town mentioned in Cracow Documents
Exporting route from Cracow
Importing route to Cracow
Source: Consularia Cracoviensia ; Najstarsze księgi; kodeks dypl.m.Krakowa.
NORTH SEA
BALTIC SEA
Gdańsk
Toruń
Wrocław
Bruges
CRACOW
Lwów
Bardejov
Nuremburg
Suceava
from Astrakhan
from Asterbad
BLACK SEA
ADRIATIC SEA
from Asterbad
Constantinople
0 40 80 120 160 200 miles
0 80 160 240 320 kms

18. Cracow's spice trade, 1390–1500

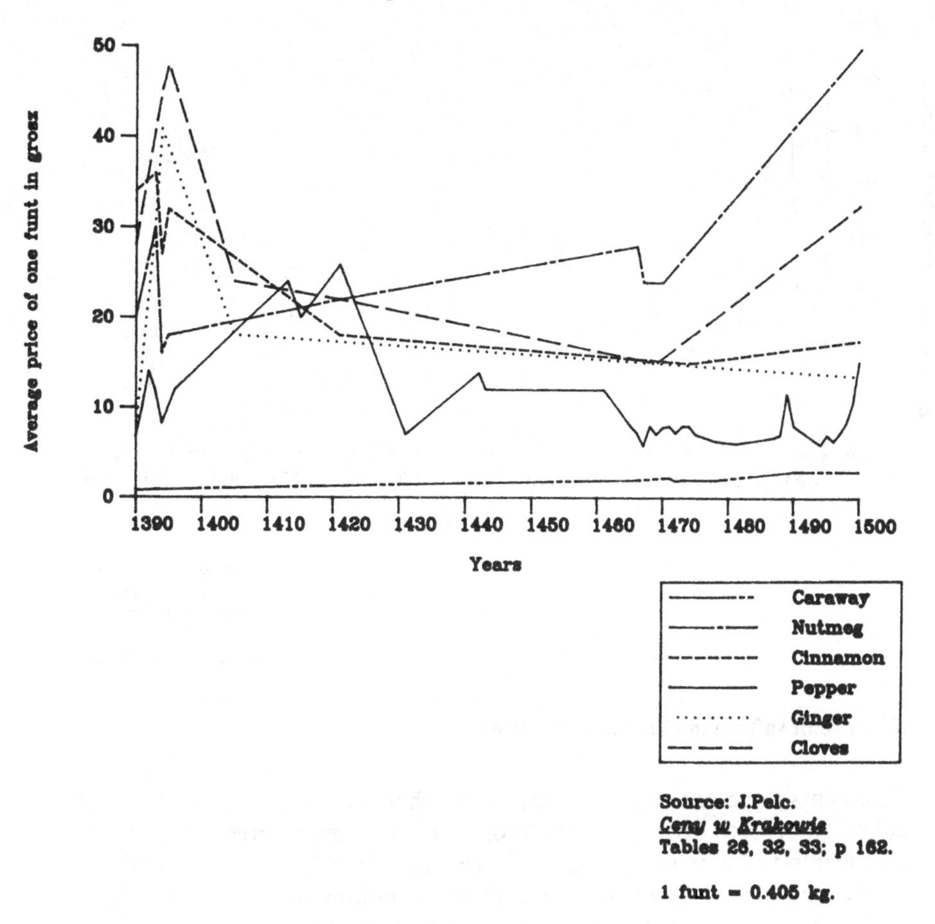

19. Price of certain spices in Cracow, 1390–1500

oriental spices from the Black Sea region moving northward through Transylvania and Hungary to Poland in the fourteenth century.[304]

Of the various types of seasoning, often cited in documents as either 'spices' or 'aroma', pepper seems to have attracted the most attention. Prior to the fifteenth century, pepper was regarded as a rarity. In Europe it was literally worth its weight in gold, and was often used for methods of payment.[305] In Cracow its price rose steadily during the first two decades of the fifteenth century (Fig. 19), only to decrease with an erratic pattern until the end of the century when military conflict between Turkey and Venice had wide repercussions on pepper prices.[306] Cracow merchants purchased pepper from Lwów; much of the spice trade in that city was in the hands of Armenian merchants who purchased large quantities from Caffa.[307] No

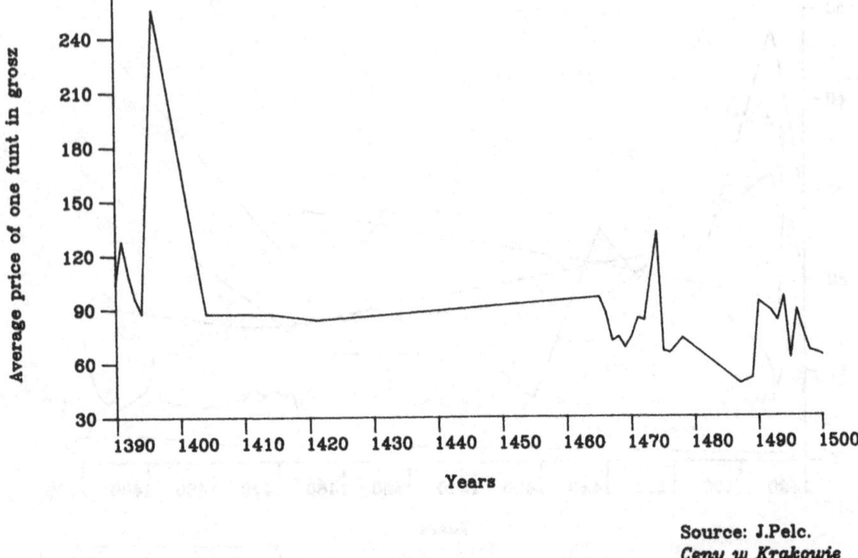

20. Price of saffron in Cracow, 1390–1500.

documents indicate the profit Cracow merchants made on pepper but they did re-export it to other parts of Europe. Other spices referred to in Cracow documents as coming from the Black Sea coast included cloves, which were used extensively in medicines, as well as in cooking and aromatic drinks; ginger, cinnamon, caraway seeds and nutmeg and were also in demand.[308] A document of 1432 gives precise details on the minimum amounts of each to be sold in Cracow market.[309] Saffron, a popular ingredient in many medieval sauce recipes, was the most costly of all herbs sold in Cracow (Fig. 20), coming from the saffron crocus; only the stigmas of the flowers were used and it took about 75,000 flowers to make one pound of saffron.[310] Consequently Cracow's merchants only bought small quantities.[311]

Continued warfare during the fifteenth century greatly hampered Cracow's transit trade in spices and other oriental goods. The wars between Poland and the Teutonic Order impaired communications between the Baltic coast and inland centres of Polish trade. Cut off from Cracow and Lwów, the Prussians now obtained their eastern goods from Venice by way of Bruges;[312] the spice accounts from 1419 to 1434 in Bruges clearly show this trend.[313] Cracow now had to find an alternative outlet for her spice trade, namely via Silesia, reaching the German market through Nuremburg, and beyond to Flanders.[314] Instead of importing spices from Hungary, Cracow's traders began exporting them to towns in Slovakia like Bardejov.[315] The source areas for spices were also changing; during the second half of the fifteenth century trade with the Black Sea region entered upon a period of decline. The disintegration of the Khanate of Kipchak and the rapid advance of the Turks brought about an entirely new situation. The Italian settlements in the Crimea fell in the 1470s[316] to the Ottoman advance followed by the Moldavian emporia of Citatea Alba and Chilia in the 1480s;[317] the Black Sea trade dwindled away and with it Cracow's supplies of oriental wares. Future spice provisions would have to come from western Europe,[318] until such a time that amicable relations between Poland and the Porte made restoration of commodity exchange a viable proposition.

(iii) Salt

Salt was fundamental in the medieval economy. It was needed for curing fish and meat, for making butter as well as for seasoning; salt was produced in many parts of the European continent, but only in modest quantities. Coastal salt pans could be constructed anywhere that had enough fuel, e.g., peat, to help evaporation, whilst inland rock-salt deposits were mined. Locally produced salt was often impure and of poor quality, unable to compete with the finer output of the Lüneburg brine springs, or the natural sea-salt found south of the Loire estuary in France.[319] Cracow's salt trade was entirely dependent on the rock-salt deposits mined at nearby Bochnia and Wieliczka.[320]

From earliest times salt must have been one of the rarer minerals in Poland and an expensive article in the medieval economy.[321] The nearest known large deposits to Wieliczka/Bochnia were in Kujavia to the west of Toruń, and around Hołyń Kałusz south of Lwów. During the second half of the thirteenth century salt mining was reformed by King Bolesław Wstydliwy, who placed a controller (żupnik) over each mine responsible for salt production and sale.[322] A dearth of known salt deposits in Silesia provided Cracow with an early market, successfully competing with Halle salt in Wrocław.[323] Like Silesia, Slovakia also lacked known salt beds; the mineral was acquired from Cracow even during the reign of Bela IV (1235–70).[324] For Cracow's traders, however, the great ordinance issued by King Kazimierz Wielki in 1368 on its saltworks at Wieliczka was most significant.[325] Besides codifying all Polish mining customs it gave Cracow merchants a monopoly in salt selling over a wide area stretching from Mazovia, Rus, Silesia, Lithuania and Lublin.[326] Growth in the popularity of Cracow salt on the Silesian market was soon experienced; in 1387 it fetched over a third more in price than equivalent amounts from Halle.[327] Clear evidence of Cracow's salt sales in Slovakia at mid-century are found in documents from

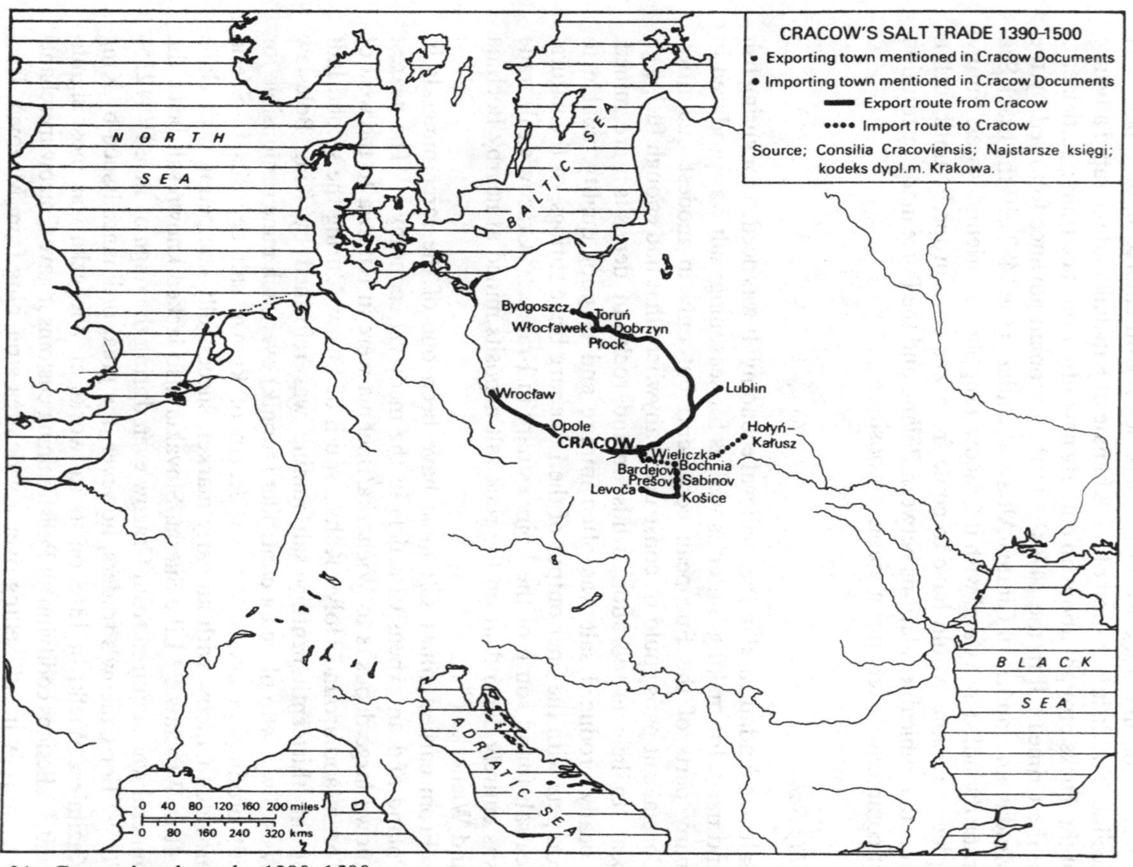
CRACOW'S SALT TRADE 1390-1500
Exporting town mentioned in Cracow Documents
Importing town mentioned in Cracow Documents
Export route from Cracow
Import route to Cracow
Source; Consilia Cracoviensis; Najstarsze księgi; kodeks dypl.m. Krakowa.
NORTH SEA
BALTIC SEA
BLACK SEA
ADRIATIC SEA
Bydgoszcz
Toruń
Włocławek
Dobrzyn
Płock
Lublin
Wrocław
Opole
CRACOW
Hołyń-Kałusz
Wieliczka
Bochnia
Bardejov
Prešov
Sabinov
Levoča
Košice
0 40 80 120 160 200 miles
0 80 160 240 320 kms

21. Cracow's salt trade, 1390–1500

Šariš and Liptov dated 1354.[328] Cracow's merchants, however, only managed to capture the markets of places in southern and central Poland, the northern part relying on supplies of French salt from Brittany organized through the Hanseatic League (Fig. 21).[329]

The dawn of the fifteenth century saw some opposition on the part of the Hungarian king, Sigismund, to the importation of Polish salt from Cracow. His political fears of Polish claims to the Hungarian throne led to a decline in commercial contacts between the two at the turn of the century; already in 1398 he complained about 'all the merchants arriving from Poland',[330] whilst in 1405 he issued an ordinance limiting the importation of Polish salt,[331] which was again renewed in 1439.[332] His attempts to create an Hungarian salt monopoly seem to have failed, for documentary evidence for Slovakian town archives record obtaining supplies from Cracow merchants who arrived from Bochnia and Wieliczka.[333] After his death in 1437, his successor Queen Barbara was more openly disposed to salt imports, particularly from Bochnia.[334] The main markets for Cracow's salt were the towns of Bardejov[335] and to a lesser extent, Levoča, Sabin, Prešov, Zemplín, Abov, Boršod, Užská, Šariš, etc.[336] although they had to compete with supplies from Marmoroš in Transylvania.[337] Nevertheless, Cracow salt still dominated the markets of northern Hungary to such an extent that in 1483 Queen Beatrice stated in one document that the whole of the country relied on foreign and particularly Polish supplies.[338]

The two other outlet areas for Cracow's salt sales were Silesia and Little Poland. Evidence suggests that throughout the fifteenth century salt from Cracow held its market in Silesia. Most of it was transported to Wrocław via Krzepice, referred to in documents as 'the salt road'; on one occasion, however, the Silesian rulers wanted to stop the free import of Polish supplies and tried to close the frontier in 1437, probably due to pressure from Hungary, but the event passed without long-term effects.[339] As Kutrzeba has stated 'Silesia was an important market for Polish salt and much of it was organized by Cracow merchants.'[340] These included such people as Antoni de Florentis, who obtained a lease on the Wieliczka mine in 1434[341] and Mikołaj Serafin who did likewise in 1456, paying a substantial sum to the king for the privilege.[342] Nonetheless, such traders had to compete in the Silesian market with Marmoroš supplies from Transylvania, just as they did in Slovakia.[343] Fortunately, such worries were not experienced in Little Poland where the city's merchants had unrestricted access to the domestic salt market.[344] Gradually, other towns like Kazimierz, Oświęcim and Chęciny obtained the right to sell salt, thus breaking Cracow's monopoly.[345] This was further exacerbated in 1454 when individual nobles received the privilege of buying and trading in the commodity.[346] To some extent this event was alleviated for Cracow's traders with the permission at mid-century to transport salt down the Vistula to Dobrzyń, Płock,

Włocławek, Bydgoszcz and even Toruń in Greater Poland.[347] Thanks to the efforts of Wieliczka's mine controllers, such as Serafin and de Sancto Romulo, not only were local supplies sent to Greater Poland, but also purchases made from Russian mines and re-exported down the Vistula.[348]

Cheaper raw materials

Luckily for Cracow her economy was not solely dependent on trade in precious raw materials; otherwise the decline of these products would have had a disastrous effect on the city's economy, through reasons quite beyond her control. Fortunately the flourishing states of Silesian, Slovakian and more distant Flemish towns, which were among Cracow's best customers during this period, were not built exclusively on trade in high-quality merchandise; in practice much the greater part of the goods which were traded consisted of skins, wax, livestock products, fish, wood and cereals – in fact the cheaper commodities. As Postan maintains, commercial dealings in luxury articles 'were something of an exception, for the main articles of northern trade were bulkier and cheaper necessities of life. Its main, and certainly the most permanent, branch was traffic in food.'[349]

It should be remembered that the Russian, Polish and Lithuanian markets produced several important commodities which, ever since the beginning of the Middle Ages, were distributed on a growing scale by international trade to the markets of central and western Europe. Van Houtte has stated 'the main support of German trade came from the Baltic and consisted of cereals from the coastal plain from Mecklenburgh to Prussia, of ores from the Carpathians brought to the coast along the Oder or the Vistula and, finally, of the produce of the tremendous forest zone of Russia, not only timber and its by-products ashes, tar and pitch, but also the yield of forest-hunting and -collecting'.[350] Such products as flax, hemp, tallow, wax, skins (leather) and furs were transported to Germany and the Low Countries for, as Attman believes, 'Eastern Europe had a virtual monopoly of these products.'[351] Pounds further reiterates this impression concerning Bruges for 'Its imports consisted mainly of the goods . . . required by its Flemish hinterland',[352] and 'These commodities were shipped to Flanders mostly by merchants of Reval, Riga, Danzig, Stettin, Rostock, Lubeck'[353] and, of course, either directly or indirectly, by traders from Cracow.

(i) Skins

Skins were exported both in their raw and treated forms and had an international market in the Middle Ages. They were a commodity that attracted merchants from both East and West Europe; source areas for Western Europe included Bohemia, Moravia, Poland and Lithuania, whilst

Russian skins and leather had been well known on the European market for a long time. There appeared to be two types of skins – the ordinary hides from cattle, sheep, horses and goats, and the rarer skins from wolf, marten, fox, lynx and other wild animals living in the forest areas. Certainly thirteenth-century customs returns for Pomerania refer to skins being sold in small quantities for export including not only horse, cattle and even wolf pelts, but also beaver.[354] The tolls at Szczecin during the reign of Barnim I (1250–60) record the skins of ox, horse, cow, goat, sheep, lamb, beaver, otter, fox and marten.[355]

At the beginning of the fifteenth century the main Polish centre for the skin trade was Poznań; Rutkowski has estimated that about 50,000 pelts were imported annually to the West from Poland, mainly via Poznań.[356] The skins were usually designated as 'Russian', 'Lithuanian' or 'tymczami' (i.e. smaller-sized ox or cow skins). After Poznań, Cracow was probably the second most important centre for the Polish skin trade, but did not enjoy the strategic position of the former city on the main trade route from Russia/Lithuania to Germany and the West; it did, however, have the nearby forested foothills of the Carpathian Mountains to draw on for supplies, and even beyond to the Hungarian plain and Moldavia.

The increased political stability of Europe between the thirteenth and sixteenth centuries had allowed greater economic growth and the development of a richer society and higher culture. Towns in Poland had formed their own handicraft organisations to cope with the growing demand for more sophisticated commodities. Evidence of the leather trade is found in several Polish towns during the Middle Ages including Opole, Gdańsk, Międzyrzecz, Warsaw, Wrocław, Kalisz, Płock and of course Cracow.[357] As capital city, Cracow found there was also demand from furriers not only to produce the goat and sheepskins, or modest furs of peasant wear, but ceremonial wares which represented the insignias of wealth and standing.

A great fillip to the trade in skins and furs for Cracow came at the end of the fourteenth century with the political union of Poland and Lithuania. Henceforth the city's merchants had easy access to one of Europe's largest fur-producing regions,[358] utilising trade contacts in Lublin first evolved through the Cracow patriarchy around 1400.[359] Małowist believes that 'it was the relatively low price of these furs which enabled them to be exported in large quantities towards the west'.[360] The other area of importance for Cracow merchants was Hungary. In 1402 one of the city's merchants, Slepkogil sent 35,000 'black Hungarian skins' via his agents to Flanders;[361] above all, however, furs were sent to Cracow from Hungary, especially weasel and wolf.[362] Furthermore, the area around the Polish–Hungarian border was noted for its skin exports, particularly in the form of sheepskin coats, whose price in the city remained fairly constant throughout the

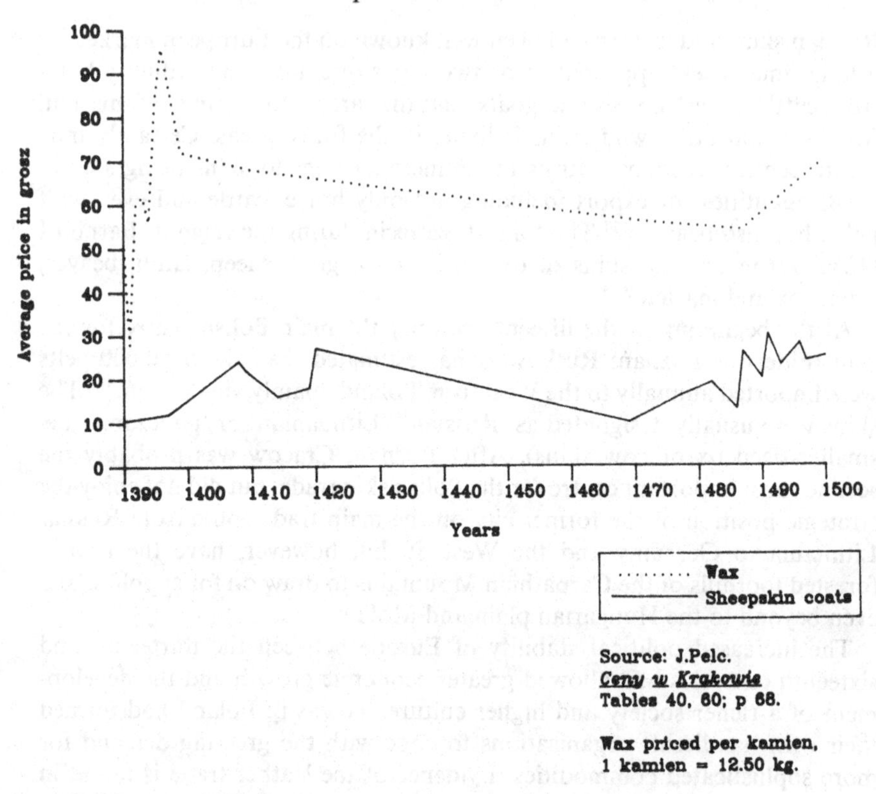

22. Average price of wax and sheepskin coats in Cracow, 1390–1500.

fifteenth century (Fig. 22). Leather fancy goods were sent from Hungary via Cracow to the rest of Poland,[363] whilst in the opposite direction leather and fur products from Little Poland were found in the markets of Hungary (Fig. 23). For example, in 1488, the wife of the Hungarian king requested a Cracow merchant to buy her a fur coat in the city for the winter.[364]

The leather, furrier and tanning industries developed firstly as rural occupations which invaded the towns of Europe only in the Later Middle Ages. During the fifteenth century Cracow acquired numerous specialists in these various branches of manufacture; for example, residence in the city was granted to furriers from Koštice and Kežmarok on several occasions between 1425 and 1500.[365] The process of tanning skins and hides and conversion to articles of clothing and of everyday use consisted essentially in removing the epidermis and hair, as well as flesh particles, and in so doing treating the hide so that it would not putrefy or decay. This involved use of various chemicals, most important of which was tannin, derived primarily from the bark and galls of the oak tree. The process gives off an offensive

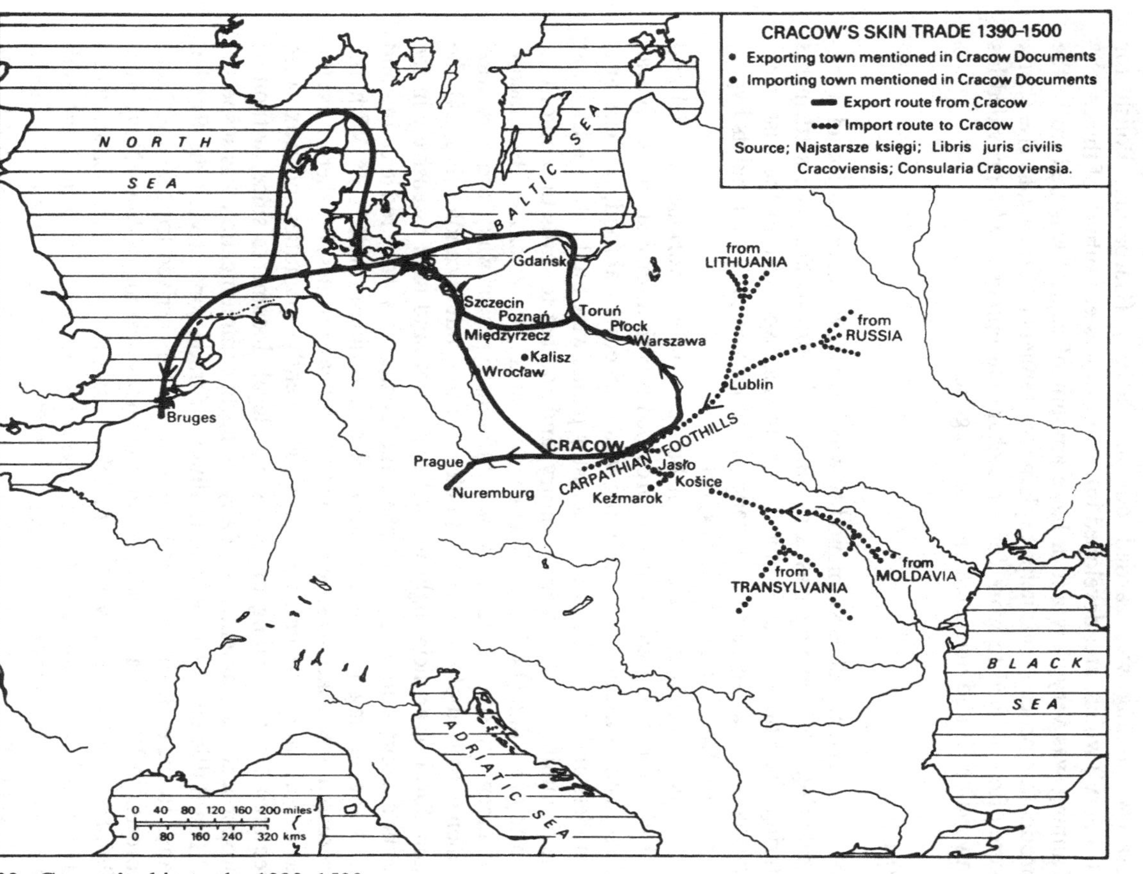
CRACOW'S SKIN TRADE 1390-1500
Exporting town mentioned in Cracow Documents
Importing town mentioned in Cracow Documents
Export route from Cracow
Import route to Cracow
Source; Najstarsze księgi; Libris juris civilis Cracoviensis; Consularia Cracoviensia.
NORTH SEA
BALTIC SEA
BLACK SEA
ADRIATIC SEA
Gdańsk
Szczecin
Poznań
Toruń
Płock
Warszawa
Międzyrzecz
Kalisz
Wrocław
Lublin
from LITHUANIA
from RUSSIA
Bruges
CRACOW
CARPATHIAN FOOTHILLS
Prague
Nuremburg
Jasło
Košice
Kežmarok
from TRANSYLVANIA
from MOLDAVIA
0 40 80 120 160 200 miles
0 80 160 240 320 kms

23. Cracow's skin trade, 1390–1500.

smell, consumes large quantities of water, and produces a highly polluted effluent. Urban tanneries were usually situated by a river bank, as far as possible from human habitation and water supply. This helps explain why in Cracow tanning was carried out in Garbary (Garbarnia, Polish for 'Tannery') which later developed into a north-western suburb of the city.[366] Tanning was very slow and a large amount of capital was tied up in the production process; this helps explain why tanners in Cracow, as elsewhere, were often obliged to work for a merchant entrepreneur who provided his raw materials and marketed his finished goods.

(ii) Wax

Wax was another cheap commodity which Cracow merchants found worth exploiting. Bee-keeping was a significant feature of Slavonic settlements during the Middle Ages; some Polish towns, especially in the east, were located in areas where an abundance of honey was produced and its byproduct provided enough wax for export.

It was collected in special town buildings where workmen melted and kneaded the wax, weighed it and provided it with a town stamp. Only wax thus ensured against counterfeiting was transported abroad, to Germany, Flanders and Italy. Such wax-processing plants were established in Lwów at the beginning of the fifteenth century, and another towards the end of that century in Vilna (1492) which was known as a 'zaboynicza'.[367] In Cracow, documentary sources refer to two different types of wax; 'ordinary wax' which consisted of beeswax with tallow fat to produce candles (Fig. 24), the main form of artificial light in medieval times, and used extensively in the home and city's numerous churches; and 'chancellery wax', red in colour and used for sealing purposes in official and business matters.[368] With the increased organisation associated with European business life there arose the need for more documentation. Consequently quantities of wax were required for the sealing of letters and legal documents since, without a proper seal, the authenticity of any medieval communication was suspect.

The Baltic coast had a long tradition of trade contact with the Russian lands to the east, where goods like wax and furs arrived from the northern forests.[369] Wax was produced in the areas around Pskov, Novgorod, Polotsk, Smolensk and Vitebsk, together with the large area which stretched from Smolensk southwards towards Kiev and Lwów; during the fourteenth century noted sources of wax were around Novgorod, Rjazan and Riga as well as Lithuania.[370] Some of the wax came along the main route connecting Vilna with Cracow via Lublin[371] whilst other source areas for the city were Moldavia[372] and Hungary.[373] Evidence suggests that the wax trade never assumed the proportions of such merchandise as minerals or salt, but a steady import to Cracow continued during the fifteenth century and beyond.

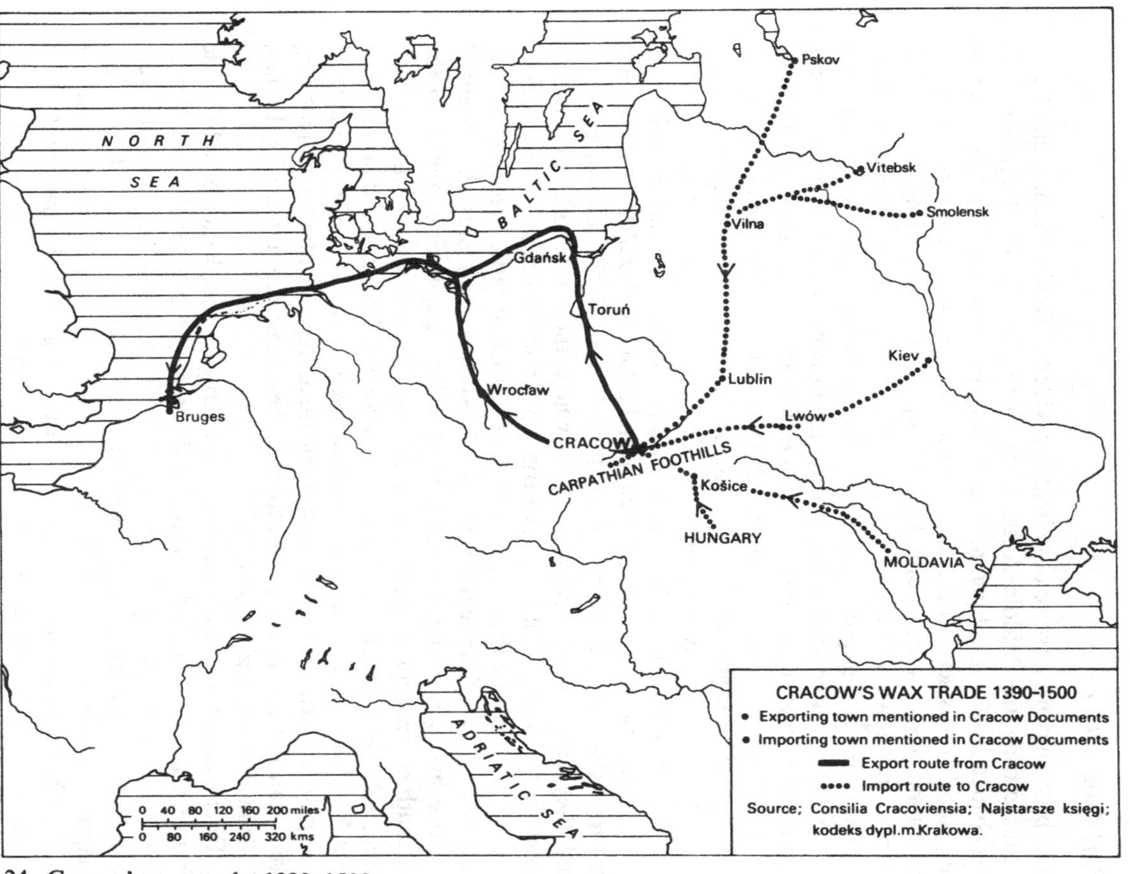
NORTH SEA
BALTIC SEA
ADRIATIC SEA
Pskov
Vitebsk
Smolensk
Vilna
Gdańsk
Toruń
Kiev
Lublin
Wrocław
Lwów
Bruges
CRACOW
CARPATHIAN FOOTHILLS
Košice
HUNGARY
MOLDAVIA
0 40 80 120 160 200 miles
0 80 160 240 320 kms
CRACOW'S WAX TRADE 1390-1500
Exporting town mentioned in Cracow Documents
Importing town mentioned in Cracow Documents
Export route from Cracow
Import route to Cracow
Source; Consilia Cracoviensia; Najstarsze księgi; kodeks dypl.m.Krakowa.

24. Cracow's wax trade, 1390–1500.

Wax surplus to local needs was then re-exported to Silesia,[374] and both overland and by river to Gdańsk[375] where it eventually found its way to Flanders and the west.[376]

(iii) Livestock products

The breeding and nurture of animals provided a livelihood for many Europeans during the Middle Ages; from them they obtained certain basic necessities of life – clothing, meat and dairy products for themselves, whilst any surplus was sent to the local market. In eastern Europe, not only the mountain pastures were suitable for sheep- and cattle-rearing but also the steppe and downland areas. Large numbers of animals, especially cattle, reared in the less populous parts, like Polish Galicia, the Hungarian Plain and Moldavia, were driven westwards along the main trade routes to supply the growing cities of central and western Europe. For Cracow, where many of the east/west routes converged, livestock and their products became of increasing importance towards the end of the Middle Ages. The transit trade in cattle was probably of greatest significance.

Małowist has written, 'It is not known for certain when larger scale trade developed in Silesia for cattle driven from Podolia and Moldavia to the west. It surely started in the fourteenth century, but only in the following century did they arrive in Wrocław ... from where they were taken further to western Germany, Bohemia, Austria and northern Italy.'[377] Moreover, according to Eberle there was 'a dynamic growth of cattle breeding towards the end of the fourteenth century'.[378] Attman maintains that 'Central Europe had a growing demand for cattle for slaughtering, and from the fifteenth century an increasing production of oxen developed in the Lwów area, in Poland (round Lublin), in the Ukraine, in Hungary and in Moldavia and Wallachia (from the last two areas even after the Turkish conquest). From the eastern production areas large quantities of cattle for slaughtering were transported via the Polish towns. They came through Lwów, Lublin, Cracow, Poznań and Breslau.'[379]

Documentary evidence suggests that Cracow merchants bought cattle from Carpathian foothill towns like Rzeszów,[380] Krosno,[381] Jasło,[382] or farther afield in Lublin,[383] or Jarosław.[384] By 1440 live animals were already an important transit article in the city's commerce.[385] One of the major source areas was Moldavia; the Polish/Moldavian trade treaty of 1408[386] had opened up this trade which was organised by merchants from Seret and Suceava, sending their bullocks and steers to Lwów and on via Bochnia to Cracow, from whence they were sent to Silesia.[387] Cattle in much smaller quantities were also sent to Cracow from Slovakia.[388]

Other types of livestock seemed to have figured much less prominently in Cracow's trade. According to live animal costs recorded in Cracow's

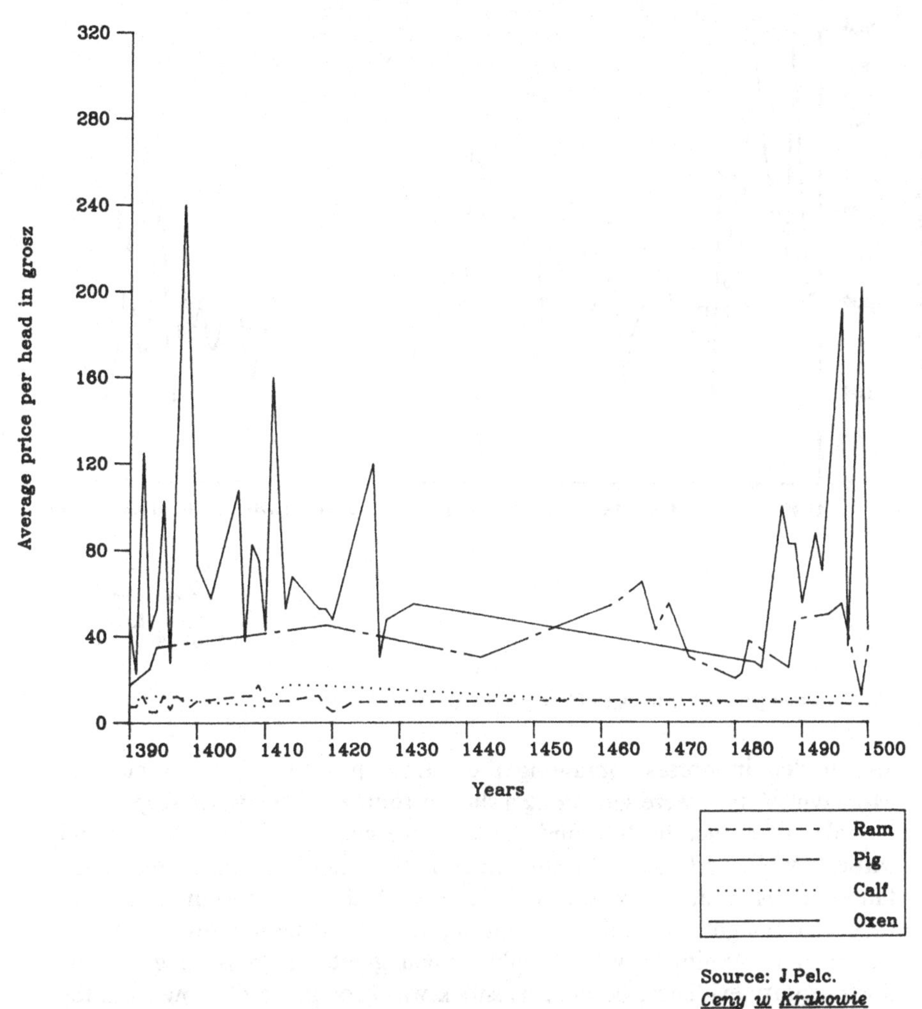

25. Price of livestock in Cracow, 1390–1500.

documents, horses fetched the highest prices throughout the fifteenth century (Figs. 25, 26). Oxen, although lower priced, reflected the same erratic changes between years as horses; calves, pigs and rams were more evenly priced, with little change in value over the whole period. A large number of horses were imported into Poland from Hungary. They were sent to fairs at towns like Jasło,[389] Biecz, Gorlice, Żmigród, Miejsce Piastowe, Krosno, Sanok, Zagórze[390] etc., but particularly from Bardejov.[391] Lwów merchants

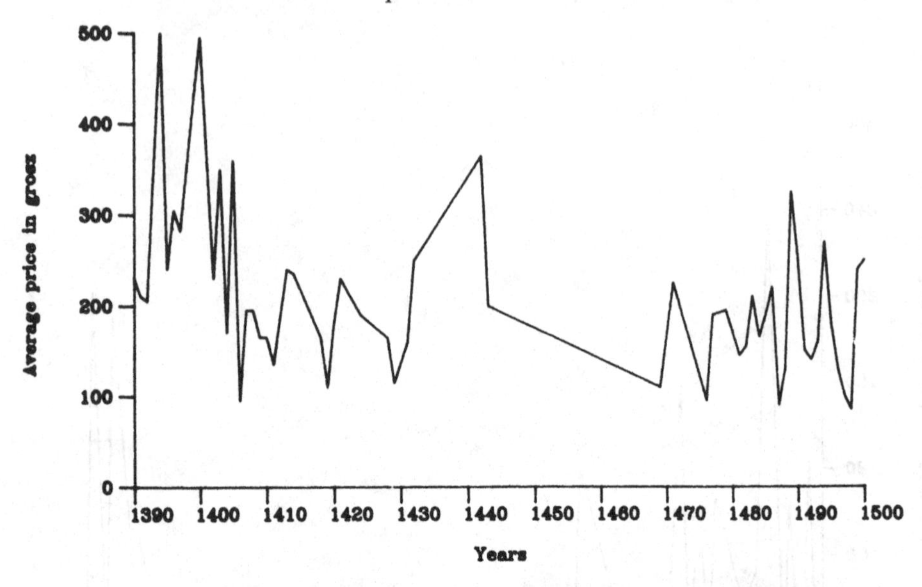

26. Average price of horses in Cracow, 1390–1500.

also traded in horses, obtaining them from parts of Transylvania and Moldavia;[392] they were sent along a shorter route to Cracow (via Ropczyce), in order to reduce feeding and maintenance costs.[393] Besides horses and cattle, sows[394] and goats[395] were sent from Slovakia to Poland often to the fair at Sanok,[396] and may eventually have reached the Cracow market where they were bought for local consumption, or sent further to Silesia (Fig. 27). Trade in live animals was probably much greater than is suggested by documentary evidence, because livestock was brought to Cracow from the hinterland, often without any formal transaction being signed.

(iv) Fish

The quantity of fish consumed in the Middle Ages was, according to Ball, 'very high in relation to that of meat, both for economic and religious reasons'.[397] This statement is further confirmed by Postan, for 'the consumption of fish in the Middle Ages was high, and sea fisheries were many. Some fish was caught in and off all the estuaries of Europe and along its sea coasts.' For Cracow, so far inland, this may not appear important, but 'throughout the greater part of the Middle Ages, and certainly from the thirteenth century onwards, by far the most important of the fishing grounds of

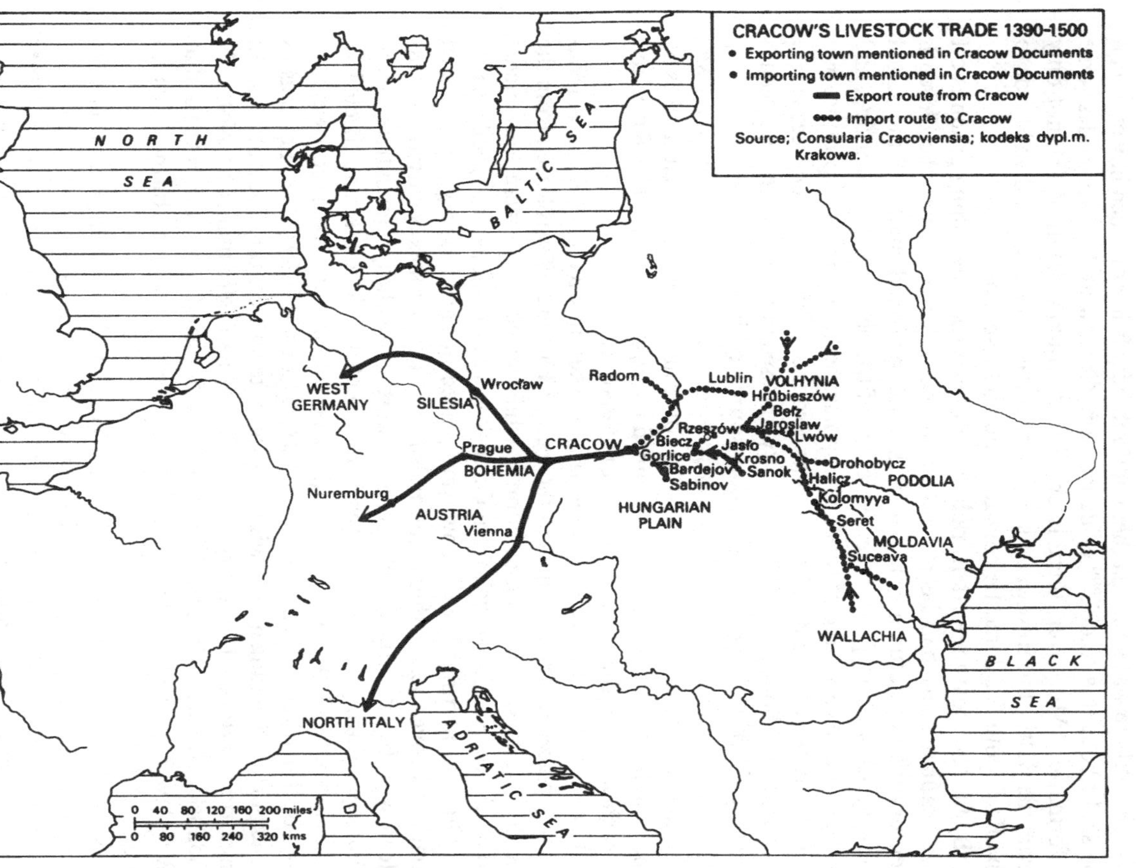
CRACOW'S LIVESTOCK TRADE 1390-1500
• Exporting town mentioned in Cracow Documents
• Importing town mentioned in Cracow Documents
Export route from Cracow
Import route to Cracow
Source; Consularia Cracoviensia; kodeks dypl.m. Krakowa.
NORTH SEA
BALTIC SEA
WEST GERMANY
SILESIA
Wrocław
Prague
BOHEMIA
CRACOW
Nuremburg
AUSTRIA
Vienna
Radom
Lublin
VOLHYNIA
Hrubieszów
Bełz
Rzeszów
Jarosław
Lwów
Biecz
Jasło
Krosno
Gorlice
Bardejov
Sanok
Sabinov
Drohobycz
Halicz
PODOLIA
Kolomyya
Seret
MOLDAVIA
Suceava
HUNGARIAN PLAIN
WALLACHIA
BLACK SEA
NORTH ITALY
ADRIATIC SEA
0 40 80 120 160 200 miles
0 80 160 240 320 kms

27. Cracow's livestock trade, 1390–1500.

Europe, and also the busiest centre of the curing industry and of the herring trade were the Baltic fisheries of Skania'.[398] Although Małowist laments, with reference to products sent to Poland from the lands of the Teutonic Knights, 'there was amongst them undoubtedly fish, not much however, can be said about this trade',[399] evidence from Cracow archives provides some information. Moreover, there is proof that Cracow obtained fish not only from the Baltic, but freshwater species from the rivers of Poland and lands to the East, both for domestic consumption and re-export.

By far the most important trade was in herrings. The fishing grounds off the Scanian coast had been a noted source of herring since the eleventh century;[400] the autumn fish fair at Skanör provided much of the salted herring sold in Europe, for they were one of the best fish for preserving, had a high value as a source of food, as popular medicine, and as a product of a major industry,[401] until their decline around 1425 when they ceased to frequent the Sound off Scania.[402] By the second half of the thirteenth century, herrings were an important source of income for such ports as Szczecin and Gryfice,[403] whilst towards the end of the fifteenth century it was Gdańsk that became the chief importer, over half from North Sea catches by the Dutch.[404] Cracow therefore obtained its supplies via the Oder River and Wrocław, or up the Vistula River from Gdańsk.

In Cracow there was a special herring market, in which particular emphasis was placed on the freshness of the fish.[405] The market was leased from the castle as it lay within its jurisdiction[406] whilst special pools were constructed to hold live fish. Salted fish from Gdańsk were also subject to specific rules. In Gdańsk, herrings for Cracow were placed in barrels and given a special mark to prove their authenticity. Falsification sometimes occurred and was dealt with by city officials.[407] Herrings were quite cheap, but suffered from shortages during the troubles with the Teutonic Knights, as reflected in prices around 1420 (Fig. 28). Not all the herring imports to Cracow were for the local market; evidence from the tariff records of 1393 show herrings as re-exported to Hungary,[408] whilst in the fifteenth century they became an important article of trade over the Carpathian Mountains.[409] Not only Gdańsk provided Cracow with herrings; a dispute in 1452 reveals that Wrocław supplied the city, some of the fish being sent further east to Russia.[410]

Salted eels were sent to Cracow from Toruń to be consumed in the city, or sent farther inland to places like Levoča in Slovakia. On occasions there were complaints about the freshness, which led to disputes in court.[411] Other evidence of the fish trade refers to specialities often ordered by the king's court;[412] for example, in 1395 beluga roe was obtained from a merchant in Košice, which had been sent from the Black Sea.[413] Lwów merchants sent salted pike and sturgeon to Cracow, which had been bought in Brailov, the main Moldavian centre for fish, coming northward on the route via Suceava

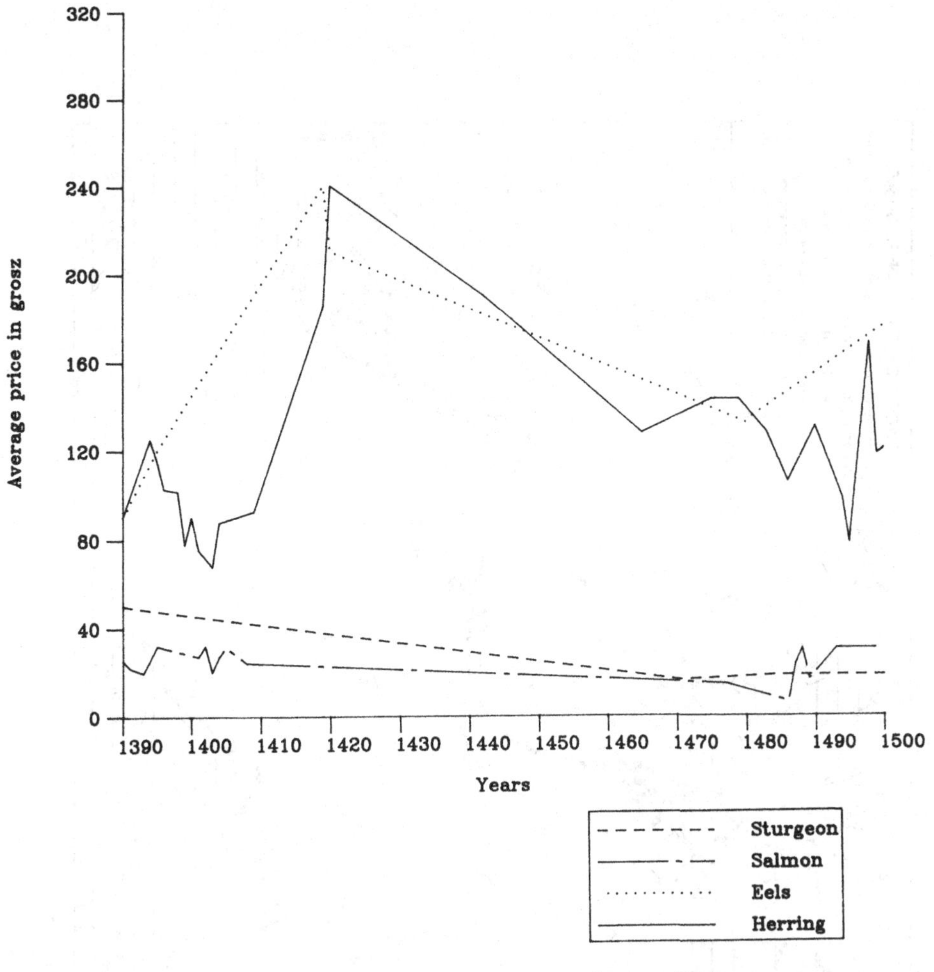

Source: J.Pelc.
Ceny w Krakowie
Tables 22: p 38–39, 161–2.

Salmon price per fish.
Sturgeon per kamień.
Herring and Eels per barrel.
1 barrel = 272 litres (60 gallons).

28. Price of certain fish in Cracow, 1390–1500.

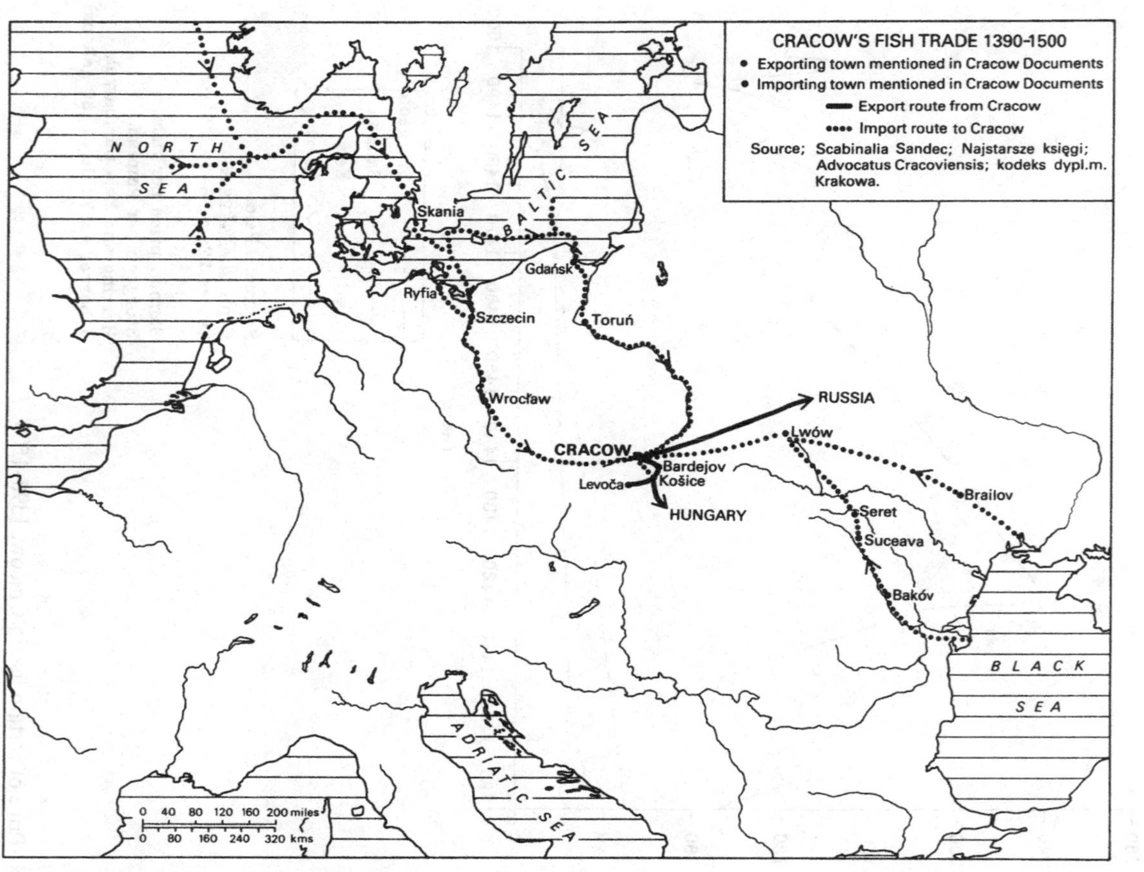
CRACOW'S FISH TRADE 1390-1500
Exporting town mentioned in Cracow Documents
Importing town mentioned in Cracow Documents
Export route from Cracow
Import route to Cracow
Source; Scabinalia Sandec; Najstarsze księgi; Advocatus Cracoviensis; kodeks dypl.m. Krakowa.
NORTH SEA
BALTIC SEA
Skania
Gdańsk
Ryfia
Szczecin
Toruń
Wrocław
CRACOW
RUSSIA
Lwów
Bardejov
Levoča
Košice
HUNGARY
Brailov
Seret
Suceava
Bakóv
BLACK SEA
ADRIATIC SEA
0 40 80 120 160 200 miles
0 80 160 240 320 kms

29 Cracow's fish trade, 1390–1500.

and Baków[414] (Fig. 29). Amongst other varieties mentioned in Cracow documents one finds examples of salmon, ordinary and crucian carp, perch, stockfish (i.e. cod) and different types of pike.[415]

(v) Wood

One of the most important basic raw materials during the Middle Ages was timber. Most medieval town building, outside the Mediterranean region, was predominantly wood, whilst it also formed the main ingredient for shipbuilding and certain types of armour such as bowstaves. Timber resources became increasingly scarce as countries, particularly in western Europe, grew in population, and their inhabitants pushed back the forests in favour of fields and pastures. The forest resources of Eastern Europe, especially coniferous timber, became increasingly available to Western Europe with the opening up of Baltic trade during the fourteenth century. From the mid-fourteenth century onwards 'eastern timber shipped from the Baltic and more especially from Danzig, all but ousted from the western markets the other types of 'white' timber. Pine, yew and fir of Baltic origin and occasionally some birch ... became one of the main articles of Hanseatic imports into this country [England].'[416] Moreover, during 'the first two decades of the fifteenth century, the wholesale prices of Baltic timber ... were exceptionally low in Poland, very high in England'.[417]

For Cracow, one of her great advantages in the wood trade was the River Vistula (Fig. 30). The bulkiness of timber meant it was an article best suited to water-borne traffic and, by way of the Polish rivers, was sent to the Baltic ports of Gdańsk and Königsberg. Postan speculates that 'The bow-staves which won the battles of Crécy and Agincourt probably came from the Carpathian mountains and were shipped to England through Hungary and Prussia.'[418] Furthermore, 'At Leipzig and Breslau, and at Cracow, Poznań and Lwów in Poland, south German merchants found a positive welcome', for their 'establishment of a trade route from west to east ... enabled them to tap the produce of the Polish interior as it descended the rivers. By the end of the fifteenth century ... The gravitational pull of the new route was sufficient to lead Breslau to quit the Hanseatic League in 1474, and to encourage ... some of its exports of grain and timber to the south-west in return for Flemish and Italian luxuries.'[419]

Cracow's proximity to the Carpathian foothills gave her merchants easy access to timber sources. The wood was floated down the Dunajec and Vistula rivers and, although mainly in the hands of merchants from Nowy Sącz, according to Korzon, Cracow was also involved in the trade.[420] Dense forests extended along the whole of the sub-Carpathian region; trees from this area were particularly sought after for ship-masts e.g., deal (*schneidedielen:dyl*),[421] and also staves for making barrels (*klappholz:klepki*), and

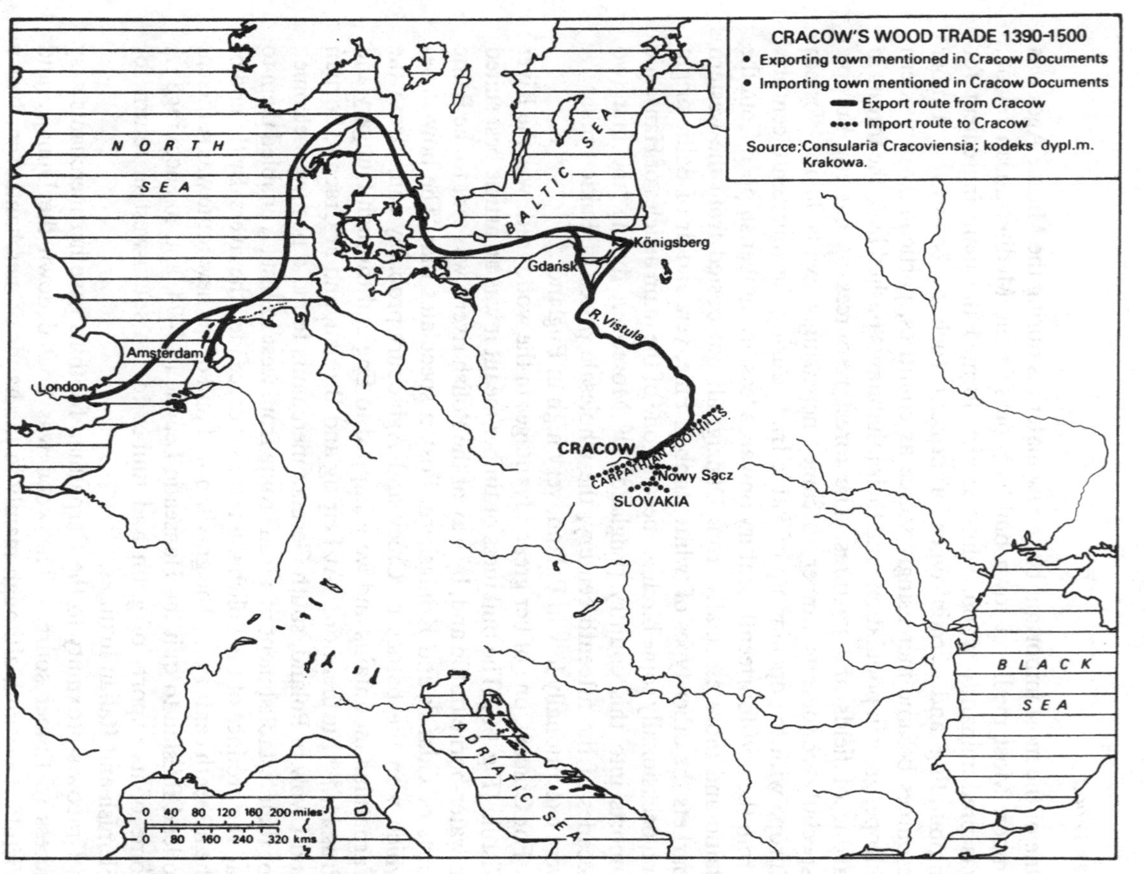

30. Cracow's wood trade, 1390–1500.

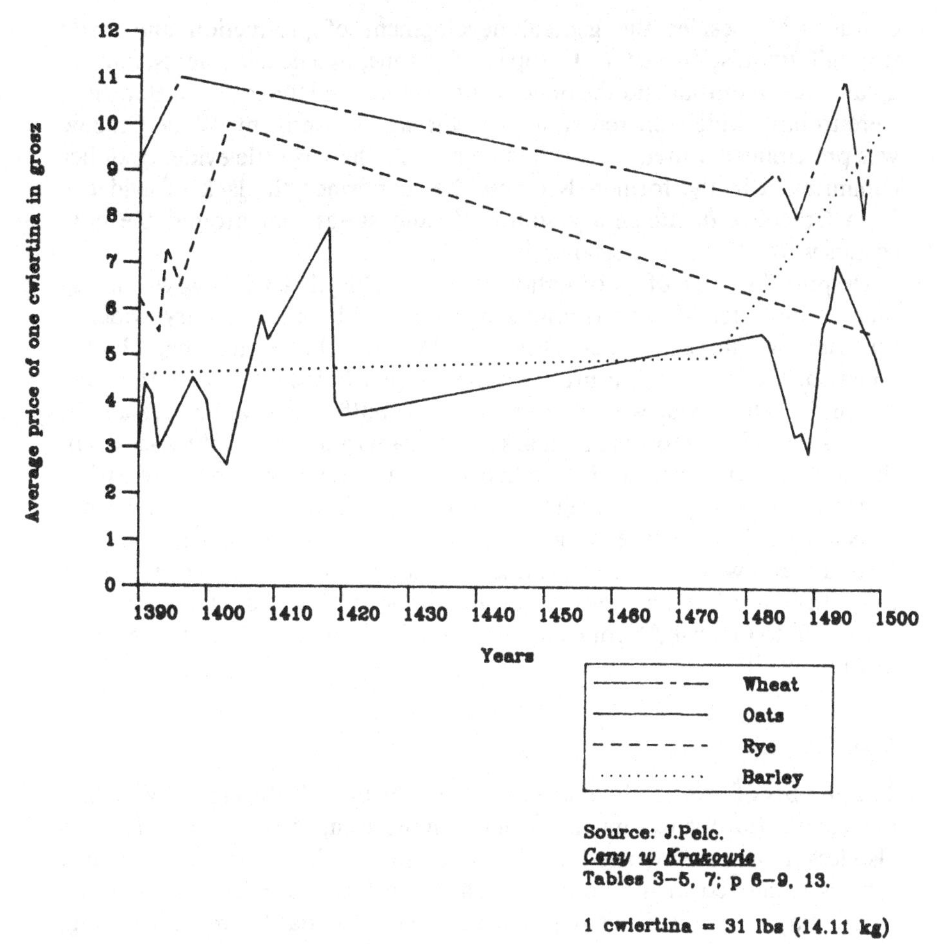

31. Price of certain cereals in Cracow, 1390–1500.

sawn boards (*Wagenschoss*:*Wańczos*) for wainscot, the split oakwood especially favoured as shipbuilding material. A wood particularly prized from this area was yew (*eybenholz*/*bogenhola*:*cis*) which was sent from Cracow and Nowy Sącz direct to Gdańsk,[422] where it was exported to England for making its famous archery bows,[423] and also to the Netherlands.[424] Besides timber, Cracow also appears to have imported wooden utensils from Slovakia,[425] probably to be used locally for domestic purposes.

(*vi*) *Cereals*

Apart from timber, grain exports began to play an increasingly important role for Poland during the second half of the fourteenth and fifteenth centuries.[426] Besides the general development of production and trade throughout most of Eastern Europe at this time, two developments, namely greater cereal output and the opening up of mines in Hungary and Bohemia, were to have wider repercussions over the whole continent. While Cracow was pre-eminently involved in the latter trade, there is little evidence of her commitment in the former. Kutrzeba has confirmed the lack of evidence from Cracow's documentary sources[427] and states that around the city, cereals were of little importance.[428]

Despite this lack of information it is probable that Cracow was a significant importer of various grains, as suggested by documentary evidence on prices assembled by Pelc (Fig. 31). Małowist has stated that, 'In the course of the fourteenth century not only Great Poland, but even the most distant Polish provinces ... show an increase both in cultivated area and in the number of little towns and markets. Peasants and lesser nobles settled in the towns and engaged the peasants to increase their productivity.'[429] Cracow, as the capital city, grew accordingly and there must have been a constant need to procure sufficient grain for its inhabitants. The region of Little Poland was capable of producing grain crops; towards the end of the fifteenth century sufficient was being grown for surplus stocks to be exported via Toruń to Gdańsk,[430] for example by Jan Tarnowski, *voivode* (wojewoda) of Cracow.[431]

Manufactured goods

The number of people who made their living by manufacturing or by trading in medieval Europe was much smaller than those engaged on the land; it was also less than those involved in food processing or the production of simple articles fashioned in the home and based on local supplies such as fibre, wood, clay or metal. Four requirements were essential for manufacturing, besides adequate sources of raw material, namely manual skills in the use of tools and a knowledge of how to combine or treat the various components, access to more elaborate equipment than the average person, sufficient quantities of power and finally, though not always, fuel supplies, especially when the application of heat was required. As Hay has pointed out, 'The volume of manufactured goods which was traded from one end of Europe to the other was impressively great. But the goods that were bought and sold were made without exception by very small producers.'[432] Specialised production was always relatively more important in the larger towns, and aimed to supply a very much wider market. For cities like Cracow, with its extensive market and transit trade, this in turn necessitated a community of merchants, and a whole complex of craftsmen; these were supported by a large tertiary employment (builders, butchers, bakers, servants etc.) whose task was to satisfy the material needs of the specialised workers. Many of the most sought after manufacturers were obviously the product of craftsmen; jewellery and the fine work of goldsmiths, illuminated manuscripts, manuscript books and in Cracow, towards the end of the fifteenth century, printed books, together with the manufacture of armour and weapons. Nevertheless, it was from transit trade in manufactured goods that Cracow secured its prosperity and from which it obtained considerable financial benefit.

(i) Textiles

Clothes more than anything else responded in the Middle Ages, as they do now, to the lust for variety and change; innovation and the attraction of foreign apparel led to the importation of textiles and acceptance of new alien styles throughout much of medieval eastern Europe. Cloth manufactured in the West became, just like the trade in metals, a major branch of Cracow's commercial activity. In the opposite direction oriental textiles were also in demand in the West, whilst the city's merchants were involved, to a lesser extent, in the sale of various types of Polish cloth.

The major source of western textiles for Cracow was Flanders and, on a smaller scale, England and Italy. There is reliable evidence that the capital of Poland maintained commercial relations with fourteenth-century Flanders; documents from 1342[433] and 1364[434] confirm that quantities of textiles and readymade clothing were being imported. Flemish cloth overshadowed all other European textiles,[435] and by the end of the thirteenth century it reached the more remote parts of the continent.[436] Evidence from the second half of the fourteenth century illustrates how Cracow merchants imported cloth from the principal textile centres – Bruges, Ypres, Ghent, Malines, Brabant, Herenthals and Wallonia, in some cases in an unfinished state, to be finished locally.[437] The most expensive cloth came from Brussels;[438] manufacturing centres most frequently mentioned were Tirlemont[439] and Ypres,[440] whilst other places included Malines,[441] Louvain,[442] Dinant,[443] Lier,[444] Courtrai,[445] Herenthals[446] and Enghien[447] (Fig. 32). In the middle of the fourteenth century the English cloth industry made a sudden leap forward which brought it to its peak at the turn of the fourteenth and fifteenth centuries. This is reflected in Cracow's documents at this time with evidence of imports from Beverley,[448] London[449] (Fig. 33), Kersey,[450] Colchester, and the county of Kent.[451] Together with the Low Countries and England, Northern Italy was one of the most important concentrations of the clothing industry during the Middle Ages. Cracow's links with Italy were enhanced by Papal connections,[452] whilst Ptaśnik claims that 'Already in the fourteenth century Italian cloth was sold in Cracow.'[453] In 1373 a merchant is recorded as selling Italian cloth and silks in the city,[454] whilst Małowist states that merchants from Venice, Florence, Genoa and Bologna were engaged, or their agents, in selling particularly Florentine cloth in the city.[455]

375

CCLXII. 1364. die 24. mensis Octobris.
De venditione piscium, sebi, cretae, picis rerumque victualium. — De mercede tonsoris pannorum.

Von dem forkowffe der fissche an der Stroze gewilkort feria quinta post vndecim M virginum 1364.

§ 1. Welch man ader ynwoner fische von anderswo her brenget czw der strozen, der zal mit den fisschen bleiben steen fir gantcze tage, vnd auff itczliche neste fast tage zal her aws itczlichem hutfasse einen czober, den man an der Stroza fleget czw haben, mit fisschen off den margkt tragen vnd forkowffen: vnd zo her dy fir tage vorbrocht hot, zo zal denne der zelbige, der dy fissche brocht hot, zal nymandes anders dy fissche vorkowffen ack den, dy dy stroze halden vnd vorczinsen. Zunder ein ynwoner, der dy fissche brenget, ab der dy fissche nicht vorkowffen möchte, als oben geschriben stet, denne mag her zy yn einen helder seteczen yn sulcher weyze, das her auff einen ideren fastagk aws dem helder czwene czober fissche off den margt trage vnd dy zelber vorkowffe vnd sy nicht den vorkowfferen weyberen ader mannen vorkowffe.

§ 2. *Von fisschen, dy do oberbleyben.* Mer ist beschlossen, das vorbas von der Strozen zal man tragen dy fissche ack yn czoberen awff den margt czw vorkowffen, vnd nicht yn plachten, noch mit emeren, noch keyne fissche zal man wider tragen von dem margkthe an dy Stroze, sunder dy teubner sullen yn dy czegel abschneyden off dem margkte.

§ 3. Vnd wer dy obengeschriben artikel nicht halden wird, vnd wirt erfunden ein obertreter, der zal der Stad ein schock busse czw geben, vnd dor czw dy fissche vorliren, vnd der Hutman an der strozen, der den obertreter nicht melden wörde vnd das vorschwige, den zal man yn dem stocke acht tage bussen.

§ 4. *Von forkawff vnsleth, kreyde ader pech etc.* Nymanndt zal kawffen vnsleth, kreyde, pech, Smer ader derley is yn zeinem hawsze als vnsleth sullen kawffen lichtcziher vnd andere, dy mit arbeten, vnd nicht dy forkoffler auf den forkowff vor der woge.

§ 5. *Was man den gewant Scherern czw lone geben zal.* Von dem Scherlon des gewandes ist alzo beschlossen, das dy scherer nicht mer nemem zallen von der elen

	Brwkesch	czw XII hellern
	florenczesch	
	Eyprisch	
von	Mechlesch	czw 1/2 gr.
	Herntalesch	
	Balbarth	
	gemeyn eyprisch	

Englesch czw VI hellern vnd vom landtuch czw fir hellern von ider elen vnd nicht anders.

§ 6. *Von vorkawff essender ware.* Nymandt zal vorkowff treiben

376

off dem margkte mit essender ware allerley bey I firdung busse vnd vorlust der ware, vnd das sullen dy tewbner besen vnd das stroffen.

Behem fol. 219.

CCLXIII. Sine anno, die et mense.
De pensa maiori et minori.

§ 1. *Was man yn der Woge den wegeren vnd der Stad geben von der wogen zal.* Wenne man hundert centner bley ader kopper wegen lest, zo nympt man von einem mitpurger, der do burgerrecht hot, 3 gr. vnd von einem gast 6 gr.

§ 2. *Von einem stuck bley.* Zo erkein burger ader gast ein stuck bley ader mer lest czw der woge brengen, zo ist der ynwoner ader gast von einem ideren stücke ein halben grosschen schuldig czw geben, welcher yn czwe teil zal geteilt werden, nemlich halb der Stad vnd halb den wog knechten, dy vnder der woge erbeten.

§ 3. *Wenne man bley weg ladet.* Zo das bley, das vor der wogen hot gelegen lang ader kurcz, durch einen burger ader gast zal weggefurt werden, zo ist der, der das bley weg ladet, zo her ein burger ist, das wog gelt als oben schuldig czw czalen, nemlich von C ct. 3 gr. yst her ein gast, zo gibet her 6 gr. ys sey kurcz ader lang do gewest ader gelegen.

§ 4. *Von bley das man czw schroth.* Wenne aber man sulch bley lest schroten ader czw hawen, zo zal man von einem stuk ein gr. beczalen, von welchen man zal geben II heller (den knechten) vnd das obrige der stad.

§ 5. *Von eyzen vnd ander dingk czw wegen was man geben zal.* Wenne man eyzen zal wegen ader andere krom dingk als ballen vnd sulch dingk, zo gibet man von acht centner ein grosschen. Von XII centner fele I gr. ys zey ein burger ader ein gast, wenne yn keynem woge lon hot der burger fortel, alleyn am bley vnd kopper czw wegen, sust is alles gleich.

§ 6. *Von der kleynen wogen.* So ein purger lest einen lapis ader einen halben ader mynner vngeferlich wegen, zo zal her von einem idern sulchen gewicht ader wegen eyn heller geben. So is aber ein gast ist vnd lest wegen als oben aws gedrucket ist, zo zal her von sulchem wegen czwene heller geben, is werde gewegen, was is zey.

§ 7. *Von zilber czw wegen.* Zo ymandt burger ader gast lest zilber wegen yn der Stad brengadem ader woge, zo gibet der is wegen lest, von einer idern margk.

Behem f. 220 v.

Wilkirz ten nie ma daty, być nawet może, że z poprzednim wilkirzem Nr. CCLXII jedną całość stanowi. Kładziemy go jednak osobno dlatego, że przepisy nim objęte do jednego przedmiotu się odnoszą, że nie zostają w żadnym związku z przepisami wilkirza poprzedniego, niemniej, że w kódeksie Behma obа te wilkirze znacznem pustem miejscem są przedzielone. Wprawdzie trudno odgadnąć, czy te puste miejsca u Behma nie były pierwotnie przeznaczone, aby w nie wpisywać

32. Document: Flemish cloth in Cracow, October 1364.
Kodeks dyplomatyczny miasta Krakowa, Vol. II (edt. F. Piekosiński) Kraków, 1882, Document 262 dated October 1364.
Evidence of Flemish cloth for sale in Cracow.

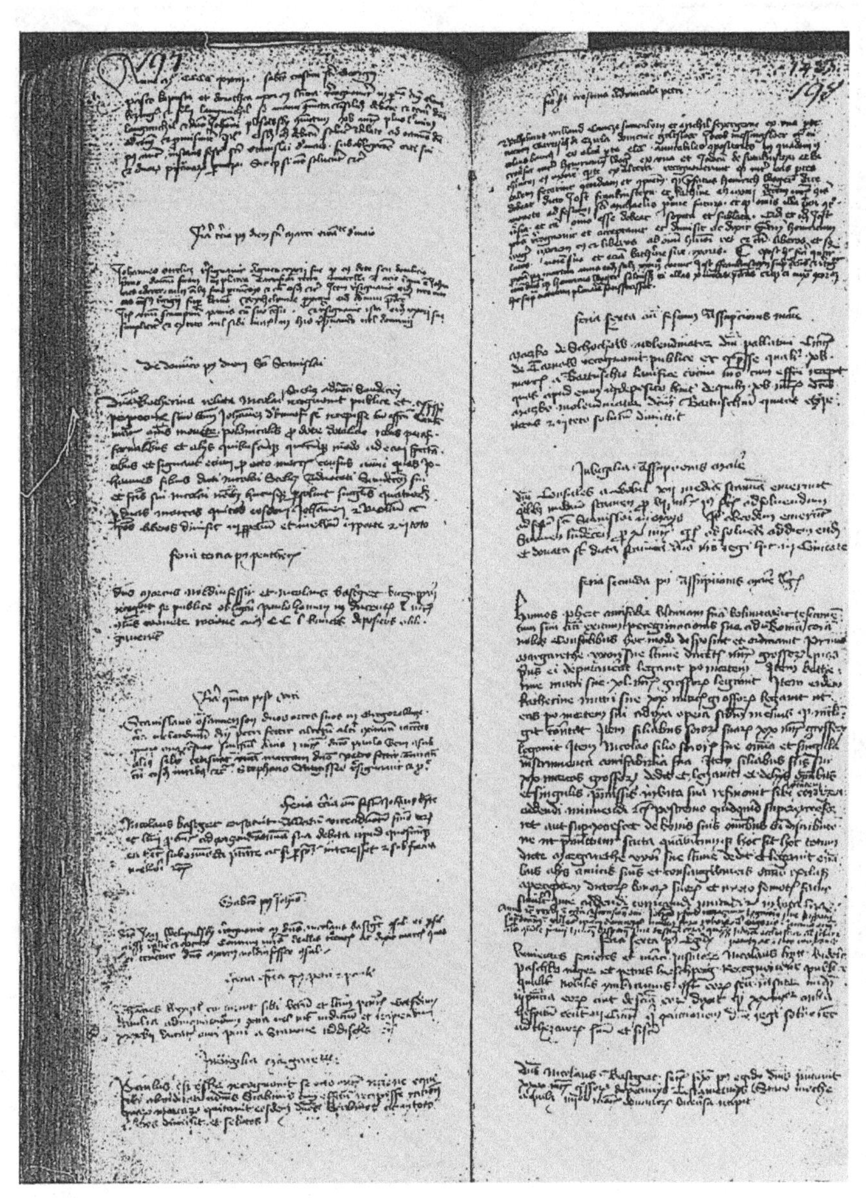

33. Document: English cloth in Cracow, 1423.

Wojewódzkie Archiwum Państwowe w Krakowie: *Consularia Cracoviensia*, Vol. II, p. 198; M/s 42. Document dated 1423 refers to purchase of English cloth 'stamen Lundencense' in Cracow.

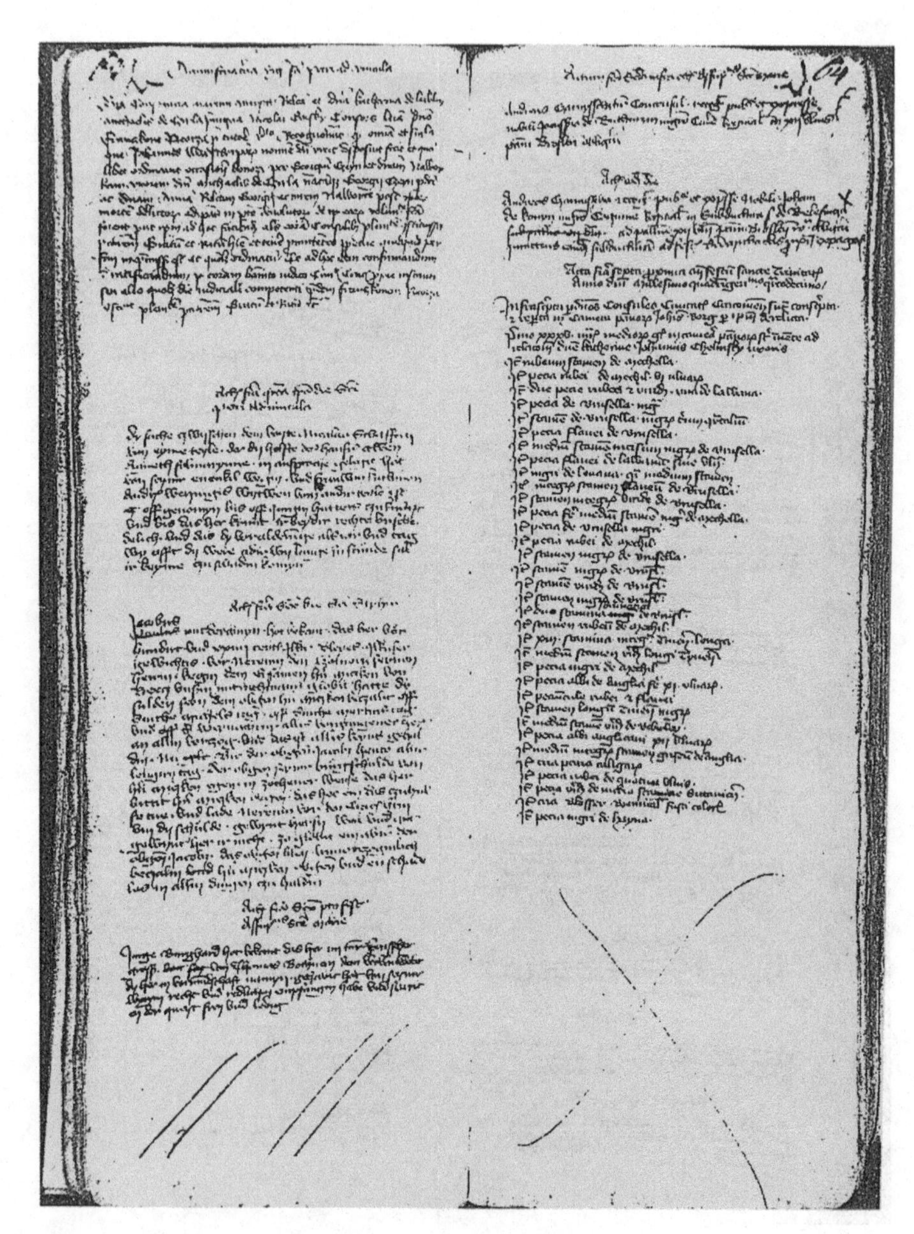

34. Document: Shop inventory of Cracow merchant, 1414.

Source: Wojewódzkie Archiwum Państwowe w Krakowie: *Consularia Cracoviensia*, Vol. II, p. 64; Ms. 428. A shop inventory dated 1414, of a Cracow merchant, Jan Borg. English cloth is mentioned twice as "pecia albi de Anglia", and "pecia albi Anglicani".

Table 3. *Shop inventory of a Cracow cloth merchant, 1414*

Origin of cloth	Colour	Amount	Origin of cloth	Colour	Amount	
Malines	Red	1 warp	Brussels	Black	3	warps
Malines	Black	½ "	Brussels	Green	1	"
Louvain	Red	½ "	Brussels	Brown	2	"
Louvain	Green	½ "	Malines	Red	1	"
Louvain	Black	½ "	Tirlemont	? (Long)	13	"
Brussels	Black	¾ "	Tirlemont	Green/Long	½	"
Brussels	Gold	½ "	Malines	Black	½	"
Brussels	Black	½ " in pieces	England	White + small amount red & gold	½	"
Louvain	Gold	½ "	Tirlemont	Black/Long	1	"
Louvain	Black	½ "	Beverley	Green	½	"
Brussels	Gold	1 "	England	White	½	"
Brussels	Green	1 "	England	Grey	½	"
Malines	Black	½ "	Zittau	Red & Green	Small amount	
Brussels	Black	½ "	Hainaut	Black	½	warps
Malines	Red	½ "	Bohemia	? (Woolly)	1	"
				TOTAL	34¾	warps

Source: Wojewódzkie Archiwum Państwowe w Krakowie, *Consularia Cracoviensia*, Vol. II, p. 64.

Chance survival of a shop inventory of one merchant in Cracow, Jan Borg (see Table 3), gives some insight into the types, origins and colours of cloth for sale in the city during the early part of the fifteenth century (Fig. 34). Certain limited deductions can be made from this table. Firstly, cloth from Brussels seemed most popular, mentioned in nine of the thirty items (30 per cent), and provided 10¼ warps of the total (30 per cent). Secondly, Tirlemont provided nearly 40 per cent of the total warpage, although only recorded twice. Thirdly, the most desired colour appears to have been black, mentioned eleven times (38 per cent) with a total of 8¾ warps (25 per cent), even though two items failed to record the colour. Finally, white and grey were only found amongst the English cloth, and may have been dyed different colours at a later stage of their purchase, yet still preserving the known quality of English woollen goods[456] (Fig. 35).

The peak of Cracow's textile transit trade appears to have been around the turn of the fourteenth and fifteenth centuries (Fig. 36). The fullest information is for the period 1390 to 1405 when a total of 228,100 warps[457] were sent to Cracow by foreign merchants, an average of 16,293 warps annually. This

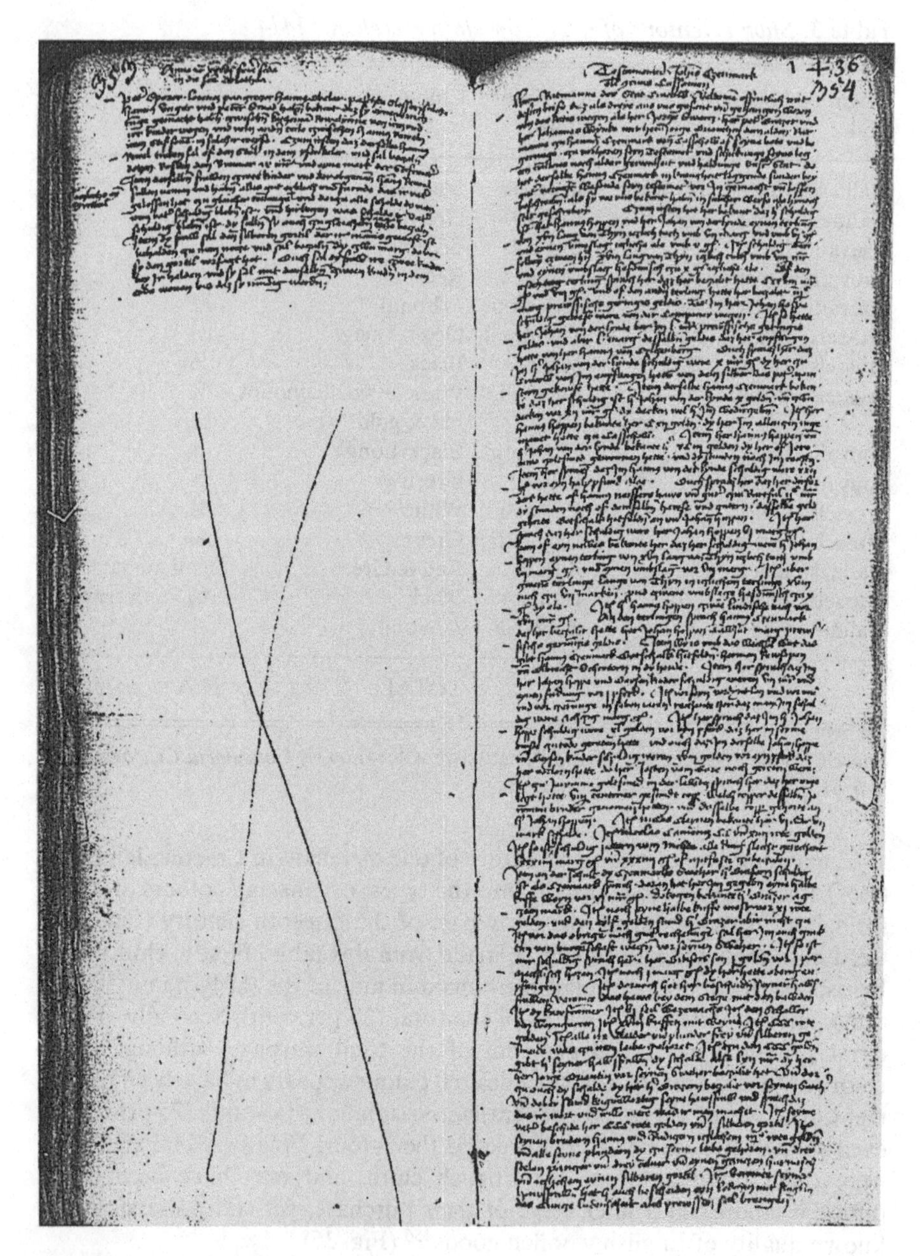

35. Document: English cloth in Cracow, 1436.

Source: Wojewódzkie Archiwum Państwowe w Krakowie: *Consularia Cracoviensia*, Vol. II, p. 354; Ms. 428. Document dated 1436, refers to purchase of English cloth by Cracow merchant, John Czenmark, and mentioned in his will.

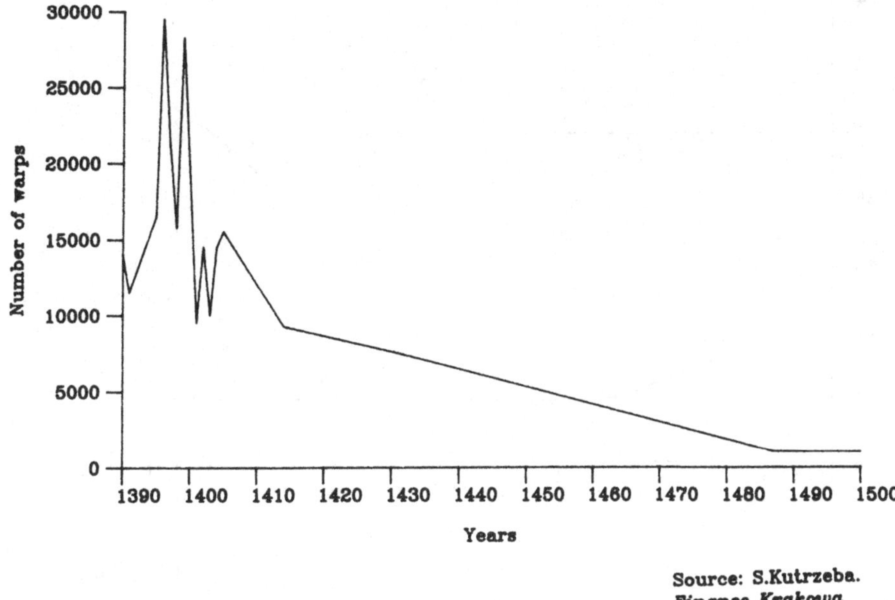

36. Transit trade in cloth through Cracow, 1390–1500.

total must have provided the city with considerable financial remuneration as all foreign merchants were liable to customs tariffs, only Cracow traders being exempt. The fifteenth century was to experience an overall decline in Cracow's cloth imports; disruptions caused by conflicts with the Teutonic Knights in the early decades was followed by a decline in direct trade contacts with Flanders after 1450, especially in cloth; further frustration was created by the Prussian War (1454–66) which disturbed cloth imports so that by 1487 the total was only a sixteenth of the average for the years 1390–1405. This can be partly explained through rises in customs duty by the city, but also a fall-off in trade with her best customers, namely, Hungary, Silesia and lands to the east.

The lands to the south-east and west of Cracow provided important outlets for the city's transit cloth trade, as well as a source for some of the more exquisite oriental textiles. Hungary provided the major market for Cracow's transit role in cloth; earliest evidence of this commerce dates from the thirteenth century;[458] by 1310, it was one of the main articles sent from the city over the Carpathian Mountains.[459] The most frequently mentioned

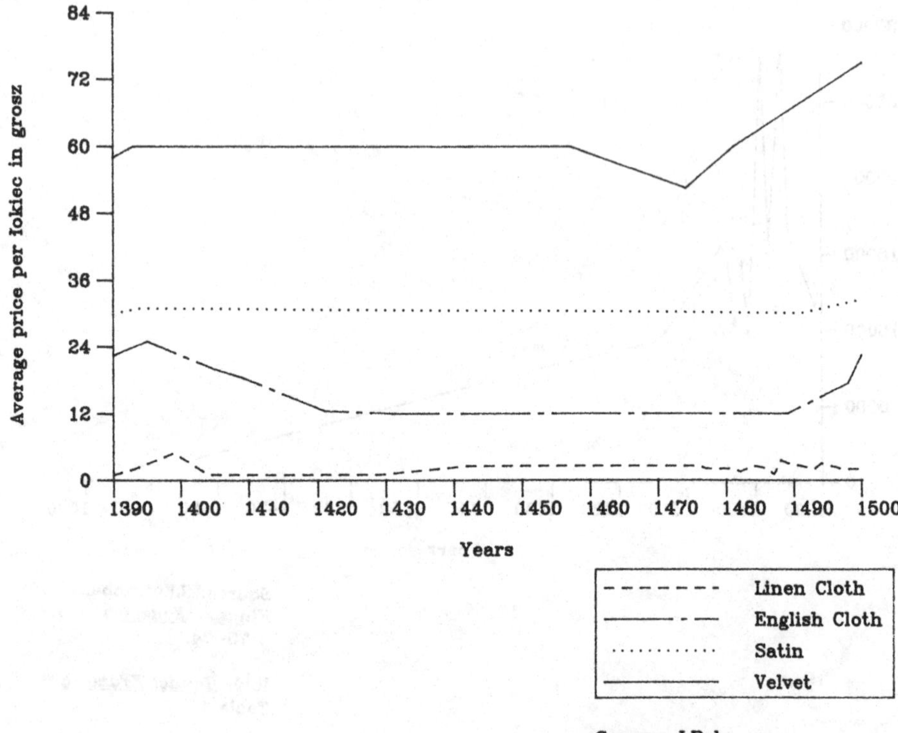

37. Price of certain textiles in Cracow, 1390–1500.

Flemish cloth sent to Hungary was from Tirlemont[460] followed by Malines,[461] Herenthals,[462] Termonde, Brussels, Ypres and Arras.[463] Highly prized was material from England, especially London,[464] and from Cologne,[465] but Cracow also sent cheaper varieties of Silesian[466] and Polish[467] cloth. Although most of the cloth was destined for the nearest Hungarian towns,[468] Kežmarok, Levoča, Bardejov,[469] Varín (near Žilina), and particularly Košice,[470] it was sent much further to Várad (Oradea) and beyond.[471] Certain Cracow merchants specialised in this trade including Jerzy Morsztyn, Jan Sweidniczer, Piotr Huenerman and others.[472]

Lwów was an important centre for Cracow merchants, both for buying and selling cloth. In Lwów they were able to purchase various types of oriental textiles – 'Istanbul' silk, scarlet, demi-satin, velvet, damask, muslin and muslinet,[473] which arrived in the city via Moldavia, from the East. Such material was then transported back to Cracow for its local market, or to be sent further down the Vistula to Gdańsk and Western Europe.[474] In the opposite direction Cracow traders sent Flemish, Italian and English cloth to such cities as Lwów,[475] the fairs at Jarosław,[476] or to the more distant lands of Moldavia and Multenia,[477] where they supplied towns like Suceava, Belgorod and Kiliya with textiles from Bruges, London and Florence, together with local varieties of cloth e.g., from Krosno.[478] Distance and quality are clearly reflected in price differential at Cracow, where velvet, satin and English cloth cost more than Polish linen material, over the whole period from 1390 to 1500 (Fig. 37).

Another source of both foreign and local cloth for Cracow was Silesia; a flourishing trade existed between the city and Wrocław except when the two centres were involved in trade wars (1350–60; 1440–90). The Silesian capital maintained lively commercial relations with Flanders in the thirteenth and early fourteenth centuries, when great quantities of cloth were being imported.[479] Only part of the Flemish cloth imports was absorbed by local buyers; some of the material found its way to Cracow for further export to the east.[480] Silesia also produced its own varieties of cloth, and although not having the same quality as its Flemish or English counterparts, nevertheless was bought by Cracow merchants. Transactions most frequently mention Görlitz (Zgorzelec) cloth (Fig. 38), especially its quality white material,[481] together with 'szleftucha' (sky-blue cloth) from Zittau.[482] Green cloth from the Wrocław workshops, and from Świdnica and Opava,[483] are also mentioned, but as it was not customary to record trade agreements in ledger books, so evidence is sporadic and accidental; nevertheless, examples quoted offer some illustrations of commercial movement in the Later Middle Ages. Certainly the customs tariffs collected at Będzin on the Silesia/Little Poland border suggest that cloth was the most important commodity travelling along the Wrocław–Cracow route.[484] It is noticeable, however, that from about 1440 onwards Silesian cloth is less frequently recorded in Cracow documents, corresponding to the intensification of trade hostilities between the capitals of Silesia and Little Poland at that time.[485]

Finally Cracow was involved in the production and trade of Polish domestic cloth. Małowist has stated that 'It is difficult today to state precisely the geographical position of the principal regions of cloth manufacture in Poland in the fourteenth and fifteenth centuries',[486] but it was important in the western part of Great Poland, along the Silesian border (e.g. Międzyrzecz, Kościan, Wschowa, Koźmin). Poznań became the main textile centre for the whole region,[487] whilst Toruń was its equivalent in Prussia.[488] In Mazovia the main production centres were at Płock, Wyszogród, Sierpc and Płońsk. The textile industry in Little Poland developed along the foothills of the Carpathian Mountains, where adequate supplies of wood led to the growth of cloth manufacture at Biecz, Lelów and other small towns, as well as in the capital, Cracow.[489]

38. Document: Cloth from Cracow to Košice, 1406.

Source: Wojewódzkie Archiwum Państwowe w Krakowie: *Consularia Cracoviensia*, Vol. I, p. 231; Ms. 427. In 1406, a Cracow merchant, Peter Huenerman, sent 110 warps of Zgorzelec (Görlitz) cloth from Cracow to Košice in Slovakia (N. Hungary).

Several varieties of cloth were manufactured in Cracow (bloser, selbfar, loden (or fleecy cloth) and szotte)[490] (Fig. 39), and in Kazimierz,[491] which satisfied local demand, any surplus being exported. Perhaps more important for Cracow was the attraction it had for other town manufacturers. Merchants from places like Nowy Sącz[492] and Krosno[493] and, to a lesser extent, Biecz and Lelów, who were unable to trade in large quantities there, brought cloth by the cartload to Cracow for sale; in this way they were able to receive the benefit of better organisation and contacts. The dominance of Cracow in the cloth trade of Little Poland eventually led to resistance by the smaller towns who wished to limit the method of sale;[494] the conflict was finally solved in 1449 with the king's intervention and support for Cracow merchants.[495] Henceforward, there was no reference to further disputes and the city's supremacy in the cloth trade of Little Poland remained supreme.

In general, Polish cloth merchants looked for markets to the north and west – to Prussia, Flanders and Novgorod in the search for greater profit, avoiding outlets in Little Poland. Only the poorer quality cloth, of which Cracow's own variety was one, had to look southward for sales; Hungary offered a substantial market. Cloth from Cracow and Biecz appear to have been the most popular varieties sent to Hungary, especially Košice[496] and Bardejov;[497] from the latter place they were also exported further south into Transylvania.[498] In 1460 a company was formed consisting of Cracow, Biecz and Nowy Sącz merchants, who wished to monopolise the Polish cloth trade in Hungary through contacts in Bardejov;[499] cartloads of linen cloth and yarn were sent to places like Bardejov, Stropkov and Humenné from small Polish production centres like Krosno, Dukla, Muszyna, Grzybów, Sącz, Tarnów, Jasło, Pilzno, Strzyżów, Ropczyce, Wielopole, Ciężkowice and Czchów.[500] Although cloth from Little Poland had small demand in the west, examples of Krosno material being sent to Silesia via Cracow are recorded[501] whilst some linen cloth was sent down the Vistula River to Gdańsk[502] (Fig. 40).

(ii) Wine and beer

Air and water are indispensable: nothing else is. There is, of course, much else that is highly desirable, but not indispensable: wine and beer for instance. Beer was the ordinary person's drink in the Middle Ages and large quantities were consumed for nourishment as well as pleasure; wine was the everyday beverage of the more affluent members of society, save in the grape-growing areas of Europe where it was cheap. Nevertheless, Postan has observed that 'Historians do not know enough about medieval consumption in different places and at different social levels to be able to judge to what extent wine and beer were true substitutes',[503] but it appears that 'Wine cost much more than beer, even local wine, and artisans took it only on special

392

erste krewcze. Und vorbas obir daz erste kreucze von eynir scheiben VI heller.

§ 3. Item von I centner ysin I hellir bis an das erste kreucze. Und vorbas obir daz krueze II heller von I centner.

Ms. Cons. I, A, str. 69. — Monum. IV, pars II, p. 142.

CCLXXXVI. 1396. die 18. mensis Octobris.

Statutum de pretio mercium, annonae, operumque manu factorum.

Nota, quod Anno domini MCCCCXIII in Annali foro festi Sancti Stanislai in Mayo infrascripta sunt proclamata et in cedula conscripta ab antiquo et in pretorio inuenta, cuius tenor talis est:

Validi ac nobiles viri domini Petrassius Rpiszka Curie, Iacussius coquine Magistri, Nawogius de lankawa, Petrassius de Sczekoczin Subdapifer Serenissime Principis domine nostre graciose Regine Polonie, Borko de Tschczenecz et Andreas de Sprowa per ipsam dominam Reginam ad infrascripta specialiter deputati, Cum circumspectis Consulibus Ciuitatis Cracouiensis de vniuersis et singulis infrascriptis vendibilibus et emptibilibus tam in terra quam ciuitate Cracouiensi constitutis, melioribus quibus potuerunt modis, saluo tamen meliori Consilio Concordarunt, composuerunt ac Amicabiliter pro publico communitatis conseruando conuictu ordinare statuerunt.

Primo mensura tritici pro VII grossis. — Siliginis mensura pro V gr. — Auene mensura pro II gr. — Order mensura pro IIII gr. — Pisarum coretum pro II½ gr. — Milij coretum pro III gr. — Papaueris coretum pro IIII gr. — Canapi coretum pro II gr. — Cepium coretum pro I. gr. — Duos bonos pullos pro I gr. — Caponem pro I gr. — Duo oua modo pro denario, post pascha tamen pro vno denario, — Auca viuens pro XVI denarijs, deplumata pro XII den. — Aneta pro ½ gr. — Porcellus bonus pro I½ gr. communis pro I gr. — Aues III pro I den. — Szemenuszky alias Brochfogil VIII pro I gr. — Asperiolus excoreatus pro III den. — Lepus pro II gr. — Capriolus pro VIII gr. — Castratus pro VI gr. viuus, Excoriatus pro V gros. Communis vero pro IIII gr. — Vitulus saginatus pro VI gr. — Sepi vnus lapis pro VIII gr. — Aruine lapis pro X gr. — Segiminis lapis pro X gr. — Lane Autumpnalis lapis pro X scotis, — Lane yemalis lapis pro VIII scotis, — Lane agnine lapis pro ½ marca, et non debet comisceri cum alia lana; — Cutis major bouina pro XIIII gr. sed comunis, quantum valere poterit; — Cutis castratina rasa pro I½ gr. non rasa pro II gr. — Cutis hurina melior pro III gr. — Scilindriorum centum pro I½ gr. — Asserum serratorum I pro I gr. — Asserum scilindriorum sexagena pro VI gr. — Tegularum bonarum sexagena pro X gr. — Lattenstebe I de bonis pro ½ gr. — Lignum scindibile alias sneydholcz pro VIII gr. melius, — Canale vltra XXX vlnas pro XII gr. — Balista pro I mrc. et I gr. — Statua baliste pro IIII gr. — Amentum pro I gr. — Nux pro I gr. — Calcaria com-

393

munia non debent fieri meliora quam pro II gr. — pro Iuuenibus I par calcariorum pro I gr. — et pro militibus I par pro VI gr., qui autem meliora habere voluerit, emat pro quanto potest. — Cerdones de maiore cute recipiant III gr. — de leuiori uero II gr., de minoribus I½ gr. — De Castratiuis hurinis et alijs paruis de qualibet ½ gr. — Ocree dominorum pro XII gr. famulorum pro VIII gr. — Item Iuuenum et kmethonum pro VII gr. — Par Botorum pro IIII gr. — par Sotularium nigrorum pro II gr. militarium scilicet, — par communium sotularium pro I½ gr. — par subsolearum non vnctarum pro I gr. — par subsolearum pro ocreis pro I½ gr. — par coopertorij baliste pro III gr. — frenum dilatum est, quousque sciatur vera taxa. — Par streparum communium pro II gr. meliorum pro III gr. — Par strepalium pro II gr. — De Babato nouo ½ gr. de antiquo sufferando IIII den. — Sella pro curru pro VI gr. — Sella communis pro Balistario pro XII gr. — Sella pro milite decupreata in capite et retro in circumferencijs pro XX gr. — faretra tecta cum taxo pro VI gr. — Alie vere faretre, quanto leuius possunt emi, — Cingulus baliste communis cum vnco pro IIII gr. — Vncus per se pro II gr. — Item quando septem libre piperis dantur pro I marca, tunc libra in Institis debet dari pro VIII gr. — Quando libra croci emitur pro I marca et VIII gr. soluet in Institа I sexagenam, — Parchanus bonus pro I mrc. — Libra bambicis pro III gr. — Tele bone de Sueuia vlna pro III gr. — Dupli alias czwelich vlna pro I½ gr. — De Snidone alias golcz vlna pro I gr. — De Bressella curti melioris vlna pro XX gr. — Longi de Thyn vlna pro XIIII gr. — Curthi de Thyn vlna pro IX gr. — De Mechil boni vlna pro XVII gr. — De lira vlna pro IX scot. — Curti de Louana boni vlna pro VIII scot. — De Anglia vlna pro XIIII gr. — De Cortir boni vlna pro XII gr. — Kirsingi vlnam pro III gr. — Edingi vlna pro VIII gr. — Girinsberg vlna pro VIII gr. — Strigoniensis vlna pro V gr. — Swidnicensis vlna pro IIII gr. — Longi Wratislauiensis vlna pro V gr. Curti uero pro IIII gr. — Albi et grisei Cracouiensis vlna pro II gr. cuiuslibet et IIII den. — Bloser Cracouiensis vlna pro II½ gr., Item selbfar pro III gr. — Loden Cracouien. alias kosmethe vlna pro II gr. — Szotte Cracouien. vlna pro III gr. — Dillirmundisch vlna pro XV gr. — Settauiensis boni vlna pro VI gr. — Harras vlna pro III gr. — Ficus, Amigdala, vuepasse etc. modo statui non possunt, quia pro presenti non adducuntur, sed cum venerint, statuentur iuxta formam.

Actum feria quarta die sancti Luce Ewangeliste.

Domini: Clemens Vicecangellarius, Petrassius Rpischka, Nauogius de Lankawa nomine domini nostri Regis Polonie:

Primo currus non ferratus pro dominabus pro ½ marca, — Item currus cursorius pro II equis pro I½ mrc. cum coopertura, — Item pro IIII equis pro II mrc. cum opertura, — Currus vectorum pro VII fertonibus, — Currus kmethonum de II equis lignorum pro II gr. et quando ducitur cum duobus bobus pro I½ gr. — Item currus Radicum pro I½ gr. — Item octuale carbonum pro II gr. medietas pro I gr. — Item pro vibracione gladij et vagina IIII gr. sed pro va-

39. Document: Foreign and domestic woollen cloth sold in Cracow, October, 1396.

Kodeks dyplomatyczny miasta Krakowa, Vol. II, (edt. F. Piekosiński), Kraków, 1582, Document 286, dated October, 1396. Evidence of woollen cloth for sale in Cracow, both of local and foreign origin.

394

ginis gladij II gr. — Item pro lopusa subductiua de Parchano V gr. de sericea VIII gr. — qui autem per modum alcioris suture habere voluerit, conueniat pro quanto potest; — Item de tunica ampla longa de peciebus simplici VI gr. et de subducta cum panno aut tela soluet octo gr. — Item de tunica IIII^or peciarum non subducta IIII gr. de subducta VI gr. qui autem (vergare) zottare (s) aut aliter contra communem consuetudinem sibi fieri iusserit, conueniat pro quanto potest. — Item de caligis cum operturis I½ gr. de bono panno, sed de simplici pro I gr. — Item Capucium subductum pro II gr. sed simplum pro I gr. — Item de pallio subducto VI gr. de simplo IIII gr. — Pellicium longum album de cutibus terrestribus pro I mrc. Nigrum uero pro I sexagena. — Rotarum frustra circumferencialia I solidum pro I gr. — Pixis rote III pro I gr. — vnum plaustrum de cantubus in numero XXXVI solidos pro I fertone, tres scilicet solidos pro vno gr. — Quatuor rotas vectori pro IX scot. — pro IIII^or equis currus dominorum pro IIII gr. Rote in paruo curru aque vel Rinwagin pro I fert. — frenum militare circumcusum vndique cum electro antelam et postelam communem pro X scot. — frenum clientare circumcusum lamnis ferreis cum omnibus attinencijs pro XV gr. — frenum cum stanno circumcuso et omnia attinencia pro XII gr. — Circumcusum cum ferro simplici similiter pro XII gr. — frenum simplex de coreo cum duobus tractorijs et omnibus attinencijs pro VIII gr. — frenum circumcusum paruis fibulis quinque lineis cum omnibus attinencijs pro ½ marca, frenum vectorum pro II gr. — Capistrum pro I½ gr.

Cons. Lib. 1536, pag. 38.

Powyższe ustanowienie cen towarów i ziemiopłodów, zapublikowane dopiero podczas jarmarku majowego r. 1413, daleko wcześniejszy ma początek. Widocznem jest nawet, iż ustanowieni przez królową (Jadwigę) do oznaczenia cen powyższych komisarze, po dwakroć do pracy zasiadali. Pierwszy wykaz cen nosi datę ze środy, z dnia św. Łukasza ewangielisty; drugi jest bez daty. Jakkolwiek rok nie jest w żadnym wykazie podany, przecież dokładnie oznaczony dzień z uwzględnieniem urzędów przez komisarzy piastowanych, dozwala rok ten z całą ścisłością oznaczyć. Mianowicie przypada u schyłku XIV wieku, kiedy właśnie komisarzy owych w innych pomnikach historycznych spotykamy, św. Łukasz ewangielista na środę tylko w latach 1391, 1396 i 1402. Otóż pierwszy zaraz z wyznaczonych przez królową komisarzy, Pietrasz Rpiszka, marszałek dworu królowej, nie piastował jeszcze w r. 1391 tego urzędu, marszałkiem królowej był bowiem podówczas Krystyn (1389, 1390, 1393 i 1394. Przezdziecki: Życie domowe Jadwigi i Jagiełły, 12, 17, 36, 50 i 73); zaś w r. 1401 Rpiszka już tylko kasztelanem wiślickim jest tytułowany (Rzysz. Mucz. I, 272). Pozostaje nam więc zatem tylko rok 1396. Jakoż spotykamy istotnie Pietrasza Rpiszkę w roku 1396 marszałkiem dworu królowej (Monum. IV, pars II, 150, 162); inni też komisarze mniej więcej na ten sam czas przypadają. I tak występuje Iakusz z Boturzyna jako kuchmistrz królowej w latach 1394, 1395 i 1397 (Przezdziecki l. c. 53, 60, kodeksu niniejszego Cz. I, N. LXXXVIII), a Nawój z Łękawy jako podskarbi królowej w latach 1393, 1394 i 1395 (Przezdziecki l. c. Monum. IV, pars II, 119); Borek z Trzcieńca wraz z pomienionym Nawojem z Łękawy świadkują razem d. 1 grudnia 1395 w Krakowie (Część I niniejszego kodeksu N. LXXXII); nawet Klemens podkanclerzy, który już tylko przy układaniu drugiego wykazu cen brał udział, również w latach 1393, 1397 i 1401 występuje (Kod. Mog. 98, — Rzysz. Mucz. I, 272 i III, 352).

39. (*cont.*)

occasions.'[504] Certainly wine was widely distributed in medieval Europe: the vine was cultivated even in unlikely northern areas until the specialisation of certain great regions enabled them to supply a higher-grade product wherever needed. Attempts at viticulture in the more climatically marginal areas of the continent were gradually abandoned as the Middle Ages progressed and northern Europe began to rely on imports from more favoured regions such as the lands of the Mediterranean. As a result, very large quantities of wine were involved in the export trade.[505] In some parts of Europe, where supplies of wine became more difficult to obtain, there was compensation from augmented production of beer, especially in parts of the Netherlands,[506] Germany[507] and Baltic countries.[508]

In Poland, place-name evidence has suggested the presence of vineyards in Greater Poland and around Cracow during the Early Middle Ages;[509] it was probably closely associated with the introduction of Christianity to the country during the reign of Mieszko I.[510] The fourteenth century saw the

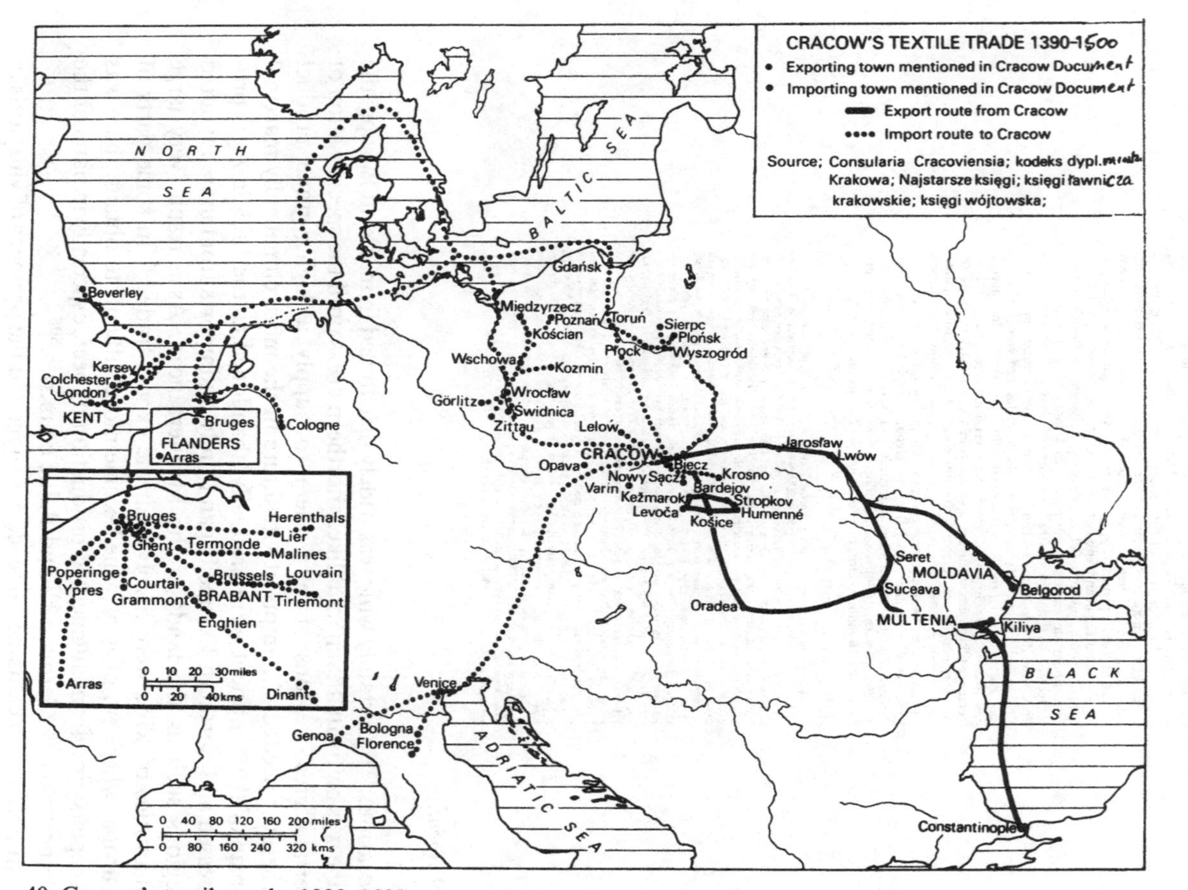

40 Cracow's textile trade, 1390–1500.

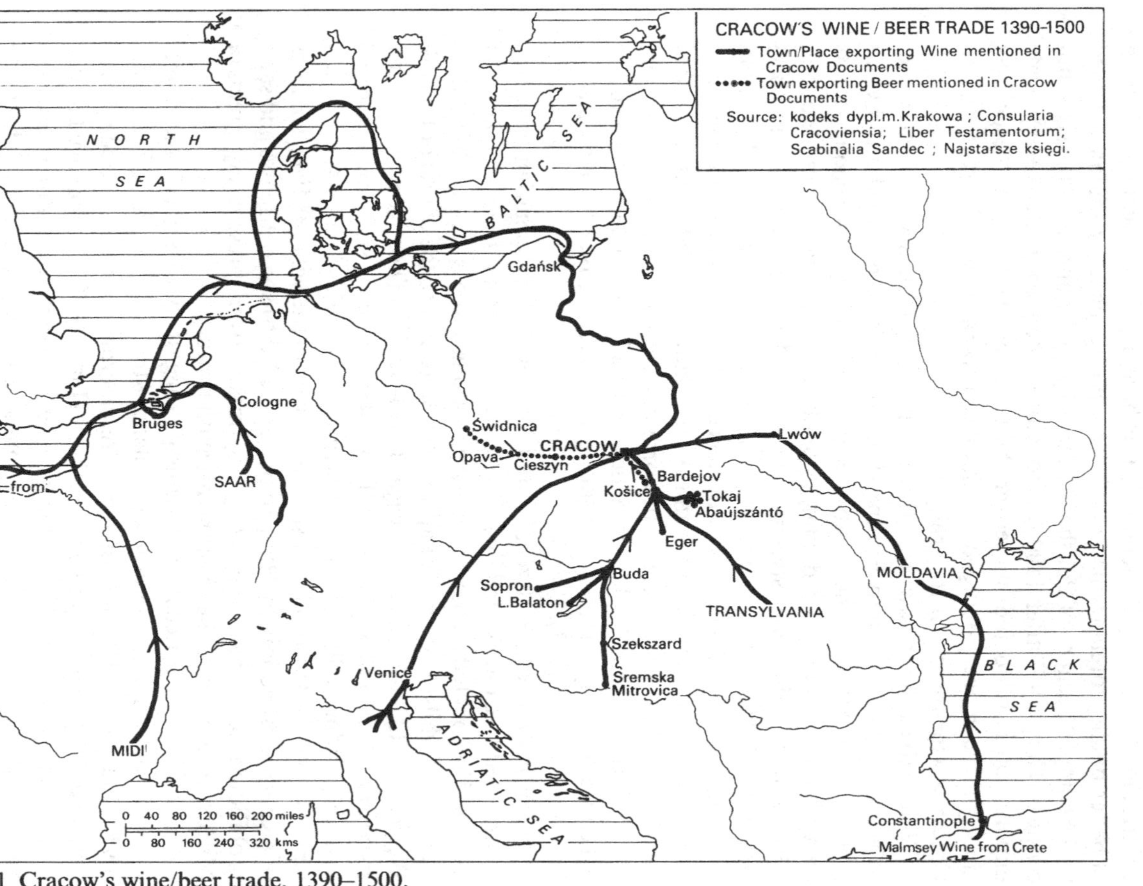
CRACOW'S WINE / BEER TRADE 1390-1500
Town/Place exporting Wine mentioned in Cracow Documents
Town exporting Beer mentioned in Cracow Documents
Source: kodeks dypl.m.Krakowa ; Consularia Cracoviensia; Liber Testamentorum; Scabinalia Sandec ; Najstarsze księgi.
NORTH SEA
BALTIC SEA
Gdańsk
Cologne
Bruges
SAAR
from
Świdnica
Opava
Cieszyn
CRACOW
Lwów
Bardejov
Košice
Tokaj
Abaújszántó
Eger
Buda
Sopron
L.Balaton
TRANSYLVANIA
MOLDAVIA
Szekszard
Sremska Mitrovica
Venice
MIDI
ADRIATIC SEA
BLACK SEA
Constantinople
Malmsey Wine from Crete
0 40 80 120 160 200 miles
0 80 160 240 320 kms

41 Cracow's wine/beer trade, 1390–1500.

peak of Polish viticultural development but after 1400 it began to decline.[511] In spite of new vineyards planted in and around such centres as Sandomierz, Poznań, Płock and Cracow, other factors both climatological and social, contributed to loss of production.[512] Furthermore the gradual introduction of better-quality foreign wines into Poland during the Later Middle Ages led to the complete eclipse of Polish viticulture by the beginning of the sixteenth century. The period after 1500 was to witness the great era of foreign wine importation into Poland; moreover the taste for alien varieties, especially amongst the more privileged medieval aristocracy, was in part due to Cracow's wine-merchant fraternity.

Wine arrived in Cracow from three main areas; Hungary, Moldavia and, to a lesser extent, western Europe (Fig. 41). Poland provided an obvious market for Hungarian wines, firstly because it did not produce enough for its own religious and other needs and, secondly, the other major possible outlet, Austria, had its domestic Styrian varieties.[513] The largest Polish consumers were the king's court, the magnates and the rich city dwellers. Hungarian wines appear from documentary evidence to have been of two basic types: mountain wines from Zemplín and Abaújszantó (referred to as 'zieleniaki samorodne' and 'maślacze asu' in north-eastern Hungary (Hegyalja)), and included the famous 'Tokaj' varieties; lowland wines (called 'piaskowe') which were mainly light wines from Eger, Szekszard, Buda, Lake Balaton, Sopron etc.[514] (Fig. 42).

According to both Kutrzeba[515] and Dąbrowski[516] Hungarian wine was transported to Cracow already in the thirteenth century. Certainly the tariff lists of 1310 include Hungarian wines amongst the articles imported to Cracow,[517] whilst vintage 'Tokaj' varieties were sent to the courts of Ludwik Węgierski (1370–82) and Jadwiga (1384–99).[518] The accounts of her successor, Władysław Jagiełło (1386–1434) record such purchases,[519] and the trading edict of Jan Olbracht (1492–1501) dated 1498 allowed Cracow merchants to collect a special duty from the wine.[520] Even so, fears of a Cracow monopoly by Hungarian traders led to some disputes and sales prohibition on certain types (e.g. 'Syrmska' from present-day Sremska Mitrovica) towards the end of the fifteenth century.[521]

Until the sixteenth century documents suggest that Cracow's merchants only sent wine in small quantities to Poland. Most wine was bought in Bardejov[522] and Košice;[523] some of the more enterprising Cracow traders even obtained their own vineyards in northern Hungary[524] and employed agents to conduct the business.[525] Purchases were made from the autumn (after the harvest) right up to Christmas; these remained in Hungary until late spring when the fear of heavy frosts and snow-blocked routes receded,[526] and rivers, like the Dunajec, were once more open to navigation.[527] Another fear came from wine dilution; on occasions this was detected by wine controllers in Sącz and led to legal disputes.[528] Merchants from the Polish foothill towns

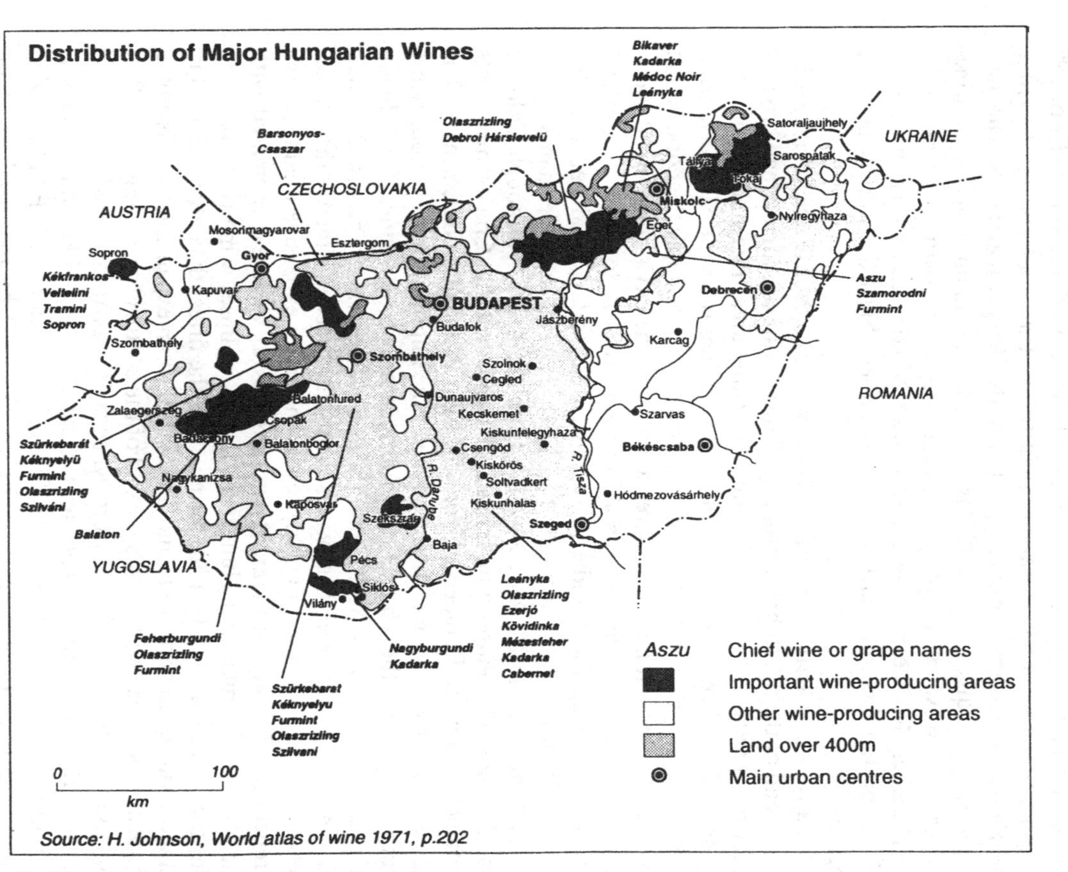
Distribution of Major Hungarian Wines
Bikaver
Kadarka
Médoc Noir
Leányka
Olaszrizling
Debroi Hárslevelü
Barsonyos-
Csaszar
CZECHOSLOVAKIA
AUSTRIA
UKRAINE
ROMANIA
YUGOSLAVIA
Satoraljaujhely
Sarospatak
Tokaj
Miskolc
Eger
Nyiregyhaza
Aszu
Szamorodni
Furmint
Debrecen
Karcag
Mosonmagyarovar
Esztergom
Gyor
Sopron
Kékfrankos
Veltelini
Tramini
Sopron
Kapuvar
BUDAPEST
Jászberény
Budafok
Szombathely
Szolnok
Cegled
Dunaujvaros
Kecskemet
Kiskunfelegyhaza
Csengöd
Kiskörös
Soltvadkert
Kiskunhalas
Szeged
R. Danube
R. Tisza
Szarvas
Békéscsaba
Hódmezovásárhely
Balatonfured
Csopak
Zalaegerszeg
Badacsony
Balatonboglor
Nagykanizsa
Kaposvar
Szekszard
Baja
Pécs
Siklós
Villány
Szürkebarát
Kéknyelyü
Furmint
Olaszrizling
Szilváni
Balaton
Feherburgundi
Olaszrizling
Furmint
Szürkebarat
Kéknyelyu
Furmint
Olaszrizling
Szilvani
Nagyburgundi
Kadarka
Leányka
Olaszrizling
Ezerjó
Kövidinka
Mézesfeher
Kadarka
Cabernet
Aszu
Chief wine or grape names
Important wine-producing areas
Other wine-producing areas
Land over 400m
Main urban centres
0
100
km
Source: H. Johnson, World atlas of wine 1971, p.202

42 Distribution of major Hungarian wines.

(Biecz, Krosno etc.) also obtained Hungarian wines which may later have been sent on to Cracow market.[529]

Much less documentary evidence is available on Moldavian wine imports to Cracow; they arrived in the city either from Lwów or Košice. Certainly products from the Moldavian vineyards were well known in Lwów,[530] together with Greek wines (malmsey) in transit through Moldavia.[531] After the Fall of Constantinople (1453) it was the Jewish merchants who first exported large quantities of goods, such as rice, alum, spices and wine, through to Lwów, whilst others sent them as far as Toruń via Cracow.[532] Even earlier, Košice had been a source of Moldavian and Translyvanian wines for the Cracow market, due to well-established contacts in these regions.[533]

Finally, wines arrived in Cracow from various west European vineyards. By the end of the thirteenth century specialised wine production had brought great prosperity to Gascony, Poitou, the Rhine area, Burgundy[534] and the Rhône valley; there was also a lively trade in wines from the Mediterranean. Some of these wines were to reach Cracow, largely to supply the nobility and royal court. Most numerous reference amongst the city's commercial transactions was Rhine wine. From Alsace northwards to Bonn, Rhineland viticulture during the Middle Ages, as it does today, produced wine for export. Cracow's town accounts frequently refer to two German wines – Rhine (reńskie) and Ruwer ('Riwuła') from the Saar;[535] although Cologne was the main wine emporium, it has been confirmed that Cracow merchants obtained them in Flanders,[536] or through Prussian middlemen.[537] The Pomeranian towns were certainly active in the importation of various wines from Bruges; at the end of the fourteenth century wines from Spain (Granada) and France (Midi), and Malmsey (Greece) were recorded in their customs lists.[538] Some of these wines eventually reached Cracow; their prices between 1393 and 1409 (Fig. 43) reveal the more expensive quality of the French and Romanian (Transylvania:Moldavia) varieties, compared with the cheaper German and Hungarian kinds. Malmsey wine (Fig. 44) from Turkey, Greece and Italy, had, for much of the fifteenth century, arrived in Cracow via the eastern route and Lwów; Ottoman expansion had disrupted some of this trade and towards the end of the century, according to Lauffer, malmsey arrived in Gdańsk from whence it was transported southward particularly by Armenian and Jewish merchants into Poland.[539] This is certainly reflected in higher prices for malmsey after 1485. Finally Italian wines appear to have been sent direct to Cracow by the usual route via Villach–Vienna–Oświęcim;[540] they were rarely mentioned in documents, probably because some years the Alps placed an insuperable barrier to the transport of Italian wines to parts of northern Europe.

Beer, according to Postan, 'made from hops was the product of the Middle Ages in decline'.[541] Certainly in northern Europe from the mid-fourteenth

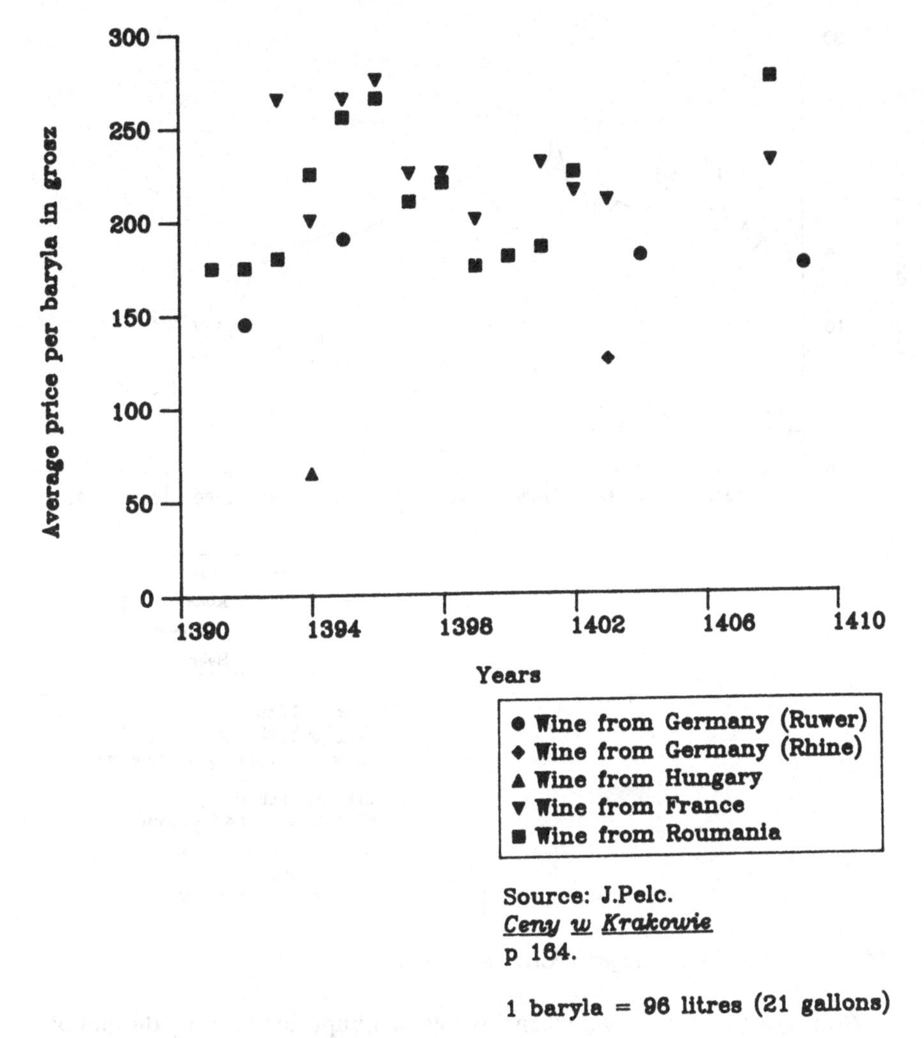

43 Price of certain wines in Cracow, 1390–1410.

century onwards it was brewed in ever-increasing quantities; reasons for this are open to conjecture, but it may have been due to changing habits, greater difficulty in obtaining imported wines because various wars interrupted trade, increased consumption of salted bread and fish by the artisan classes which worked up a thirst,[542] or simply that beer was gradually replacing ales and other drinks of a humbler kind like mead. In the fifteenth century wine consumption in Holland declined at the expense of beer production,[543] whilst beer was exported from north Germany to Scandinavia and Prussia in large quantities.[544]

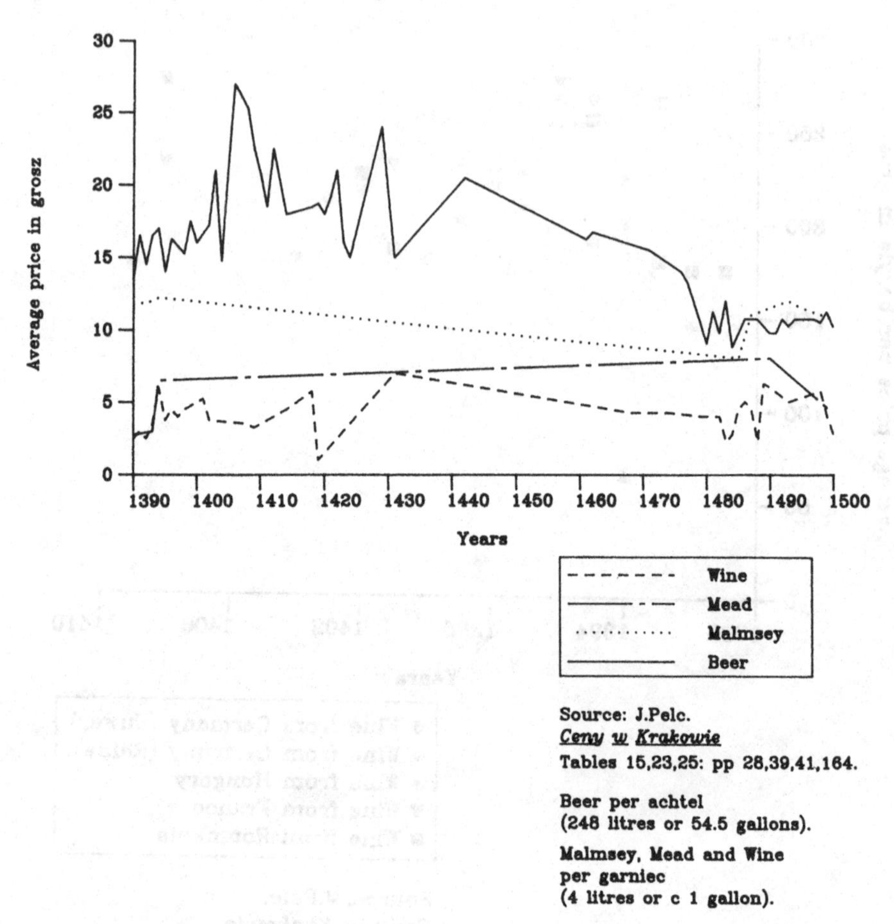

44 Price of certain beverages in Cracow, 1390–1500.

Although it is not known when Cracow first imported beer, by the end of the fourteenth century the Silesian breweries at Świdnica (Schweidnitz) seem to have had a monopoly in the city.[545] The first extant documentary evidence dates from 1447 when Cracow merchants imported quantities of beer from Świdnica and Cieszyn (Teschen).[546] The other main source of foreign beer was Bardejov in northern Hungary, which exported it to the towns of southern Poland. Documentary evidence from the mid-fifteenth century supports this claim.[547] Pressure at this time to protect the home market by Cracow's domestic brewers, led the king, Kazimierz Jagiellończyk, in 1456, to forbid the import of foreign beer into the city.[548] This ordinance was to last for nearly half a century and its impact in lowering the price of beer in the city is clearly seen in Figure 44. In 1501 it was repealed by King Jan

Olbracht, when once again Świdnica took over a monopoly in Cracow's beer supplies.[549]

Miscellaneous goods

Besides textiles, wine and beer, other products worthy of mention were the various manufactured articles which had their places in the city's trade structure. Such products were the result of increased manufacturing activity in many parts of Europe; within the urban centres craftsmen became organised into confraternities or guilds during the twelfth and thirteenth centuries. These were organisations of men who practised the same craft. Each craft guild had its own special ordinances and privileges, and in Cracow these were conferred by the City Council or granted by the king. A revealing insight into the city's workshops at the beginning of the sixteenth century is obtained from Balthazar Behem, who produced a book containing illustrations of Cracow's craft guilds, partly by realistic miniature paintings of the workshop interiors, and partly by symbols.[550] These depicted a movement, begun during the fifteenth century, of intense artisan activity characteristic of many Polish towns,[551] which must have involved considerable investment over a prolonged period of time. By 1505, Cracow had guilds not only producing for local consumption (bakers, blacksmiths, carpenters, coopers, dressmakers, cobblers and tanners) but manufacturing articles for more distant trade (bell making,[552] goldsmiths,[553] sword and knife manufacture,[554] needle making, painters and potters).[555]

Considerable emphasis in Cracow was placed on the transit trade and local sale of metal manufactured articles, known in documents as 'Nuremburg goods' from their obvious place of origin. They had an excellent reputation with a wide market throughout Europe;[556] Nuremburg 'Tand' (merceria), especially knives, were preeminent in their time.[557] Because Nuremburg, unlike many other cities, did not find her main markets in the immediate vicinity her products had to be designed to suit international taste and demands; hence the enormous variety. In 1365 King Kaziemierz Wielki gave Nuremburg merchants trade privileges throughout his territory;[558] both they and their metal wares are continually mentioned in Cracow documents throughout much of the fifteenth century.[559] In turn, Cracow traders then re-exported some of these wares, for example to markets in Hungary,[560] and along the Vistula River.[561]

Cracow merchants were also involved in jewellery goods. These were always in demand, not only from the royal court and nobility, but also from the richer citizens of various towns. For example, Cracow merchants supplied places like Jarosław with corals, pearls and crystal[562] together with gold and silver goods from Leipzig.[563] In 1464 one of Cracow's merchants sent a ruby weighing 250 carats (5 dekagrams) to England.[564] Competition for the Hungarian market, however, came via Balkan merchants, particularly from Dubrovnik.[565] Finally, of the other luxury manufactured goods arriving in Cracow mention should be made of the various oriental carpets which were sent through Lwów.[566] Of the more mundane, everyday products there seems to have been some importance attached to the trade in glass.[567]

Commodity prices

In the fourteenth and fifteenth centuries money began to play a greater part in the economic life of eastern Europe. Throughout the whole of Europe the period between 1150 and 1350 was one of growth and maturing economies. Unfortunately, all times of expansion come to an end, and the medieval era was no exception. For at least a century after 1350, less favourable circumstances prevailed for agriculture, industry and commerce. While wages rose and taxes grew heavier, prices stagnated, or declined, and some commercial enterprises became less profitable. Therefore, just as the twelfth and thirteenth centuries were followed by steep price rises all over Europe, so was the decline of the fourteenth and fifteenth centuries accompanied by price changes no less continuous but, according to Postan, 'possibly less spectacular'.[568] Certainly, the process of devaluation was 'already under way before 1440, continued well after 1750 and indeed down to the present day'.[569]

It is now opportune to question to what extent Cracow's own commodity prices reflected these general trends evident throughout Europe up to 1500. A word of caution is needed here for it should be remembered that it is not easy to find out what the prices really were in the past or, once discovered, to interpret the information correctly. There are many difficulties to overcome including changes in the intrinsic value of coins, like the grosz (groat), regional differences in weights and measures (the kersey cloths sold in Poland, for example, varied from 32 to 37 ells), and fluctuations according to the season. Therefore information gleaned from documentary evidence should be regarded as mainly for the record rather than possible interpretations for a wider sphere.

In Cracow, as in other parts of Poland, control on the quality of goods was in the hands of the guilds; similarly they supervised weights, measures and prices. Price catalogues (cenniki) – well known in the Middle Ages – were concerned with the control and regulation of prices; these were superintended by the city council and Polish government authorities. Effectively, this meant the local 'Wojewoda' and the city elders.[570] Special attention was paid by the king's price commission on articles brought to the city market from the surrounding rural area (e.g. cereals, cattle, poultry, wool, etc.), on handicraft goods manufactured in the towns, and on products from abroad, especially cloth and spices.[571] The town council paid particular heed to the quality and price of soap[572] and beer,[573] whilst comparisons were made with other Polish cities such as Poznań.[574]

Longer term price movements in Cracow from the end of the fourteenth century until 1500, have been graphed by commodity (see Figs 15, 19, 20, 22, 25, 26, 28, 31, 37, 43, 44), thanks to systematic research on the history of prices in the city carried out by J. Pelc, part of a larger project on several Polish cities.[575] The earliest evidence on prices in Cracow dates from the second half of the fourteenth century. The overall impression is one of a general tendency to rise during the first half of the fifteenth century, then a decline for the next fifty years. The impact of new silver sources into the economies of Europe do not appear to have had any more spectacular effect on the regions nearer to the point of origin than elsewhere. As Postan states, 'it is difficult to read into the regional differences between the price series in Cracow, Alsace, Holland and Norway'.[576] Nevertheless, within Cracow individual products obviously differed from the overall price trend for the city.

Foodstuffs in Cracow tended to have higher prices *circa* 1400 than at the end of the fifteenth century. One of the best examples was peas, which in 1389 cost 4.72 grosz per bushel, in 1419, 6 grosz, but only 4 grosz by 1500.[577] Similar trends in Ghent and Cracow reveal 'The smoothness of the latter hides the very considerable fluctuations resulting from the ravages of weather and war.'[578] Certainly meteorological conditions were reflected in foodstuff prices; for example, 1440 was a very bad winter in Poland and prices rose. A wet summer was experienced in 1456, whilst flooding of the Vistula in 1486 and particularly in 1495 (when it lasted six months around Cracow)[579] led to unstable price conditions. Prices of mining and manufacturing goods also mirrored the military and trade wars, which involved Cracow. For example, demand for lead rose sharply in Cracow during the second and third decades of the fifteenth century, coinciding with Poland's conflict with the Teutonic Knights; textile imports were also affected by trade wars with Wrocław and Toruń throughout much of the fifteenth century when not only transit trade in cloth declined, but textile prices in Cracow stagnated until the more peaceful conditions after 1490. Increased European conflict with the Ottoman Turks towards the end of the century was also echoed in higher prices for some spices in the Cracow market place.

Trade treaties and foreign merchants

From the thirteenth century onwards the merchants of the Later Middle Ages lived under the protection of their own town or city. Reciprocal guarantees were often given between urban centres and the protection of their merchants was an object of policy. The merchant operating abroad was therefore encouraged to settle in other places by the granting of privileges agreed in trade treaties drawn up between individual commercial centres. A trader was therefore able to operate freely within this framework and did so throughout most of Europe, with the exception of the English and, to a lesser extent, the north Germans, whose own communities adopted a more disciplined form of organisation which theoretically restricted their activity both at home and abroad.[580]

One of Cracow's oldest trading partners was Košice. The latter's attractive geographical position on the borders of the Carpathian Mountains and Hungarian Plain helps explain its commercial importance, for it controlled routes leading south to Transylvania, and trade connections to the north and west. During the Middle Ages it was a walled town covering 45 ha. with a population of some 2,500 inhabitants.[581] The town first signed a trade treaty with Cracow in 1324[582] in which it was agreed to help each other in commercial matters. The excellent political and economic relations between Poland and Hungary during the reigns of Ludwik Węgierski (1370–82) and Jadwiga (1384–99) encouraged the formulation of a second treaty between Cracow and Košice towards the end of the fourteenth century. This followed more general treaties between the town and Poland in 1375[583] and 1378.[584]

In 1394 members of Cracow's City Council went personally to Košice and, on 25 February, agreed to the following treaty known as a 'reciprocal pact' (pactum mutuum):

Clause I. Košice merchants have full rights to come with their goods through Cracow to all other countries.
Clause II. However, at the present time, no authorisation will be given for Košice merchants to use trade routes leading through Cracow to Prussia.
Clause III. Cracow merchants are allowed to bring their goods to Košice and unload them according to earlier rights and customs.
Clause IV. If either side does not wish to continue upholding this agreement, four months' notice must be previously given, so that each merchant can arrange to send his goods to, or from, his home.
Clause V. Both sides agree to help over the problem of debtors.[585]

Similar trade treaties probably existed with other important commercial centres, but the general rule seems to have been that the nearer and more important emporia, like Lwów, Wrocław, Toruń etc., conducted so many trade wars with Cracow that such agreements would rarely have been concluded. More distant cities fared better; for example, Nuremburg merchants in the fourteenth and fifteenth centuries made pacts for themselves and their countrymen to trade in a number of European states including Poland. These agreements were concerned mainly with the reduction and fixing of customs duties, relief from other commercial restraints, and promises of safe conduct. Moreover, her merchants got protection from reprisals, and overcharging which, although they appear of minor importance, provided considerable advantages in the market competition of the Later Middle Ages.[586] From Cracow's viewpoint, an agreement signed in 1365 between the Polish king, Kazimierz Wielki and Nuremburg, guaranteed the latter free trade as far as Lwów; from then on Poland, south of a line from Poznań to Warsaw, became the commercial domain of her trade for several centuries.[587]

Another interesting aspect of Cracow's commercial activity was the city's attitude towards foreign merchants, and curtailment of their retail selling opportunities. A body of rules was designed to cover two main fields of merchant activity; firstly, to prevent a foreign merchant, through collusion with a Cracow trader, benefiting from the various commercial privileges given the latter. Secondly, foreign merchants were subject to various burdens and requirements, e.g. the staple laws, which limited their commercial ventures. Nevertheless, merchants from abroad found ways of circumventing these restrictions. One of the more common methods was to obtain citizenship of some other Polish town, which entitled a merchant to privileges accorded to domestic traders; concurrently he could still utilise trade contacts with his homeland, and continue to own family property abroad.[588] Another popular practice for a foreign merchant was to join a local trade partnership and thus avoid all the nocuous rules on customs duty, stapling, selling limitations, etc. In Cracow, laws against this latter *modus operandi* were so openly flouted that stiffer financial penalties were introduced[589] and remained operative until the end of the Middle Ages.[590]

Cracow's city council was very rigorous and strict with regard to awarding foreigners citizenship (*civis Cracoviensis*), particularly merchants. This was quite understandable, for someone obtaining this right acquired the same freedom to trade as local merchants and was excused other commercial restrictions applicable to merchants from abroad. In order to obtain Cracow citizenship a person had to have real estate in the city, be married (bachelors were seen as unreliable, for nothing tied them to the city and they could easily leave the country taking their accumulated financial gains with them), present evidence to the council of being 'of good birth', give information on their place of origin, and offer a certain sum of money in payment.[591] As a result of such harsh restrictions Cracow inhabitants of foreign extraction had to remain for long periods guests within the city; many, like ordinary artisans, and artists in the service of the king, on a permanent basis.

Those inhabitants lucky enough to receive citizenship were duly recorded in the *Libris Iuris Civilis* which gives a picture of these privileged people for over three and a half centuries (mid-fifteenth to early eighteenth century). Although the number of foreign merchants gaining this honour was rather limited until the end of the Middle Ages, other supplementary archival

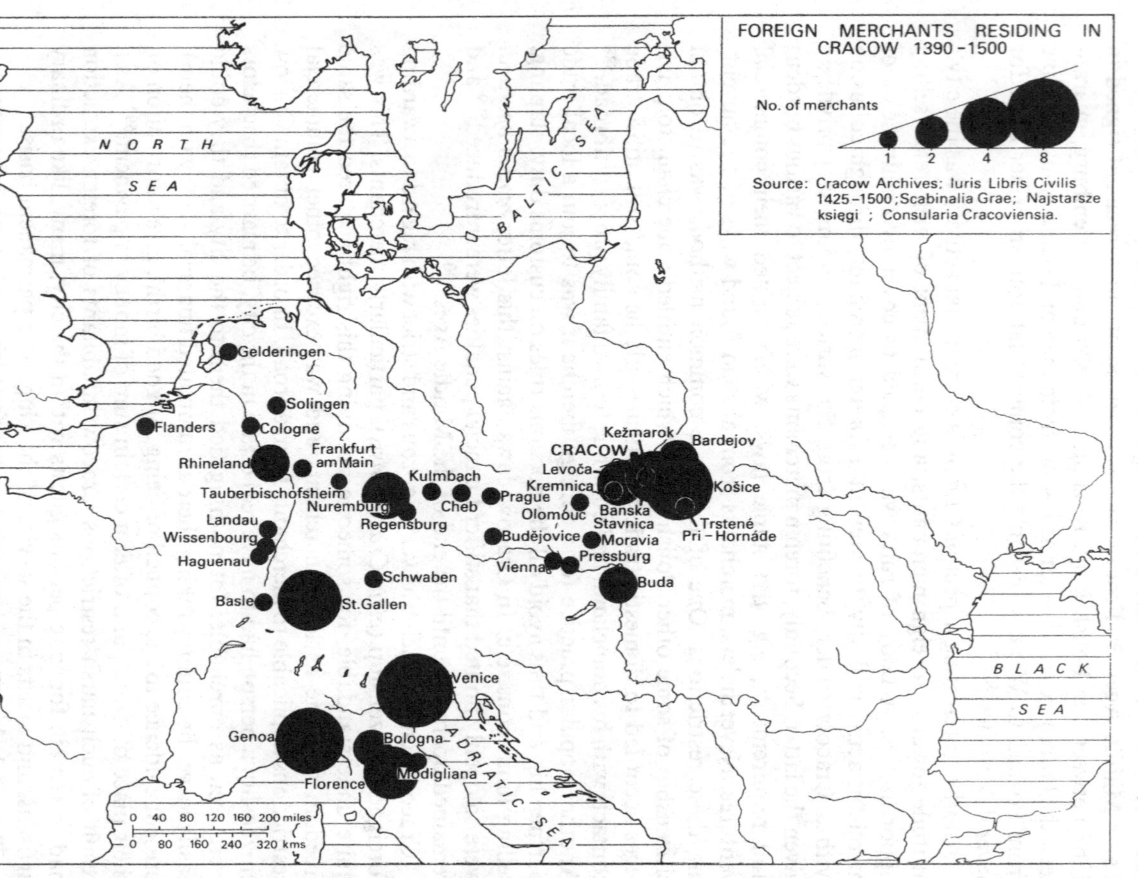
FOREIGN MERCHANTS RESIDING IN CRACOW 1390 -1500
No. of merchants
1
2
4
8
Source: Cracow Archives; Iuris Libris Civilis 1425-1500; Scabinalia Grae; Najstarsze księgi ; Consularia Cracoviensia.
NORTH SEA
BALTIC SEA
BLACK SEA
ADRIATIC SEA
Gelderingen
Solingen
Flanders
Cologne
Rhineland
Frankfurt am Main
Tauberbischofsheim
Kulmbach
Nuremburg
Cheb
Prague
Regensburg
Landau
Wissenbourg
Haguenau
Basle
St.Gallen
Schwaben
Budějovice
Vienna
Pressburg
Moravia
Olomouc
Kremnica
Banska Stavnica
Levoča
CRACOW
Kežmarok
Bardejov
Košice
Trstené Pri - Hornáde
Buda
Venice
Genoa
Bologna
Florence
Modigliana
0 40 80 120 160 200 miles
0 80 160 240 320 kms

45 Foreign merchants residing in Cracow, 1390–1500.

information on foreign traders does allow their place of origin to be plotted (Fig. 45). The commercial importance of any city can be measured by its power to attract foreign nationals to live within its walls. One can see, therefore, the magnetism of Cracow's emporium for many merchants from south and west Europe, particularly Italy and north-west Germany, as well as countries nearer at hand such as Hungary, Bohemia and Austria.

Conclusion

An attempt has been made here to utilise primary and secondary source material to interpret the commercial activity of one city, and its links with Europe and beyond. Documentary evidence from the fourteenth and fifteenth centuries has confirmed that the settlement below the Wawel Castle operated in trade connections over a wide area during the Later Middle Ages. Cracow proved to be an important link in the chain for transporting the products of underdeveloped regions like Slovakia and Moldavia, to the more developed areas of western and southern Europe; the significance of the Low Countries, especially Bruges, in this latter process was paramount. It is hoped that the preceding pages have illustrated how Cracow was a significant centre for men living from trade, the commodities they handled, the routes they followed, the markets served and the effective, but not over elaborate way business was conducted and organised.

Although, commercially, Cracow did not rank amongst the great medieval emporia like Venice, Nuremburg etc., it certainly deserved a position on the second rung of the European commercial urban hierarchy in the Middle Ages. The repeated recurrence of East Slovakia, Moldavia, Prussia and north Germany in the trade patterns of the city tend to suggest that Cracow's merchants were part of an effective large-scale organisation for inland commerce. Her traders were concerned with far more than just local traffic, whilst at a national level she was Poland's major business centre, as well as capital. The city was able to participate in the most profitable part of distant or international commerce which helped it to develop and prosper as a trading centre during the Later Middle Ages. After 1500 events, both political and commercial, quite beyond her control of sphere of influence, were to bring fundamental changes to the city's previously exalted position and to have lasting effects on her future development.

6

The political situation of Cracow, 1500–1795

There are many signs that the years around 1500 saw a new age dawning not only in Europe but also in other parts of the world. This date is still regarded as a convenient and satisfactory point at which to draw a line between the Later Middle Ages and the Early Modern Period of European history; the boundary remains in spite of a recent well-justified tendency to emphasise the essential unit of a period extending from the fifteenth to the seventeenth or eighteenth centuries.[1] The 'early modern' period is a term which refers to a process applied to an era when the modern Atlantic world emerged from the confined Europe of the Middle Ages dominated by tradition, agrarian life and superstitious practices; but it was a process that occurred at various times in different countries. In England it happened rapidly, but in eastern Europe the procedure was far from complete by 1800.[2] In the East, in Russia and, to a similar extent in Poland, east Prussia and Hungary, the agriculturally based society remained little disturbed by new political and economic developments, such that the traditional social order continued largely intact and scarcely challenged even at the end of the eighteenth century. For Cracow, as an important trade emporium, the years after 1500 were to prove a testing time; Polish commercial policy did not prove sagacious enough to meet these new conditions, and thus led to a decline of trade exchange and the general prosperity of the country's urban centres.

Changes in Western Europe

The sixteenth century dawned to new political situations in many parts of western Europe. The rise of the nation-state headed in many countries by an absolute monarch was unquestionably one of the salient features which dominated the history of the Early Modern Period. England under its Tudor monarchs rose in status and significance as a European power. France, in spite of many vicissitudes, steadily became more important under its Valois and Bourbon kings. Politically the most significant factor was the growth of the Spanish Empire; in 1400 Spain was still a divided land, but in the course of a century it became united under one ruler, obtained new dominions and wealth and was a force in Europe. From 1519 the House of Hapsburg ruled Spain, the Holy Roman Empire and the Netherlands and was on the point of consolidating its control over almost the whole of Italy. The political unity of the empire ruled by Charles V (1519–56) was to be a potent force for greater economic and financial integration.[3] Its political strength emerged from the fact that within its borders were concentrated most of the economically advanced centres of western Europe, further enhanced by the precious metals from Mexico and Peru channelled into the European economy through Spain.

The discovery of new ocean routes to the East and the American continent effected a reorientation in European commerce. By 1490, the Portuguese explorations of the west coast of Africa had passed the Cape of Good Hope, thanks to the efforts of Bartholomeu Dias who accidentally reached Mossel Bay in the Indian Ocean in 1487. Thus the way lay open to the trade of India, dazzling in its variety; in the early summer of 1498 Vasco da Gama made his epochal voyage ending at the great spice port of Calicut and a year later the first cargoes of pepper and spices reached Lisbon directly from India. This furthered the desire of the Portuguese to control the profitable trade in these commodities. Columbus' voyage in 1492, under the auspices of Ferdinand and Isabella of Spain, to the New World was a major event in world history. After many difficulties, he sighted Watling Island in the Bahamas, later reaching northern Cuba which he firmly believed to be the mainland of Asia. Before Columbus made his last voyage in 1502 others were already continuing the work of exploration, which was to prove the basis for colonisation and trade. For example, precisely in 1500 Pedro Alvares Cabral made the first landfall on the coast of Brazil.

These two events changed the geographical values of European lands. The Baltic and the Mediterranean lost their former centrality and supremacy in European trade. Moreover, the geopolitical roles within the Mediterranean were once more reversed. Spain and Constantinople, whose decay as centres of civilisation had permitted the rise of Italy in the eleventh century, again became seats of powerful states. After about 1500, the Italian city-states had increasingly to rely on local resources, and ceased to exert a dominant political influence on the European continent; however, in the definition of *bon goût* in art, architecture, music, manners and, to a considerable extent, religion, they remained supreme.[4] Nevertheless, in the new oceanic world those states which had more westerly situations, with sea-boards fronting the Atlantic and the North Sea, enjoyed geographical advantages in relation to the sea routes to the Indies and the Americas.

It was a fortunate circumstance for at least some of these states that the new opportunities came at a time of political consolidation; certainly in

Spain, Portugal and England the growth of royal power and national unity favoured economic advance. The Mediterranean and Baltic seas, although losing their importance, nevertheless continued to play an active part in European commerce. The former still distributed its own local products within and outside its shores, whilst the latter continued to supply western Europe with essential raw materials like timber and grain. Even so, relationships between the area bordering the North Sea and the Baltic were also undergoing fundamental changes in their trading pattern. The Hanseatic League had already passed the peak of its influence as the provider of a major link between the eastern Baltic and the north-west sea-board of Europe. Similarly, what affected the Hanse was also felt by Cracow for, although they were never close associates, these external events were experienced by them both. Industrial change in western Europe was an important factor in this situation. The Flemish cloth industry had suffered severely from the political conditions of the fifteenth century, while the increasing competition of English cloth took English merchants and ships into the Baltic; perhaps even more alarming inside the Baltic was the very real threat the Dutch made on Hanseatic supremacy. Unfortunately, the League's over-aggressive defence of its interests, especially in the Baltic, led to an exposure of its own lack of unity, and the Dutch became welcome visitors, both in Germany and the east Baltic.

Perhaps, finally, one should mention that the easy dominion of the high seas lured west Europeans, especially those living on or near the Atlantic coasts, to transoceanic adventures. A horde of explorers, missionaries, and merchants therefore continued to bring new commodities and ideas back for the attention of their less intrepid contemporaries throughout much of the sixteenth and seventeenth centuries. New wealth, information, techniques and concepts flooded into Europe as a result and the scale of European activities in the Atlantic, Indian and Pacific oceans continued to increase at a rapid rate. Even for Cracow, somewhat distant from all this activity, the sixteenth century in many other ways than the purely geographical, was to see a slow process of change. Absorption and adoption of commercial techniques, sometimes radically new in approach, together with the independent evolution of solutions for places and countries in similar economic circumstances, were to be a hallmark of the Early Modern Period. In turn, such changes were to spill out through Poland to the lands farther east, and their approval could only lead to an intensification of commercial competition for Cracow itself.

Events in eastern and south-eastern Europe

These events basically affected the political life of western Europe, but Cracow also had to take account of events in eastern and south-eastern

Europe, situated as she was on the borders between East and West. Moreover, through the Union of Mielnik (1501), Poland and Lithuania were once again united[5] after temporary separation following the death of Kazimierz IV (Jagiellończyk) in 1492. This union first under Aleksander (1501–6) and later ruled by Zygmunt I Stary (the Old) 1506–48, and Zygmunt August (1548–72)[6] has been called the 'Golden Age' of the Polish–Lithuanian Empire; geographically, it saw the extension of political boundaries far to the east and south. This political unit, referred to variously as the 'Polish Commonwealth' or the 'Commonwealth of the Gentry', stretched from the Baltic Sea eastward to Smolensk in the north, and to the Khanate of Crimea[7] around the Sea of Azov in the south; within it was contained a large part of the Dnieper valley and its tributaries (Fig. 46). Whilst Poland directly controlled the smaller eastern areas of Red Ruthenia, Chełm and Podolia, the Lithuanians administered Byelorussia, Black Ruthenia, Polensia in the northern part and Volhynia and the Ukraine in the south.[8]

During the second half of the seventeenth and early eighteenth centuries the Commonwealth began to show signs of decline. After 1572 the so-called 'Silver Age' suffered the vagaries of European wars,[9] such as the Thirty Years War of 1618–48 in spite of bright interludes of conquest exemplified by Stefan I Batory (Stephen Báthory), 1576–86.[10] The downward momentum continued with the 'Iron Era' (1648–97) under the reigns of Jan Kazimierz (1648–68) and Michał (Korybut) Wiśniowiecki (1669–73) with recurring struggles both along the northern and eastern borders; perhaps some equilibrium was restored with the exploits of Jan III Sobieski (1674–96) for his liberation of Vienna in 1683 has been interpreted as a crucial factor in the development of European history.[11] The eighteenth century saw Poland experience new waves of war and catastrophe during the reigns of the two Augustus – II, Mocny (the Strong) (1697–1733) and III (1733–63); finally, the reign of Stanisław II August Poniatowski (1764–95) – last king of Poland – was characterised by attempts at reform[12] but led eventually to the partition of the country, in which the former commonwealth was divided between the Russian, Prussian and Austrian Empires.[13]

Underlying many of the political events influencing the life of the Commonwealth was the growing power of two empires to the east and south, namely the Ottoman/Turkish[14] and Muscovy/Russian Empires.[15] Certainly Poland was seen as the last outpost of western civilisation during the Early Modern Period, holding the line against Islam and the Muscovite schismatics. Poland, like neighbouring Hungary, was seen by western Europe as the bulwark (antemurale) against the barbarian nations to the south and east. For Cracow, the victorious progress of the Ottoman power, manifested in the conquest of Constantinople by the Turks, was of greater political significance than events further east. Ottoman successes at Kaffa (1475), Kiliya and Akkerman (1484) effectively meant that Poland and, in particular,

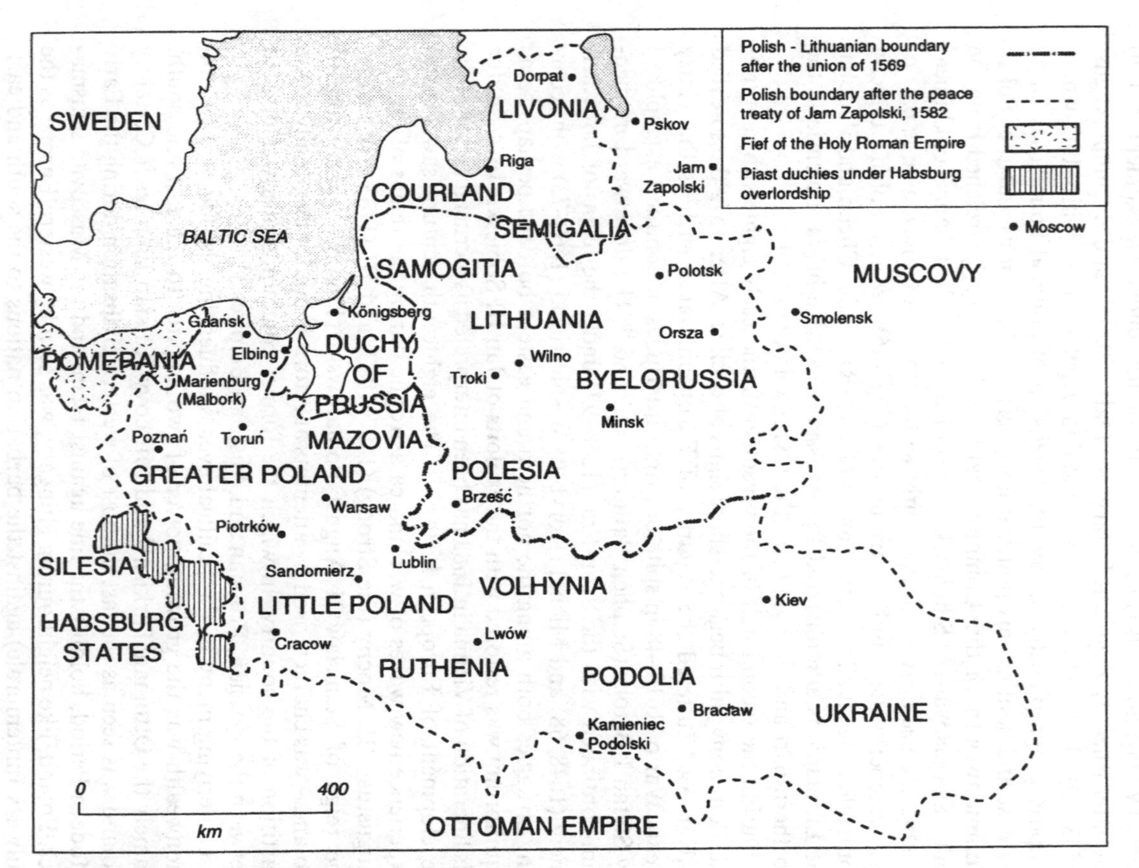

46 The Polish/Lithuanian Commonwealth, 1569.

Cracow and Lwów, was cut off entirely from trade with the Orient. The question therefore arose as to whether the city's merchants would have the sagacity to adapt to these changing political conditions. Before one can attempt to answer such a question there is need to outline the major factors behind Ottoman expansion in Europe.

Turkish policy was always founded more on expediency than a far-sighted diplomacy, whilst Turkish aggression was usually a response to external pressure or foreign alliances. Thus prolonged war was often astonishingly indecisive. Of the four themes which governed Turkish policy[16] it was probably opposition to the Hapsburgs in eastern Europe and, to a lesser extent, in the Mediterranean, that had greatest significance for Cracow. The Ottoman Sultan could be relied upon to harass the Hapsburg army in Hungary and the Hapsburg fleet in the Mediterranean. Even so the large-scale military and political operations on which the Ottoman empire embarked after the fifteenth century were made possible by the development of commercial and economic life in the empire, and by the increase in state revenue that followed. In fact from Cracow's viewpoint, the political order established by the Ottomans produced conditions of safety, provided a link between remote areas and brought about an economic integration of the region as a whole. Foreign merchants from non-Muslim lands such as Cracow were allowed to trade under treaties of capitulation (in Turkish, *amān-nāme*, literally document of mercy) provided they paid a 5 per cent customs tariff on goods.[17] During peaceful periods, therefore, trade flourished; in Poland a strong political element consisting of the gentry and those magnates opposed to Hapsburg absolutism, realised the importance of diverting Poland from a dangerous war with Turkey in the interests of Vienna. Preservation of peace along the country's southern borders was essential for commercial prosperity; that peace was broken only once during the reign of the last Jagiellonian kings (i.e. up to 1572) when Ottoman troops, with a strong Tartar contingent, advanced all the way through Little Poland in 1524, under the leadership of their Sultan (Sulaimān I the Magnificent, 1520–66) but Cracow seems to have avoided the worst of this onslaught. Poland found it expedient to conclude an agreement of 'eternal peace' which was signed with Turkey in 1533.[18]

Unfortunately, this agreement did not prevent the Muslims from threatening Poland's southern boundaries from bases in Hungary – a threat that was to last for about 140 years. In fact 1526 was a momentous year for Christendom. The Turkish Wars with Hungary had been going on intermittently for many years, with first one side gaining the advantage, then the other, but no decisive operations had taken place recently. In 1526 Sulaimān again invaded Hungary and, in the battle of Mohács, the Hungarians were totally defeated; 20,000 soldiers including the King were killed.[19]

Poland remained aloof from any anti-Ottoman league throughout much of the sixteenth century, considering such a course to be the best means of preserving her own position between the Hapsburg realm and Muscovy. This basic tenet of Poland's policy to avoid direct military engagement with the preponderant military power of the Ottoman empire was severely tested during the last years of the century. Pro-Ottoman Tartar forays into Poland had repeatedly strained relations between the two countries; Turkish incursions into Moldavia, Wallachia and Hungary,[20] critical areas for Cracow's trade-route connections with the Balkans, were warily watched by the Polish rulers. Similarly, the Sultan became increasingly aware of the damage caused by Cossack marauding parties from the south-eastern part of the Commonwealth, who ventured into the Ottoman dominions of the Black Sea coast in search of loot, in the form of booty from Turkish galleys.[21] Poland's political brinkmanship and readiness for war with the Ottomans over such events was often severely tested, but in some way open conflict was averted.

During the early years of the seventeenth century relations between Poland and Turkey deteriorated. Poland's policy of neutrality finally broke down when Zygmunt III Waza (1587–1632) supported ideas of eastward expansion for the Commonwealth, combined with continued Cossack raids on Black Sea ports like Varna. The former agreement of 'eternal peace' was revoked by the bellicose Sultan Osmān II (1618–22) who embarked upon war with Poland; in 1621 negotiations for an armistice were quickly concluded and the ensuing peace treaty restored the political boundaries present during Jagiellonian times. The frontier followed the Dniester river, whilst Chocim, an important trade centre for Cracow merchants, was to remain under Turkish control, both sides agreeing to respect each other's territory; in practice, the treaty did nothing to prevent continued incursions of Tartars or Cossacks into opposing lands.[22]

After a lapse of forty years the peace of Poland's southern borders was broken once more with an Ottoman invasion bent on reducing the whole of Poland to Ottoman vassalage. In 1672, Sultan Mehmed IV (the Hunter, 1648–87) sent a large army commanded by grand vizier Köprülü Fazil Ahmed Pasha into Poland, conquered the fortress and trade centre of Kamieniec Podolski, and advanced as far as Lwów; besides disrupting all commercial contacts through this area, in 1672–73 the Poles lost Podolia and two-thirds of the Ukraine. Such ignominy was not accepted lightly and the following year Jan Sobieski reversed these encroachments, which gave him the accolade 'vanquisher of the Turks'.[23] In spite of further success against the Ottomans in 1675–76 he was unable to regain the south-eastern regions of the republic; ironically Turkish claims to the Ukraine were not maintained, the area eventually becoming part of the Russian Empire in 1681 after the Turkish army had left it in ruins. Although the right bank of the Dnieper and Podolia were returned to Poland, and the Turks withdrew from the contested area, the foundations were being laid for Russia as a new great European power.[24]

The eighteenth century saw the gradual withdrawal of Ottoman influence in Europe, whilst Austrian troops little by little regained Hungary for Christendom. For Poland, and Cracow in particular, it was the growing might of their eastern neighbour, Russia that was to have increasing influence on political events after 1500. At the beginning of the sixteenth century the united Polish–Lithuanian monarchy seemed to be entering a period of growing wealth and steady progress, having firmly asserted the union against all foreign foes. However, the Muscovite Empire was growing irresistibly in strength under its determined, absolute rulers. Even so, as İnalcik has observed, until the end of the first quarter of the sixteenth century, the Ottoman Porte looked on Muscovy as an element in the balance of power in the north and considered Moscow as its natural political ally against Lithuania–Poland.[25] By the middle decades of that century, however, circumstances had changed and it became clear that an impending conflict was looming between the Ottoman and Muscovite states. At stake was hegemony over the vast territories into which the former empire of the Golden Horde had disbanded, including the commercially viable area of the Crimea.

Early warning signs came at the beginning of the period under review, for Tsar Ivan III (the Great, 1440–1505) attacked Lithuania (1550–3) and, with the aid of Crimean Tartars, captured all the principalities on the Oka and Desna rivers as far as the middle Dnieper River. Russian advances were not curtailed during King Zygmunt I's reign,[26] either through sacrifice of the territory of Smolensk (1514), or with successes inflicted upon the Russians by the united Polish–Lithuanian power in military conflicts at Orsza in 1514;[27] further encroachments continued during the reign of Zygmunt Augustus, last of the Polish Jagiellonians. For twenty-four years Tsar Ivan IV (the Terrible, 1530–84) had led a war of conquest in order to obtain the remaining lands belonging to the Teutonic Order in Livonia between 1558 and 1582, and was partially successful. Such aggression prompted Zygmunt Augustus to enter an agreement with the last Grand Master of the Livonian Order, Gotthard von Kettler; this treaty, known as the Union of Vilna, 1561,[28] directly integrated Livonia into Poland while Courland (a secular duchy) was subjected to Polish authority. In retaliation and suitably antagonised, the Tsar during the 1560s invaded Poland and conquered lands around Połock on the Dvina River. Pressurised by Russian threats and the fact that there was no heir apparent to the Polish throne, the government (Sejm), combined all the former Crown Lands (Lithuania, Royal Prussia and Livonia) to create a unified and indivisible republic – the Union of Lublin (1569).[29] This, in turn, generated a feeling of greater unification in the Commonwealth; the Polish population began to integrate the nobility, the burghers and a large part of the peasantry in the Lithuanian–Ruthenian regions to the east. Migration, particularly between 1540 and 1580, from the heavily populated region of Mazovia led to resettlement in the largely forested area of west Lithuania, especially in Podlasie; noticeably by 1580 there were over 150,000 Jews among the numerous nationalities of the Polish Commonwealth, more than anywhere in Europe, and the largest community in the diaspora.[30]

For Polish Jewry the sixteenth century was a golden age, and its numbers may have risen from about 50,000 to 500,000 between 1500 and 1650.[31] Unfortunately, many of its professions, of moneylender, tax-farmer, estate-manager, were not such as to endear it to the common man, any more than in old Spain; while, as there, its presence hindered the rise of a native middle class.[32] Christian burghers in the Polish cities and some of the Polish nobility resented Jewish competition; in Cracow, the adjacent suburb of Kazimierz had a Jewish community which was one of the most important in Poland. Nevertheless, during the sixteenth century, Jews moved from western Poland to the east and south, for economic competition and anti-Jewish feeling in the older cities of Poland was offset by the opening up of large tracts of land owned by the Polish nobility in the south-east.

Such settlement was partially encouraged by the successes of Stephen Batory against the Russians, a constant threat to the stability of the Commonwealth's eastern lands. His defeat of Tsar Ivan IV in three brilliant campaigns between 1578 and 1581 forced the Russians to return the occupied territories of Livonia and Połock by 1582. This triumph retarded for more than a century Muscovite hopes for an outlet to the Baltic, and encouraged Batory to devise plans to conquer Moscow itself and create an immense Christian empire in eastern Europe, with the ultimate aim of driving the Turks from Constantinople; such limitless personal ambitions were never upheld by the Polish nobility, even before Batory's untimely death in 1586.[33]

It was not until the middle decades of the seventeenth century that serious threats of Russian invasion again worried the Polish state. The reign of the last Vasa king in Poland, Jan Kazimierz (1648–68) is remembered in his country's history as one of bloody wars and political disasters. The epoch began with a rebellion of Ukrainian Cossacks; they heavily defeated the Polish army in 1648 and 1651 and in the compact of Pereyaslavl, 1654, recognised the authority of the Russian ruler Tsar Alexis Mikhaylovich (1629–76). Flushed with success Tsar Alexis, supported by Cossack troops, attacked Poland, conquering in 1654–55 almost the entire eastern part of the country, as far as the important emporium of Lwów. Further encroachments were averted due to the military success of the Swedes who, in the summer of 1655, invaded the Commonwealth; Sweden took advantage of Poland's predicament and marched into the country, receiving en route some support from local aristocrats. Russia, fearing the rise of Swedish power, withdrew from Poland; concessions were made to the Poles in a peace treaty, among which was an agreement to form a joint campaign against Sweden[34] – such are the vagaries of war.

The eighteenth century saw two major events in Polish/Russian political relations. The whole century was characterised by foreign powers interfering in the internal affairs of Poland, often encouraged by the Poles themselves. The country's armed forces appeared unable to suppress a Cossack uprising in 1702; appeals were made to Russian troops to quell the disturbance which was effectively achieved.[35] Such favours were not easily forgotten by Russia, further enhanced by Tsar Peter I's (the Great, 1672–1725) aid in securing for the king (Augustus II Mocny, 1697–1733) his position on the Polish throne. Attempts by Prussia to acquire territory along the Baltic coast in 1715 met with some success until the intervention of Peter the Great. Negotiations between Poland and Prussia proceeded with difficulty, until Peter, who resided in Gdańsk at the time, frustrated further Prussian advances by ordering 18,000 Russian troops into Poland in 1716, and who stayed on effectively to control the southern Baltic coast. By the treaty of Warsaw (1717)[36] Peter the Great thus assumed the role of protector of the territorial integrity of the Polish state; more profoundly this event meant that from then on Poland was to all intents and purposes a protectorate of the Russian government – a relationship that was to last until the dissolution of Poland in 1795.

The second event concerned the intervention of Catherine the Great (1729–96) of Russia, in Poland's internal affairs. Like one of his predecessors, Augustus II Mocny, the last king of Poland, Stanisław II August Poniatowski (1764–95) gained the Polish throne with the aid of 14,000 Russian troops, and financial support from Catherine;[37] such an act tended to uphold the view that Poland's internal affairs were an exclusive Russian preserve. Nevertheless, introduction of political reforms by the new Polish ruler appeared to arouse the displeasure of the Empress of Russia, and his plans never bore fruition. Religious turmoil in Poland also enabled Catherine to meddle in the discord surrounding Poland's internal problems.

The period between 1716 and 1768 had by Polish standards, been a fairly peaceful era in which the country's population had risen to nearly eleven and a half million. Part of this growth resulted from steady immigration both from the German lands to the west, and from Russia, mostly composed of refugees escaping from the system of enslaved peasantry. By 1770, Poland's population contained not only a large Jewish element scattered throughout the country[38] (more than half the world's Jews lived here at that time), but about a million dissidents, half Protestant, half Orthodox. The predominantly Roman Catholic population had witnessed laws passed in 1717 and 1733 which restricted the secular and spiritual rights of the dissidents i.e. non-Catholic inhabitants; opposition to this situation by Catherine led to a repeal of these laws, only to be met by a Catholic uprising (with Turkish support).[39] Four years of civil war (1768–72) eventually saw Russian troops again act as the decisive factor in Poland's internal affairs, with the suppression of Catholic ambitions. Unfortunately, this event indirectly led to Prussian demands for the partition of Poland, the first of which occurred in 1772.[40]

Although Cracow itself was not directly involved in the maelstrom of Poland's political affairs throughout much of the period under review, it has been necessary to outline the major events generated by the Russian and Ottoman Empires, in order to obtain a better understanding of the vicissitudes of trade experienced with these regions during this time. Even with this political backdrop however, the picture is not yet complete. This can only be achieved when the component parts created by the sweeping changes of the European Wars of the seventeenth century, are fitted into the overall scene.

Cracow and the European Wars

Religious intolerance had been a feature unknown to Polish life until the early years of the seventeenth century.[41] It made its appearance during the reign of King Stefan Batory's successor, Sigismund Wasa, heir to the Swedish throne, who also became King of Poland. The Catholic church provided Poland with a sense of national unity which the divided Protestant faith could not, against enemies of other religious beliefs. As Batory died without issue in 1586, the election of Sigismund seemed to favour the preservation of this Catholic tradition; he was a nephew of the last Jagiellonian king and brought up as a Catholic. The king-maker role had been played by Jan Zamojski, Grand Hetman of Poland, who favoured, along with his gentry party, the Swedish heir to the Polish throne. Together, they saw Sweden as a possible useful ally in the Baltic – that one goal of Polish Imperial policy.[42]

A rival faction supported Maximilian of Austria, but the House of Hapsburg was seen by Zamojski as a mortal danger to the future of Poland. Nevertheless, Maximilian's candidature was predictable, for the Hapsburgs saw the rise of the gentry in Poland as a warning to themselves, a spur to their efforts to make sure of their power at home. To the Hapsburg nobility it was an incitement to thwart their own gentry and, by winning the Polish throne they might be able to suppress such developments, as well as enhance their position in Europe. Zamojski's one political creed was to keep Poland free from Austrian influence, and the succession to Batory led to passions so extreme, that it was necessary to have recourse to arms.[43]

Maximilian was encouraged by his Polish associates, particularly the Zborowski noble family, to enter Poland quickly with his armed followers, which he did in late September 1587.[44] By mid-October he had reached Cracow along with about 8,000 supporters.[45] Zamojski, in the name of the new king, fortified the city, which lay under siege for the first time during the Early Modern Period. Maximilian delayed his attack on Cracow in the hope that his actions would impress the Polish nation; he summoned a meeting of the Polish nobility at the nearby town of Mogiła, but attendance was poor.[46] The Archduke, therefore, again resorted to force in an attempt to achieve his objective, and assailed the walls of the city, only to be repulsed with heavy losses. Maximilian retired to Silesia in December to await reinforcements for a march on Greater Poland; before this could be organised, Zamojski attacked his position at Byczyna in Silesia in January 1588. The Archduke suffered a crushing defeat, was captured and imprisoned for a prolonged period in Zamość and, in the ensuing Treaty of Będzin[47] between the Polish Grand Hetman and the Austrian Emperor, Rudolph II, all Hapsburg claims to the Polish throne were renounced. Cracow's inhabitants could once more return to the peace of their everyday lives, whilst the chance that the Hapsburgs might become rulers of the Polish–Lithuanian realms passed for ever.[48]

The early years of the seventeenth century proved troublesome times for the Swedish king of Poland. His plan to limit the powers of Parliament with the help of Austria and to suppress Protestantism by force led to civil war. Nicholas Zebrzydowski, Palatinate of Cracow, and successor to Jan Zamojski, gathered around him a group of dissidents, who objected to the king's plans.[49] In 1607, the malcontents under Zebrzydowski went into open rebellion at Sandomierz, but were decisively beaten by the royal army. The failure of the rebellion meant the Polish Constitution remained unchanged, so that Poland did not pass through a period of absolutist reaction, unlike England under the Tudors and several continental states during the Early Modern Period.[50]

The reign of Sigismund survived this national calamity, but the wars with Sweden proved more long-lasting. Sweden objected to Sigismund's claim to the Swedish throne made at the beginning of the century, which entangled Poland in a dispute which lasted for thirty-five years and continued more or less uninterruptedly for over half a century under his two successors from the same House of Vasa. These wars were intended to settle the issues whether Sweden was to gain a foothold on the southern shore of the Baltic, and so convert this sea into a Swedish lake or, conversely Poland was to extend her influence over the Baltic coast to the north-east, and thus strengthen her position as a maritime power in that part of Scandinavia. Poland's early successes included control of Latvia, but the territory of Estonia remained inconclusively decided.

Poland's international situation entered a difficult phase early in the third decade of the century. Although Sigismund's policy during the Thirty Years

War (1618–1648) was to remain neutral at all costs, it brought the country little profit. Poland's involvement in suppressing a massive Turkish invasion along her more southerly borders allowed the Swedes under Gustav II Aldolphus (1594–1632) to occupy Livonia and capture Riga, along with other important Baltic ports belonging to Prussia, confirmed in the Peace of Altmark.[51] Moreover, the hoped for reunion of Silesia with Poland as a result of the latter's neutrality did not materialize; even worse Poland's distinctly pro-Austrian and anti-Protestant neutrality had not prevented a conflict with that champion of Protestantism, Gustav II Aldolphus so that by the end of Sigismund's reign in 1632,[52] the whole of Poland's Baltic shore was still at stake. In fact the gradual process of isolating Poland from its vital outlet to the sea had irrevocably begun. For Cracow, distanced from these events, the effects were not so rapidly felt, but conflicts along Poland's southern and northern borders were certain to curtail the free flow of commerce on which the city was so dependent.

A far more direct influence on the city came with the invasion of the Polish and Lithuanian lands by Charles X Gustav (1622–60) of Sweden, who launched the War of the North in 1655. Poland's involvement in that year with trouble in its eastern lands, including continued exposure to Cossack intrusions, undermined the country's ability to meet challenges elsewhere, especially in the Baltic. The Swedish king was made aware of Poland's vulnerability by a Polish magnate Hieronim Radziejowski[53] who, after a private quarrel with Poland's ruler Jan Kazimierz (1648–68), fled to Stockholm and advised Charles X Gustav to attack a weakened Poland. The treacherous attitude of other magnates in Greater Poland enabled the Swedes, helped by their veteran soldiers from the Thirty Years War, to occupy, in August 1655, large parts of Poland[54] with little local resistance. The Swedish ruler, having therefore already obtained Polish Prussia and Pomerania as part of his plans for a *dominium maris Baltici*, extended his ambitions, with the aid of 30,000 troops, to control the whole of Poland. By September 9 he had captured Warsaw;[55] he marched southward and, on 25, the siege of Cracow began.[56]

Experience had warned the city's inhabitants always to be prepared for war;[57] nevertheless, fires and pestilence, especially during 1651–52,[58] had reduced Cracow's population and, together with those who fled to the highlands and forests[59] left about 5,000 citizens to face the Swedish force. The overwhelming odds of 12,000 Swedish soldiers quickly led to capitulation and Cracow faced a dark period in her history. Amongst the many impositions placed on the city was a compulsory system of taxes on all merchandise brought into its precincts,[60] which discouraged even local trade. The oppressive conditions continued until 1657. In that year Jerzy Rákóczi, ruler of Transylvania, Moldavia and Hungary, was approached by the Swedes for help;[61] 40,500 of Rákóczi's troops marched across the Tatra

Mountains and helped occupy Cracow for Charles; it was not until 30 August that King Jan Kazimierz at the head of Austrian and Polish troops finally forced all occupying forces to leave the city.[62] Not only Cracow but also the surrounding villages and small towns presented a sad sight;[63] many lay in ruins, razed to the ground by the occupying forces. Those settlements still standing contained few inhabitants, whilst Cracow's walls suffered severe damage; inside the city the churches were vandalised, houses damaged by cannon artillery, and even the Wawel had, in parts, been destroyed.[64]

Cracow's fears of further occupation were not over. On the death of the Polish king Jan III Sobieski (1674–96), several candidates presented themselves for the vacant Polish throne. The successful competitor was Augustus II Mocny (the Strong), 1697–1733, under whose reign Poland experienced a new wave of wars and catastrophies.[65] His alliances with Peter I the Great of Russia and King Frederick IV of Denmark against Sweden led to the Northern War of 1700–21. The Swedish king, Charles XII (1697–1718) repelled Denmark and Russia in 1700 and marched from Livonia to Poland in 1701–2.[66] The Swedes occupied Warsaw in May 1702[67] and, as in 1655, once more marched on Cracow, which they captured without a shot being fired after the battle of Kliszów in July of that year.[68] The second Swedish invasion proved even more disastrous than its predecessor; Wawel Castle was burnt down, but it was the city's inhabitants who suffered most from the constant presence of foreign troops.[69] Charles XII squeezed over 60,000 thalers from Cracow in retributions, causing severe poverty amongst the local population. Merchants and craft guilds were even forced to relinquish their precious insignias.[70] If this was not enough, in 1703 a strong wind swept through the city causing considerable damage.[71] In 1705, the Swedes under General Stromberg billeted over a thousand soldiers in the city,[72] and again in 1709 it was subjected to heavy Swedish harassment and control.[73] In 1711 Russian units entered the city and occupied the Wawel castle for the next five years.[74] The Swedes may have departed never to return, but further occupation of the city was to take place during the 1730s.

Russia's relations with Poland became increasingly important as the eighteenth century progressed. For Russia, Poland was the essential land link with the West, and Russian influence became a predominant feature of Polish history, especially after the battle of Poltava in 1709.[75] Its growth was further illustrated by the War of the Polish Succession (1733–35);[76] during these years Russia, with Austrian support, imposed on the Polish nobility a king (Augustus III, Elector of Saxony) whom the majority obviously did not want, at the expense of another claimant (Stanisław Leszczyński, father-in-law of Louis XV of France), whom most of the nobles were prepared to accept. Stanisław Leszczyński actually travelled from France to Warsaw, where the Polish Diet proclaimed him king in September 1733; Austria and Russia, however, feared this might be followed by French interference in

Poland and decided to support the eventual ruler August III (1733–63).[77] The noble families remained divided on the possible methods of saving their country; the Czartoryski family supported ideas of internal regeneration, while the Potockis aimed to shake off the foreign yoke by armed force.[78] In addition, there was a strong party among the Polish nobles who were jealous of the influence and power of such families as the Czartoryskis, and particularly the latter's ideas on parliamentary reform. Such internal disagreements eventually led the Czartoryskis to actually extend their hands to Russia for help;[79] perhaps unwittingly the first step had been taken towards the situation which ended the existence of Poland as a nation.[80]

Against this political background, Cracow remained a detached observer for much of the Polish Succession War, as the main theatre of operations was located in the northern part of the country.[81] However, in 1735 Russian troops under General Lascy were quartered in the city[82] during their march across Poland to support August III in defence of the country from possible attack by France and Stanisław Leszczyński's supporters. In fact, by the thirties of the eighteenth century, the stationing of large Russian military forces in Poland had become a permanent institution; Russian troops marched through the country on their expeditions into Germany and Turkey;[83] Cracow's strategic situation on an important routeway proved an obvious halting place; in 1749, 30,000 Russian troops marched through Cracow on their way to Silesia.[84] Moreover, not only was the city subject to a deluge of foreign soldiers moving through her precincts with all its disrupting circumstances but, in 1736, Cracow suffered an inundation of a more natural kind; in that year the R. Vistula overflowed its banks[85] causing considerable damage. It is not surprising therefore, that, by the middle of the eighteenth century, Cracow showed signs of stress and decline; its inhabitants numbered barely 20,000,[86] whilst commerce, craftsmenship, learning and the arts were virtually at a standstill.

The future was to prove little better for the city; Cracow found herself in a country whose army was ridiculously small in proportion to its real size and the length of its exposed land frontiers, and was poorly equipped and trained. During the Seven Years War (1756–63),[87] Poland was made into a thoroughfare and even a battlefield by the various combatant foreign armies. Between 1757 and 1763 much of the country was occupied and plundered by Russian troops en route to the battlegrounds of Brandenburg and Pomerania, which left Poland exposed to retaliatory raids by the forces of Frederick II (the Great) of Prussia.[88] Cracow was not to escape these calamities. Whilst the territories of Gdańsk, Toruń and Poznań were used as operational bases by the Russian army against Prussia, Cracow was utilised by Russia's allies, Austria, in the struggle with Prussia for control of Silesia. In 1759, General Laudon marched his Austrian troops through the city[89] during one of the Silesian campaigns, whilst Russian troops were also quartered there for a short while in that year en route from the battle grounds of the north.[90]

The Seven Years War continued for another four years when, in 1763, the way to peace finally materialised with the Treaty of Hubertusburg.[91] The European wars, whether based on religious intolerance, or territorial greed, had proved disastrous for Poland and her cities in particular. Cracow, the ancient capital of the country, had degenerated into a dull provincial town by the end of August III's reign in 1763. His successor King Stanisław II August Poniatowski (1764–95) was to rule during a period of reform and genuine national regeneration, in which Cracow was to see a partial revival of her former glory. Unfortunately, it was to end with the partitions of the country and ultimate disappearance of Poland from the map of Europe – a catastrophe which could not have been averted. Cracow, however, had already experienced her own demise as capital of Poland; this came with the transfer of this position to Warsaw, early in the seventeenth century.

The capital moves to Warsaw

There has been some disagreement in Western geographical literature as to the exact date when the capital of Poland was transferred from Cracow to Warsaw. Dickinson states it was 'after 1595, when Warsaw became the seat of kings, who attracted courts, Parliament and nobility in their train';[92] Osborne maintains that 'Kraków succeeded Poznań as capital ... until Warsaw became capital at the end of the 16th century'.[93] Mellor has written, 'In 1596 Zygmunt III had moved the capital from Kraków to Warsaw because the latter was geographically more central for the state at that date.'[94] Davies points out that 'By the 17th century the rise of oceanic commerce and the dwindling of the Asiatic overland trade had crippled the commercial importance of the city [Cracow]. Warsaw became the political capital.'[95] Various encyclopedias have also given different dates; for example, a French source gives 1596 as the date of transfer so that the official capital would be nearer the Baltic Sea.[96] An Italian citation prefers 1595[97] and a Spanish one gives two dates – 18 March 1595 under the section on Warsaw,[98] and 1610 for Cracow.[99] Similarly, an American source describes how Warsaw became capital after a fire in Cracow (i.e. 1595),[100] but that Cracow remained capital until 1609 when the king's court was transferred to Warsaw.[101] A major German encyclopedia also has two contradictory dates, one stating 1596[102] and the other 1609;[103] a Swedish reference declares that although the Polish capital and king's court was in Warsaw in the sixteenth century,[104] another tome gives Cracow as capital until 1610.[105] The *Encyclopedia Britannica* categorically states that on 16 March 1596 the Polish king decided to have his capital and court in Warsaw.[106]

The Soviet account appears more accurate. Under the heading 'Cracow', it maintains that at the end of the sixteenth century, as a result of the union between Poland and Lithuania, Warsaw was a more centralised location for capital, to which the king brought his court;[107] under 'Warsaw' it asserts that in 1596 the official residence of the king was moved from Cracow to Warsaw.[108] Even Polish publications have this event taking place between 1595 and 1611.[109] The Ukrainian encyclopedia states that Cracow was Poland's capital up to 1609,[110] but the section on 'Warsaw' states that with the joining of Mazovia to Poland in 1550, it became capital of the Polish state.[111] An Hungarian version puts forward 1610 as the year the capital moved from Cracow to Warsaw,[112] but when describing the latter city in another volume, 1573 is given as the year it took the mantle of Polish capital.[113] A Romanian source affirms the date as 1603 under 'Cracow',[114] but for 'Warsaw' gives 1587 as the year it became capital city.[115] Clearly there is a need for a greater precision in determining what actually happened.

It is more than possible that the movement of Poland's capital from Cracow to Warsaw was part of a long-drawn-out process. No specific date can be attributed to this event. No contemporary chronicler from Sigismund III's reign is known to have recorded the move.[116] Some nineteenth-century Polish scholars believed the transference of the king and his court to Warsaw in 1596 signalled the permanent movement of the capital from Cracow,[117] but evidence is not conclusive. It appears the king decided to leave Cracow for Warsaw to attend a sitting of the Parliament (Sejm) and it was necessary for the court to accompany him there; also a disastrous fire in the Wawel Castle in 1595[118] must have provided difficulties of accommodation, whilst in Sweden contemporary manoeuvring for a successor to the throne demanded the king's closer connections with parties supporting his candidature.[119] A document dated 1598, in which the king gave the city of Warsaw increased privileges, particularly for trading purposes, has also been interpreted as supporting the view that Warsaw was the new capital, but again the argument is unconvincing.[120] Both 1596 and 1598 therefore lack the necessary credibility for positive proof.

The year 1609 suggests a more probable transfer date. It has been noted that this was the last time the king was in Cracow.[121] He had resided in the restored Wawel Castle between 1607 and 1609, only making journeys to Warsaw for parliamentary reasons. However, on 28 May 1609 he left Cracow for Lithuania. The phenomenal rise of the Muscovite State in recent years and the ambitions of their Grand Princes to rule over all the Russian lands, began to provide a direct threat to Polish supremacy in the East. In February 1609 the Muscovite army had signed an alliance with Charles IX of Sweden[122] which precipitated Sigismund's departure from Cracow; he was never to return, an event prophetically noted by a contemporary Jesuit chronicler.[123] In September 1609, the king advanced with his troops to Smolensk and its strongly garrisoned fortress; failure to capture it outright led the Polish forces to settle down to a two-year siege.[124] By mid-June 1611 Sigismund had forced the garrison to surrender and he returned in triumph to Warsaw, where he took up residence in the recently restored Royal Castle.[125]

In spite of such compelling facts the actual transfer of the capital to Warsaw is, according to Małecki, an 'open question'.[126] Certainly, no official act for the transfer was ever passed; also if Cracow lost its place as the king's residence in 1609, but he did not live in Warsaw until 1611, this hardly implies that for two years Poland was without a capital. It is also interesting that parliamentary sessions throughout the seventeenth and eighteenth centuries still referred to the 'capital city of Cracow'.[127] Even as late as 1789 Parliament alluded to Cracow, 'as capital of the Polish kingdom';[128] moreover, in 1792 Warsaw was still never mentioned in parliamentary debates as the capital, only the city containing the king's residence.[129] In 1793, Cracow, for the first time, lost its title of capital city[130] and ranked second to Warsaw, but above Lublin, in official pronouncements by the state.[131] Perhaps another indicator of change concerned the coronation ceremony. Up to 1637 this had always been held in Cracow, but in that year the coronation of Queen Cecilia Renata took place in Warsaw.[132] Others were held there in 1670[133] and 1764,[134] but each time the coronation regalia were sent from Cracow, and returned after the ceremony.

Much of this discussion really depends on what is understood by the term 'capital'. In the present-day context it is usually synonymous with location of the seat of government,[135] but in Poland's feudal society this definition did not apply. Warsaw was only seen as the senatorial residence of the king and his immediate advisors; after 1717 this august body obtained a larger membership, but not until the second half of the eighteenth century was government, in the sense of a more modern political instrument, applicable to the Polish state. This came with the creation of new administrative machinery, central ministries and other trappings associated with changes in the political system.

All these above facts tend to suggest that the transfer of power was a gradual evolutionary process, and the loss of Cracow's character as Poland's capital city only emerged slowly over time. Admittedly, one may isolate critical dates in the chronological process, which hindsight has shown were of significance in this development (Table 4).

Other factors may also be proposed that could have contributed to the locational change of the capital city. The expansion of the Polish state and accompanying boundary changes in the fifteenth and sixteenth centuries meant that Cracow was becoming increasingly peripheral in relation to the centre of the country. This was clearly demonstrated in 1569 following the union of Poland and Lithuania;[136] Cracow suddenly found herself tucked away in the south-western corner of the state, whilst Warsaw unexpectedly

Table 4. *Important dates in the loss of Cracow's capital status*

Year	Event
1550	Last time the ceremony of feudal homage held in Cracow.
1559	King Sigismund August permanently left his residence in Wawel Castle.
1569	Warsaw chosen for holding future Parliamentary (Sejm) meetings.
1573	Warsaw assigned as electing place for future Polish kings.
1578	The establishment of the Coronation Court in Piotrków and Lublin.
1603	The last Parliament (Sejm) to be held in Cracow.
1609	King Sigismund III leaves Cracow for the last time.
1637	First coronation outside Cracow.
1697	Final performance of the Coronation Parliament in Cracow.
1734	The last coronation (August III) in Cracow.
1734	The last king to be buried in Wawel Castle.
1734	The final Coronation Parliament planned to be held in Cracow (did not take place).
1765	The Coronation Archives transferred to Warsaw.
1787	The last visit of a Polish king (Stanisław August Poniatowski) to Cracow.
1795	Coronation treasures stolen by Prussian army.

Source: J. M. Małecki, 'Kiedy i dlaczego Kraków przestał być stolicą Polski', *Rocznik Krakowski*, Vol. XLIV, Kraków, 1973, p. 34.

obtained a more centralised location. Indicators demanding a more northerly political capital were also appearing; growth in the importance of the Baltic Sea, Poland's annexation of Latvia, the increasing danger of war with the growing Muscovite state, would all be better served by a political centre nearer to these areas of possible conflict.

Such events were, in turn, to have a negative effect on Cracow. The city's significance in the economic life of feudal Poland noticeably declined after the mid-sixteenth century. Changes in the importance of various trade routes, growth of the estate system run by noble families and the emphasis on the grain exports through the Baltic ports to western Europe, were pointers towards the emergence of the more northerly regions of the country. These were to become the most economically viable areas. At the same time, the decline of the old routes for the Eastern trade, and rise of the Ottoman Empire, meant Cracow's merchants saw themselves outflanked as the commercial gravity of Poland moved northwards. Coupled with these external events, the removal of the king's court to Warsaw, with all its commercial advantages and profit for the city's richer merchants, proved an internal blow to the city's future prosperity. No longer was there a need for the richer city merchants, who had specialised in the costly luxury trade associated with the ruler's court; great merchant families like the Boners and Wierzyneks found themselves eclipsed by events far beyond their control.[137] Their fate, however, was but a precursor to that awaiting Poland itself just two centuries later.

Finis Poloniae

Nine months after the death of August III in 1763, a new king, Stanisław August Poniatowski was elected to the Polish throne, the first native holder of that title for over sixty years.[138] Formerly a diplomat in St Petersburg, he gained his regal position thanks to the help of Russian troops, support from Catherine II (the Great),[139] and co-operation of Frederick II (the Great) of Prussia.[140] Both Catherine and Frederick found it in their interests to keep Poland weak, while the other European powers stood idly by; the Poles themselves could do little to resist the combined willpower of St Petersburg and Potsdam and suggests the inability of Poland to control, or even influence her own destiny during the latter years of the eighteenth century.[141]

In spite of his weak character and complete subservience to the wishes of Catherine II, Stanisław August was a man of intelligence, considerable education and well-founded appreciation of culture. Under the influence of the Czartoryskis, and using their great wealth, he was able to bring about their liberal political reforming ideas, and extend the scope of Polish literature, science and the arts under the new regime; under the King's protection the press and the theatre were to develop, whilst groups of scholars and writers enjoyed the support of royal favour.[142] However, foreign intrigue was never far from the domestic scene; Catherine's first open intervention was to try and assert Russian power in Poland, only two years after Stanisław Augustus' coronation, with the demand for equality of rights and freedom of worship for members of the Orthodox Church. Contemporary Polish public opinion meant that such an action would not be tolerated by the Catholic majority, and the government (Sejm) promptly rejected the ideas in 1766.[143] The following year Orthodox dissidents formed their own Confederation of Radom and Catherine II sent troops into Poland to support them; such interference in Poland's internal affairs led many of the nobility to unite in their disapproval of such Russian action; in 1768 they formed the Confederation of Bar and appealed to France for help. France took the opportunity to incite Turkey to aid a Catholic uprising, for the Ottoman Empire was already alarmed by Russia's intervention in Poland. Turkey's declaration of war on Russia led to four years of fighting (1768–72), which threatened to involve both Austria and France, but this was finally diverted and Russian troops were able to gain the upper hand over the Bar confederates.[144] For Cracow, together with nearby Tyniec, this conflict meant another period of sustained siege, in which the Bar confederates defended themselves in Wawel Castle for three months, the town was once more devastated and the inhabitants had to pay financial retributions.[145] The conflict also directly led to the First Partition of Poland in 1772.[146]

The idea of a partition of Poland, rather than wholesale annexation by Russia, was the joint work of other jealous neighbours, namely Prussia and Austria,[147] who both looked with dismay at the advance of Russian power in Poland and Turkey. Anticipation of Russia's intentions led Austria and Prussia to occupy portions of their border territory with Poland, and then approach Catherine, who was not strong enough to resist their joint pressure, with a *fait accompli*. In the treaty of August 1772,[148] Prussia annexed Gdańsk Pomerania (excluding Gdańsk itself), thereby uniting its East Prussian and Brandenburg possessions, Austria seized the southern territories, and Russia the eastern White Russian region. Thus Poland was deprived of about a third of her territory and four million inhabitants and, even more important, left with only a narrow corridor for access to the Baltic Sea (Fig. 47). The treaty was generally accepted by the rest of Europe in silence, although public opinion in some countries was aroused.[149]

Although Russia received the largest areal share of territory (92,000 km^2)[150] it was Prussia with the smallest amount (36,000 km^2) who obtained the most valuable possessions, due to the vital economic importance of the Vistula River outlet to the Baltic Sea. In comparison Russia only acquired a comparatively poor and thinly populated area. The Austrian annexation of 83,000 km^2 covered the area of Galicia and stretched northward to the banks of the R. Vistula, upstream from Sandomierz; it contained, however, some rich and densely populated areas, as well as the valuable salt mines of Bochnia and Wieliczka.[151] Thus while Cracow remained in Poland, the city's suburb of Kazimierz on the eastern bank of the Vistula passed to Austria.[152]

For Cracow, the years immediately after the First Partition proved difficult. The city now found itself as a frontier settlement cut off from previous markets in the Carpathian foothills and further east in Russia.[153] Even worse, the Austrian Emperor, Joseph II, founded a rival commercial centre on the Vistula River opposite Cracow called Josephstadt (present-day Podgórze).[154] This port siphoned off much of the transit trade activity on which Cracow's commercial prosperity had been so dependent. Even the return of Kazimierz suburb to Cracow in 1776 did little to improve the situation,[155] and many of the merchants and richer magnates sold their residences in the city and moved elsewhere, particularly to Warsaw.[156] Any hopes of Cracow developing increased trading links with ports on the Baltic coast also suffered a blow from the First Partition. Now that Prussia controlled this coast, she had approximately four-fifths of Poland's total foreign trade going through her territory, i.e. along the River Vistula; by levying enormous customs duties on this commerce, Prussia was able to

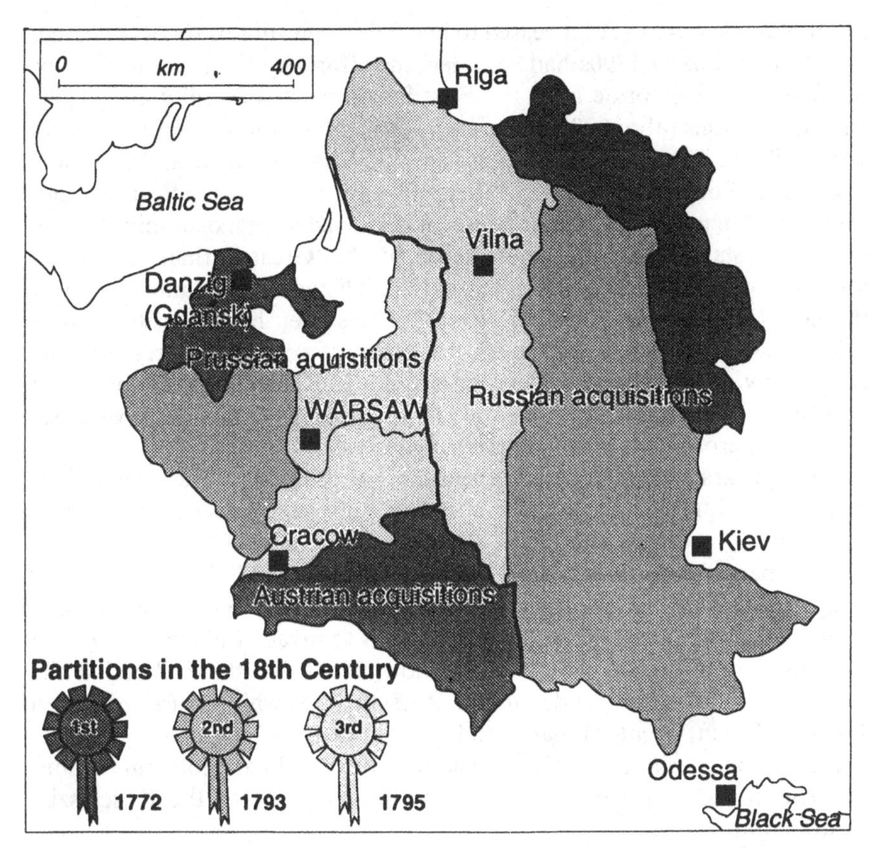

47. Poland: the three partitions.

obtain an important new source of income.[157] Hopes in Cracow were raised in 1787 when King Stanisław August visited the city; he promised an annual subsidy of 300 ducats to help Cracow's financial situation.[158] However, what remained of the former Polish territories was by this time so poor that the allowance never materialised.[159] Although the city tried to develop port facilities for long-distance trade in salt and wheat, the Russo-Swedish war (1788–90) prohibited commerce through the Baltic ports;[160] any commercial development, therefore, had to be restricted to the city's local hinterland and within the reduced area of the Polish state.

The 1790s were to witness further harmful partitions of the country. As Reddaway has aptly stated, 'All the world knows that the Poland of the First Partition endured for barely twenty years. The three-power dismemberment, proclaimed in 1772, accepted by Poland in September 1773, and finally ratified by the Diet in 1775, was transformed beyond recognition in 1793. The two-power dismemberment of that year gave the Republic its death-wound, and in 1795 it ceased to live.'[161] In spite of this tragic outcome, the 1780s and early 1790s had seen genuine efforts by the Poles to recover something of their pride after the First Partition; it was still a country of large dimensions (about 224,400 miles^2) with a population of approximately seven million. The shock of the First Partition caused political and economic reforms in Poland,[162] in which attempts were made to eradicate certain weaknesses in public life and organise an improved system of administration within the state. The task was achieved by the 'Great Parliament' over a four-year period lasting from 1788 to 1792. Of the various improvements, the new Constitution signed in May 1791 has been hailed as its greatest achievement;[163] this altered the whole system of government, restored the rule of law, gave urban dwellers more rights and improved the lot of the peasants. Unfortunately, it came too late to preserve Poland's independence from the encroachments of neighbouring states.

The appearance of a new Constitution led directly to the Second Partition;[164] it differed from the first in that only two powers, Prussia and Russia were involved. The Russo-Prussian treaty of 1793 (the Second Partition) emerged after Catherine II considered the revised Polish Constitution dangerous to the existence and continuation of her own governmental system, and therefore ordered Russian troops to invade Poland in 1792; the reforms were destroyed by force. This now meant that the Red and White Ruthenian lands came under direct Russian rule, while Prussia received Gdańsk, Toruń, Greater Poland, and part of Mazovia.[165] Cracow, as part of Little Poland, was not directly affected by the Second Partition, but obvious ferment in the city was to manifest itself in 1794, with the Kościuszko insurrection.

Tadeusz Kościuszko (1746–1817), born into a family of the lesser nobility, had left Poland in 1776, to fight for American independence.[166] He returned to Poland in 1785 and, after distinguished service in the Polish army, he was elected the leader of a new patriotic party which based its membership on those opposed to the Second Partition. On 24 March 1794, Kościuszko launched an armed insurrection from the Market Square in Cracow, which soon gained nationwide support.[167] Mobilisation of over 15,000 men helped create early successes; at Racławice, a village north of Cracow, his army defeated a small Russian force, which encouraged him to march on Warsaw and set up a provisional Polish government. The decision of Frederick William, King of Prussia, to join Russia in crushing the uprising made defeat inevitable; ultimately Kościuszko was defeated by better-equipped Russian troops at the Battle of Maciejowice in October 1794; Warsaw also capitulated at the end of the year. At the same time, Russia and Austria, later joined by Prussia, agreed to the Third Partition of Poland in 1795.[168] Russia annexed all the territory east of the Niemen and Bug rivers, Austria obtained almost the whole of Little Poland, including Cracow, and Prussia received the remaining lands, together with Warsaw (Fig. 47). Stanisław August II abdicated in 1795 and to all intents and purposes the Republic of Poland was erased from the map of Europe for the next one hundred and twenty-three years.

For Cracow, the euphoria generated by the Kościuszko insurrection of early 1794 was soon replaced by despair as Prussian troops advanced and occupied the city after its capitulation on July 15 – and whose presence was to last until January 1796.[169] Then Austrian soldiers replaced them, for Cracow lay within the boundaries of Hapsburg rule according to the Third Partition – this stretched along a line from Pilica, the middle Vistula to the River Bug.[170] Cracow's destiny now lay within the hands of Austria. The outlook was bleak, for its streets 'looked like heaps of ruins'[171] and its inhabitants, who numbered less than 10,000,[172] were to be subjected by the Austrian government to a strong campaign of Germanisation; the German language was to be officially used in all offices and schools.[173] The path from the happier, wealthier days of the early sixteenth to the late eighteenth century had certainly been a hard and long one, whilst the immediate years of the dawning nineteenth century offered little hope for a brighter future.

Conclusion

Political events in the Early Modern Period provided Cracow with new problems for her commerce. The discovery of sea routes to Asia and America by the Spanish and Portuguese diverted the old trade routes from the Mediterranean and Black Seas. The wars of Europe meant that Poland was subjected to mass movement of foreign troops across her soil; the Swedish campaigns around the middle of the seventeenth century with their sieges, plunderings and fires, created terrible havoc in the countryside. Urban centres, like Cracow, did not escape – the siege of 1655 dealt a hard blow to the welfare of the city and brought about its economic decline.

Turkish foreign policy and ambitions caused disturbance not only in the Balkans but also in eastern and western Europe, Poland often finding the need to walk a political tightrope as a result of sudden incursions; for Cracow, future trade lay in the harder earned markets of the West. Further, political danger lurked in the growing aspirations of the Muscovite/Russian state as it pushed its boundaries and influence slowly westward. Poland, clamped between Swedish, Prussian, Russian and Austrian claims on her territory, was gradually reduced in size; for Cracow this meant a shrinkage of her former market areas and greater concentration on national and local commerce.

In spite of these difficult times created by the political patchwork of events, Cracow may have been able to maintain its previous high level of living standards, culture, influence and the like by means of the accumulated resources of centuries. However, King Sigismund III ultimately lowered the city's importance by transferring his residence to Warsaw together with the numerous trappings that accompanied the royal presence.

The eighteenth century was not to prove a happier era for the city. Wars, and Cracow's strategic position brought new misfortunes; soldiers, Swedish, Saxon, Russian and Polish alternately occupied the city, leaving it poorer each time. As the century progressed the whole country grew more destitute and the First and Second Partitions only aided the eventual downfall, in which Cracow now played the role of a small provincial centre on the edge of a dying state. Even the war of independence in 1794, under the aspiring leadership of Kościuszko and begun in Cracow itself, only helped sound the death knell of the Polish state. The following year Cracow awoke to find herself allotted to a foreign power, Austria. The lasting fear of her dwindling population had been realised. Even worse, Cracow was now tied to the Hapsburg Empire and no longer allowed such a free hand in exploiting her own commercial advantages – a severe blow to any centre dependent on the flow of commerce for its own livelihood.

7

The commerce of Cracow, 1500–1795

The period known as the 'Golden Age of Cracow' covered the first half of the sixteenth century and roughly coincided with the reign of King Zygmunt Stary (Sigismund the Old) from 1506 to 1548. The vigorous development of intellectual life in the city, combined with foreign contacts and a flourishing artistic life was supported by commercial prosperity.[1] This was reflected on a larger scale within Poland, expressing itself in the wealth of the country and a favourable growth in foreign trade.[2] Raw materials such as grain, timber, livestock, hemp, flax and potash led to a lively export trade, especially with the Low Countries and England; the production of raw materials, however, was in the hands of the propertied Polish nobility who were gradually becoming farming entrepreneurs. Here one sees the first hint of looming disaster for Cracow, which helped precipitate the city's decline. Not only in Poland but also in other parts of Eastern Europe,[3] the nobility were taking advantage of the rural origins for raw materials such as grain and wresting control of them from urban merchants; in Poland the nobility decisively asserted its dominance over the towns which led to a steady flow of restrictions on the wealth and authority of such places as Cracow. The city also suffered from external factors; the great geographical discoveries had opened up new sea and ocean trade routes, while the old land routes, on which Cracow was so dependent, gradually diminished in importance.[4] Moreover, following the Union of Poland and Lithuania in 1569, Cracow found itself tucked away in the south-west corner of Poland and had to cede its prime position in Poland to Warsaw.[5] Such events, combined with the ravages of war during the seventeenth century, meant the prosperity of its merchants declined. As the eighteenth century progressed and the clouds of depression gradually lifted, the shape of things to come for Cracow was becoming fairly clear; gone were the days of profitable long-distance transit trade, to be replaced by greater emphasis on the import/export of individual commodities, combined with increasing dependence on Polish and local market areas.[6] The question therefore arises as to how Cracow adapted to these changing conditions, not only in the spatial contraction of its trading sphere but also in the commodities involved.

Trade decline in expensive raw materials

(i) Spices and other colonial goods

Much has been written on the impact of new geographical discoveries and conquests by the Portuguese in the East Indies at the beginning of the Early Modern Period together with their importance for the spice trade.[7] Similarly, military successes by the Ottoman Empire, not only in Syria and Egypt but also in the Balkan Peninsula, have been seen as influential in altering the trade pattern of routes supplying spices to the growing European markets.[8] Just as Venice and Genoa had been the main centres for the spice trade in the earlier centuries, after 1500 the spice market moved to Lisbon, later to Antwerp[9] and by the end of the sixteenth century, to Amsterdam.[10] How were these changes to affect Poland, and Cracow in particular?

Admittedly, one of the previously most important products to decline in profitability in Cracow's trade structure was spices, but this was no overnight phenomenon. With the fall of Kaffa, Kiliya and Belgorod to the Turks in the late fifteenth century the effect on Polish purchases of eastern merchandise was quickly felt by cities like Cracow. The Ottomans realised this situation and treaties signed between the Turks and Poland in 1490 and 1494 allowed free trade to continue between them.[11] More serious for Cracow was the sharpened tension between itself and Lwów for transit trade through Poland to Western Europe. Onerous tariff customs imposed on Cracow by Lwów in 1507 and 1509[12] seriously affected the profits obtained on eastern goods by the city's merchants, but the flow of products, like spices, continued. As Rutowski baldly states, 'Spices, southern fruits and oil came [to Poland] through Danzig and from Turkey',[13] whilst Małowist adds, with reference to the fifteenth and sixteenth centuries, 'the valuable products ... [from Turkey] cotton and silk fabrics, spices, for which there was then a great demand in Poland',[14] suggests that absolute decline was not yet at hand.

In spite of the opening up of the Cape route to India and Portugal's role as the 'Grocer King' it was still quicker for places like Cracow to import spices though the Balkan Peninsula or via Lwów.[15] In fact one may draw a demarcation line through Cracow–Lublin–Brześć which separated spices coming to Poland from Venice, Antwerp via Gdańsk and those arriving through the Ottoman lands, via Wallachia, Moldavia, Transylvania, or Kamieniec Podolski and Lwów[16] (Fig. 48). Koczy has shown that spices arriving in Poznań from Goa via Lisbon[17] and Gdańsk took over six months, whilst those coming via Lwów travelled half the distance and were twice as

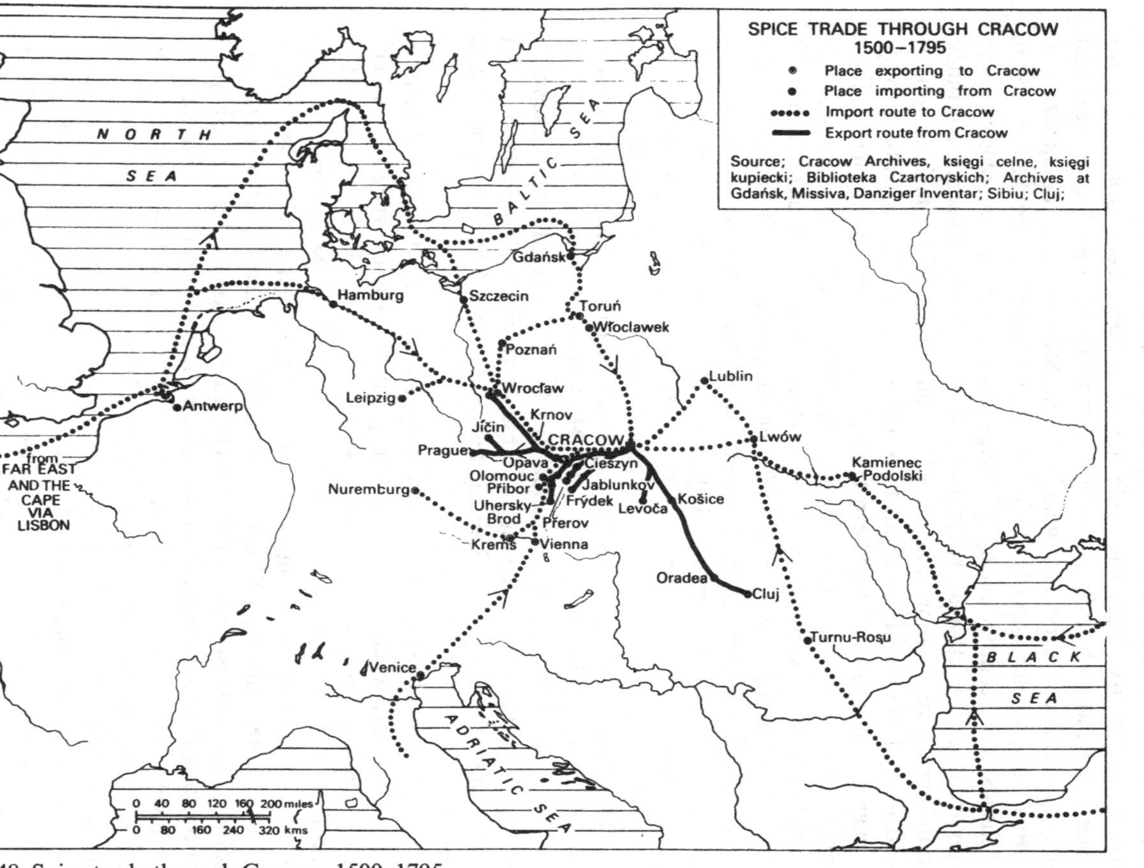
SPICE TRADE THROUGH CRACOW
1500–1795
Place exporting to Cracow
Place importing from Cracow
Import route to Cracow
Export route from Cracow
Source; Cracow Archives, księgi celne, księgi kupiecki; Biblioteka Czartoryskich; Archives at Gdańsk, Missiva, Danziger Inventar; Sibiu; Cluj;
NORTH SEA
BALTIC SEA
BLACK SEA
ADRIATIC SEA
from FAR EAST AND THE CAPE VIA LISBON
Antwerp
Hamburg
Leipzig
Nuremburg
Szczecin
Gdańsk
Toruń
Włocławek
Poznań
Wrocław
Krnov
Jičin
Prague
CRACOW
Opava
Olomouc
Přibor
Uhersky Brod
Přerov
Krems
Vienna
Cieszyn
Jablunkov
Frýdek
Levoča
Košice
Oradea
Cluj
Lublin
Lwów
Kamienec Podolski
Turnu-Roșu
Venice
0 40 80 120 160 200 miles
0 80 160 240 320 kms

48 Spice trade through Cracow, 1500–1795.

cheap.[18] Cracow's position on the margins of both marketing spheres meant that during the first few decades of the sixteenth century spices, along with other colonial goods, better described as tropical and subtropical products, arrived in the city. They were referred to in documents as *merces* or *res aromaticae* and included various southern fruits such as figs, oranges and lemons, which came to Cracow mainly from Mediterranean countries.

Evidence of spice imports to Cracow from Turkey during the early years of the sixteenth century exists[19] although changes were taking place; pepper imports from the Ottoman lands declined, in preference to sources brought from Venice via Vienna and Wrocław, but other spices such as ginger, saffron and cinnamon continued to arrive in the city, usually through Armenian merchants.[20] Southern fruits, however, were from the Mediterranean, particular Italy, and organised in Cracow by Italian traders mainly from Venice.[21] Towards the end of the sixteenth century spices rarely arrived in Cracow from Lwów or Lublin,[22] but increasingly from Wrocław – the main suppliers – and other Silesian towns; seasoning included saffron, ginger, cloves, pepper, almonds, bay leaves and caraway seeds, together with Mediterranean oranges, raisins, figs, olives, and diverse types of sugar.[23] Throughout much of the sixteenth century the various spices and colonial goods arrived from numerous directions. Comparison of the customs books for 1538/9 and 1584 reflect this theme. Herbs from Tartary came via Lwów and sent on to Wrocław; cinnamon was further relayed from Cracow to Prague;[24] sometimes merchants in Cracow received consignments direct from Nuremburg (sugar), Leipzig (almonds) and Wrocław (spices and rice).[25]

Comparison of earlier and later customs books from the sixteenth century does, however, reveal interesting changes in the use of trade routes. The dominant route for colonial goods (spices and southern fruits) *circa* 1540 was from Italy via Vienna; by the last two decades of the century they came mostly via Gdańsk. This suggests the growing popularity of sea as opposed to land routes, although Wrocław continued as Cracow's other supplier of colonial goods; these arrived either via southern Germany for the early years or through Hamburg and Szczecin[26] during the later period. The importance of Lwów as Cracow's major supplier of colonial goods decreased as the century wore on, with only Tartar herbs and Turkish saffron being noted in the customs books for the closing years. In direct contrast, one gradually sees as the century progressed the rise of Gdańsk as Cracow's major supplier. The customs books for Włocławek on the Lower Vistula rarely mention spices being sent on to Cracow during the first half of the sixteenth century,[27] whilst in 1537 a document was sent by Toruń merchants to Gdańsk complaining that Cracow traders were buying pepper in Wrocław and not from them,[28] a theme reiterated in 1544.[29] Many other spices were increasingly being sent from Gdańsk to Cracow as the century progressed,[30] including ginger, cinnamon, nutmeg, mace, cloves and saffron; southern fruits also figured in this trade with examples of figs, raisins, almonds, olives, citrus fruits, olive oil and citric acid. Rice was transported in small quantities whilst there seems to have been a growing market for sugar[31] and, in later years, sugar-coated products, e.g. orange peel.[32]

The seventeenth century saw a strengthening of Gdańsk's position as chief supplier of spices and colonial merchandise to Cracow. For example, Gdańsk was the largest market for pepper imports to Poland; as Jeannin states, 'The market for pepper at Poznań was smaller than that of the ports of Danzig and Lübeck, and there was an enormous tenfold increase in imports at Danzig between 1538 and 1634.'[33] Małecki's analysis of merchandise imported by Cracow from Gdańsk during the first half of the seventeenth century reflects this development with consistently large quantities involved during the first and third more peaceful periods of this troubled century.[34] Supplies to Cracow via Gdańsk were partly in the hands of various consortia, of which the Fugger enterprise was a notable example,[35] or increasingly in the seventeenth century by the Dutch and, to a lesser extent, the English.[36] Other spices such as ginger and cinnamon arrived in Cracow via Gdańsk during the first half of the seventeenth century but in much smaller quantities than pepper;[37] perhaps this may be explained by the fact that pepper was widely used as a seasoning in Polish cooking and also for meat packing and curing illnesses.[38] Of the other colonial goods through Gdańsk, the most popular appear to have been raisins, olives, lemons, aniseed and of course sugar.[39]

Cracow also obtained various spices and colonial goods during the early part of the century direct from Italy;[40] the major condiments and fruits were brought to Cracow straight from central Italy by way of Venice, Krems and Vienna and in smaller quantities through Olomouc. Oranges were very popular amongst the upper echelons of Polish society, becoming an extremely fashionable fruit during the reign of King Zygmunt (Sigismund) III Waza (1587–1632), along with other types of citruses famed for their curative properties.[41] Of the spices, pepper, saffron and Tartar herbs appear to have been in most demand, whilst other delicacies included Parmesan cheese (via Vienna) and olives.[42] Finally, there is evidence that Cracow was still obtaining some of her spice supplies from the East through Lwów during the early part of the seventeenth century, particularly pepper and, to a lesser extent, saffron and ginger.[43]

During the second half of the seventeenth century documentary evidence suggests that Cracow continued to import certain spices through the Ottoman Empire mainly via agents in Transylvania. The customs registers for Turnu-Roşu reveal that quantities of saffron,[44] ginger,[45] sugar[46] and rice were sent to Poland, especially Cracow, during the 1680s and 1690s. This helps support the idea that the Turks continued to allow trade with the

Infidel. Netta has noted that 'the Ottomans monopolised certain products and exploited both the producers and the merchants in the consumers' interest';[47] nevertheless, the Turks were no merchants and neither were they commercial intermediaries. This role was practised for them by foreigners. Further, it appears 'they were not interested in the commercial income some conquered regions enjoyed',[48] which may help to explain the reason for commercial exchange with Poland long after the Ottoman occupation of the Balkans.

The eighteenth century was to witness a decline in Cracow's role in longer distance international trade and concentration on local and regional markets.[49] Spices and other colonial goods continued to be imported mainly from western Europe, through Gdańsk,[50] largely to satisfy the city's home market and towns in the immediate vicinity. There were no longer opportunities for extensive profits from the re-export of such goods, probably due to increased competition from other Polish cities such as Poznań[51] and Silesian,[52] and German towns, who now successfully competed in areas like Upper Hungary, Cracow's major spice export market in the sixteenth and seventeenth centuries.

The earlier trade relations between Cracow and Hungary took a further stage in their development during the sixteenth century. The Ottoman expansion into southern and central Hungary after the Battle of Mohács in 1526 gradually cut off Upper Hungary from southern trade routes which had previously brought merchandise, including spices and colonial goods, from the Mediterranean and Levantine markets.[53] Merchants from Upper Hungary, especially the Slovakian towns and those still lying within the Monarchy, now looked to Cracow for their supplies of spices, etc. The transformation came around mid-century, and is clearly evident from the customs books of such towns as Košice (Kassa). Granasztói has shown how, around 1555, spices began arriving in the town mainly from Cracow, but also from Germany via Wrocław in Silesia.[54] The most important spices sent from Cracow to Upper Hungary were pepper and ginger, exported in small quantities in sacks[55] and transported with other general goods referred to in documents as *res institoriales*. During the first half of the seventeenth century Cracow's spice trade developed particularly strong links with certain Slovakian towns like Levoča,[56] although the effects of the Thirty Years War and the Swedish–Polish war tended to reduce Cracow's significance as a source for spices. Gradually these towns began to look elsewhere, e.g. Vienna and Wrocław, for future supplies. Eastern and Southern Moravian towns also sought spices on the Cracow market. Merchants from places like Opava, Krnov, Těšín, Jablunkov, Ostrava, Frýdek, Fryštát, Příbor, Rožnov, Valašské Meziříčí, Jičín, Přerov, Olomouc and Uherský Brod, are all noted in the customs books as having visited Cracow for the purchase of spices.[57] Again pepper appears to have been most important, followed by ginger, saffron, cloves and bay leaves; less important spices included aniseed, brought by Armenian merchants from Lwów to Cracow, and mace.[58] Other colonial goods included raisins, and sugar, described as a luxury article used on occasions to replace honey.[59]

Besides Slovakia and Moravia, the Princedom of Transylvania also provided Cracow with an export market for her spices. Transylvania appears to have played a dual role in Cracow's spice trade, merchants from some towns exporting them to the city, others, like Cluj, importing certain varieties of seasoning.[60] The time factor, however, reveals that whereas Transylvania was largely exporting spices to Cracow in the late seventeenth century,[61] the customs books for a century earlier clearly illustrate that towns like Oradea[62] obtained certain spices – saffron (from Aragon, Catalonia), cloves, ginger, fruits such as oranges, and nuts (almonds) – from merchants in Cracow.

Finally mention should be made of spices/colonial goods prices in Cracow in relation to the general European trend during the Early Modern Period. There is one economic occurrence which affects all developing economies – inflation. When historians discovered this phenomenon in the sixteenth and early seventeenth centuries, they called it, rather exaggeratedly, the 'price revolution'. This was contrasted with the long period of static or declining prices of the Later Middle Ages. To people living in the sixteenth century this inflation was not immediately obvious as prices, especially of foodstuffs, fluctuated extensively according to season and type of harvest.[63] Over the century as a whole there was a general upward trend of between 2 and 3 per cent annually; by the middle of the sixteenth century the cumulative effect of prices was becoming evident throughout Europe.[64] Spices and colonial goods were no more immune from inflation than other merchandise; perhaps more so because of their long-haul characteristics and the impact of new geographical discoveries.[65] Nevertheless spice prices, for example, did suffer fluctuations, as witnessed around 1600 when a fall in consumer prices led to a dramatic surge in European consumption.[66]

In Poland, and more generally in central Europe, the price revolution was closely related to its counterpart in western Europe.[67] Graphs for certain spices and colonial goods (Figs 49, 50) in Cracow show price increases in the sixteenth century commensurate with the general European trend. More particularly the drop in pepper prices throughout Europe after 1600 is also clearly visible from Cracow's documentary sources,[68] but the city's pepper prices remained higher than in other Polish towns.[69] After 1650 both graphs illustrate the steeper increase and greater fluctuations in prices, again evident in many other parts of Europe.[70] In Cracow, merchants during the early decades of the eighteenth century had difficulty coping with such vacillations, many of them continually having to obtain long extensions on credit facilities for, as Hundert has written, 'The rate of inflation during that

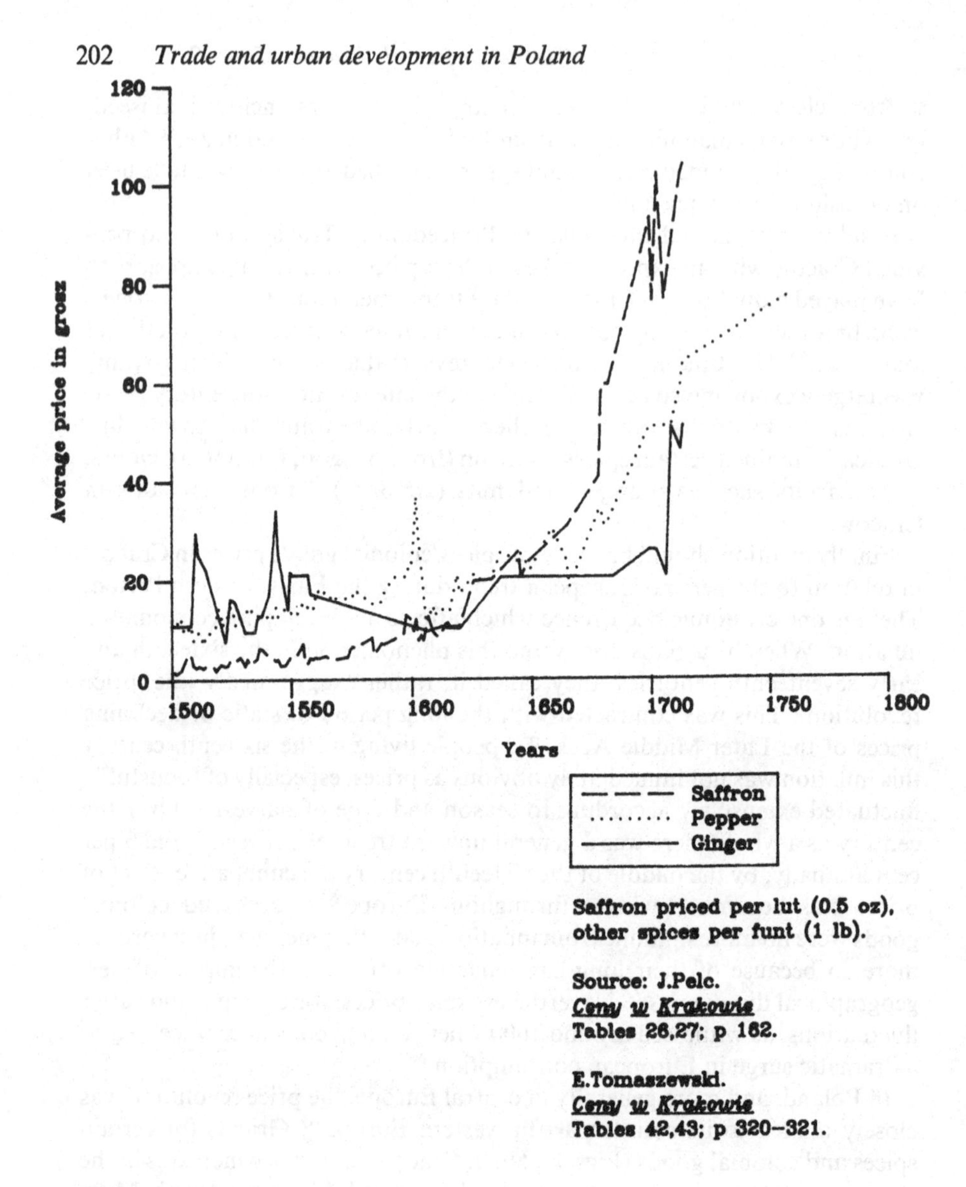

49 Price of certain spices in Cracow, 16th–18th centuries.

period, though it cannot be calculated with precision, was extremely high.'[71] This perhaps helps to explain why spices in particular and, to a lesser extent, other colonial goods, declined as transit or re-export commodities in the city's commercial structure during the later years of the Early Modern Period, in contrast to the first half of the sixteenth century, when Cracow was enjoying its 'Golden Age'.

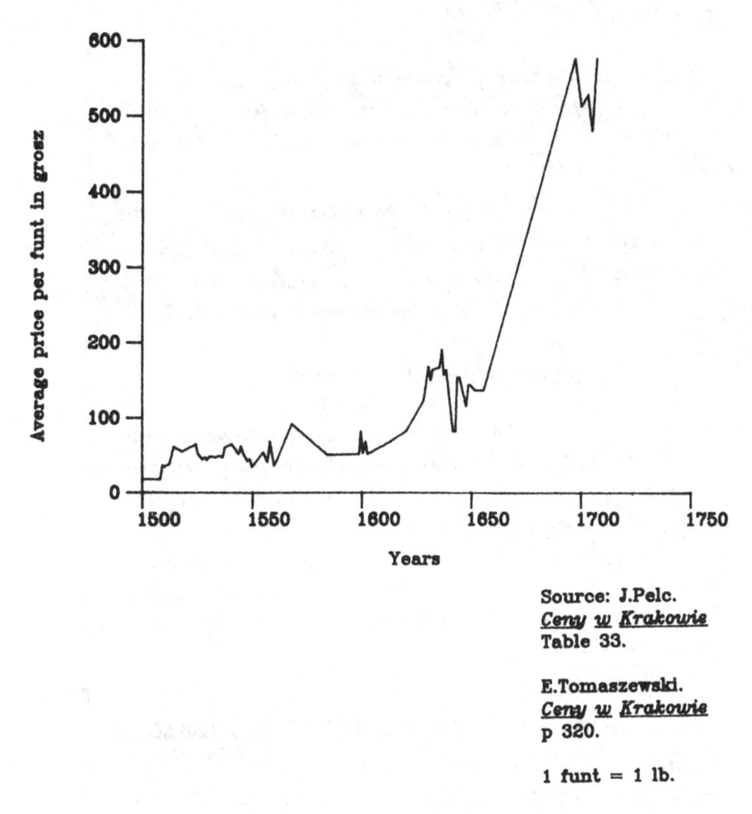

50 Price of Cinnamon in Cracow, 16th–18th centuries.

(ii) Salt

Salt, like food and clothing, was needed everywhere. For Cracow the salt trade was another product to suffer decline as a profitable export commodity after 1500, but this was not an immediate process. Salt continued to be in demand from the surrounding countries and regions throughout the sixteenth and seventeenth centuries. The rock-salt mines of Wieliczka and Bochnia near Cracow were the greatest industrial enterprises in Poland well into the eighteenth century;[72] these and the small salt-producing centres of Red Ruthenia (Przemyśl, Drohobycz, Stara Sól), were sovereign monopolies, the revenue covering expenses at the royal court. Everything was done to promote domestic production and to protect the home market from threats of river salt competition from outside. In the eyes of Cracow's merchants, working for a state monopoly must have provided added impetus for trade, and lasted until 1772 (First Partition) when Cracow's mines were appropriated by the Austrian monarchy; they became part of the Austrian

Fol. 15

N: 14. Eleemosynarius Regalis

Additus est ergo Eleemosynarius alter qui Vocatur Regius qui Seorsum ab aliis pauperibus circa propriam mensam sedet, potitur hisce ferculis, quibus Officiales [illegible]. Habet pro quolibet die quatuor panes [illegible] cum [illegible]. Diebus [illegible] [illegible], solum datis panes pro prandio

Item datur eidem pro prandio et cenae duo santeri panis Cerevisiae. Item feriis [illegible] et diebus Sabbathinis datur ei portiones casei et consueti santanuli cerevisiae cum duobus panibus. Nil amplius habet in Zuppis nec pecuniis aliquibus consolatur sicuti hi priores. Pro his Eleemosynis sibi largitis, tenetur Deum rogare pro felici Statu Sacrae [illegible] et Domini Zupparii.

N: 15. Praebendarius Altarista Zuppae

Quorsum necesse [illegible] res agitur nutritum [illegible] dono non posse ~~[illegible]~~ habere in manibus. His Praebendarius fruebatur expensis quotidianis de die indicta Regiis Cera ad Altare sibi dabatur propter quod et ipsum Altare vocatur Zuppae. Quartualiter Salis qualibet Septimana potiebatur. Merito ergo [illegible] in Regiis [illegible] cedere deberet tanquam in Provisorem huius altaris. Quo pacto istud [illegible] revidendum esset Privilegium huius Altaris. Nunc tantum prandium et cenam habet nullum aliud dignerium in pane et Cerevisia habet nec candelis sepalibus potitur.

N: 16. Tantum est de Eleemosynis et Salariis Vielicien.

Descriptio Salariorum et omnium Digeriorum sub tenuta Gnosi Dni Ioannis Bonner videlicet in peccuniys Panno Officialibus. et Familiae Zuppae Vielicien. expensas habentibus.

Vice Zupparius

Quemadmodum Officia alio aliud [illegible], ita et quae sunt eiusdem attinentiae et alia debent esse maiora et nobiliora ceteris. Quid enim Officiatori gloriosius quid felicius, quam ipsam intueri ceteris recte imperantem, sibi commissa debite regentem, et [illegible] [illegible] [illegible] actiones ferentem. Idem non solo commodo parere, vi et astu oblationes extorquere, mala dissipare et [illegible] ad bonam gubernationem decet feliciter attentare, fines debito subiecta sibi et credita Ordinarie Iniurias, si quae delatae fuerint, et Montanis et Gazarii domosue continentiis iustis rationibus disponere. ViceZupparius enim Similitudo sive imago Zupparii est, cui ceteri Officiales parere debent. Sed proh dolor temulentia et ignavia obruti debiles in his omnibus reddimur. Nec mirum: non postratu enim [illegible]

Source: Muzeum Żup Krakowskich Wieliczka: *Commissiones Annorum*, 1518; Fol. 15, Bonner.

51 Document: Boner mine leases, 1518 and 1614.

Komora Czyzowska.

Komora Gembalinskie.

Szyb Boner

Komora Morsztyn.

Source: Museum Żup Krakowskich Wieliczka: *Commissiones Annorum*, 1614 Szyb Bona Komora Morsztyn, p. 184.

state monopoly which has been described as the 'brightest jewel in the possession of the Hofkammer'.[73]

The 'golden age' of Poland in the sixteenth century also coincided with the period of greatest significance for Cracow's salt mines. Increasing production has recently been noted by Keckowa; 'Information dating from the late XVth and early XVIth century shows an output of approximately 12,000–13,000 tons. Towards the end of the XVIth and first half of the XVIIth centuries, the output rose to approximately 30,000 tons, and in the mid-XVII century to about 37,000 tons. The second half of the XVIIth century and the start of the XVIIIth were marked by a considerable decline in output.'[74] It appears that for much of the eighteenth century the state limited production to about 30,000 tons annually until the Austrian takeover in 1772.

The profits to be had from selling salt attracted not only local merchants from Wieliczka,[75] but Cracow itself; well-known families such as the Morsztyns held leases at the mines over several generations, as did the Betmans, Sapiehas and Boners (Fig. 51).[76] Besides wealthy Cracow burghers,[77] foreigners, particularly from Germany and Italy, had mine tenancies and sometimes made large fortunes for themselves from commerce in salt.[78] The Italian connection was especially strong;[79] all foreign merchants had received an unexpected filip in their activities when, in 1565, a new Polish law stated that 'no goods could be freely exported over the Crown's frontier by merchants – only foreigners'.[80] This complemented the acts of 1507 and 1538 when foreign merchants were allowed into Polish markets not only as wholesalers but also as retailers.[81] This meant merchants from abroad now took up residence in cities like Cracow and used contacts in their homeland to foster merchandise exchange. Italians were particularly active in the salt trade;[82] one Gaspar Gucci had trade links with Poznań, Lublin, Lwów and Vilna, acting as a middleman for trade with Italian and German towns.[83] He had been asked by the King (Zygmunt Stary) to arrange for salt to be sent from Cracow to Toruń and Prussia due to devastation of coastal salt-pans during wartime.[84]

Transport costs were a major factor in this trade, some governments even assisting in the movement of salt.[85] Cracow merchants utilised both land and river routes for transporting this commodity. The obvious river route was along the Vistula and reliable documentary sources from the first half of the sixteenth century suggest the cost of movement (e.g. in 1527) from Cracow to Warsaw was about half the initial purchasing price.[86] By mid-century transport costs had risen further, while the price of salt in Wieliczka remained fairly stable throughout the century.[87] It was not until the eighteenth century, however, that some form of transport organisation by independent owners emerged; previously all river transport was co-ordinated by the king's representatives.[88] Nevertheless, as Rybarski states 'It is important to

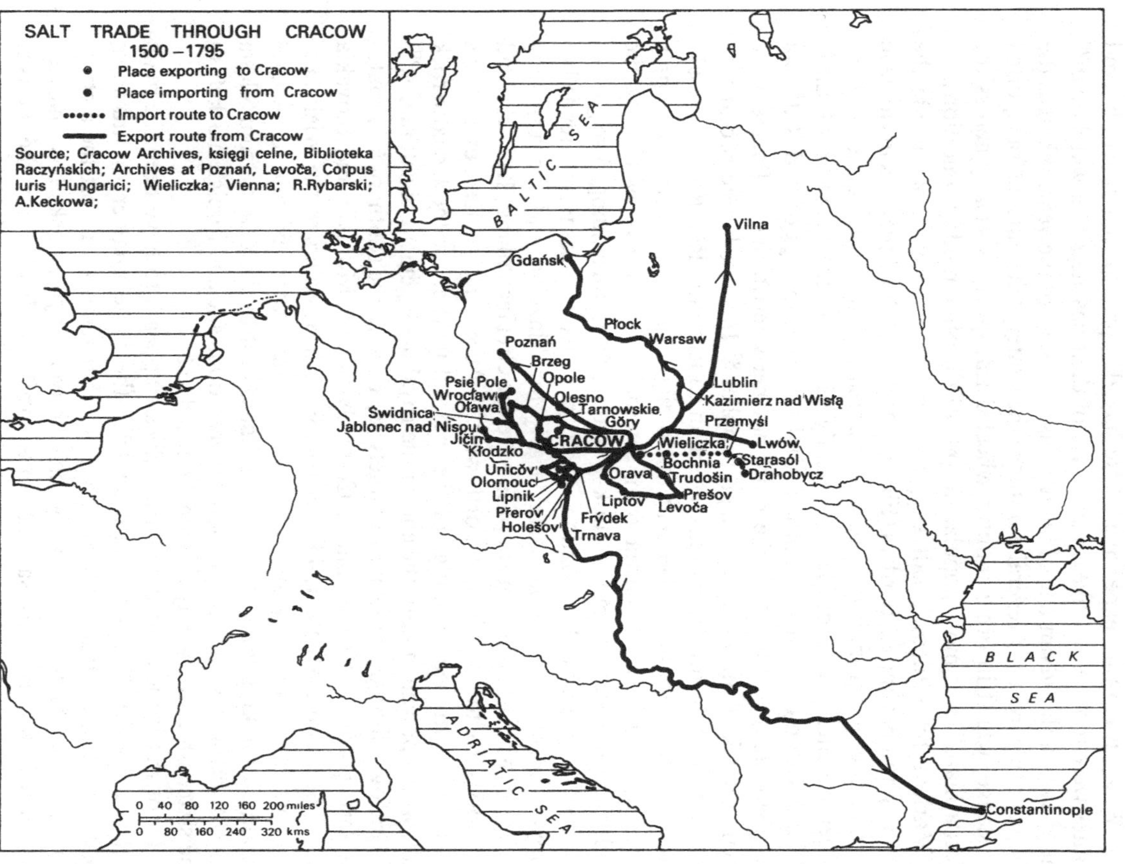
SALT TRADE THROUGH CRACOW
1500 – 1795
Place exporting to Cracow
Place importing from Cracow
Import route to Cracow
Export route from Cracow
Source; Cracow Archives, księgi celne, Biblioteka Raczyńskich; Archives at Poznań, Levoča, Corpus Iuris Hungarici; Wieliczka; Vienna; R.Rybarski; A.Keckowa;
BALTIC SEA
Vilna
Gdańsk
Płock
Warsaw
Poznań
Brzeg
Opole
Psie Pole
Wrocław
Oława
Olesno
Tarnowskie Góry
Świdnica
Jablonec nad Nisou
Jičín
Kłodzko
CRACOW
Lublin
Kazimierz nad Wisłą
Przemyśl
Wieliczka
Lwów
Bochnia
Starasól
Drahobycz
Trudošin
Orava
Prešov
Levoča
Liptov
Frýdek
Trnava
Uničov
Olomouc
Lipnik
Přerov
Holešov
BLACK SEA
ADRIATIC SEA
Constantinople
0 40 80 120 160 200 miles
0 80 160 240 320 kms

52 Salt trade through Cracow, 1500–1795.

remember that river transport costs were not as great as on land. Actually getting the salt blocks from Wieliczka to the banks of the Vistula at Cracow, a short distance compared with the boat trip to the warehouse at Kazimierz nad Wisłą, was more expensive.'[89] He has also proved that generally land routes were more expensive, irrespective of their nearness to a warehouse,[90] but they had the advantage of being in use for most of the year. Whereas the Vistula often froze for several months in the winter, salt could be transported by sledge, whilst in other months wheeled transport, especially across the Carpathians[91] was more popular in spite of the often muddy conditions.

Markets for Cracow's salt supplies were to be found both inside the country and beyond its borders (Fig. 52). Within Poland there seems to have been a strict geographical division of purchase areas. Up to the end of the fifteenth century Cracow tended mainly to supply the province of Little Poland within the domestic market, whilst the salt springs of Red Ruthenia supplied areas further east; the more northerly territories of Poland were then dependent on local or foreign supplies.[92] Increased state production encouraged these northern areas to obtain Cracow and Red Ruthenian salt, thus making them less dependent on foreign sources. Attempts to supply Greater Poland and Kujawia from internal markets was evident by the mid-fifteenth century, but it took a state decree forbidding foreign salt imports[93] to activate the situation. Thereon, Cracow mines were intended to supply the middle Vistula basin largely for Mazovia and the adjoining Podlasie, whilst Ruthenian supplies were for Greater Poland, Kujawia, and territories to the north and north-west.[94] According to Keckowa, 'This division determined the range of the internal market for the Cracow rock salt, which prevailed until the end of the royal management of salt mines.'[95]

Cracow's foreign trade in salt largely lay beyond the southern and south-western borders of the country. Certainly Silesia, Moravia and Slovakia were important markets. Evidence from French travellers to the Wieliczka/Bochnia mines in 1574 and 1585 confirm the export of salt to 'several neighbouring kingdoms'[96] and the purchase of it by merchants from Silesia and Moravia.[97] Some comparison between domestic and foreign markets in the sixteenth century may be seen from Table 5.

Throughout the sixteenth century Silesia appears to have been the main salt market for Cracow. Unfortunately, the large number of unknown destinations for salt movement abroad in 1591 tends to distort the picture, but, even so, Silesia remains the chief buyer. Wyrozumski has analysed documents referring to salt customs tariffs at Będzin on the Polish–Silesian border during the last decade of that century, which confirm the importance of the Silesian market for Cracow's salt merchants.[98] According to customs books in Cracow from the 1590s[99] the main Upper Silesian destinations were Tarnowskie Góry, Piekary, Gliwice, Pyskowice, Bytom, Pszczyna and around Opole;[100] in Lower Silesia, it was Wrocław, Strzelin,

Table 5. *Domestic and foreign markets for Cracow's salt supplies, 1519, 1538/9, 1584*

1519			1538/1539			1584		
Destination	No. of Horses*	% total	Destination	No. of Horses	% total	Destination	No. of Horses	% total
Domestic market	7,335	65.6	Domestic market	3,767	49.2	Domestic market	18,661,	54.1
Foreign markets	3,354	30.0	Foreign markets	3,111	40.6	Foreign markets	15,241	44.3
of which:			of which:			of which:		
Silesia	2,911	86.7	Silesia	2,242	72.1	Silesia	3,279	21.6
Moravia	302	9.0	Moravia	234	7.5	Moravia	190	1.3
Hungary	141	4.3	Hungary	50	1.6	Hungary	40	0.2
			Unknown	585	18.8	Unknown	11,732	76.9
Unknown	485	4.4	Unknown	779	10.2	Unknown	558	1.6
TOTAL	11,174	100%	TOTAL	7,657	100%	TOTAL	34,460	100%

* It is accepted that the average load for one horse was 5 cetnar (i.e. 500 kgs), or the load of one waggon. See R. Rybarski, *Handel i polityka handlowa Polski w XVI wieku*, Vol. 2, Poznań, 1929, pp. 332–40.
Source: Ibid. pp. 178–9; 184–5.

Brzeg, Nysa, Kłodzko, Ząbkowice, Świdnica, Ząbice, Oława and Psie Pole that were the main receiving centres.

Moravia appears to have been much less significant. Perhaps it should be remembered that much of the salt trade was in the hands of less wealthy Cracow traders, the powerful rich merchants specialising in more profitable commodities. This may explain the rather limited distances for the salt market, the traders usually organising business through their 'Furman' or waggon driver. Salt was therefore transported to the nearest markets at the border, e.g. Olomouc, Jičín, Hranice and Lipník.[101] Nevertheless, a document dated 1557 confirms the willingness of King Ferdinand I to allow Polish salt to be sold in an area between Ostrava and Holešov (near Kroměříž).[102]

Finally, salt from Cracow went to Upper Hungary. During the Middle Ages there was a long history of objections on the part of the Hungarian monarchy to the importation of Polish salt; the fifteenth century had seen laws passed specifically forbidding it from entering the country (1405, 1417, 1439, 1458, 1464)[103] in an attempt to protect their own state monopoly. This theme continued into the sixteenth century with a mandate published in 1518–19 which forbad, under threat of confiscation, the importation of foreign salt.[104] Northern Hungary was particularly affected by this decree as there was only one small salt mine in the whole area, at Sóvár;[105] in general, Hungarian salt production was insufficient to supply the whole population. Several towns in Northern Hungary (Slovakia) had managed to obtain permission to import foreign salt during the later part of the fifteenth century,[106] in particular Prešov which had the right to import (mainly Polish) salt, a privilege its citizens enjoyed until 1616.[107] It was not until the personal intervention of King Zygmunt August in 1550 over salt disputes that the Hungarian government gave official permission for the importation of Polish sources,[108] most of which came from Wieliczka. Such action may have been prompted by the Ottoman occupation of southern and central parts of Hungary, the increasing difficulty of obtaining supplies from their main mine at Marmaroš (Maramureş),[109] and general inaccessibility of Transylvanian salt.[110]

Małecki maintains that there is no information on the sending of salt from Cracow to Hungary[111] and suggests Hungarian/Slovakian merchants purchased their supplies at Nowy Sącz or Nowy Targ. He disagrees with Pieradzka's view that salt could not be bought direct from Wieliczka[112] but had first to be sent to Cracow. Whether the city exported salt to Northern Hungary by direct or indirect methods, evidence exists to prove that trade took place. The main route lay through the Carpathians via Tvrdošin, Stará Kubín to Ružomberok; Nowy Targ lay on this route and in 1593 obtained warehousing rights for salt and lead.[113] Throughout the second half of the sixteenth century there appears to have been an increasing demand for

Polish salt in the towns of the Spiš region, especially Levoča.[114] In 1581, the Polish king, Stefan Batory (1576–86) issued a mandate whereby the thirteen towns of the L'ubovňa district could purchase salt from Nowy Sącz and Wieliczka, provided they observed the obligatory three-day stay in the former town.[115]

The deepening economic crisis in Poland during the first half of the seventeenth century naturally affected Cracow's trading sphere; during the first quarter of the century commerce remained lively, but around 1630 this stopped, with overall decline in the second half of that century. The salt trade was also affected; nevertheless, if the description of the French traveller Le Laboureur's visit to Wieliczka during the 1640s is to be believed this was far from true. He states 'The salt from Wieliczka is bought by Poles, Silesians, Moravians, Austrians and Germans. The Hungarian rivers of Orawa [Orava] and Wagus [Váh] are utilised for transport purposes. Two of the salt columns [*bałwans*], measuring perhaps six feet long and four feet wide, are carried on one raft, which quickly transports them to Trnava, about five miles from Pressburg [Bratislava]; there they arrive at the Danube, not only for sending to the whole of Hungary, but even Turkey, the salt going as far as Istanbul.'[116]

Documentary evidence certainly supports the importance of Cracow's salt trade with neighbouring areas over the Carpathians during the first three decades of the seventeenth century. In 1606 large quantities of Polish salt were sent to the main Bohemian warehouse at Jablonec nad Nisou, some of this presumably coming from Wieliczka.[117] Even the great merchant Jan Thurzo had an interest in salt purchases at this time.[118] During the 1620s, carters delivering Slovakian handicraft goods to Cracow often made the return journey with salt supplies. For example, in 1624 customs books reveal that salt was sent to Jičín, Opava, Krnov, Ostrava, Příbor, Frýdek, Lipník, Uničov, Olomouc, Fryštát, Hlučín, Přerov and Mirov.[119] Later examples may also be found of purchases throughout the seventeenth century.[120] On the domestic front Cracow's salt trade suffered from a shrinking market to the north and north-east of the city, faced with increasing competition from Baltic port supplies[121] and further complicated by the vagaries of war in the later part of the century.

The second half of the seventeenth and beginning of the eighteenth centuries were marked by a considerable decline in output at Cracow's salt mines.[122] Wars, internal unrest and natural disasters did little to improve the situation. Certain years also suffered from the illegal importation of foreign supplies to some parts of Poland, notably Mazovia and Podlasie, in spite of stiff legal penalties and close guarding of frontiers.[123] For Cracow, some improvement in the salt trade took place in the second half of the century, as seen from Table 6.

Prior to 1778 there appears to have been lively trade in Cracow, with 1775

Table 6. *Salt transported from Wieliczka to Cracow 1750–1785 (No. of cart loads)*

Year	1750	1755	1760	1765	1770	1775	1780	1785
No. of cart loads	2,596	1,396	3,181	883	3,601	4,736	119	3,674
Approx. weight (in tons)	1,298	698	1,591	441	1,800	2,368	60	1,832

Source: Wojewódzkie Archiwum Państwowe w Krakowie, *Varia Cracoviensis, Akta i Rachunki do handlu solą (1588–1794)*, sygn. 2991.

having the highest amount of salt sent through the city during the second half of the century. Production was temporarily affected in 1778–79 after Maria Theresa passed a law allowing a new salt warehouse to operate in Galicia, the only place where the commodity could be obtained. In turn, this led to a rise in both contraband and prices in Poland but, fortunately for Cracow, the law was rescinded by her son Joseph II between 1781 and 1783.[124] Such action helps explain the phenomenal drop in salt movement to the city in 1780, and subsequent rise five years later.

The main receiving areas for Cracow's salt appear to have been the northern part of the Cracow region and Greater Poland via the Vistula river route, whilst quantities travelled by land to Prussia and Silesia.[125] Opportunities for exporting salt to Hungary during the eighteenth century were severely hampered by protectionist policies on the part of the Hapsburg government, particularly after an edict of 1721.[126] Salt from Wieliczka still reached towns in Slovakia and other parts of Hungary, but it met with increasing competition, especially from Transylvanian[127] and Crimean sources.[128]

The price of salt in Cracow between 1500 and 1795 suggests three phases of development (Fig. 53). During the whole of the sixteenth century prices seem to have been very stable; Rybarski has illustrated how a *bałwan* (1,075 kg), cost 4 złotys, 3 grosz in 1499 and only one grosz more in 1588.[129] Prior to 1500, both wholesale and retail salt was sold in Wieliczka, to allcomers; after it became a state monopoly, price differentials appeared between the nobility, ordinary Polish merchants and foreigners.[130] The average price for a barrel of salt (390 kg),[131] however, remained stable until after 1600, when the disturbed political situation of the seventeenth century began to reflect changes; early Polish involvement in the Thirty Years War (*circa* 1620) and the period 1647–61 (Cossack and Swedish Wars) reflect higher prices, particularly in 1656 when Wieliczka was partially destroyed.[132] The final phase began around 1710 followed by price-rise fluctuation on a much steeper scale than previously and lasting to the end of the

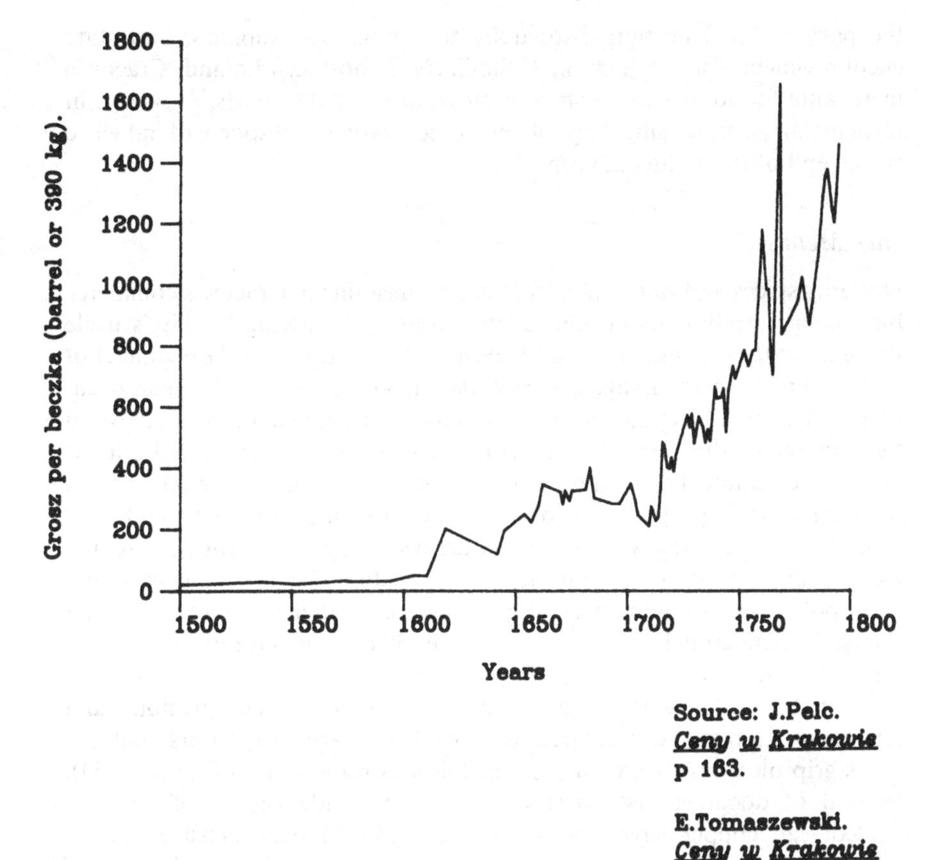

53 Average price of salt in Cracow, 1500–1795.

century. The years 1766–70 show a phenomenal price rise, which perhaps prompted the increased production seen in Table 6. The impact of the 1778 law concerning the new Galician warehouse and its abolition in 1781–83 is also reflected in the price pattern, thereafter rising further and making Cracow's salt merchants even less competitive in their traditional foreign markets by the end of the century.[133]

Throughout the period 1500–1795 the main centres abroad for Cracow's salt were in Northern Hungary (Slovakia), Silesia and, to a lesser extent, Moravia and Bohemia; on the domestic front it was Little Poland and Mazovia that provided the main outlets. Prior to 1500, Slovakia had greater importance as a market for salt than Silesia, but between 1500 and 1700 this situation seems to have been reversed. The main reason was opposition on the part of the Hapsburg Monarchy to foreign salt supplies, and state encouragement for production.[134] Similarly, in northern Poland, Cracow's merchants found growing competition from the Baltic ports,[135] which, in turn, meant a further shrinkage of her former commercial sphere of influence by the end of the eighteenth century.

(iii) Metals

Not only spices and salt gradually lost their meaning in Cracow's commerce but mineral trading also declined; they were still entering the city's trade throughout the sixteenth and early seventeenth centuries but the character of their commerce had changed. Partly, the answer lay in the Ottoman occupation of some former source areas, which now depended, unfortunately, on the consent of the Turkish authorities, who realised their value both for armament manufacture and minting of coinage. Another factor was the rise of international companies involved in the mineral trade; they could raise more easily than Cracow's merchants the necessary capital for new mining technology, a costly item with increasing inflation. Finally, Cracow's near monopoly of certain minerals in the Middle Ages had now been broken, and the city's merchants had to compete for markets in the face of much wider international rivalry.

Nevertheless, a wide range of minerals, precious/semi-precious and ferrous/non-ferrous still entered Cracow from various quarters, but the city's grip on transit trade in these metals was no longer so firm (Fig. 54). Perusal of documentary sources indicates the wide variety of minerals traded: lead, copper, silver and gold constituted the more expensive variants whilst sulphur, saltpetre, iron, tin, calamine and vitriol formed the bulk of cheaper deposits. For Cracow, the first half of the sixteenth century was to be the zenith of its involvement not only in mineral trading, but also in metallurgical processing, being third in importance after Lwów and Vilna in Poland at that time.[136] Małowist has stressed the value of 'exports of the products of Polish mining and handicrafts'[137] for this period, but the overall prominence of Poland's position in these industries began to decline after about 1560.[138] According to Piekosiński, 'the export of lead completely ended in the seventeenth century' for Cracow,[139] whilst Topolski laments with reference to southern Poland that 'in metallurgy' there was 'a sharp decline in the seventeenth century, worsened further by growing imports and a decrease in artisanship in the towns'.[140] Some revival has been noted by Rusiński, for 'transport in the first half of the seventeenth century comprised also small deliveries of minerals (copper, iron, sulphur)', but, 'In the second half of the eighteenth century they increased again and iron ore from the frontier regions of Greater Poland was exported';[141] for Cracow it was a case of too little, too late.

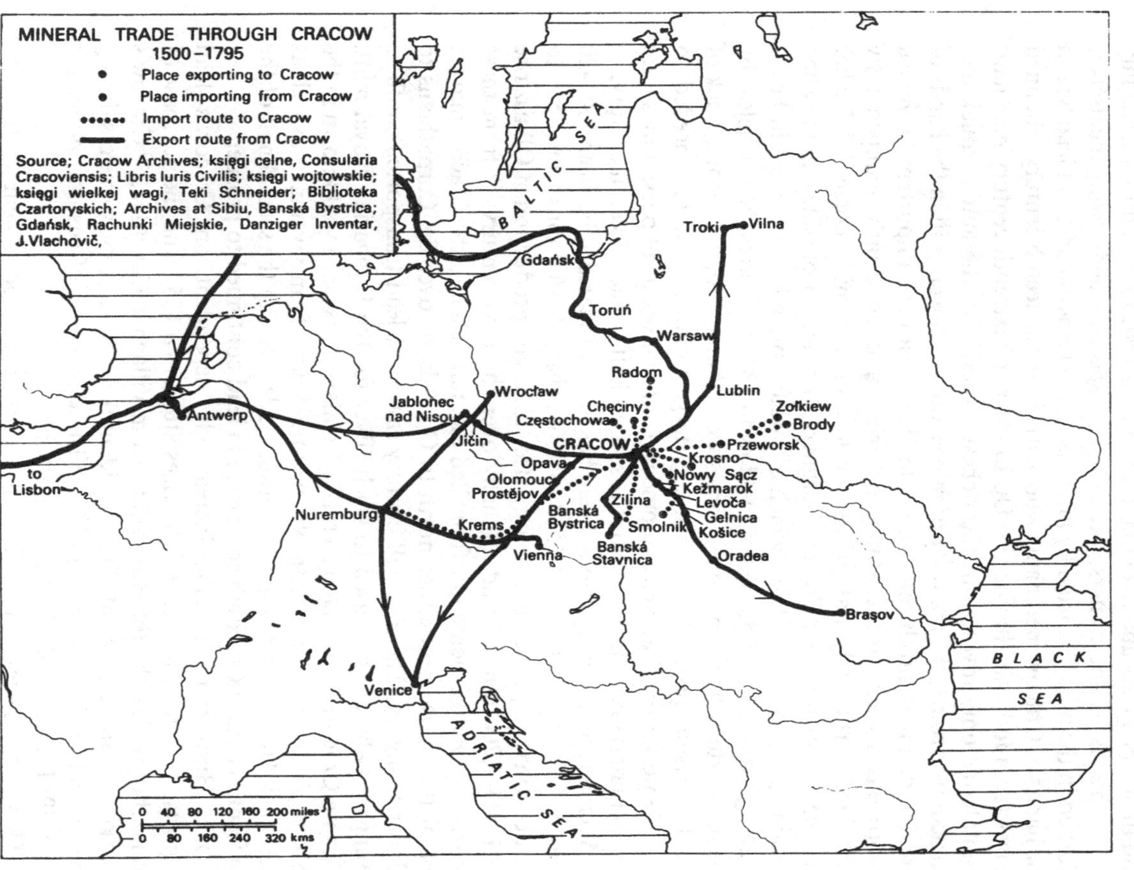
MINERAL TRADE THROUGH CRACOW
1500–1795
Place exporting to Cracow
Place importing from Cracow
Import route to Cracow
Export route from Cracow
Source; Cracow Archives; księgi celne, Consularia Cracoviensis; Libris Iuris Civilis; księgi wojtowskie; księgi wielkej wagi, Teki Schneider; Biblioteka Czartoryskich; Archives at Sibiu, Banská Bystrica; Gdańsk, Rachunki Miejskie, Danziger Inventar, J.Vlachović,
BALTIC SEA
Troki
Vilna
Gdańsk
Toruń
Warsaw
Radom
Lublin
Wrocław
Jablonec nad Nisou
Chęciny
Częstochowa
Zołkiew
Brody
Antwerp
Jičin
CRACOW
Przeworsk
Krosno
Opava
Olomouc
Prostějov
Nowy Sącz
Kežmarok
Levoča
Gelnica
Košice
Zilina
Banská Bystrica
Smolnik
Banská Stavnica
to Lisbon
Nuremburg
Krems
Vienna
Oradea
Brașov
BLACK SEA
Venice
ADRIATIC SEA
0 40 80 120 160 200 miles
0 80 160 240 320 kms

54 Mineral trade through Cracow, 1500–1795.

Lead was the major mineral exported from Cracow during the sixteenth century. The mining of Polish lead and silver was mainly concentrated in Olkusz and Chęciny,[142] the latter some 100 km to the north of Cracow and the former only about 40 km to the north-west. Exploitation was mainly in the hands of the miners, organised in special guilds, who themselves worked in the mines assisted by hired labour. The actual trade in lead was in the hands of Cracow merchants who, early in the sixteenth century, obtained leases from individual mines.[143] Up to the mid-sixteenth century merchant capital was only invested in the richest mines, with the aim of quick and immediate profits; the majority of mines, however, were still in the hands of small individual producers, often unable to afford important technical improvements e.g. for draining galleries, and constantly hampered by inflation, particularly oat prices for feeding the mine horses.[144] Richer merchants, such as the Cracow burghers, brought in outside capital when obtaining quantities of lead, which allowed some mines to invest in better drainage systems and deeper shafts down to 40–50 metres. Nevertheless by the mid-sixteenth century many mines were abandoned, not through lack of ore but rising production costs, e.g. large number of horses needed to provide power for drainage equipment. Higher investment possibilities were not to materialise until well into the second half of the sixteenth century.

Average annual production of lead has been estimated at between 100–420 tons in Olkusz for the years 1520 to 1530, rising to 700 tons from 1540 to mid-century. Improved technology after 1550 saw production at Olkusz rise to 1,000–1,200 tons of lead annually over the next century;[145] from these sources at Olkusz merchants funnelled the minerals into the Cracow market for further movement either north to Gdańsk, or over the Carpathians to Upper Hungary. The first half century saw most lead transported south to Hungary, where it was used to smelt silver for the Hungarian Royal mint. Certain Cracow families (e.g. Tratkop, Haller)[146] became well known in the lead trade and had links with Nuremburg[147] and Transylvania[148] as well as the north Hungarian towns and Lithuania.[149] During the second half of the sixteenth century Cracow's commerce in lead continued to prosper.

The Hungarians became increasingly dependent on Cracow's lead exports; in spite of higher Polish prices after *circa* 1550, and the purchase of some cheaper lead by the Hungarians from Villach in Corinthia,[150] Cracow's market remained buoyant. This is evident from documentary sources during the second half of the century,[151] whilst comparison of customs books for 1538/39, 1585/86 and 1591/92 support this fact.

These figures suggest that about half of the lead exports went to Hungary, in particular the main silver smelting centres of Slovakia.[152] Although the sending of minerals to Turkey was strictly forbidden, some of Cracow's lead exports were still reaching the semi-independent parts of Transylvania at the end of the century.[153] Far less was sent down the Vistula to Gdańsk;[154] in

Table 7. *Lead exports from Cracow in the sixteenth century*

	1538/39		1585/86		1591/92	
	Cetnars	%	Cetnars	%	Cetnars	%
Total amount recorded in Cracow	4,597 (4,60 tons)	100	22,600½ (2,262 tons)	100	3,208 (320 tons)	100
of which to Hungary: arriving in:	2,105 (210 tons)	45.7	13,335½ (1,334 tons)	59.0	1,573 (157 tons)	49.0
Levoča	64 tons	(30.4)				
Banská Štiavnica	59 tons	(28.3)				
Kežmarok	55 tons	(26.2)				
of which to Gdańsk via Vistula R.:	1,100 (110 tons)	23.9	1,179 (118 tons)	5.3	100 (10 tons)	3.2
of which to Moravia/Silesia:			7,885 (790 tons)	34.8	359 (36 tons)	11.2
of which to Austria:					262 (26 tons)	8.1
Unknown destination	1,392 (140 tons)	30.4	201 (20 tons)	0.8	914 (91 tons)	28.5

Source: R. Rybarski, *Handel i polityka ... op. cit.*, Vol. II, p. 184; Ibid., Vol. I, pp. 140–2; Wojewódzkie Archiwum Państwowe w Krakowie, *Księgi celne*, rkp 2116 k.15 v.

1588 lead from Cracow was sent to Spain via Gdańsk, perhaps coinciding with demands for supplying the needs of the Spanish Armada. Moravia and Silesia also provided outlets for Cracow merchants but this seemed to decline in importance towards the end of the century.[155] Finally, it was also sent to Austria, especially Vienna and Krems, often in the same consignment as wax.[156]

Trade in lead from Cracow, particularly to Hungary and Gdańsk, continued during the first half of the seventeenth century; decline after mid-century was largely due to flooding problems and poor maintenance of the Olkusz mines. During the early years of the century lead was transported over the Carpathians to Slovakia,[157] and even further to Oradea in Transylvania.[158] Increasing use of the Vistula was made by Cracow merchants for sending lead to Gdańsk as seen in Table 8. According to Bogucka, between

Table 8. *Lead exports from Cracow to Gdańsk 1600–1650*

Cracow Merchant	Years operating	No. of separate years recorded	Total amount		Annual average	
			cetnars	tons	cetnars	tons
M. Gerstman	1612–1616	four	15,030	1,503	3,757½	375
W. Celesta	1605–1641	eighteen	53,325	5,332	2,962½	296
J. Celesta	1634–1641	five	13,905	1,390	2,781	278
W. Gerstman	1617	one	2,250	225	2,250	225
M. Szembek	1609–1612	three	6,660	666	2,220	222
Z. Celesta	1651	one	1,350	135	1,350	135
G. Wizemberg	1605–1649	four	3,150	315	787½	79
K. Kestner	1632–1641	five	3,863	386	772½	77
K. Celesta	1649–1650	two	1,260	126	630	63
K. Pszonka	1650–1651	one	400	40	400	40
Z. Hipolit	1648–1651	three	1,097	110	365½	37
J. Celesta	1651	one	209	21	209	21
Total amount			102,499 cetnars	10,249 tons	18,485½ cetnars	1,848 tons

Source: Calculated from Table 18, H. Obuchowska-Pysiowa, *Handel wiślany w pierwszej połowie XVII wieku* (Ossolineum), Wrocław-Warszawa-Kraków, 1964, p. 144.

1608 and 1618, 4,391 cetnars (440 tons) of lead was sent from Gdańsk to the Iberian peninsula,[159] some of which may have resulted from consignments sent by the above merchants, or other sources recorded in Cracow documents.[160]

A partial revival of the lead trade took place during the second half of the eighteenth century with evidence of several purchases sent from the Olkusz mines to Cracow *circa* 1750–60;[161] some was re-exported from the city towards Gdańsk, Toruń and Warsaw,[162] but by the 1780s the overall decline of lead mining became complete.[163]

Silver and gold also entered Cracow's metal trade. The very nature of these precious metals often meant their movement and trade were shrouded in secrecy. Whilst Hoszowski maintains that 'there was practically no production of silver or gold in Poland'[164] it is known that both minerals were mined at Olkusz and in Lower Silesia.[165] Moreover, sources in Hungary were in decline for, according to Nef, 'Beginning in the late twenties of the sixteenth century the Turkish invasion and the growing dislike of the local population for the German capitalists who had gained partial control of the local enterprises, apparently brought about some decline in the output of silver at Neusohl and most other Hungarian mines.'[166] The existence of gold and silver smiths in Cracow during the first half of the sixteenth century[167] meant that such metals must have been brought to the city; furthermore, it is known that the Fugger and Thurzo families were involved in silver-mining enterprises[168] and must have sent quantities to Cracow either for domestic use or re-export,[169] e.g. to Vilna.

Domestic sources of other, less expensive, minerals, also entered Cracow's trade structure in the sixteenth century and similarly were to decline in importance after 1700; sulphur, calamine and iron were the most significant. Although it has been stated that 'The export of sulphur had always been small and without importance',[170] more recent research has shown this to be rather overstated. The earlier impetus of sulphur mining at Swoszowice, near Cracow, continued during the Early Modern Period. There was certainly a local demand for sulphur in Cracow, especially during the years 1508, 1543, 1555 and 1591, when the worst plagues were recorded in the city;[171] there was also national and international markets for the city's sulphur supplies. In 1581, 90 per cent of the sulphur produced went for export abroad[172] usually down the Vistula to Gdańsk,[173] but also south to Transylvania.[174] The first half of the seventeenth century saw continued development; particularly via Warsaw to Gdańsk (Table 9).

The other outlet during the seventeenth century was Upper Hungary,[175] particularly Slovakia, but here it was in direct competition from Sicilian supplies.[176] Silesia was also a market for Cracow's sulphur mines.[177] By the mid-eighteenth century the Swoszowice complex was run down and neglected and Cracow merchants had to buy their supplies from Gdańsk, Toruń and Opava;[178] sulphur mining was revived at Swoszowice under Austrian rule in the 1790s,[179] but the profits were to go to Vienna not the merchants of Cracow.

Calamine, previously imported into Poland during the earlier centuries, began to be exploited by Cracow merchants around Chrzanów, 35 km west of Cracow, at the end of the sixteenth century. Its importance as an ingredient for brass founding led to increased trade in the seventeenth century; not only was the powder used locally by metallurgical works, e.g. in Cracow,[180] but also it was in demand by Gdańsk merchants for export to western Europe.[181] The peak years of local production in the mid-seventeenth century appear to have coincided with considerable quantities stored in Cracow's Vistula warehouses for further delivery to Gdańsk.[182] Decline in the eighteenth century meant a return by Cracow traders to earlier sources, notably around Tenczyn.[183]

Domestic supplies of Polish iron ore were not utilised in Cracow until the sixteenth century. The extraction of iron ore and production of iron was carried on by small undertakings, scattered in many parts of the country, mostly near Częstochowa, Radom and Chęciny.[184] In 1511, King Zygmunt I granted a royal privilege for mining iron ore near Chęciny[185]

Table 9. *Export of sulphur from Cracow to Gdańsk via Warsaw, 1589–1636*

	1589	1597	1607	1617	1632	1636
Annual amount	26 cetnars 4 faska	3 faska	465 kamień	268.2 kamień	992.3 kamień	223.5 kamień
Tonnage equivalent	2.08	0.29	6.04	3.5	13.0	2.9

Sources: Calculated from R. Rybarski, *Handel i polityka* Vol. I, p. 146; Wojewódzkie Archiwum Państwowe w Krakowie, *Księgi Celne*, for 1589 and 1597; H. Obuchowska-Pysiowa, *Handel wiślany w pierwszej połowie XVII wieku* (Ossolineum), Wrocław-Warszawa-Kraków, 1964, Table 40, p. 164.

and, by the end of the century, several villages in this area were engaged in its extraction;[186] unfortunately, this was insufficient for Cracow's needs, and throughout the sixteenth century the city was very reliant on Hungarian ore from the Slovakian mines,[187] sent over the Carpathians often to the iron foundry at Nowy Sącz,[188] whence it arrived in Cracow in a malleable form. It was in great demand in the city largely for the manufacture of various metal goods[189] (e.g. knives) some of which were then exported abroad. In other cases, the iron was dispatched in loads down the Vistula to Gdańsk,[190] or eastward to Lithuania and Belorussia.[191] The seventeenth century witnessed further iron imports over the Carpathians via places like Krosno,[192] eventually arriving on the Cracow market; after 1640, however, iron and iron goods were supplied to the Baltic coastal towns almost entirely by the Dutch and Swedes,[193] relieving Cracow of her northern markets. Although ironworking had long been carried out in the Holy Cross mountains, near Kielce, in central Poland, the methods used remained simple and output small until the last decades of the eighteenth century.[194] Cracow's metal handicraft industry, therefore, still had to rely on imports of worked metal during the second half of the century, often supplied by Opava merchants.[195] Transit trade in the ore through Cracow had, however, long ago suffered decline.

Of the other minerals entering Cracow's commercial sphere, perhaps mention should be made of saltpetre. Its attraction for military purposes, meant a steady demand from places like Transylvania[196] and Gdańsk in the sixteenth century[197] and, though usually transported in small quantities,[198] continued to be in demand in the eighteenth century.[199] Production was concentrated in eastern Poland (Brody, Żółkiew, Przeworsk) whence it was sent to Cracow, largely by Jewish merchants.[200] Part of the saltpetre consignments were then exported from Cracow to places farther west, e.g. Wrocław[201] probably for re-sale on the German market.

The purchase of metal ore and its refinement was a business in the hands of Cracow's patricians. Individuals and firms combined to invest in the opening of ore mines and to finance metal purchases and then sold the crude material to the owners of liquation works for smelting in the case of copper and lead, or forges for iron. Besides the purchase of domestic ore supplies, Cracow was dependent for certain minerals on mining sources abroad, of which by far the most important was copper. This metal, in demand for both bronze and brass and for use in coinage, was produced mainly in Slovakia and at Mansfield in the Harz Mountains.[202] From these sources it was distributed to Italy and north-western Europe. For Cracow, the former area, Slovakia, was the main source of her copper ore.

On 15 November 1494, Jacob Fugger of Ausgberg[203] and the Cracow patrician Johann Thurzo[204] signed a deed of partnership upon which was based their joint exploitation of the North Hungarian copper mines at

Table 10. *Fugger transit trade in copper through Cracow, 1512–1517*

	1512	1513	1514	1515	1517
Amount	2,200 cetnars	5,103 cetnars	9,337 cetnars	2,439 cetnars	16,357 cetnars
Tonnage equivalent	220 tons	510 tons	935½ tons	544 tons	1,636 tons

Source: Calculated from J. Ptaśnik, 'Przedsiębiorstwa kopalniane krakowian i nazwiązanie stosunków z Fuggerami w początku XVI wieku', in J. Ptaśnik (ed.), *Obrazki z przeszłości Krakowa*, Seria 1 (Biblioteka krakowska No. 21), Kraków, 1902, pp. 77–8.

Banská Bystrica.[205] This alliance of technical ingenuity and financial resource proved highly successful; by 1500, mines around Banská Bystrica had been extended and new ones opened up further east at Smolnik[206] and Gelnica.[207] As the accumulated wealth of the Fugger Empire was poured into the enterprise, return on investments increased and further expansion took place. Although Cracow no longer had a monopoly position in the copper trade, and had to compete with places like Wrocław for its control,[208] nevertheless it was a major transit point en route to the Baltic coast. Some idea of Fugger's trade during the early years is illustrated in Table 10. The Fugger/Thurzo enterprise was only one, though probably the largest, that sent copper through Cracow. Others existed throughout the sixteenth century, e.g. Linke Langnauer and Haug,[209] each paying money into Cracow's coffers for the privilege of using the city as a transit centre.[210] Thus, although the full profits from copper did not remain in Cracow, as in the former monopolistic days, at least the city was partly able to benefit from these links in international trade. Of course, there were always companies who tried to avoid paying Cracow's copper taxes,[211] whilst others were freed of tax commitments; this was especially so if the copper was for the King's court, for building purposes in the city, or armament manufacture.[212] Therefore, figures recorded in the customs tariff books, e.g. 1538/39, 1584 and 1593,[213] were probably much smaller than actual amounts of copper entering the city.

Copper imports from Hungary gradually lost their importance for Cracow in the following century. Even during the second half of the sixteenth century some of the earlier impetus had been lost when exports from the Slovakian mines were increasingly sent via Silesia and Bohemia to Hamburg.[214] The copper trade did not disappear entirely in Cracow thanks to the efforts of such entrepreneurs as Wolfgang Paler[215] during the first three decades of the century, together with some of the merchants previously mentioned (Table 11) as active in the lead trade.

According to Kulczykowski[216] the copper trade was still lively during the first half of the eighteenth century, but then declined completely. The second half of the century saw very small amounts enter the town, largely for local use by Cracow boiler-makers.[217]

The main export market for copper travelling through Cracow in the sixteenth century was Gdańsk, en route for the major copper emporium at Antwerp.[218] Copper was sometimes transported entirely overland on the route from Banská Bystrica via Wrocław and along the 'Hohe Landstrasse' to Leipzig, where the Fuggers had a separating plant;[219] but the route from Slovakia via Cieszyn, Cracow, Sandomierz, Toruń to Gdańsk, utilising the R. Vistula was often preferred, because the water route was cheaper.[220] Documentary sources prove the use of this route early in the sixteenth century[221] whilst figures for the period 1495–1529 suggest that peak move-

Table 11. *Copper exports from Cracow to Gdańsk, 1605–1651*

Cracow merchant	Years operating	No. of separate years recorded	Total amount		Annual average	
			cetnars	tons	cetnars	tons
W. Celesta	1605–1641	sixteen	5,322	532	332.6	33¼
G. Wizemberg	1605–1649	one	210	21	210	21
K. Kestner	1632–1641	three	573	57¼	191	19¼
Z. Hipolit	1648–1651	three	1,128	113	376	37½
J. Celesta	1651	one	181	18	181	18¼
W. Orseti	1650	one	395	39½	395	39½
J. Furmankowicz	1648–1650	two	494	49½	247	24½
TOTAL			8,303	830¼	1,932.6	193¼

Source: Calculated from Table 18, H. Obuchowska-Pysiowa, *Handel wiślany w pierwszej połowie XVII wieku* (Ossolineum), Wrocław-Warszawa-Kraków, 1964, p. 144.

Table 12. *Copper exported from Cracow via Gdańsk to Western Europe, 1495–1529*

	1495–1504	1507–1509	1510–1513	1514–1516	1517–1529
Amount in cetnars	24,000	36,000	77,000	50,000	46,000
Tonnage equivalent	2,400	3,600	7,700	5,000	4,600

Source: A. Divéky, *Felsö Magyarország kereskedelmi összeköttetése Lengyelországgal fölog a XVI–XVII században*, Budapest, 1905, p. 329.

ment was reached at the end of the first decade of the sixteenth century (Table 12). It should be noted, however, that much of this trade was in the hands of non-Cracow merchants, particularly traders from Toruń.[222] Others were agents acting for the Fuggers, who organised the movement of copper through Cracow and Gdańsk to Portugal,[223] Spain[224] and other west European countries including Denmark[225] and Holland.[226] This situation continued until the dissolution of the Fugger agency in Cracow in 1548,[227] although the stream of copper sent through Cracow did not dry up immediately. However, as the second half of the sixteenth century progressed less and less copper from Hungary came through the city.[228]

Copper from Cracow was also sent to Lithuania in the sixteenth century,[229] especially to Troki[230] and Vilna.[231] Other consignments to places like Lublin may also have been en route to the Lithuanian markets,[232] but both the size of loads and frequency were infinitesimal compared with copper movements to Gdańsk and western Europe.

Even at the end of the sixteenth century copper was still being sent from Slovakia via Cracow to Gdańsk, but changes in the export routes were clearly visible with preference given via Wrocław to the German towns.[233] Some respite for Cracow was given by the trading activities of Wolfgang Paler during the first two and a half decades of the seventeenth century (Table 13). Beside Paler's near monopoly in Cracow's copper trade with Gdańsk, Table 13 also reveals that over 40 per cent of his copper exports from Slovakia reached Gdańsk. Little of this amount now seemed to be reaching the Iberian peninsula for, according to Bogucka, copper exports there only totalled 18 cetnars in 1612, none for 1613/14, 5 in 1615, 56⅓ in 1617 and 36 cetnars in 1618.[234] Between 1625 and 1650 copper movement from Cracow to Gdańsk declined, largely due to increasing amounts now sent from Slovakia to western Europe via Wrocław-Frankfurt and the Oder and Elbe rivers to Hamburg.[235] Copper is last recorded as sent from Cracow to Gdańsk in 1649;[236] after that the eclipse of this transit trade was complete.[237]

Of the remaining minerals imported to Cracow from abroad mention

Table 13. *Copper exports through Cracow to Gdańsk, 1612–1624*

Year recorded	Copper sent from Slovakia to Cracow by Paler		Copper sent from Cracow to Gdańsk			
			Through Paler		By other merchants	
	Cetnars	Tonnage Equiv.	Cetnars	Tonnage Equiv.	Cetnars	Tonnage Equiv.
1612	358	36	288	29	40	4
1613	748	75	178	18	–	–
1614	703	70	–	–	–	–
1615	1,752	175	782	78	–	–
1616	3,087	309	1,444	144½	–	–
1617	1,584	158½	968	97	–	–
1618	321	32	121	12	–	–
1619	4,026	402½	2,123	212	50	5
1623	2,370	237	391	39	–	–
1624	124	12½	–	–	–	–
Total	15,073	1,507½	6,295	629½	90	9
% of copper sent by Paler		98.60%		1.40%		

Source: Calculated from J. M. Małecki, *Związki handlowe miast polskich z Gdańskiem w XVI i pierwszej połowie XVII wieku*, (P.A.N.), odd. w Krakowie (Prace Komisji Nauk Historycznych No. 20) Wrocław-Warszawa-Kraków, 1968, p. 136.

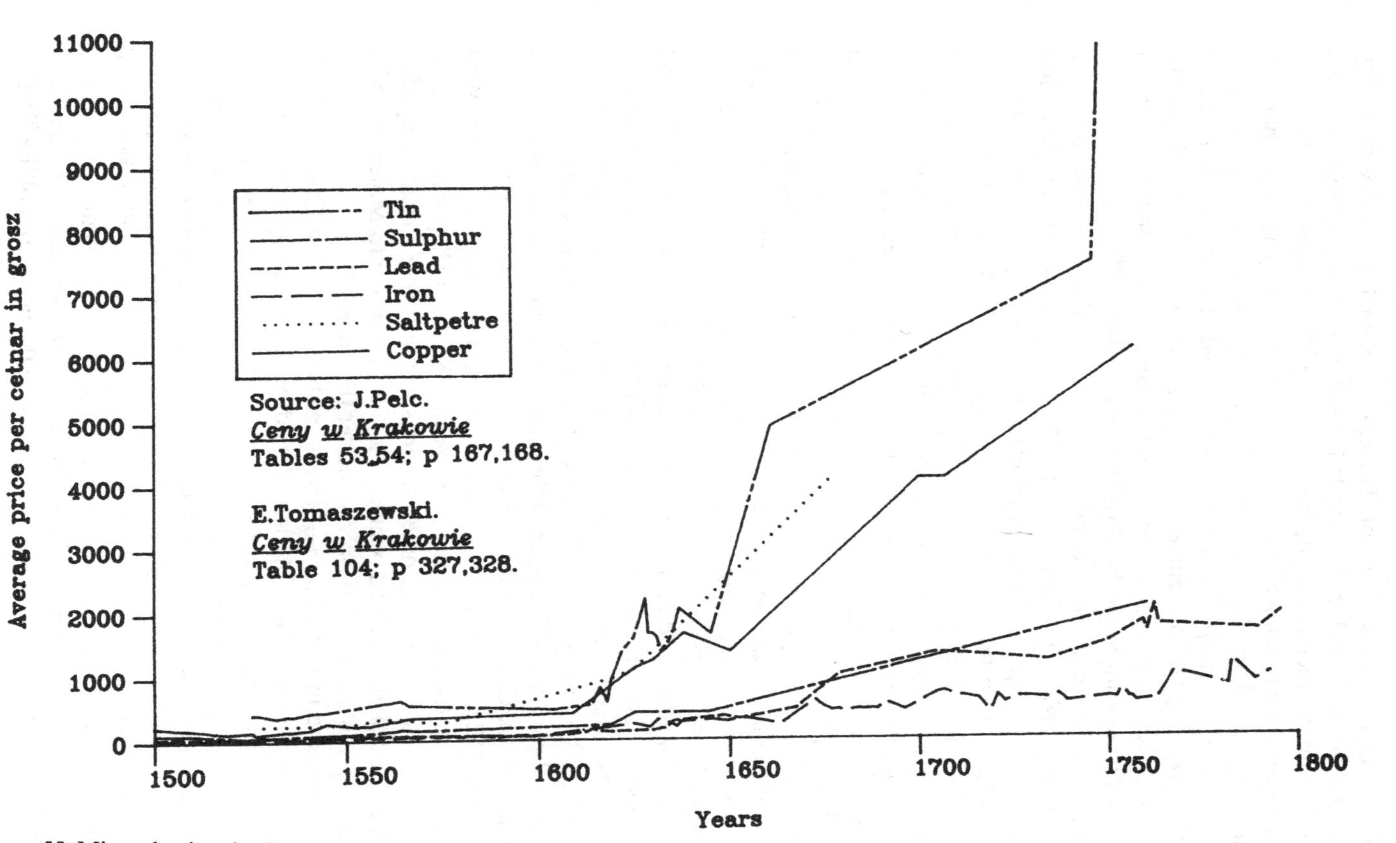

55 Mineral prices in Cracow, 1500–1795.

should be made of tin and vitriol. Tin was in demand in the city mainly by metal craftsmen[238] and was imported, usually in small quantities, from Silesia[239] and, to a lesser extent, from Germany[240] and Moravia.[241] Surplus amounts in the city sometimes led to re-export, mainly to other parts of Poland and northern Hungary.[242] Vitriol, referred to in documents as *Kupherwasser* and *Koperwasser*[243] was imported from northern Hungary together with copper consignments.[244] Its general application for the metal-working trade meant steady demand in Cracow, whilst in some years small quantities were sent down the Vistula river to Gdańsk.[245]

Mineral prices over the period 1500 to 1795 (Fig. 55) show dramatic changes after 1600, especially up to the Thirty Years War. As Vlachovič has stated, 'The attempts at obtaining higher prices for copper were begun as early as 1599, even for copper of inferior quality';[246] Braudel and Spooner have shown that 'with the beginning of the sixteenth century, we see a sharp rise in copper prices [in western Europe],[247] and after that . . . there is a trend towards equilibrium and a steady market', reflected in Cracow's figures for the sixteenth century and supported by the general evidence for west and central Europe collected by Hoszowski.[248] The decline of Cracow's heavy involvement in the copper trade after 1650 saw a rapid rise in the city's prices after mid-century; it also meant less income for the city council from taxes on minerals which, according to Pieradzka, were 'some years yielding nothing'.[249] Iron, a rival to copper in everyday life, and another metal for coinage, showed a steadier price rise than some of the other metals. Braudel and Spooner have discussed iron prices with reference to western Europe and maintain that 'prices were steady with a marked – though late – upswing after 1750'.[250] Cracow seems also to fit that trend, but Lwów deviates from this general pattern.[251] The price of tin shows it was the most expensive metal arriving on the Cracow market, a cetnar costing twenty-six times more in 1748 than 1525; this compares very unfavourably with some other West European countries, for example, 'Between 1480 and 1740 prices of English tin only doubled.'[252] Of the other minerals, sulphur rose considerably after 1600, a trend similar in lead, whilst calamine, according to Molenda, had a pattern resembling lead but only a quarter the price.[253]

The decrease in importance of spices, metals and salt in Cracow's trade structure during the Early Modern Period meant it had to find alternative commodities to trade in; the results from these changes were to cause a reorientation in the city's commercial sphere. This meant a decline in the intensive exploitation of transit trade, particularly the more expensive raw materials, coupled with a need to adopt more extensive trading connections over a wide area in the import/export of cheaper merchandise. Perhaps of necessity this was a wise move; Kula has shown that in Poland, as probably elsewhere, luxury goods gradually became more accessible in the Early Modern Period. What did become dearer were the necessities of life.[254] This was particularly the case in eighteenth-century Europe, where luxury began to spread and established itself widely among the middle classes,[255] whilst demand for basic goods, like the population, continued to grow.

Trade in cheaper raw materials

With the decline of the more expensive raw materials in Cracow's commerce, particularly after 1650, there was a greater emphasis placed on cheaper unprocessed merchandise – skins, wax and livestock, all of which were produced in large quantities throughout parts of Eastern Europe and had a ready sale in the West. Pounds has commented on sixteenth-century European commerce for 'There was change also in the pattern of trade ... the internal routes of Europe were very far from being neglected; indeed, the volume of goods transported over them increased ... In eastern Europe new towns sprang up, and a few grew rich on the profits of trade in grain, metals and forest products.'[256] It now remains to be seen how Cracow's trade structure adapted to these changed conditions and what commodities replaced diminishing profits from such vendibles as spices, copper and lead.

(i) Skins and furs

The close of the Middle Ages saw a new chapter open in the historical geography of Cracow's trade. In the sixteenth and seventeenth centuries its development became mainly dependent on the general processes of economic expansion in Poland and on those transformations which affected the main tendencies of European commerce. The trade route leading from the Black Sea through Poland to Western Europe had lost its significance, which inevitably weakened the leading position of Little Poland, and Cracow in particular, in certain branches of Polish commerce. Certainly the new geography of European trade was to make northern and central Poland increasingly significant areas in East–West commercial exchange, but in some commodities Cracow was more than able to hold its own position. While cloth, hardware, metal and metal goods were in the main traded within Poland and the East, the exchange structure with the West favoured the export of cattle, hides, furs, wool and even feathers.[257] Cracow seized the opportunity to develop its leather/tanning industry, in direct competition with its main rival Poznań,[258] now better located in terms of East–West links. For Cracow, it was now not enough just to trade in skins and furs, it also had to offer semi-processed or manufactured items for sale; this in turn led to a strengthening of the artisan element in the city.[259]

Cracow imported skins from various sources (Fig. 56); some came from Silesia (Wrocław),[260] from Russia and Lithuania, usually through Lublin,[261] from Moldavia, via Lwów[262] or, as Małecki has shown,[263] from

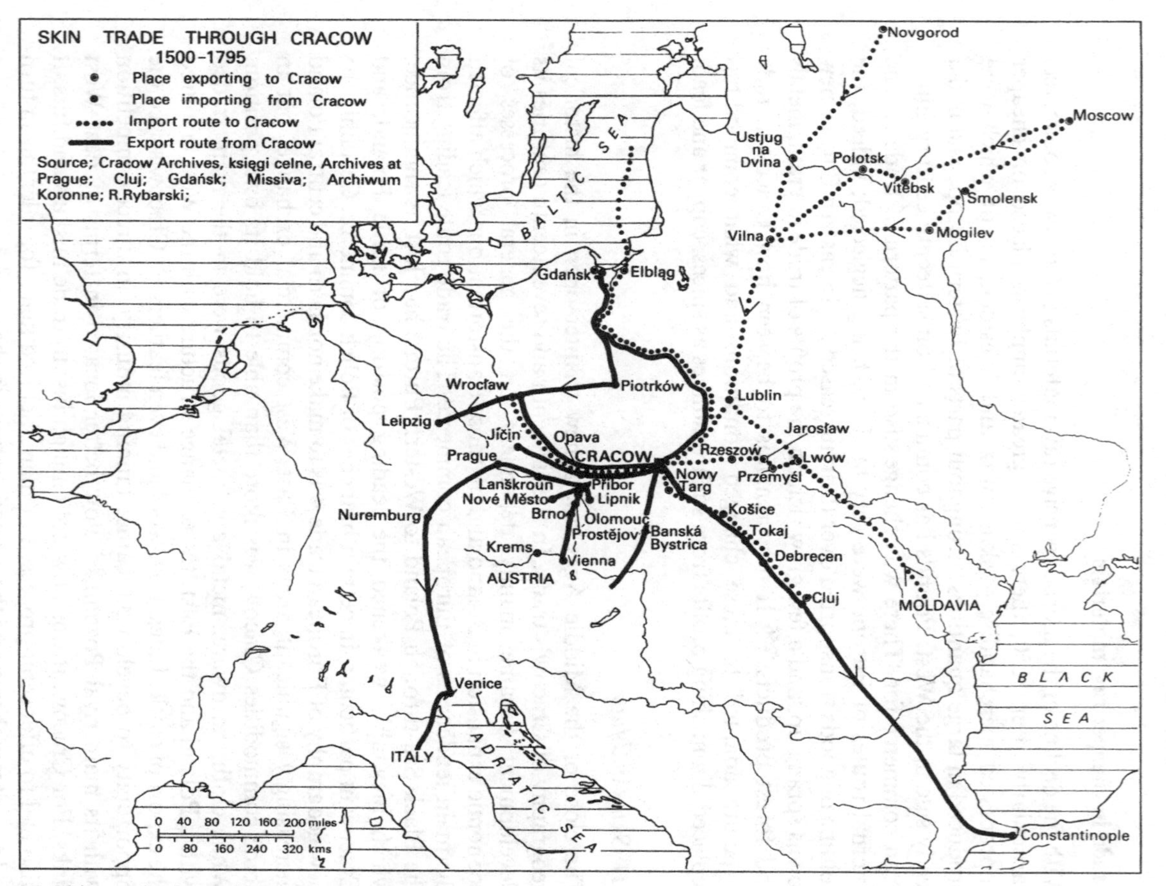

56 Skin trade through Cracow, 1500–1795.

Table 14. *Raw cattle hides exported from Cracow to Moravia in the sixteenth century*

Importing centre	1509/10	1533/34	1549/50	1584	1591–1593
Brno	36		4,253		
Jičín		909	113		
Lanškroun		702			
Nové Město		3,140			
Olomouc	876	200		2,335	
Opava			2,393	4,859	2,363
Prostějov		2,270	600		
Příbor				1,222	
Moravia (general)					870
Total	912	6,221	7,359	8,416	3,233

Sources: R. Rybarski, *Handel i polityka*, Vol. II, pp. 175; p. 194; Wojewódzkie Archiwum Państwowe w Krakowie, *Księgi celne*, rkp. 2116, 2117; F. Hejl, 'Český obchod na Krakovském trhu po Bílé Hoře' *Sborník Prace Filosofické Fakultety Brněnské University*, Vol. 10, (řada historická č. 8), Brno, 1961, p. 240.

the local region, and Greater Poland.[264] Besides the larger skins of oxen, cow, calf and heifer, the first half of the sixteenth century saw increasing demand for small skins; sheep, lambs, goats, etc., often based on local supplies. Markets for Cracow's skins and leather goods were as varied as their places of origin. Moravia appears to have imported skins from Cracow throughout the sixteenth century, especially cattle hides, as seen in Table 14. In the peak year, 1584, a third of all Cracow's exports of raw cattle hide went to Moravia.[265] Decline in the 1590s may have been due to an increasing amount of cured or worked leather being sold by the city's workshops. In Bohemia, Prague was Cracow's main export centre for skins.[266]

The city's customs books from the end of the sixteenth century reveal that Wrocław was the main destination for skins, followed by Nuremburg (chamois) and Austria (Vienna and Krems).[267] Other places importing varied skins and leather goods from Cracow included Gdańsk,[268] Leipzig[269] and, to a much lesser extent, Transylvania, Slovakia[270] and Italy.[271]

The fur trade was also important in sixteenth-century Cracow. Perhaps of less significance than the skin trade, nevertheless, Cracow's furriers bought pelts and furs from a wide area of north-east Europe. Russia and Lithuania emerge as the main sources of supply. Russian furs have always had a high market reputation throughout the world, the more expensive varieties such as sable, ermine, marten, mink etc. coming from the great trading centres, like Ustjug on the Dvina river, Novgorod, Smolensk, Vitebsk and

Polotsk;[272] they arrived in Cracow usually via Lublin.[273] The decline of the Hanseatic League did not signal the fall of Lithuanian trade; many of her merchants now came overland through Mogilev–Vilna–Warsaw–Poznań–Leipzig,[274] some bringing not only the more expensive pelts but also cheaper varieties of grey/red squirrels, rabbit and hare, beaver and badger.[275] In spite of competition from Poznań, Cracow's furriers had lively connections with Lithuania and Russia, but the turnover, according to Małecki, was small.[276] Customs returns from the end of the sixteenth century refer to quantities of squirrel, marten, fox and rabbit (collectively called *Nowogrodkowa*),[277] whilst Rybarski has found similar amounts for the early part of the century.[278] Pelts also arrived from Transylvania,[279] and Lwów[280] whilst, towards the end of the century, furriers in Cracow obtained some supplies from the Baltic region via Gdańsk.[281]

Outlets for Cracow's fur trade were mainly to the west and south in the sixteenth century. Most furs seem to have left for Wrocław and Prague, with business largely in the hands of Jewish traders[282] who had connections further west in Germany. A small, though profitable market was found in northern Hungary; Slovakian merchants came to Cracow for the purchase of Lithuanian/Russian sable, lynx, marten, fox, etc. These traders arrived chiefly from the mining towns where expensive luxury goods had a ready sale amongst the rich mine-owning classes.[283] Likewise, Polish fur traders, especially from Cracow, visited these mining areas, often spoiling the commercial interests of local furriers; their frequency reached such a stage in Banská Bystrica that a statute issued by the local council forbad Polish merchants from selling fox, marten and beaver furs in the town.[284] Finally, some of Cracow's furs found markets amongst the ruling elite of the Ottoman Empire,[285] largely due to the commercial activities of Andrej Halkokondil, a merchant from Constantinople, who supplied Sultans Selim I and Süleman I during the last two decades of the sixteenth century with sable, fox and squirrel furs, principally through his Polish connections.[286]

The seventeenth century was to witness further development of Cracow's skin and fur trade, at least until the 1640s, when the ravages of war began to disrupt commerce until the last three decades of the century. The basic source areas remained as before; cattle skins were sent to Cracow largely overland from Moldavia and via Lwów.[287] This meant the city's furriers could not benefit from the cheaper water transport afforded skins from Podolia and the Ukraine which profited from easier river connections, particularly along the Vistula[288] and San rivers.[289] The early part of the century also saw increasing competition for Cracow from Toruń, which now began to import large quantities of furs and skins from Lithuania, and Belorussia;[290] some came to Cracow from Moscow,[291] via Lublin. Skin imports from Elbląg,[292] in these early years, could have been in transit from more northerly ports like Turku.[293] Revitalised trade in the 1670s and

Table 15. *Export of skins from Cracow to Gdańsk, 1633–1647*

Year	Sent by H. Thiring		Sent by Gdańsk merchants		Sent by other Cracow merchants	Total
	By Land	By River	By Land	By River	By Land	
1633					1,205	1,205
1636					1,100	1,100
1637			800			800
1643	1,315		1,025			2,340
1644	500	1,840				2,340
1645	2,100	2,100				4,200
1646	400	6,071		527		6,998
1647		8,325				8,325
Total	4,315	18,336	1,825	527	2,305	27,308
	22,651		2,352			

Sources: Calculated from Wojewódzkie Archiwum Państwowe w Krakowie, *Księgi celne*, rkp. 2145; 2146; 2157–2161; J. M. Małecki, *Związki handlowe miast polskich z Gdańskiem w XVI i pierwszej połowie XVII wieku*, (P.A.N.), Wrocław–Warszawa–Kraków, 1968, Tables XVII, A.B.

1680s experienced renewed imports from the Ottoman-occupied lands – fine skins being sent through the Turnu–Roşu customs post for Poland, and Cracow in particular[294] probably originating in Bulgaria.[295]

Skin and fur exports from Cracow during the first half of the seventeenth century were predominantly destined for western markets. Bohemia and Moravia imported consignments of raw cattle hides as in the sixteenth century, but not in such great quantities. For example, customs returns for 1624 show a total of 1,417 hides sent there, 40 per cent designated only as 'to Moravia', whilst a further third went to Lipník.[296] Many of these skins, as with Cracow's exports to Slovakia,[297] were destined for Austrian markets. Such movement disappeared in the war-torn 1650s and 1660s, with no goods of any description being sent to Transylvania or Bohemia in the 1660 customs register; little appears to have changed even by 1688.[298] Gdańsk was also an important outlet for Cracow's skin dealers, of whom H. Thiring[299] was the chief merchant in the 1640s as seen in Table 15. This illustrates not only that Thiring sent over 80 per cent of all Cracow's skins to Gdańsk during this period but, more generally, one sees the attraction of cheaper water transport (69 per cent) over land routes (31 per cent).

Another outlet for Cracow's furs and skins was Italy, particularly during the early years of the century. Obuchowska-Pysiowa states that 'Leather and

Table 16. *Leather and furs exported from Cracow to Italy in 1604*

Origin of merchant	Pieces of Leather				Number of Furs	
	Ox Hide	Calf Leather	Russian Leather	Chamois Leather	Grey Squirrels	Fox
Cracow	1,955	2,512	400			
Italian in Cracow	2,217		1,832	677	5,950	
Intermediary town				52,380		121
Venice			240			
Total	4,172	2,512	2,472	53,057	5,950	121

Sources: Wojewódzkie Archiwum Państwowe w Krakowie, *Księgi celne*, rkp. 2126 (1604); H. Obuchowska-Pysiowa, 'Trade between Cracow and Italy from the Customs-House Registers of 1604', *Journal of European Economic History*, Vol. 9, No. 3, Rome 1980, Table 15, p. 649.

furs exported to Italy [in 1604] were partly of home production, from areas near Jarosław, Przemyśl, Rzeszów and Cracow, partly from the East, reaching Cracow through Lublin.'[300] These are seen in Table 16.
It is clear from this table that chamois (suede) leather was most important, and sent from Cracow via Nuremburg to Italy, whilst grey squirrel pelts took second place. Fox furs consisted of 120 white from Russia and one red from Germany.

Eighteenth-century Cracow seemed gradually to lose its importance in the international commerce of skins and furs; locally and regionally the tanning industry preserved its significance, still having the most important and largest guild in the city. This was partly explained by the increasing resident population, and the large demand for meat; residue skins, especially from cattle, were then supplied to the city's leather workers, or to nearby tanneries in places like Pińczów, Bochnia and Biała. Ox, cow and calf hides seemed in plentiful supply but on occasions ram skins had to be imported from Hungary via Nowy Targ, for example in 1763.[301] During the last quarter of the eighteenth century a two-way traffic developed with the Carpathian foothill region; raw skins would be sent from local villages to Cracow for tanning; some of these were then returned, the rest going for export (Table 17).

Quantities of the prepared skins were sent from Cracow to Bohemia (Opava), Silesia (Wrocław) or northward to Piotrków and the main textile centres of Greater Poland.[302] Some local Jewish merchants had connections in Wrocław, and Skoczów (Silesia), Gdańsk and even in Frankfurt.[303] Nevertheless, the general decline, according to Attman, of Russian sources

Table 17. *Cracow's skin trade 1775–1785 (in cart loads)*

	1775	1780	1785
To Cracow	24	101½	60
From Cracow	6	41	84
Total	30	142½	144

Source: Wojewódzkie Archiwum Państwowe w Krakowie, *Varia Cracoviensis*, Akta i Rachunki, (1588–1794), rkp. 2991.

for skins, leather and furs to western Europe,[304] the growing importance of England and Scotland in supplying skins and hides to the Baltic,[305] and the weak demand for Polish leather goods in south-east Europe,[306] all helped force Cracow's merchants to increasingly concentrate their efforts on providing skins and furs for the local region.

Prices for skins (Fig. 57) showed a continual rise throughout much of the period under review; the gradual increase in skin prices for the sixteenth century rose dramatically the following century, particularly in the 1650s as a result of the Swedish incursions. Although some types, e.g. heifer, dropped in value after 1663, the general upward movement continued well into the eighteenth century. Interesting details for one year, 1604, have been published by Obuchowska-Pysiowa; the 'Price of oxhide [raw] was 15 grosz; dressed 1 złoty; calf leather – 5½ grosz; Russian leather – 15 grosz; Chamois leather – 7½ grosz; grey squirrel – 3 grosz; white fox – 7½ grosz; red fox – 1 złoty.'[307] This at least gives some idea of comparative values placed on certain skins and furs in the early sixteenth century, before the political and military events of that century created price rises of enormous proportions.

(ii) Wax and honey

There was always a great demand for Polish wax in West European markets and, according to Rybarski, Cracow was the country's main exporter.[308] Both inside Poland and beyond there was a need for wax in the form of candles for the large number of churches that existed and to supply the requirements of numerous court residences.[309] Crude wax came to Cracow from various sources (Fig. 58); important were local/regional supplies from the Carpathian foothills and forested areas of the Little Poland Highlands,[310] but the major source came from Lwów[311] and Podolia. Wax was also an important export from Russia and continued to hold this position until the seventeenth century.[312] In 1555 the Council of Gdańsk referred to

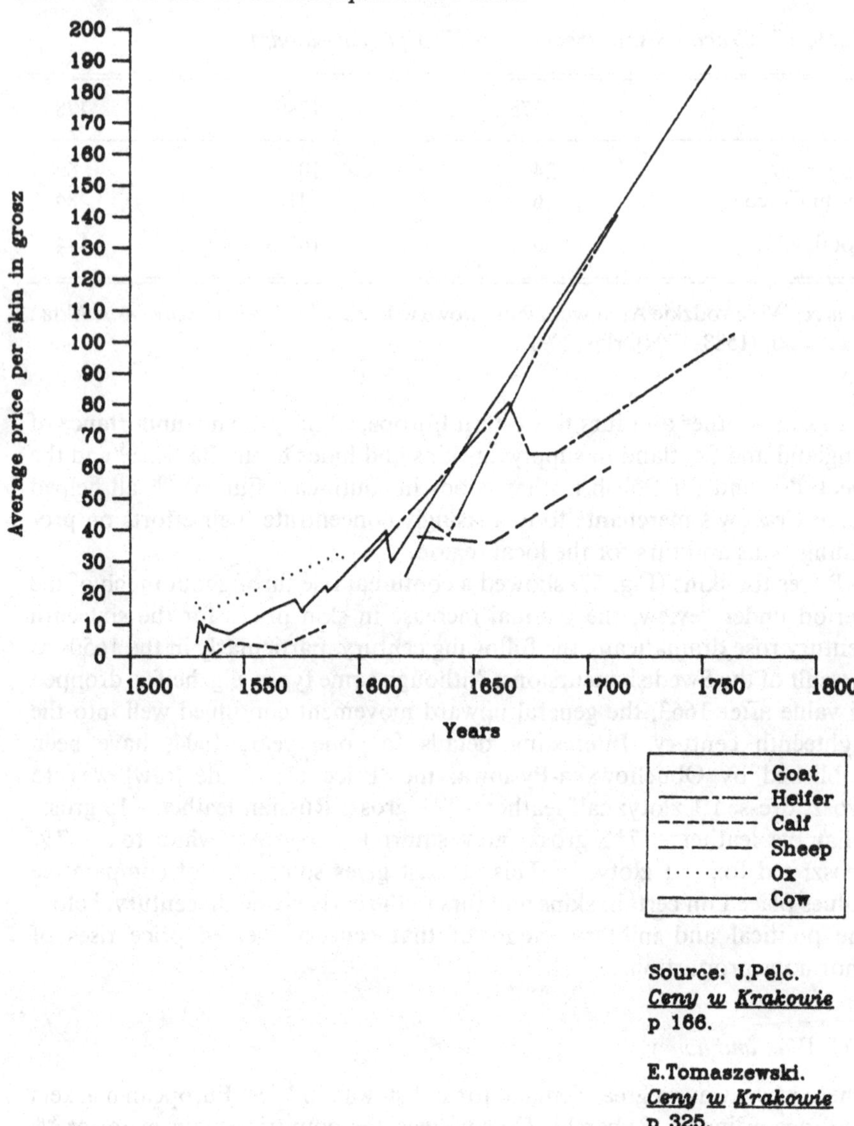

57 Average price of certain skins in Cracow, 1526–1761.

the quality of wax in sixteen Polish and Lithuanian towns; according to this document the production zone stretched from Minsk in the east as far as Lwów in the south-east.[313] There was also large-scale production in Lithuania, Belorussia and Galicia, in addition to other Polish provinces in the east which, according to Jeannin, 'probably provided more of the wax

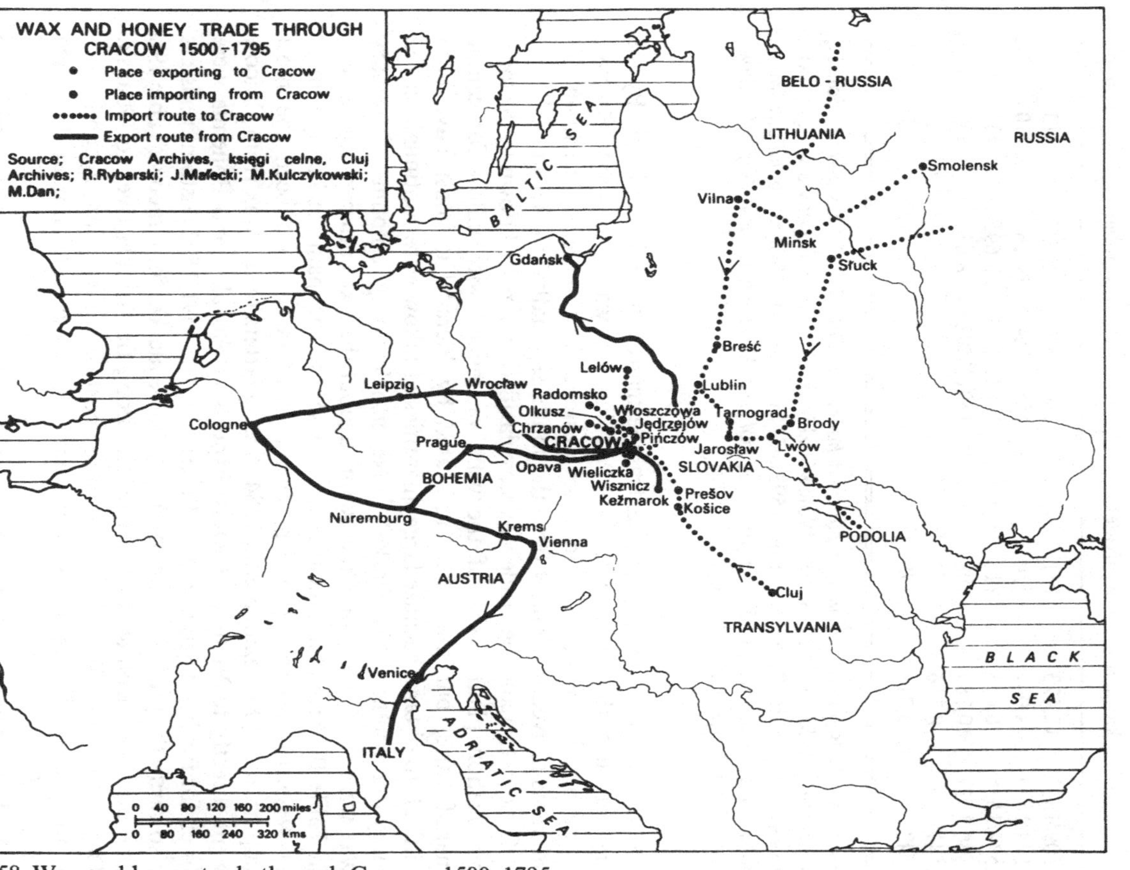

58 Wax and honey trade through Cracow, 1500–1795.

Table 18. *Wax exports from Cracow, 1509–1592*

Year	Weight in kamień	Tonnage equivalent	Year	Weight in kamień	Tonnage equivalent
1509/1510	12,123	231	1549/1550	11,188	213
1533/1534	8,041	154	1553/1554	14,169	270
1534/1535	6,035	115	1584	22,695	433
1536/1537	10,389	198	1591/1592	18,812	360
1541	8,756	167			
1542	9,313	178			

Sources: R. Rybarski, *Handel i polityka handlowa Polski w XVI stuleciu*, Vol. II, Warszawa, 1958, pp. 172–3, 198; Wojewódzkie Archiwum Państwowe w Krakowie, *Księgi celne*, rkp. 2116 (for 1591/1592); J. M. Małecki, *Związki handlowe miast polskich z Gdańskiem w XVI i pierwszej połowie XVII wieku* (P.A.N.), Wrocław–Warszawa–Kraków, 1968, pp. 121–2.

exported towards Poland and central Germany than did Russia'.[314] By the first decade of the sixteenth century Podolian wax was being sent to Cracow, by-passing Lwów in order to avoid customs duty there.[315] Much smaller amounts arrived in Cracow from the south, especially Transylvania.[316]

Once in Cracow the crude wax was clarified and melted down in the local purification plant; some of the wax was mixed with tallow for candle manufacture, but the latter, as stated by Pieradzka, 'must have been difficult to obtain, because large quantities of tallow were continually taken away to the mines at Wieliczka'.[317] Nevertheless, large amounts of ordinary wax were exported from Cracow throughout the sixteenth century, as seen in Table 18. From Cracow the wax was sent to the main marketing centres in western Europe, of which Vienna appears to have been the most important.[318] For example, in 1584 three-quarters of the wax from Cracow went directly to Vienna,[319] whilst exports to Slovakia and Bohemia were often for further destinations in Austria.[320] The 1591/1592 customs registers also verify the predominance of the Viennese market for Cracow (Table 19). The other main outlet was northward along the Vistula river to Gdańsk. Data from the Włocławek customs post for the years around mid-century show the varying involvement of Cracow merchants in this trade (Table 20).

Małecki maintains that the wax trade provided a reasonable income for Cracow merchants.[321] Certainly some of the more noted ones in the city were involved, e.g. Jan Boner,[322] whilst towards the end of the century Cracow's merchants had a near monopoly in the trade.[323] The lack of foreigners involved in wax exports suggests they had insufficient knowledge of the

Table 19. *Exports of wax from Cracow, 1591–1592*

Destination	Weight in kamień	Tonnage equivalent	%	Destination	Weight in kamień	Tonnage equivalent	%
Vienna	12,566	240	66.80	Wrocław	260	5	1.38
Krems	1,582	30	8.40	Unknown	2,506	47	13.33
Nuremburg	1,898	36	10.08				

Source: Wojewódzkie Archiwum Państwowe w Krakowie, *Księgi celne*, rkp. 2116; J. M. Małecki *Związki handlowe miast polskich z Gdańskiem w XVI i pierwszej połowie XVII wieku*, (P.A.N.) Wrocław–Warszawa–Kraków, 1968, p. 121.

Table 20. *Wax shipped through Włocławek to Gdańsk by Cracow merchants 1544–1561*

	1544	1557	1558	1560	1561
Total recorded (kamień/tons)	1,333 (23)	1,148 (22)	490 (9)	1,325 (25)	1,089 (21)
By Cracow merchants	100 (2)	140 (3)	400 (8)	150 (3)	560 (11)
% by Cracow merchants	8.18	12.19	81.63	11.32	51.42

Source: S. Kutrzeba, 'Wisła w historji gospodarczej dawnej Rzeczpospolitej Polskiej, *Geografija Wisły*, Vol. XI, Warszawa 1920, Table 3; J. M. Małecki, 'Przyczynek do dziejów handlu Gdańska w drugiej połowie XVI wieku (Związki handlowe z Krakowem), *Studia Gdańsko-Pomorskie*, Gdańsk 1964, Table 1.

Cracow market compared with local traders, who could more easily organise on-the-spot purchase of wax from their numerous rural contacts.

Closely tied to the collection and sale of wax was honey, although this never reached the proportions, or significance, attained by the wax trade. As with wax, honey was collected from the foothill areas of the Carpathian Mountains, and pastures of the Olkusz and Chrzanów region to the north-west of the city early in the century.[324] Małecki has made a detailed analysis of this trade during the last decade of the sixteenth century,[325] and shown the importance of those areas previously mentioned, together with the Little Poland Highlands (e.g., Jędrzejów, Lelów, Kurzelów, Radomsko). Honey also arrived in the city from Transylvania.[326] Although Poland seemed to have an active export trade in honey during the first part of the century, especially to Germany, Bohemia and Moravia,[327] Cracow did not play a significant role. Much of the city's production went for the Polish market, although local specialities like honeyed gingerbread (*miodownik*) was exported in small quantities to Hungary.[328] More generally, with the increasing use of sugar throughout Europe in the second half of the sixteenth century, honey exports from Poland somewhat declined.[329]

The early years of the seventeenth century saw continued importation of wax to Cracow, largely from sources to the north and east. It still remained an important export article from Russia,[330] arriving in Little Poland either from Smolensk via Minsk, Brześć and Lublin, or through Słuck to Lwów. From Lublin, Cracow's merchants also obtained Lithuanian wax sold at the annual fairs in Vilna and Minsk.[331] Small amounts also arrived from Slovakia,[332] but the largest quantities were obtained from areas within Little Poland itself. Once processed, it was sent on to Silesia (about 200 tons a year *circa* 1600),[333] some to Gdańsk,[334] and special consignments for the Italian market.[335] Less information is forthcoming on honey, but it is known, for example, that in 1631 a delivery of

Transylvanian honey was received in Cracow,[336] as well as regular small quantities from Slovakia.[337]

By the eighteenth century, Cracow had, for two hundred years, been one of Poland's main commercial centres for wax. The true significance of this trade may never be fully known as many Cracow merchants utilized their own conveyances for transport, and therefore were not noted in the city's customs registers. Even so documentary evidence suggests that the first half of the eighteenth century had not seen such thriving activity in wax sales as in the previous century, a phenomenon similarly noted in other commercial centres like Poznań.[338] As the second half of the eighteenth century progressed there appears to have been some revival of Cracow's dealings in wax, particularly in supplies from eastern Poland. Much of it came from towns like Brody and Jarosław,[339] together with two towns to the south of Cracow (Tarnogród, Wiśnicz) and three north of the Vistula river (Pińczów, Włoszczowa and Chęciny).[340] The main importer of Cracow wax was Opava,[341] the period after 1780 seeing closer commercial ties between the two places; smaller quantities went to Wrocław in Silesia and Krems in Austria.[342] Wax was often used as direct barter for iron goods, especially scythes, from these places.

Trade in honey seems to have experienced some revival in the later eighteenth century. After decline in the sixteenth century, and a very minor role in Cracow's commercial structure during the seventeenth century (especially with its former major market in Silesia)[343] a welcome renewal took place after 1750. Sugar was still a scarce luxury in eighteenth-century Cracow, honey forming the main sweetener for its growing population, together with increasing needs for the production of mead.[344] The main supply areas were similar to those for wax, and Kulczykowski has shown that of the 116 places providing Cracow with honey, over two-thirds were from urban centres;[345] much of this trade was in the hands of Jewish merchants, especially from Tarnogród.[346] The demands of Cracow's mead factory and local consumption left little for export (about 5 per cent)[347] most of it, as with wax, to Wrocław, Opava and Krems.[348]

The average price of wax in Cracow remained fairly stable throughout the sixteenth century, but experienced a steep rise with the advent of military conflict around 1620; it had reached a plateau of 52 grosz/kamień (stone) for over a century but the unstable economic conditions prevailing during the Swedish invasion set off further price rises which continued unabated well into the eighteenth century (Fig. 59).

(iii) Livestock

Since the Middle Ages there had been a strong tradition of sending livestock from the eastern parts of Europe to the West; it was a branch of commerce

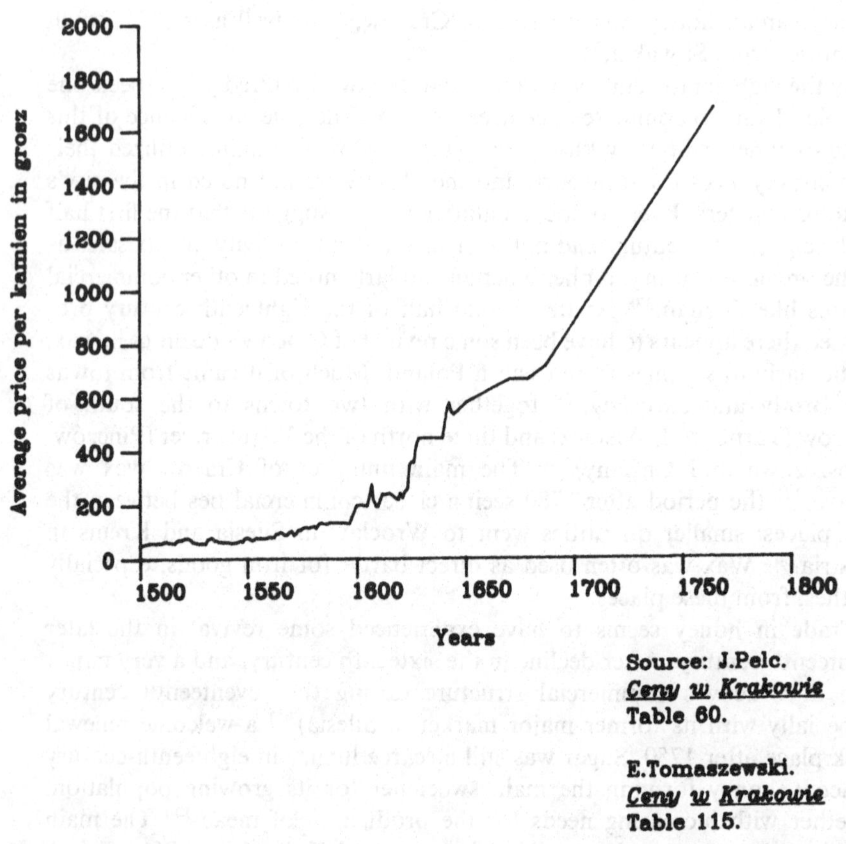

59 Average price of wax in Cracow, 1500–1761.

which remained well beyond the period under discussion here. This massive transcontinental trade consisted of thousands of cattle (mainly oxen) on the hoof, which occupied, according to Jeannin, 'far and away the most important place in the exports of the central-eastern countries'.[349] Recent studies have served to reveal in some detail the composition of this trade[350] which consisted of two major areas of movement, one north, the other south of the Carpathian Mountains. The southern route left the important breeding grounds of Wallachia/Moldavia and cattle were driven through Hungary to the West. The other main route left the Ukrainian–Moldavian Steppe and the Podolian tableland passing through Poland en route to the western markets. The principal stations on this latter route were Lwów, Lublin and Cracow, each having its own annual cattle fair.[351] At Cracow the route bifurcated (Fig. 60); one branch led through Moravia/Bohemia, usually via Plzeň, and on to Regensburg and Nuremburg; the other and most significant,

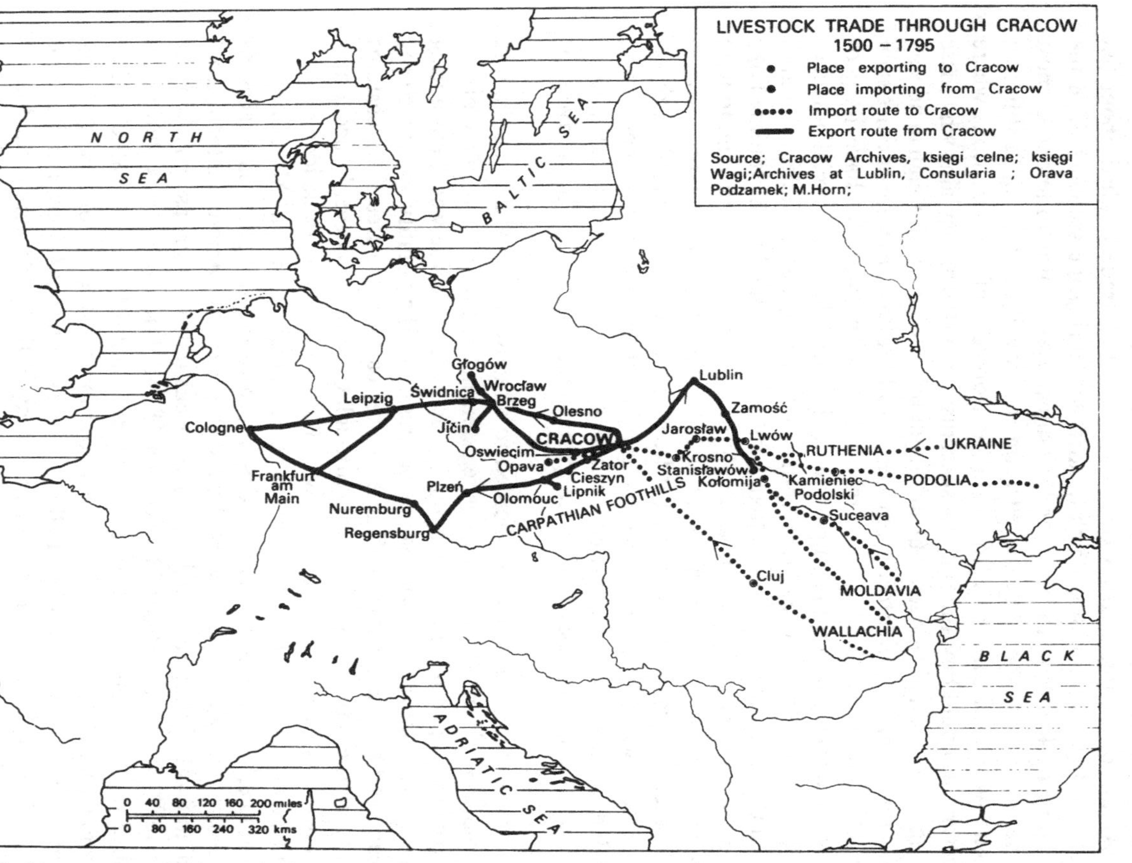

60 Livestock trade through Cracow, 1500–1795.

went through Silesia, with its important fair at Brzeg, and on to Leipzig, Frankfurt-am-Main and Cologne, supplying meat to the major urban centres of central Germany (north of the R. Main).[352]

At the beginning of the sixteenth century this trade totalled about ten thousand cattle annually, which quadrupled during the second half of the century to 40,000 head per year.[353] Transportation was different to most other exports as the cattle travelled on their own, but feeding facilities en route were necessary; this led to the development of cattle stations where merchants could buy and sell in a specialist market. This explains why these centres were usually located outside major urban centres, although in no way was this commerce divorced from the overall trading structure of cities such as Cracow. Nevertheless, for Cracow the cattle trade did have its geographical implications; in comparison with most other commercial branches it did bind the city more closely with central Europe as a whole, compared with the dominance of the north and Baltic ports found in the trading links of other commodities passing through the city. It is also interesting to note that some areas to the north-east of Cracow, such as Lithuania, provided no evidence of large exports of live cattle. It is important to remember, therefore, that when trying to define the geographical movement of commerce, different trade branches provide varying patterns of spatial distribution; these lack the precision of more permanent features such as rivers or coastlines.

The source areas for cattle herded to Cracow were mainly Podolia and Moldavia,[354] but from the customs registers in Cracow it is difficult to obtain a true trade picture. Before reaching Cracow many cattle were kept during the winter months on farm estates belonging to the nobility and were only ready for sale in the spring. These 'winter cattle' were free from customs duty, so the actual number moving through Cracow may never be accurately known. Documents from other towns, however, do give some idea of their movement towards Cracow. From the breeding grounds in Moldavia, for example,[355] the cattle were herded northward through such places as Suceava and Kołomyja towards Lwów;[356] they were wintered in the Carpathian foothills and recorded in Cracow's customs registers as arriving from there.[357] Certainly places like Krosno[358] and Jarosław[359] must have supplied the city's butchers with cattle, as did Lublin;[360] certain cattle traders from Cracow dealt with longer distance movement, of whom one Hieronimus Kriegel in the 1530s and 1540s,[361] and Hieronimus Gelhorn in the 1560s and 1570s, were active merchants.[362] Many of the cattle were herded down the San valley towards the city, and it is interesting to note that in Lwów and Cracow prices per head showed no great differential, despite the latter's nearer proximity to central European markets (Table 21).

Considerable numbers of cattle must have been herded along the more northerly route via Sandomierz and Krzepice, and therefore by-passing

Table 21. *Price of cattle per head in Lwów and Cracow during the sixteenth century (in grosz)*

Year	Average price in Lwów	Average price in Cracow
1527	72	68
1530	75	71
1531	36	90
1561	150	160
1591	157	159½

Sources: S. Hoszowski, *Ceny we Lwowie w XVI i XVII wieku*, Lwów, 1928, p. 180; J. Pelc, *Ceny w Krakowie w latach 1369–1600*, Lwów, 1935, p. 75.

Table 22. *Cattle movement through Cracow in the sixteenth century*

Year	1509/10	1519	1552	1584	1591/92
No. of cattle	3,305	4,944	4,460	5,290	4,452

Sources: R. Rybarski, *Handel i polityka handlowa Polski w XVI stuleciu*, Vol. II, Warszawa, 1958, pp. 172, 177, 193–4, 237; Wojewódzkie Archiwum Państwowe w Krakowie, *Księgi celne*, rkp. 2115, 2116, 2117.

Table 23. *Livestock registered in Cracow's customs books for 1589, 1593*

Type of livestock (per head)	1589	%	1593	%
Oxen	2,837	67.74	2,807	87.70
Cows	2	0.05	27	0.84
Heifers	292	6.97	–	–
Goats	183	4.36	169	5.27
Sheep	741	17.70	198	6.19
Unknown category	133	3.18	–	–
TOTAL	4,188	100.0	3,201	100.0

Source: Calculated from Wojewódzkie Archiwum Państwowe w Krakowie, *Księgi celne*, rkp. 2115, 2117, 2118.

Cracow, but one still finds large numbers coming to the city throughout the century (Table 22). This table reflects perhaps the more general trend in cattle trading for, according to Jeannin, 'The animals in transit through Wallachia, when added to those from Hungary, explain the record figures reached in the 1580s – nearly 200,000 head per year.'[363] It also puts into sharper focus, the fact that Cracow was handling only about 2½ per cent of this estimated grand total.

Cattle from Podolia and the Ukraine were sent from Lwów towards Silesia, passing through Cracow or further north to markets such as Łęczyca; in this case they entered Germany via Brandenburg, as did cattle from the Poznań region.[364] However, the markets of Saxony and Thuringia were supplied not only from this route but also through Silesia. During the sixteenth century two Cracow families seem to have dominated cattle exports from the city; the Schilling family[365] had strong connections with Wrocław, whilst the Gutteters were concerned with sending cattle to Nuremburg and Frankfurt-am-Main.[366]

The dominance of cattle trading in Cracow during the sixteenth century tends to obscure commerce in other live animals. The customs registers towards the end of the century reveal movement of other livestock through the city (Table 23). Again, it has to be remembered, these figures only give some idea of livestock movement because neither the nobility, nor Cracow's citizens, were subject to customs duty; in practice, therefore, the numbers would be higher; even so, the overall dominance of trade in oxen would not have been overshadowed. Horses were also a significant item, particularly for their greater versatility within society as reflected in their higher price (Fig. 61). Locally bred horses were certainly exported to Bohemia (Olomouc, Jičín, Lipník, etc.) during the last decade of the sixteenth century,[367] whilst Cracow imported horses from Transylvania[368] and Slovakia.[369] However, it would appear there was never a large trade movement in horses either to or from Cracow.[370]

According to Obuchowska-Pysiowa, 'The trade in oxen performed an important role in the economy of Cracow',[371] at the beginning of the seventeenth century. Not only did it provide a constant source of meat supply for its inhabitants, but brought in profits from abroad for merchants and money for the city treasury through customs duty. Customs registers reveal how cattle were sent to the city from Ruthenia, the Ukraine, Podolia and Moldavia, with a certain percentage being re-exported to Brzeg and Świdnica in Silesia (Table 24). On average about 400 head of cattle arrived in Cracow each month, with a peak in August and low in March and May; nearly a fifth were re-exported. It is difficult to tell how representative 1604 was for the trade, but the annual sum appears fairly commensurate with other totals in the sixteenth century (Tables 20 + 21) which suggest that about 4,000 head of cattle arrived annually in the city from the eastern breeding grounds.

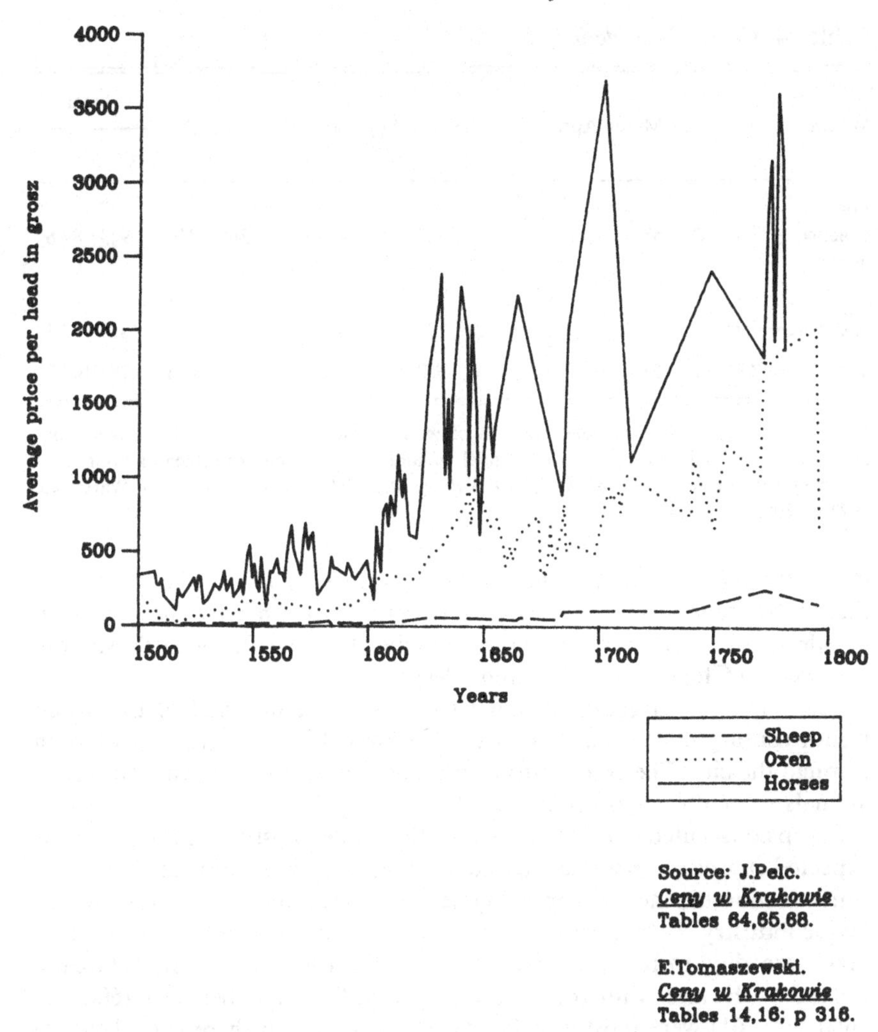

61 Average price of certain livestock in Cracow, 1500–1795.

Ruthenia was one of the chief suppliers during the first half of the seventeenth century, in which Jarosław, as in the sixteenth century, held an important position (Table 25). Despite Cracow's low ranking in this table, it should be noted that a further 29,273 head of cattle were recorded as going to Silesian and foreign fairs (8,357 of which went to Brzeg) and for 39,992 cattle no destination was noted. Some of the cattle in these two latter categories must have travelled through Cracow en route to the West. Other points of interest in Table 25 indicate firstly the major position of Jarosław

Table 24. *Cracow's trade in oxen, 1604*

Month	Jan.	Feb.	March	April	May	June	July	Aug.	Sept.	Oct.	Nov.	Dec.	Total	
													No.	%
For Cracow market	538	526	58	271	40	262	382	422	438	282	368	356	3,943	82.6
For re-export	97	–	–	–	–	–	170	483	–	–	–	79	829	17.37
Total	635	526	58	271	40	262	552	905	438	282	368	435	4,772	100.0

Sources: Calculated from Wojewódzkie Archiwum Państwowe w Krakowie, *Księgi celne*, rkp. 2126; H. Obuchowska-Pysiowa, 'Handel wołami w świetle rejestru celnego komory krakowskiej z 1604 roku', *Kwartalnik Historii Kultury Materialnej*, Vol. XXII, No. 3, Warszawa, 1973, p. 508.

in this trade (50 per cent of the Polish market went through the town), and that of the grand total (275,660), over 14 per cent are known to have gone on to Silesia and Germany, the largest number through Brzeg cattle station, followed by Głogów, Wrocław and Olesno.[372]

Trade in other livestock through Cracow in the first half of the seventeenth century continued to be overshadowed by the movement of oxen through the city. The 1604 customs register records the following variety of animals entered for tariff purposes (Table 26).

Sheep constituted over 90 per cent of this trade, most going for export; as expected it was a very seasonal commerce, largely concentrated into the summer months, most sheep being sent to Silesia to provide wool for its textile industry.[373] Interestingly, there seems to have been little mention of pigs; according to Rusiński, exports to the West were chiefly from Greater Poland.[374] Horses, with their erratic price differentials between 1600 and 1650 (Fig. 61) were mostly of Polish origin, although there is evidence of imports from the Orava region of Slovakia.[375]

In the middle of the seventeenth century significant changes began more generally to take place in East–West trade relations. On the Polish side there was a noticeable decline in the cattle trade, a factor also experienced on the other side of the Carpathians in Hungary.[376] Exports of oxen from Poland were confronted with serious difficulties after having survived the traumas of the Thirty Years War. Attman believes that 'exports could no longer compensate for the import of products – especially cloth – from the West',[377] largely due to the development of trade credit, which heralded the decline of earlier great markets in Polish urban centres. In their place the fairs at Wrocław, Frankfurt and Leipzig began to develop, the latter

Table 25. *Oxen exported from Ruthenia to Polish cattle fairs, 1601–1648*

Fair	Kamionka	Jaworów	Jarosław	Gródek	Radymno	Mościska	Gdańsk	Bełz	Luków	Zamośc	Cracow
Total oxen	20,054	20,701	103,772	20,498	8,299	1,156	3,621	879	11,896	7,521	1,074
Rank position	4	2	1	3	6	9	8	12	5	7	11

Sources: Centralne Państwowe Archiwum Historyczne U.S.S.R. we Lwowie, *Sąd Grodzki Halicki*, pp. 109–21, 354–66, 368–88, 390–3; M. Horn, 'Handel wołami na Rusi Czerwonej w pierwszej połowie XVII wieku', *Roczniki Dziejów Społecznych i Gospodarczych*, Vol. XXIV, Warszawa, 1962, p. 83.

Table 26. *Cracow's trade in livestock (excl. oxen), 1604*

Month	Jan.	Feb.	March	April	May	June	July	Aug.	Sept.	Oct.	Nov.	Dec.	Total	
													No.	%
For Cracow market:														
Cows + Heifers	5	–	1	12	132	147	16	40	4	15	16	5	393	8.35
Sheep	–	–	–	–	45	28	–	74	–	–	16	–	163	3.47
Goats	–	–	–	–	5	28	2	30	–	–	–	–	65	1.39
For re-export														
Sheep	–	–	–	–	–	926	2,100	717	–	340	–	–	4,083	86.79
Total	5	–	1	12	182	1,129	2,118	861	4	355	32	5	4,704	100.0

Sources: Calculated from Wojewódzkie Archiwum Państwowe w Krakowie, *Księgi celne*, rkp. 2126; H. Obuchowska-Pysiowa, 'Handel wołami w świetle rejestru celnego komory krakowskiej z 1604 roku', *Kwartalnik Historii Kultury Materialnej*, Vol. XXI, No. 3, Warszawa, 1973, p. 508.

dominating commercial links between on the one hand Russia/Poland, and on the other central and western Europe.[378] Cracow became increasingly isolated from these developments not only in the cattle trade, but also in the general commercial field; the Swedish occupation of the city and its aftermath only contributed further to this deteriorating situation during the second half of the century.

Indications suggest that by the eighteenth century emphasis in the cattle trade had moved northward from Little to Greater Poland; it was concentrated around Wieluń, where cattle were herded to Silesia and beyond.[379] It seems Poznań now took on Cracow's previous role in this trade for, 'Livestock commodities were characterised by the highest rates of dynamism ... The intensive development of the livestock market due to the demands created in the municipal, regional and foreign markets'.[380] Only after the first partition of Poland (1772) was the old route through Ropczyce–Bochnia–Cracow again brought into intensive use,[381] when the Austrian government established cattle markets at Oświęcim, Zator and Cieszyn. Merchants from Kleparz in Cracow were active in buying cattle from the city's neighbourhood, but much of the Silesian trade was in the hands of people from the surrounding villages of Cracow such as Płaszów.

As Cracow meat merchants/butchers were free of customs duty in the eighteenth century, figures in the city tariff registers are of little use for estimating actual demand within the urban centre, but one may safely assume that a large percentage of the cattle entering the city were for local consumption. Similarly for pigs, their numbers were an unknown quantity; for sure they were in demand, not only for meat, but also in the preparation of salted bacon and lard. The registers do, however, give some idea of the supply catchment area, pigs arriving in the city from Little Poland (Chrzanów, Wolbrom, Sędziszów) and farther afield – above all from Zamość, Jarosław and Stanisławów.[382]

Horse trading appears to have been very much a local affair, unlike such places as Poznań, where 'The horse trade, which after the 1770s increased considerably, was geared first of all to supply the needs of the Prussian army.'[383] Livestock by-products were in demand, such as tallow, for soap and candle-making, coming mainly from well-known cattle towns (Jarosław, Drohobycz, Tarnogród) and largely organised by Jewish merchants. It was also supplied by local villages around Cracow, any surplus usually going for export to Silesia.[384] Dairy products (e.g. ewe's cheese) were mainly for the local market, but some exports went along the R. Vistula to Warsaw.[385] The wool trade was of little significance.[386]

Prices of certain livestock in Cracow (Fig. 61) follow a similar pattern experienced with other products. The gradual rise in the sixteenth century was overtaken by great annual variations in the seventeenth, which continued into the following century. Horses were clearly the most expensive animal on sale in Cracow; only in 1555, 1601, 1645 and 1776 were they on average cheaper than oxen, whilst in 1767 they fetched the same price. The vicissitudes of war also help explain some of the price variations, especially in the seventeenth century; the drop in oxen prices at the end of the period under study to that similar to fifty years earlier may have been connected with the disturbed economic situation resulting from the second and third partitions; alternatively, it may have been linked with an overall European trend alluded to by Braudel and Spooner – 'What people are generally less well aware of is that the situation sketched in 1750 – large rations of bread and a little meat – which continued by and large for another century until about 1850, was itself the result of deterioration ... In the German towns the yearly consumption of meat declined from the fifteenth to the eighteenth century ... What was true of Germany was sooner or later true of Europe.'[387] Sheep, even though much lower priced, demonstrate those phases characteristic of other livestock between 1500 and 1795.

(iv) Miscellaneous goods

The above products by no means exhaust the list of goods Cracow traded in during the Early Modern Period, but perhaps highlight those of greatest financial reward for the city. There remains however, numerous other

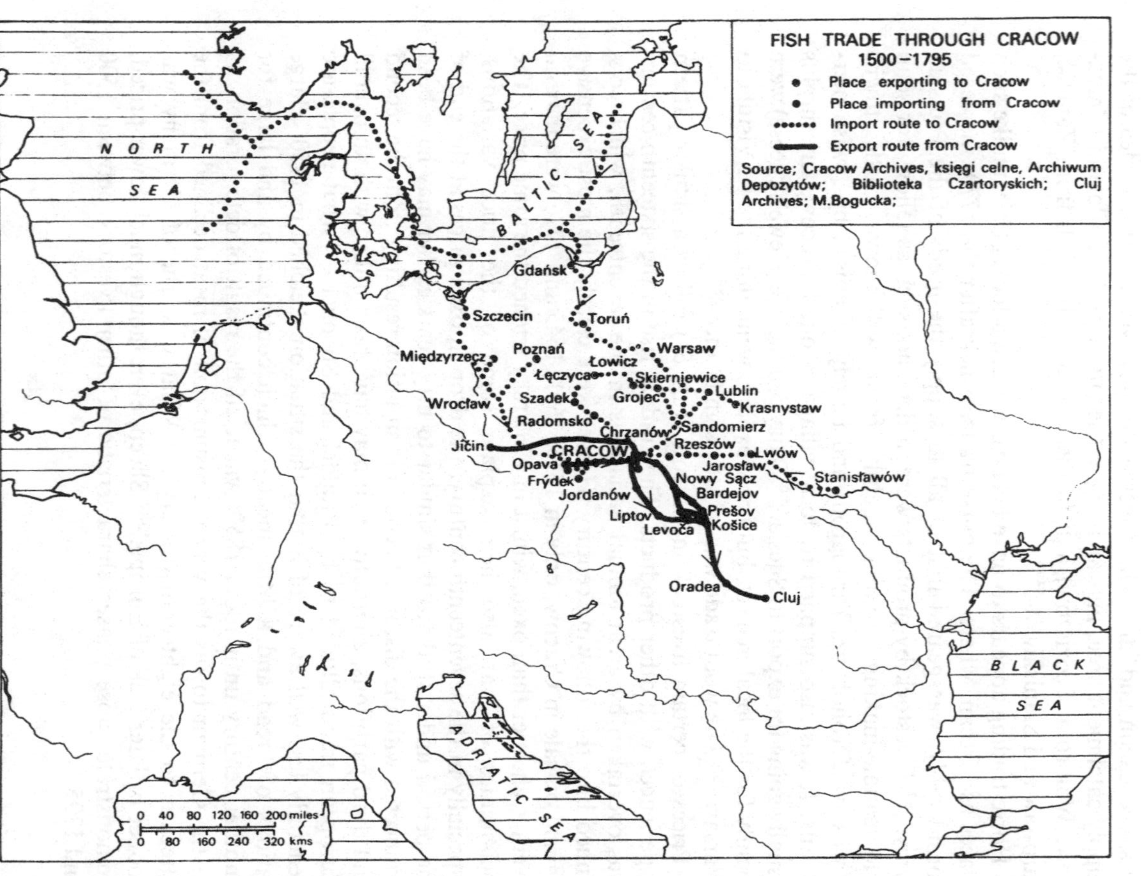
FISH TRADE THROUGH CRACOW
1500–1795
Place exporting to Cracow
Place importing from Cracow
Import route to Cracow
Export route from Cracow
Source; Cracow Archives, księgi celne, Archiwum Depozytów; Biblioteka Czartoryskich; Cluj Archives; M.Bogucka;
NORTH SEA
BALTIC SEA
BLACK SEA
ADRIATIC SEA
Gdańsk
Szczecin
Toruń
Warsaw
Międzyrzecz
Poznań
Łowicz
Łęczyca
Skierniewice
Lublin
Szadek
Grojec
Krasnystaw
Wrocław
Radomsko
Sandomierz
Chrzanów
Rzeszów
Jičin
CRACOW
Lwów
Opava
Jarosław
Frýdek
Nowy Sącz
Stanisławów
Jordanów
Bardejov
Liptov
Prešov
Levoča
Košice
Oradea
Cluj
0 40 80 120 160 200 miles
0 80 160 240 320 kms

62 Fish trade through Cracow, 1500–1795.

merchandise sent either to, or through Cracow, most notable of which were fish, cereals and wood.

(a) Fish

The compulsory fish diet of Catholic Poland was provided partly by local supplies such as carp, pike, eels, perch, trout, lamprey, salmon and other river species, but also by cod, plaice and, above all, herrings from the northern seas. The importance of herrings overshadowed all other fish imported into Poland; during the first half of the sixteenth century their importation was rather irregular[388] but after mid-century increasing amounts entered the Polish market. Much of the trade was in the hands of Baltic coastal merchants and those inhabiting the Lower Vistula valley, particularly Mazovia.[389] During the sixteenth century herrings began to prefer the North Sea for spawning, possibly due to a change in water salinity. According to Rutkowski about 60 per cent of the herring supplies came to the Baltic ports through the Sound;[390] from here herrings were only occasionally sent up river to the Polish markets, preference being given to overland routes.

Cracow was one of the major fish-marketing centres of Little Poland (Fig. 62); it possessed two centres, one in Garbary (mainly local fresh fish) the other beneath the castle with its warehousing facilities for salted and dried fish;[391] the latter served a wider area, including the Hungarian transit trade. Documentary sources suggest a steady supply of herrings from the Baltic ports, especially Gdańsk[392] to Cracow throughout the century (Table 27). The table illustrates this continuing trade, but the actual amounts involved varied considerably from year to year; it also reflects the crisis in Poland's fish trade by mid-century and consequent price increases. Pieradzka has calculated that herring prices rose by 71 per cent during the second half of the century, only becoming more stable after 1584.[393] Most of the herring catch came from Gdańsk through Prussia, but were recorded in Cracow documents[394] as having been supplied from inland route towns such as Stężyca, Warsaw and Sandomierz (river ports), Skierniewice and Łowicz.[395] Other towns included Radomsko, Szadek and Łęczyca, the latter a noted place for eels.[396]

Although most of Cracow's herrings and eels came from the northern parts of Poland, the city also imported fish from Lwów. These consisted mainly of sturgeon and various types of salted, dried fish which were transported to Cracow in special containers making it easier to identify them in the customs registers; actual varieties were more difficult to distinguish, often being referred to as 'Lwów fish'. Similar consignments were recorded from Lublin.[397]

Apart from fish consumed in the city, Cracow exported quantities surplus to requirements. The main receiving areas appear to have been the Carpa-

Table 27. *Baltic herring imports to Cracow during the sixteenth century*

Year	1509/10	1533/34	1534/35	1536/37	1541	1542	1549/50	1576	1588	1589	1593	1595	1597	1600
No. of barrels	392	1,928	1,171	1,619	302	589	264	48	660	42	11	31	38	193

Sources: Calculated from R. Rybarski, *Handel i polityka handlowa Polski w XVI stuleciu*, Vol. II, Warszawa, 1958, pp. 40, 172–3; Wojewódzkie Archiwum Państwowe w Gdańsku, *Odpisy korespondencji*: 300 R/Ee, No. 1, p. 1435; *Missiva* 300, 27/14k. 127v–128v; Wojewódzkie Archiwum Państwowe w Krakowie, *Księgi celne*, rkp. 2115–2121; H. Obuchowska-Pysiowa, *Handel wiślany w pierwszej połowie XVII wieku* (Ossolineum), Wrocław–Warszawa–Kraków, 1964, p. 149; *Regestra thelonei aquatici Wladislaviensis saeculi XVI*, S. Kutrzeba and F. Duda (eds.), Kraków 1915, p. 70; J. M. Małecki, 'Przyczynek do dziejów handlu Gdańska w drugiej połowie XVI wieku', *Studia Gdańsko-Pomorskie*, Gdańsk 1964, p. 32.

thian foothill region, northern Hungary and, to a lesser extent, Silesia. After 1541 there was a decline in fish recorded in the customs books, whilst Cracow's fish traders seem to have suffered a takeover by Hungarians and dealers from Wrocław; this was particularly apparent by the end of the century when direct contacts by these merchants began in the smaller towns of Little Poland.[398] Even so, as Małecki's study has shown, fresh fish from Cracow were transported to many of the Carpathian foothill towns including Biecz, Bochnia, Brzesko, Chrzanów, Jordanów, Nowy Sącz and Nowy Targ etc., as the 1594 customs registers have shown.[399]

The other great outlet for Cracow's fish trade was northern Hungary. Evidence reveals that throughout the sixteenth century such towns as Košice, Kežmarok, Bardejov, Levoča, Prešov, the Liptov and Orava valleys received fish consignments from Cracow.[400] Less information is forthcoming for urban centres in central Hungary. However, Dan has proved that fish was transported to Transylvania, especially herrings to towns like Cluj[401] and Oradea.[402] Certainly several well-known merchants were involved in the north Hungarian trade including Thurzo, Spies, Melchjor and the Schilling family,[403] some, like the Fuggers and Hegels, concentrating on trade with individual places like Košice and Lwów.[404] Finally, evidence suggests that trade contacts with Silesia and Moravia existed for supplying both herrings and salted fish from Cracow's fish market.[405]

In the eyes of contemporaries the seventeenth century was to see an explosion of Dutch trade in northern Europe. By 1600, according to Christensen, 'the Dutch were unquestionably the world's leading seafaring nation';[406] their trade with the Baltic continued to increase until the 1620s when the Swedish invasion of north Germany interrupted commerce. A new trading peak was reached in the 1650s followed by a slow decline towards the eighteenth century. Indirectly all this was relevant to Cracow, for the ports of the Baltic littoral constituted the largest single market for Dutch cured herring. Even in 1500 half the herring imported into Gdańsk came from Holland[407] and the volume increased over time. The herring, states Unger, 'was consumed in Baltic ports or shipped inland along the rivers of Poland and Russia. On average, some 11 per cent of the herring that entered the Baltic in the first half of the seventeenth century found its way up the Vistula as far as Warsaw.'[408]

Certainly long periods of religious fasting and skill in preserving herrings made this fish very popular; it also ensured that herrings were carried in transit throughout Poland and beyond.[409] Obuchowska-Pysiowa's research on Warsaw customs registers during the first half of the seventeenth century has shown that Cracow held thirteenth place amongst Polish cities in its herring imports;[410] this position, though not as impressive, was still significant within Little Poland. Moreover, documentary evidence suggests that the city continued to export surplus quantities of herring over the Carpathians to places like Bardejov,[411] Prešov[412] and Opava, often organised by merchants from Międzyrzecz and Šenov.[413] Salted fish also left Cracow for Cluj and Oradea in Transylvania.[414]

After 1700 Dutch involvement in the herring trade declined sharply and, according to Unger, 'The destruction of a great part of the Dutch herring-fleet by the French shortly after 1700 was surely not unconnected with this.'[415] Their heirs were the Scots and Scandinavians who were already taking over the herring trade at the end of the seventeenth and beginning of the eighteenth century, a period of instability and war. In the first half of the eighteenth century it was the Norwegians and, after 1760, the Swedes who dominated herring sales in the ports along the Baltic coast, and on whom ultimately Cracow's merchants depended for their supplies. Nevertheless, even in 1763 Dutch and Scots as well as Swedish herrings arrived in Cracow from Wrocław, Gdańsk and Toruń;[416] other fish included lamprey, salmon, cod, eels, etc.; as for Poznań, some of this fish came from the Baltic port of Szczecin.[417] Inland varieties arrived in Cracow from Lwów, Jarosław and Stanisławów, especially sturgeon,[418] but consignments from the Lower Vistula region were small in comparison due to continued conflict with Warsaw merchants over their fish prices.[419] In contrast, the Upper Vistula region and the Polish–Silesian border were important suppliers (e.g. from Kopciowice, Wielki and Mali Ochab (Austrian Silesia), Skidzynia, Grójec, Dańkowice (in the Soła valley),[420] Babice, Porąbka etc.); river fish also arrived in Cracow from the Skawa valley and places nearer to the city.

Documentary sources for the later eighteenth century reveal that Cracow's herring trade was not on such a large scale as two centuries earlier;

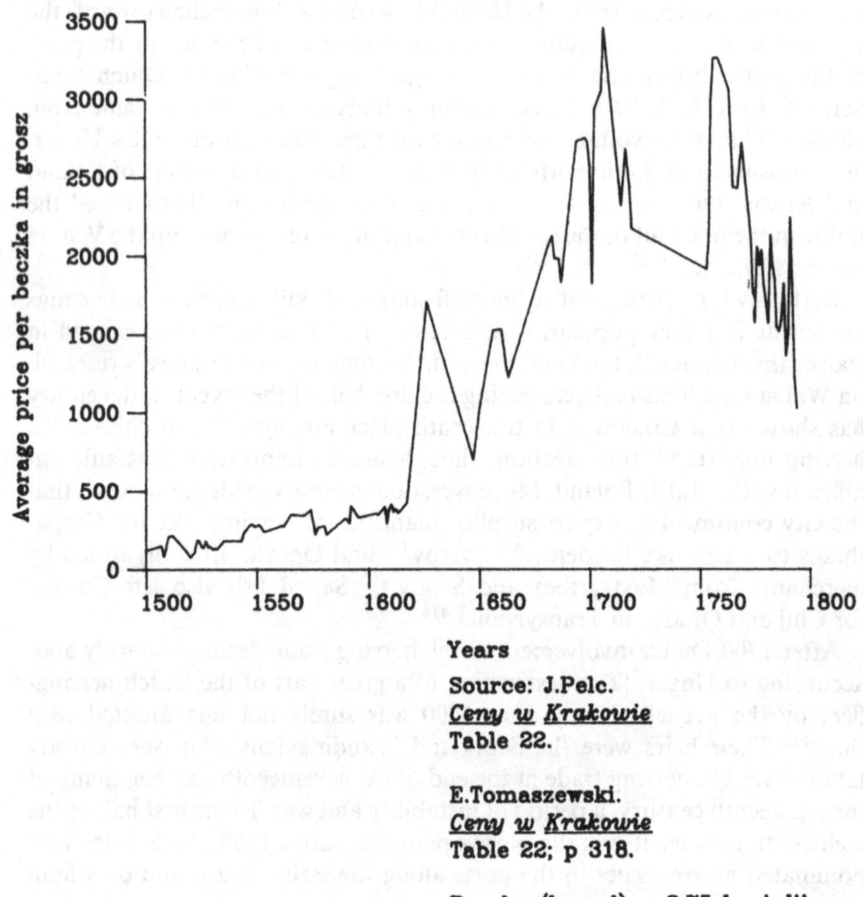

63 Price of herring in Cracow, 1500–1795.

small quantities were purchased by the city's fish traders in Wieliczka, Ropczyce, Rzeszów, Jarosław, Frýdek and Opava. Sometimes fish such as cod, smoked cod, sturgeon and lampreys arrived in consignments of colonial goods from Gdańsk and Toruń along with small amounts of cod-liver oil.[421] One thing that changed little over the centuries was the seasonal nature of the fish trade concentrated in the winter months (November–March), particularly herring following the summer and autumn catches in the North and Baltic Seas.[422]

Fish prices based on Cracow's documentary sources are rather of a sporadic nature, with several varieties having only short-run data; the exception are herrings (Fig. 63) with a time series covering the period

1500–1795. The price of herring in Cracow rose in the 1590s to a point higher than previously in the sixteenth century,[423] followed by yet another rise in the 1620s and 1630s. Prices continued to increase, peaking during the troubled war years of the 1700s; by 1720 a long-term decline set in, only disturbed by the high totals experienced during the Seven Years War. By 1793, however, they returned to a similar level found a century and a half before.

(b) Cereals

From the end of the fifteenth century the importance of grain in international commerce was steadily growing;[424] during the second half of the sixteenth and first half of the seventeenth century it totalled from 50–60 per cent of total trade between Eastern and Western Europe.[425] The major source of grain in the Baltic area at this time was Poland, which accounted for about half the total cereal exports from the Baltic. Grain production in the country was approaching the crucial point at which surpluses were being regularly obtained; thus external demand became closely matched with favourable prospects for increased internal supply. Many scholars have emphasised the significance of cereal exports in the evolution of the economic, political and social structure of Poland in the past. Weighty arguments have been put forward on this theme, most recently by Topolski and Wyczański,[426] but critical in this study is to examine its importance as a product in Cracow's commercial structure.

Documentary sources suggest that Cracow's role in the Polish grain trade was insignificant although, as a city of several thousand inhabitants, there was internal demand for grain, especially wheat for making bread. It is even questionable if many of Cracow's merchants were involved in the grain trade; according to the customs tariff returns at Włocławek, cereals were sent down the Vistula to Gdańsk by them only in 1556, 1557 and 1575.[427] Fertile cereal-growing areas lay to the east and north-east of the city, with easier direct access to the banks of the Vistula downstream of Cracow; upstream the region was less favourable for cereal growing, whilst arable land in the immediate vicinity of the city appears to have been utilised for other crops. Certainly, cereal exports from the whole Cracow region to the Baltic coast were much lower than for other parts of the country during the second half of the sixteenth century.[428]

Any cereal surpluses left over from internal needs, appear to have been sent, not to the main grain mart in Gdańsk, but westwards to Silesia;[429] often this was in the form of return cargo by carters from places like Bielsko-Biała, who had delivered linen and other textiles to the city.[430] Wheat, barley, millet and buckwheat were most frequently documented as sent to small centres like Bestwina, Komorowice, Kozy, Lipník (near Bielsko-Biała), Mikuszowice, Pszczyna and Żywiec.[431]

Cracow's involvement in the grain trade showed a marked contrast between the seventeenth and eighteenth centuries. The pattern set in the sixteenth century of Cracow's minimal involvement in cereal exports has been confirmed by Obuchowska-Pysiowa as continuing well into the following century. Her analysis of goods imported and exported along the River Vistula between 1605 and 1651 based on customs ledgers from Warsaw reveal that only 0.4 per cent of the total tonnage for all Polish localities was recorded as coming from Cracow.[432] In a much earlier study Lepszy also maintained that 'in the seventeenth century the great trade in cereals, in which Poland supplied the whole of Europe, was absent from Cracow's trading interests'.[433] The decline in vital Polish grain exports after the mid-seventeenth century has been generally analysed in terms of material and human disasters brought about by war; this stimulated a fall in agricultural profits and productivity, to which Jeannin believes should also be added a satisfactory analysis of the demand factor.[434] Whatever the outcome of such research, it is clear that Cracow's cereal trade was only symptomatic of the overall situation in the country during the latter part of this century.

In contrast, the eighteenth century was to witness the development of Cracow as an important centre for the Little Poland region. Madurowicz and Podraza have described how Cracow at this time lay on the boundary between two agricultural regions; wheat and barley production was concentrated to the north, oats and other cereals to the south.[435] The First Partition of Poland in 1772 accentuated this development. The Austrians received territory to the south of the River Vistula which effectively divided Little Poland into its two grain-producing halves. Inevitably this led to Cracow becoming the main centre for cereal exchange, and gradually developed to include markets in Austrian-controlled Galicia. After 1785, however, a series of poor harvests in Galicia led to decline in Cracow's entrepôt position in the grain trade of southern Poland.[436]

Cracow's catchment area for obtaining cereals varied on average between 22 and 65 km, although instances were recorded of up to 100 km from some of the smaller villages in the Carpathian foothills.[437] These more distant traders brought wooden carvings, Hungarian wine and dried fruits, Silesian linen and rough woollen cloth to exchange for vegetables and cereals; they came to the main grain market in Cracow at Kleparz.[438] Usually, however, grain arrived from villages nearer to Cracow, such as Proszowice, Słomniki, Koszyce, Miechów, Skała and Żywiec.[439] The most intensive period of cereal export was during the First Partition of Poland (1773–84) when the main outlet for grain to Galicia was through Cracow (Table 28). The table suggests that quite large quantities of cereals were sent annually to Galicia from Cracow, mainly transported between June and August (i.e. an average of 40 per cent for the two years).

Table 28. *Cereal exports from Cracow to Galicia, 1780, 1785*

Month/ Year	Jan.	Feb.	March	April	May	June	July	Aug.	Sept.	Oct.	Nov.	Dec.	Total
1780 Cart Load	598	1,154	63½	78	166	802	816	440	229	241	268	491½	5,347
1780 Ton equiv.	465	898	49	61	129	624	635	342	178	188	209	382	4,160
1785 Cart Load	912	844	804½	353	537	869	1,697½	1,137	331½	249	514½	729	8,978
1785 Ton equiv.	710	657	626	275	418	676	1,321	884	258	193	400	567	6,987

Sources: Wojewódzkie Archiwum Państwowe w Krakowie, *Rachunki miasta Kazimierz*, rkp. 665, 670; M. Kulczykowski, *Kraków jako ośrodek towarowy Małopolski zachodniej w drugiej połowie XVIII wieku*, (P.A.N.), Warszawa, 1963, p. 50.

The last decade before the Third Partition of Poland in 1795 saw a considerable grain crisis in Galicia due to a series of bad harvests; further, the Russo-Turkish War of 1787 restricted Polish cereal imports from Kherson,[440] which in turn led to cereal price rises in Little Poland. Moreover, the Austrian government forbad cereal exports to Galicia,[441] which forced Cracow to reorientate her grain trade once more towards the Baltic coast – Gdańsk, Elbląg and Królewiec (Königsberg).[442] Finally, it should be remembered that Cracow was also an important flour-milling centre for Little Poland[443] supplying towns like Wieliczka, Bochnia, Skawina and Myślenice.[444]

Cereal prices in Cracow between 1500 and 1795 (Figs. 64, 65) show wheat and rye as higher value crops compared with barley and oats throughout the whole period.[445] As with other product prices it is possible to differentiate between the centuries; the sixteenth was fairly stable with increasing fluctuations during the latter decades, the seventeenth rising and falling commensurate with the vicissitudes of war and harvest, the eighteenth having dramatic oscillations related to the economic and political vagaries of the time.

The survival of grain-price data throughout much of Europe over this period allows some comparison between Cracow and other areas. In southern and western Europe price rises for cereals were only moderate during the first half of the sixteenth century. After 1550, contact between Polish and other European corn markets meant similar price patterns. In Cracow, as in Gdańsk and Malbork, slight price rises in grain were already discernible in the 1540s;[446] substantial price increases for Polish grain during the second half of the century were the result of growing West European demand for

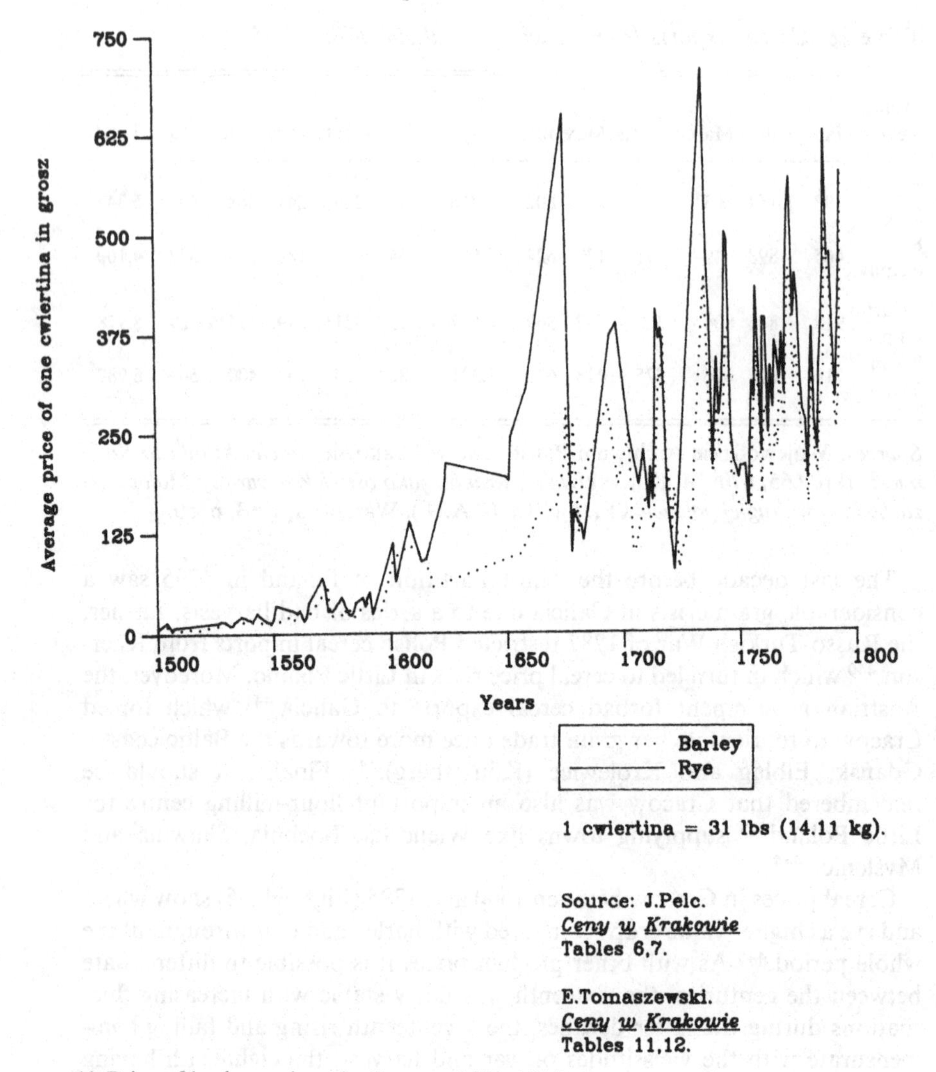

64 Price of barley and rye in Cracow, 1500–1795.

cereals combined with the general depreciation of precious metals throughout the continent.[447] The intense price rises in the seventeenth century were directly related to the devaluation of the grosz *vis-à-vis* silver content; this led to the disastrous monetary crisis of 1617–31,[448] and was reflected in Cracow's grain prices. Hoszowski has shown, for example, that 'Within the sixteenth century ... the price of oats increased fivefold in Warsaw, Cracow and Gdańsk',[449] and goes on to suggest that 'corn prices

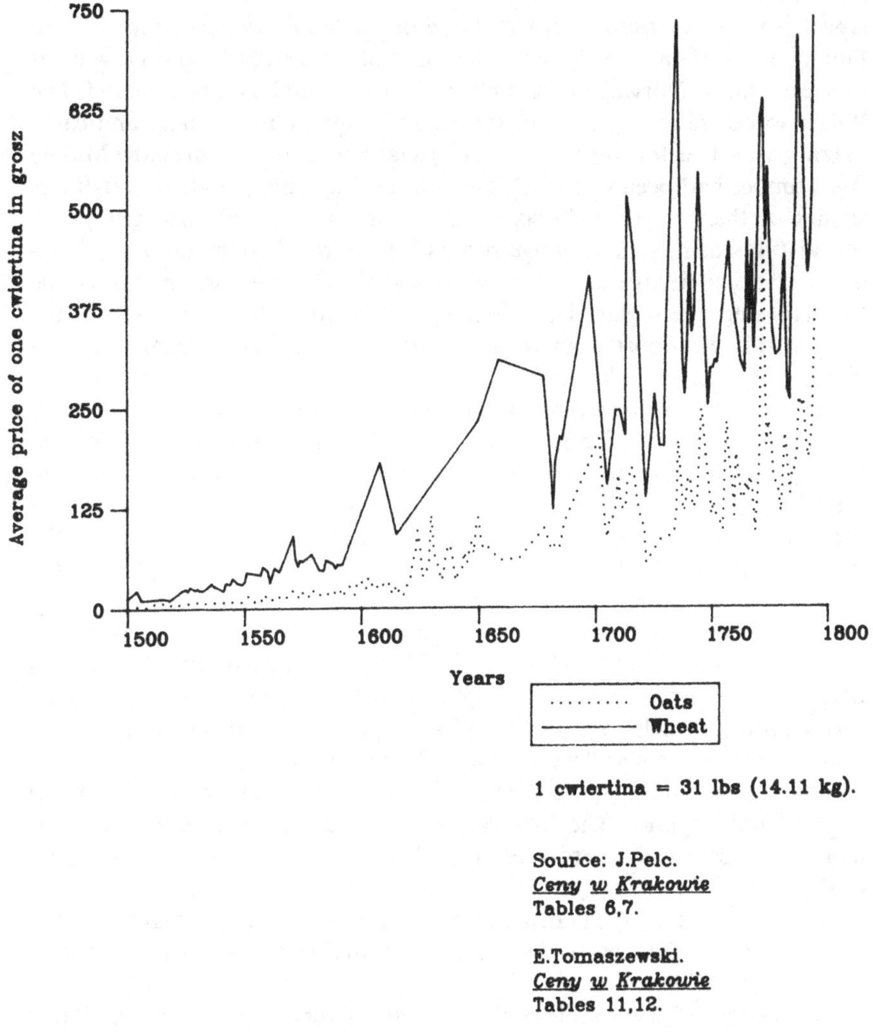

65 Price of oats and wheat in Cracow, 1500–1800.

were rising faster in Poland than the majority of western countries'.[450] The eighteenth century saw the development of Poland's domestic cereal market, which was dominated by cities on the Vistula river. Kuklińska has proved in her analysis of rye and wheat prices in Gdańsk, Warsaw and Cracow (1752–80) that a close correlation existed between these places, suggesting some measure of interdependence;[451] such results further support an earlier study by Kula[452] which suggested a tendency for price standardisation.

(c) Wood

Apart from grain, timber provided one of the largest items of export from Poland and Lithuania until the second half of the sixteenth century, when it was replaced by Norway in the timber markets of Holland and England. The Polish wood trade consisted of timber for shipbuilding, joinery and cask-making as well as forest products such as tar and charcoal. Since the Middle Ages timber had been one of Poland's principal exports and, although less significant than cereals in the sixteenth century, it still involved large areas of the Vistula basin, yielding huge profits for the royal court and large land-owners.[453] The decline of Polish timber and allied exports was probably due to increasing areas placed under arable cultivation at the expense of the forests; this was most dramatically experienced in eastern Mazovia, previously a densely wooded area.

Cracow was an important market centre for the timber trade; demand within the city itself for heating houses, building purposes (e.g. Wawel Castle),[454] constructing river craft and, to a lesser extent, for charcoal in metallurgy, must have been considerable. Further, its position on the Vistula provided a collecting point for further wood transhipment down river to the Baltic coastal ports. The main source of Cracow's timber supply was the heavily wooded valleys of the Upper Vistula, Soła, Skawa, Przemsza and, above all, Pszczynka rivers in the Carpathian foothills, and from Silesia.[455] The sixteenth-century Polish historian Marcin Kromer specifically referred to these timber imports when writing about Cracow.[456] Besides timber, other forest products including pitch[457] (for coopers, shoemakers, ropemakers), lubricants[458] and resin[459] were imported into the city.

Some timber was exported from Cracow down the Vistula to central Poland and beyond. The Włocławek customs tolls reveal that Cracow merchants sent wood to Gdańsk throughout the sixteenth century (Table 29). This table suggests Cracow merchants were most involved in yew exports, a wood in great demand throughout Europe for making bows. Other ports on the Lower Vistula received timber from Cracow, amongst them Warsaw.[460]

Exhaustion of forests close to navigable rivers in the southern Baltic during the seventeenth century, meant West European merchants turned to the northern reaches of the Baltic Sea. Lithuania and especially, Belorussia now became the chief suppliers of timber for the West, particularly through the ports of Królewiec (Königsberg) and Riga.[461] This affected Cracow's export role in the timber trade, but the import of wood for internal needs continued. The eighteenth century saw little change for Cracow in this situation, in spite of a revived demand for Polish timber abroad, particularly through Gdańsk.[462]

Local demand for wood in Cracow remained. Wood was needed as a fuel and for building purposes; large quantities were consumed by the brewing,

Table 29. *Timber exports from Cracow to Gdańsk in the sixteenth century*

Type of timber	Measure	1537	1544	1546	1555	1557	1558	1560	1561	1572	1574
Logs	Pile	–									
	Beam	4									96
Wainscot	Bundles			5		80	10				
	per 100			18½							
Yew	No. of Horses	50	56		9½			20	8		
Grained wood	per 100									2½	

Source: *Regestra thelonei aquatici Wladislaviensis saeculi XVI*, S. Kutrzeba and F. Duda (eds), Kraków, 1915, pp. 381, 385, 395, 399–400, 419–20.

Table 30. *Woodfuel imports to Cracow from Southern Little Poland 1750–1785*

Year	1750	1755	1760	1765	1770	1775	1780	1785
No. of Cartloads	1,608	1,075	1,400	2,995	3,748	1,746	1,742	1,023½
Tonnage equivalent	1,251	836	1,089	2,330	2,916	1,358	1,355	796

Sources: Calculated from Regestra thelonia civitatis Cracoviensis, rkp. 2216, 2217, 224, 2235, 2238, for relevant years; M. Kulczykowski, *Kraków jako ośrodek towarowy Małopolski zachodniej w drugiej połowie XVIII wieku*, (P.A.N.), Warszawa, 1963, p. 69.

distilling, brick-making and lime-extracting industries together with the salt mines in Wieliczka/Bochnia.[463] Cracow's position on the boundary between the forested south and arable north of Little Poland led to its significance as a timber mart. After mid-century, archival sources note wood arriving in Cracow from many surrounding villages to the south of the city, including Żarnów, Zawada, Biała, Skawa, Budzów, Jasienica, Palcza, Harbutowice, Izdebnik, Skawinka, Zachełmna, Maków, Sidzina, Rudnik, Sułkowice, Barcice and Żywiec; and such places sent particularly firewood. The overall annual totals are seen in Table 30.

After the First Partition of Poland (1772) Cracow was cut off from its southern hinterland and had to rely on other source areas; wood was floated down the left bank tributary of the Vistula river; some timber was also exported downstream from Cracow to Brzesko, Nowe Miasto Korczyn and

Warsaw.[464] Other forest by-products such as charcoal (from the Beskidy Mountains), oak bark for leather tanning, resin and potash according to documentary records assumed little importance amongst the city's imports.[465]

Timber prices in Cracow are difficult to ascertain, due to the varied nature of use for wood; the only long-run data series is for firewood, whilst timber used for building purposes only has prices for short-term periods.[466] Signs of the great price fall in the seventeenth century are clearly visible in Cracow, as in Lwów;[467] they follow a pattern found in the German cities. However, although Polish prices tended to follow the 'German' pattern (i.e. price decline by 1620–40), Lwów and Cracow escaped this decline until the early 1650s.[468] This tendency continued for over a century when, again, timber prices, and particularly building wood, rose towards the end of the eighteenth century.

Trade in manufactured goods

Sixteenth-century Europe inherited from medieval times a system of manufacturing that remained unchanged in technology and little modified in organisation and structure; this situation was to continue until the second half of the eighteenth century. From then on technological innovation, wider markets and new types of industrial organisation were to transform the manufacture of goods throughout the continent and make way for the period known as the Industrial Revolution. Such developments, however, did not appear at the same time throughout Europe; in Poland manufacturing remained small-scale and predominantly domestic almost everywhere well into the eighteenth century, and in many areas even longer. The failure to develop a system of large-scale/factory production may be attributed to the lack of mass demand for most manufactured goods, shortage of capital investment, and imperfect knowledge on ideas of management. There was no European market as it is understood today; except for a few products demanded by a limited section of society, prices remained stable for, as Pounds has noted, 'The price of pepper, or of copper, might not vary except within narrow limits between Spain and Poland, Italy and England. But for most other commodities there was a series of local markets.'[469]

Nevertheless, industrial production did expand during the sixteenth century. There was, for example, an expanding market for cloth in eastern Europe, but evidence suggests that the overall production of consumer goods failed to keep pace with the increasing population of that century, and that real incomes were falling.[470] In Poland urban life was decaying, as the gentry increasingly turned for their purchases to the merchants who exported their cereals. Conversely, there was a movement of craftsmen into Poland during the first half of the seventeenth century; these came mainly from Germany, a country troubled by the Thirty Years War. The relative peace of Poland proved attractive; Polish manufacturing, especially cloth production, was to benefit from this influx of foreign expertise.[471] Unfortunately, much of Europe suffered during the second half of the seventeenth century which in Poland turned to a depression during the period known as the 'Deluge' (*Potop*) and the country, economically weakened and politically divided, succumbed to its neighbours during the eighteenth century. Admittedly, the landed aristocracy in Poland and, to a greater extent in Bohemia/Moravia, established factories for themselves in the eighteenth century, to utilise the raw materials produced on their estates, with workshops attached to their own houses.

Given this more general background, one may now ask how Cracow fitted into this changing scheme of things; to discover which products of the European manufacturing system were to pass through the city and what changes took place in the spatial origin and destination of these goods.

(i) Textiles

The international trade of the fifteenth and sixteenth centuries experienced a growing importance of manufactured goods, particularly of cloth for use by increasing numbers of consumers. Although textiles had been an important item in medieval long-haul trade, especially in the high-quality cloths for the central European markets from Flanders and Northern Italy, they were usually delivered in small quantities at high prices. By the beginning of the Early Modern Period the international cloth market was producing larger amounts for an expanding consumer commerce. To reiterate Pounds, 'One must not, however, underrate the scale and importance of those industries [e.g. clothmaking] which had developed to supply distant markets.'[472] Poland's imports were mostly of manufactured goods. The best cloth was from Flanders, England, France and Italy and imported either through Gdańsk or overland; medium grades came from Bohemia and Moravia, and the cheapest types from Silesia and Brandenburg. Silk fabric imports from Italy and the Orient were much smaller in quantity. Superior quality linen came from Flanders and the Rhineland, cheaper varieties from Silesia.

Antwerp was the centre of the international textile trade until about 1575, and it was from here that Flemish and some cloths from England were exported; while English wool exports diminished in the sixteenth century, their cloth trade to the continent more than doubled.[473] Some of these exports were sent to Poland, mainly in exchange for grain and timber:[474] in turn these textiles filtered through to the regional and local markets of the country. Here one found the inferior, cheaper textiles from central Europe;[475] cotton and silk fabrics from the Ottoman Empire were also evident, for which there was a great demand in Poland.[476] Important urban

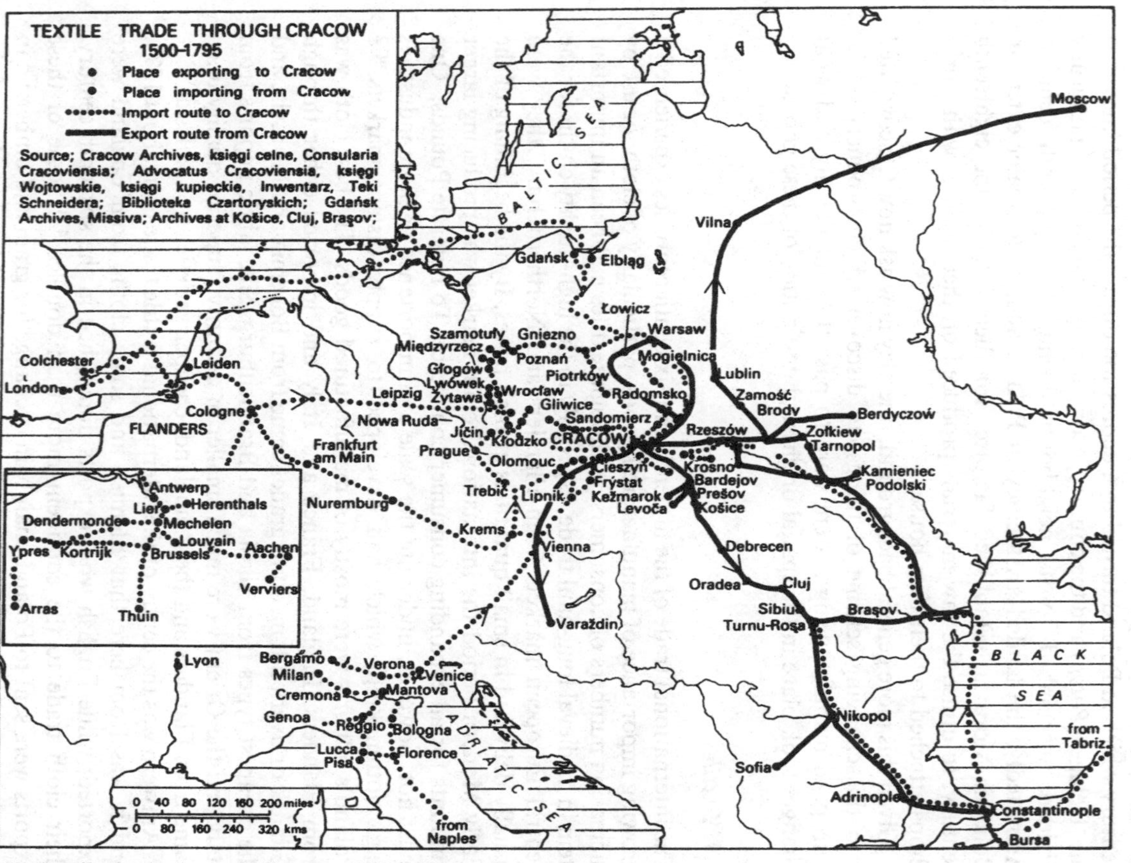

66 Textile trade through Cracow, 1500–1795.

centres, like Cracow, Poznań and Lublin, usually with their own annual fairs,[477] provided the ideal environment for textile purchase, either for the surrounding local area, or for further transit on the long-haul commercial network by textile merchants.

Cracow was an important textile trading centre in the sixteenth century (Fig. 66) although it had no significant cloth-manufacturing industry.[478] Material arrived in Cracow's Cloth Hall (*Sukiennice*) from a variety of sources. More generally, the finer quality cloths came from the distant workshops of Italy, London, Flanders and Constantinople, the cheaper from central European sources, particularly Silesia. It is perhaps worthwhile to note that in spite of changing trade routes, re-orientation of European trade from the Mediterranean to the Atlantic seaboard, and new varieties of cloth for sale, Cracow retained its position as an international textile mart in the sixteenth century; cloth provided an important turnover article in the city's commercial structure.

Quality cloths from western Europe arrived in the city either via Gdańsk or through Wrocław in Silesia. In the early sixteenth century the purchase of cloth from Flanders, London and other west European countries was largely in the hands of merchant families such as the Boners in Cracow and Popplaus in Wrocław;[479] the expensive English broadcloth, known as *Luński* in Polish documents[480] referred in the sixteenth and seventeenth centuries both to Dutch and English cloth[481] and was regarded as amongst the best imported.[482] In Cracow, the city council bought it, partly to remunerate its officials and functionaries.[483] On occasions it was called 'Gdańsk' cloth, because their merchants sold it.[484] According to Rybarski most of this cloth came through Silesia to Cracow and, to a lesser extent, via Gdańsk.[485] Besides cloth rolls, other textile products arrived in Cracow, such as Flemish tapestries for the decoration of Wawel Castle.[486]

Increasingly, during the second half of the sixteenth century, Cracow market displayed cloth from central European workshops, above all Silesia. Production was located mainly in central and lower Silesia, particularly the towns of Lwówek Śląski, Zielona Góra, Głogów, and Zgorzelec.[487] Linen cloth production (mainly for canvas) was concentrated in the Sudeten foothills, at Jawor, Gryfów Śląski, Kłodzko, around Cieszyn, and Bielsko-Biała.[488] Moravian cloth was also present on the Cracow market; most of it originated from the towns of northern and south-western Moravia. In the north-central region the most important cloth-making workshops were in Nový Jičín, Příbor, Meziříčí, Přerov, and Holešov; the main centres in south-west Moravia were at Jihlava and around Třebíč.[489] Linen cloth manufacture was most advanced in northern Moravia near the Silesian border, where adequate flax supplies could be obtained.[490] The Moravian textile industry was noted not only for cloth material, but also for

Table 31. *Cracow's linen cloth imports from Poland at the end of the sixteenth century*

Region	1589				1593				1594	
	No.of Bolts	Metre equiv.	No. Horses	Approx. Metre equiv.	No.of Bolts	Metre equiv.	No. Horses	Approx. Metre equiv.	No. loads	Approx. Metre equiv.
Biecz	18	558	27	16,200	269	8,339	29	17,400	21	27,300
Bielsko-Biała	272	8,432	–	–	890	27,590	16	9,600	16	20,800
Other	10	310	2	1,200	5	155	–	–	1	1,300
Totals	300	9,300	29	17,400	1,164	36,084	45	27,000	38	49,400
Approx. Total Metre equiv.	26,700 metres				63,084 metres				49,400 metres	

Sources: Calculated from Wojewódzkie Archiwum Państwowe w Krakowie, *Księgi celne*, rkps. 2115, 2117, 2118 for relevant years; J. M. Małecki, *Studia nad rynkiem regionalnym Krakowa w XVI wieku* (P.W.N.), Warszawa, 1963, p. 189.

other products including hats, tents, handkerchiefs, tablecloths etc.[491] Bohemian cloth mainly came to Cracow from Prague.[492]

Polish textile production was concentrated in two main weaving areas. Most of the weaving centres of Greater Poland were situated on the left bank of the River Warta, mainly within a 40-km belt stretching along the Silesian border. Here one found the major cloth-making region of Greater Poland; in the sixteenth century it covered the larger part of present-day Poznań voivodship, whilst the adjacent Kalisz voivodship formed its eastern boundary. To the south and west this region reached up to the Silesian frontier.[493] Important weaving centres included Wschowa, Leszno, Kościan and Międzyrzecz. Nearer to Cracow the main textile manufacturing area at this time was in eastern Little Poland, centred around the town of Biecz.[494] Production in the area exceeded local demand and Biecz cloth/linen material enjoyed wide appeal both within Poland and abroad. Although linen utilised local flax supplies, the cloth industry had to rely on northern and north-western Poland for its raw material base. Rybarski found that cloth from Biecz, Krosno and Stary Sącz were already noted in Cracow's customs books by 1538/39.[495]

The large number of extant Cracow documents for the latter part of the sixteenth century allow a clearer picture to emerge regarding the city's cloth imports. Polish linen cloth arrived in Cracow mainly from two regions

Table 32. *Cracow's foreign cloth imports 1591/1592*

Type of cloth	Quantity							
	Bale	Approx. Metre equiv.	Terling	Approx. Metre equiv.	Warp	Approx. Metre equiv.	Horse	Approx. Metre equiv.
1. From Silesia								
a) Silesian	13½	6,480	½	500	569	11,380	4	2,400
b) Zgorzelec	48	23,040	1	1,000	–	–	–	–
c) Lwówek	20	9,600	19	19,000	–	–	–	–
d) Opole	–	–	–	–	201	4,020	–	–
e) Brzeg	1	480	–	–	327	6,540	4	2,400
f) Kłodzko	2½	1,200	–	–	281	5,620	–	–
g) Nowa Ruda	37	17,760	–	–	55	1,100	–	–
h) Ziębice	–	–	–	–	165	3,300	–	–
2. From Moravia								
a) Moravian	–	–	–	–	468	9,360	52	31,200
3. From England								
a) London	1½	720	–	–	13	260	–	–
4. Other (origin unknown)								
a) Plain					25	500		
b) Fine	1	480						
c) Kersey	1	480	2½	2,500	22	400		
d) Unidentified	10	4,800			130	2,600	34	20,400
Totals	135½	65,040	23	23,000	2,256	45,080	94	56,400

Sources: Calculated from Wojewódzkie Archiwum Państwowe w Krakowie, *Księgi celne*, rkp. 2116; J. M. Małecki, *Studia nad rynkiem regionalnym Krakowa w XVI wieku*, (P.W.N.), Warszawa, 1963, p. 45.

(Table 31). This table shows that, on average, about 46,500 metres of Polish linen cloth arrived in Cracow annually; half of this came from the Biecz region (50.14 per cent) and nearly half from Bielsko-Biała – 47.72 per cent. The other regions (2.13 per cent) played a very insignificant role, supplies coming from around Olkusz, and Szydłów (near Kielce), or with unrecorded origin.

Cracow's cloth imports near the end of the sixteenth century suggest heavy dependence on Silesian supplies. The 1591/92 customs book supports this view (Table 32). This table emphasises Cracow's reliance on cloth imports from or via Silesia; over 60 per cent of the cloth (in metres) was produced there, and may well have been more. Unfortunately, 17 per cent of the documentary entries failed to record cloth origin.

Italian cloth was also to be found in the Cracow market place. A number of Italian merchants lived in Cracow during the sixteenth century, and sold their country's cloth by the ell in stalls behind the Cloth Hall,[496] but they were not the only foreign traders in the city. Germans, the largest group, Armenians and Jews also traded there. This fact perhaps prompted the Italian merchant, Paulo Giovanni, to write in 1565 that only the Poles failed to benefit from this commerce in textiles.[497] Further impetus in textile trading came during the reign of King Stefan Batory (1576–86) when he gave Cracow merchants the right to sell cloth 'from Italy, Venice and Mediolani' (Milan) throughout the whole country.[498] Much resentment was felt by other traders at the methods the Italians adopted in selling cloth; a controversy begun in 1581, culminated in an appeal to the king in 1586[499] who clarified the situation by giving preference to merchants holding Cracow citizenship. Some Italian trading companies in Cracow also specialised in importing their own and other western cloth, for further dispatch to the east.[500]

Cracow's strong Italian connection may have resulted from the marriage of Zygmunt I (Sigismund) to his second wife Queen Bona Sforza (1494–1557)[501]; as Tazbir states 'The arrival of Italians with Bona Sforza ... enabled the average squire, who never travelled abroad, to acquaint himself with the ways of these foreign visitors and to compare it with his own "mos polonicum".'[502] He further maintains that in general foreigners did not meet with an enthusiastic reception,[503] for there was dislike of their intellectual acumen and the manner in which they managed to acquire wealth and rank. These traits were often a source of irritation to the average Polish merchant, as testified in the above-mentioned dispute.

Finally, mention should be made of oriental cloth imports to Cracow. Eastern textiles from the Ottoman Empire including silks, silk taffetas, brocade, damask, velvet, camlet, kaftans, and various other types of silk fabric arrived in Cracow from the distant workshops of Constantinople and other parts of Turkey.[504] The main middlemen in this trade were Armenian and Jewish merchants from Lwów,[505] who not only sold such materials in the city, but also took them in transit to Gdańsk for sale in the west.[506]

The export both of foreign and domestic Polish cloth from Cracow to the east and especially south (Hungary and Transylvania) continued in the sixteenth century. Throughout much of the Later Middle Ages, Hungary had provided Cracow's merchants with a steady outlet for its textile trade, particularly the more expensive west European cloths. During the sixteenth century there was a change in emphasis with increasing sales of cheaper Polish and other central European cloth to the Hungarian markets. Admittedly there was still demand for English and Dutch cloth from the Hungarian court and richer individuals found throughout the country[507] but Cracow's customs books suggest that it was the cheaper cloth, especially

Table 33. *Cracow's cloth exports to Hungary: 1591/1592 and 1593*

Type of cloth	1591/1592								1593							
	Bale	Metre equiv.	Ter-ling	Metre equiv.	Warp	Metre equiv.	Ell	Metre equiv.	Bale	Metre equiv.	Ter-ling	Metre equiv.	Warp	Metre equiv.	Ell	Metre equiv.
1. From Silesia																
a) Silesian					22	400							11	220		
b) Zgorzelec	32	15,360			4	80			19	9,120	½	500	14	280		
c) Lwówek	5	2,400							2	960						
d) Nowa Ruda	1	480														
e) Świebodzin					2	40										
2. From Moravia																
a) Moravian					8	160										
3. From W.Europe																
a) Dutch					10	200	10	5					16½	330	20	10
b) London					6	120										
4. Little Poland																
a) Broniów					23	480										
5. Great Poland																
a) Międzyrzecz					1	20										
6. Other (origin unknown)																
a) Plain	2	960			6	120										
b) Fine			½	500	12	240										
c) Kersey					92	1,840			1	480			45	900		
d) Unidentified	2	960	1	1,000	9	180							237	4,740		
Totals	42	20,160	1½	1,500	196	3,920	10	5	22	10,560	½	500	323½	6,470	20	10

Sources: Calculated from Wojewódzkie Archiwum Państwowe w Krakowie, *Księgi celne*, rkp. 2116; K. Pieradzka, *Handel Krakowa z Węgrami w XVI wieku* (Biblioteka Krakowska No. 87), Kraków, 1935, p. 214.

from Silesia[508] that now dominated in places like Prešov, Bardejov, Kežmarok, Varaždin, Debrecin and Košice.[509]

What factors contributed to this changing situation? Firstly, it should be remembered that in Hungary war, internal disturbances and Ottoman incursions (e.g. the Battle of Mohács, 1526) had impoverished urban centres and disrupted trade routes to such an extent that any return to former commercial prosperity was beyond possibility. Secondly, during the late fifteenth and whole of the sixteenth century, Polish cloth production saw strong development in the Carpathian foothill region (e.g. Biecz) which produced material in large quantities for the Hungarian market.[510] Thirdly, a possible reason for decline in foreign-cloth imports to Hungary and Cracow's near monopoly situation in these products, which now had to satisfy a growing home market. Fourthly, by the mid-sixteenth century price differentials between western and central European cloth was forcing Hungarian merchants to increasingly purchase the latter.[511]

Some idea of the cloth varieties sent from Cracow to Hungary during the last decade of the sixteenth century is seen in Table 33. Despite over a quarter of the total metre length having an unknown origin, Table 33 reveals that nearly 70 per cent of Cracow's cloth exports to Hungary were produced in Silesia. Only 1.5 per cent were of west European origin, whilst together cloth from Poland and Moravia only totalled a similar percentage. Such results, however, must be interpreted with caution. It should be remembered that quantities quoted in the customs books were only the minimum amounts; in reality a much larger total could have been traded, due to customs avoidance, commercial privileges etc. Given these weaknesses of Cracow's documentary sources, it seems nevertheless that the city's cloth exports to Hungary were much less in the 1590s than during the first half century, or even when compared with 1584.[512]

Linen cloth also figured in Cracow's export market but on a much smaller scale. Demand for high-quality linen cloth came from royal courts (especially for tablecloths); the expensive varieties, particularly from Cologne and Gniezno, were sent to Hungary by Cracow merchants, along with cheaper cloths from Silesia (Głogów, Bielsko-Biała and Lipník).[513] The reason only limited amounts went from Cracow to Hungary was the protectionist policy of the Magyar rulers who wished to encourage domestic production (e.g. at Bardejov);[514] even so, evidence suggests that Polish exports continued throughout the century with Cracow's merchants specialising in linen cloth from Biecz.[515]

Transylvania was another outlet for Cracow's textile transit trade. Woollen, silk and cotton fabrics were the most important goods imported into the area and had been since the end of the twelfth century.[516] They reached the Transylvanian market from Europe, the Middle East and even the Far East; although the Ottoman Empire was the major trading partner of

Table 34. *Cloth imports Cracow–Cluj, 1592–1595, 1597, 1599*

Type of cloth	Quantity in metres	% of total
Dutch or Swiss dresses	4,424	21.91
Silesian cloth	3,846	19.05
Black cloth	2,829	14.02
Veils from Flanders	2,598	12.86
Simple white cloth	2,068	10.24
Coloured cloth	1,579	7.83
Cloth from Biecz (Little Poland)	728	3.61
Cloth from Cologne (damask, atlas, Dutch)	687	3.41
Other types	1,428	7.07
	20,187	100.00

Source: Wojewódzkie Archiwum Państwowe w Krakowie, *Księgi celne*, rkps. 2116 (1592); 2117 (1593); 2118 (1594); 2120 (1597); 2021 (1599).

the Romanian Principalities (Wallachia, Moldavia and Transylvania) during the sixteenth century, evidence from Cracow's archives certainly confirms cloth imports from central and western Europe. According to customs data from the 1590s Cluj was one of the main cloth-importing centres, with over 3,000 metres of material on average annually (Table 34).

Analysis shows that the better-quality material (from Flanders and Cologne) totalled 3,285 metres (16.27 per cent), but over three-quarters was poorer quality cloth (15,474 metres: 76.65 per cent) mainly from Silesia, Poland and the poorer dress material from Western Europe which had declined in quality when compared with earlier in the century. The predominance of Silesian cloth in Cluj imports from Cracow is best illustrated for 1599 (Table 35). Thus only English cloth (*Luński*) was not manufactured in Silesia, for even Kersey was now being produced there.[517]

Cloth imports from Cracow to Oradea during the last decade of the sixteenth century are seen in Table 36. These quantities show that three-quarters of the cloth was of West European origin, with Silesian material totalling 16 per cent and Polish cloth a mere 5 per cent; the other important market for western textiles sent through Cracow was Braşov.[518]

Most of this cloth, of course, did not remain in these towns but was dispatched further into the markets of the Ottoman Empire through Nikopol to Sofia and beyond. Similarly, the Romanian Principalities were the main areas of exchange in goods travelling from Adrianople and Constantinople en route to Cracow and Western Europe.[519] Oriental textiles, particularly silk and cotton fabrics were sent from Tabriz, via Bursa and Constantinople northward to Poland, often conveyed by Polish merchants.[520]

Table 35. *Cloth imports Cracow–Cluj, 1599*

Type of cloth	Quantity in metres	% of total
Cloth from Zgorzelec	41,274	55.69
Kersey	13,813	18.64
Cloth from Wrocław	12,058	16.27
Cloth from Gorzów	5,303	7.16
English cloth	1,657	2.24
	74,105	100.00

Source: Wojewódzkie Archiwum Państwowe w Krakowie, *Księgi celne*, No. 2021 (1599).

Table 36. *Cloth imports Cracow–Oradea, 1593, 1595, 1597, 1599–1600*

Type of cloth	Quantity in metres	% of total
Dutch or Swiss dresses	822½	3.35
Silesian cloth	3,437½	14.00
Black cloth	495	2.00
Veils from Flanders	7,671	31.23
Cloth from Biecz (Little Poland)	165	0.67
Cloth from Podgórze (Little Poland)	825	3.35
Cloth from Koło (Greater Poland)	220	0.90
Cloth from Cologne (damask, satin, taffeta, silks)	10,127	41.25
Fustian cloth	110	0.45
Other types	687½	2.80
	24,560½	100.00

Source: Wojewódzkie Archiwum Państwowe w Krakowie, *Księgi celne*, rkps. 2117 (1593), 2119 (1595), 2120 (1597), 2021–2 (1599–1600).

Textiles were also sent through Cracow to lands further east. Irrespective of the changing fortunes of the Hungarian cloth trade, Cracow's merchants still had other income sources from textiles, particularly in trade connections with other Polish towns; of these Lwów and Lublin were the main importers. Lwów traders relied on the Cracow market for supplies both of local cloth and that in transit from western Europe; Lwów's city records from the sixteenth century reveal that the more expensive imports included cloth from Brussels (usually in black, yellow, green or brown shades), Thuin (black),

Ypres, Mechelen (red and black), Louvain (black and yellow), Dendermonde, Lierre, Kortrijk, Herenthals and Eysden as well as from England (London; Colchester). The simpler, cheaper kinds were from Silesia (Wrocław; Żytawa; Opole; Zgorzelec), Moravia, Prussia and Poland.[521] One Lwów firm alone was importing over 2,000 metres of cloth annually from Cracow at the end of the century.[522] Merchants from Cracow rarely took their cloth to Lwów; they transported it only as far as Rzeszów, Przemyśl, or Jarosław where they agreed contracts with their Lwów counterparts.[523]

The other important cloth centre was Lublin. Like Lwów, Lublin imported a variety of luxury material from the West, together with simpler cloths from central Europe. Lublin customs records note the import of Italian and English cloth, along with cheaper varieties from Greater Poland (Szamotuły), Mazovia (Mogielnica), Silesia (Świebodzin, Zgorzelec), and Little Poland (Bukowiec); linen cloth came from Flanders and England.[524] Some of the textiles from Cracow recorded in the Sandomierz customs books may also have been en route to the Lublin market.[525] An inventory dated 1586 of Jan Cechi, a well-known Italian merchant in Cracow, lists goods sent to Gromnik fair including large amounts of cloth and silks destined for Lublin.[526] From Lublin some of the textiles were sent further east; also on occasions merchants from Vilna even came to Cracow to purchase cloth material.[527]

Finally, and more generally, mention should be made of an influx of Jews who came to live in Cracow at the end of the sixteenth/early seventeenth centuries; they arrived from Germany, Italy, Spain and Portugal, many settling in Kazimierz. They became involved in many branches of commerce, but their knowledge of west European contacts was particularly useful in the textile trade. Links were forged with other Jews living in southern Poland and Silesia (e.g. Bielsko-Biała)[528] where local cloth purchases were made and then exported through Cracow to Hungary,[529] Moravia, Lwów, Lublin and places east.

The seventeenth century was to see changes in the textile industry throughout Europe. The making of cloth, more than some other forms of manufacture, experienced modifications in its pattern of location. Some of the medieval textile centres declined whilst prosperity came to new countries and new areas; above all, cloth manufacture became a rural industry. This was the age of the 'putting-out system', which depended on a widely scattered domestic labour supply. The regions that decayed were largely those of small-scale urban manufacture. Rural wages were often much lower than those artificially fixed in towns and expansion of the industry took place mainly in areas that adopted the new system of utilising cheaper labour in the countryside.

Western Europe, traditional source of better-quality cloth for the Polish

Table 37. *Cloth exports to the Baltic region from England and Holland, 1591–1650*

Annual average for the period	From England	From the Netherlands
1591–1600	28,700 pieces	2,600 pieces
1601–1610	32,000 pieces	6,700 pieces
1611–1620	34,000 pieces	16,200 pieces
1621–1630	29,000 pieces	30,500 pieces
1631–1640	32,600 pieces	39,800 pieces
1641–1650	20,700 pieces	33,800 pieces
Total	177,000 pieces	129,600 pieces

Source: Calculated from R. W. K. Hinton, *The Eastland Trade and the Common Weal in the Seventeenth Century*, Cambridge, 1959, pp. 227–30.

market, experienced these changes throughout the century. For example, woollen-cloth production in Venice was reduced to a tenth its former size, whilst Milan, Florence and Genoa suffered almost as badly. Some of this was due to urban–rural movement of manufacturing, whilst luxury textiles had to compete with lower-priced French and Dutch varieties.[530] In Germany continued warfare resulted in the collapse of many urban industrial centres whilst the devastated countryside found little opportunity for the new putting-out system or work. Likewise in the Netherlands, the medieval Flemish clothing towns suffered badly during the Revolt in that country; many weavers moved northward and re-established their industry around Leiden. However, the whole industry was dominated by astute merchants who dealt in finished cloth and the superiority of Dutch finishing processes (dressing, bleaching and, above all, dyeing) ensured it a continued market throughout this troubled century. In some cases this led to complaints by foreigners that the Dutch bought cloth abroad, dyed it, and exported it as their own product. England and Holland were particular rivals in this trade, the former dominating Baltic cloth imports until the end of the sixteenth century'[531] from then on the Dutch became increasingly important for the east European market, as seen in Table 37.

For Cracow the major source of Baltic cloth imports came from Gdańsk. Although occasionally hampered by personal disputes between textile merchants of the two cities,[532] western cloth in large amounts and various sizes arrived in Cracow throughout the first half of the seventeenth century.[533] Customs books' data gives some idea of this trade (Table 38). This table, however approximate it may be, does give some indication of the Gdańsk–Cracow cloth trade. For example, it suggests that during the first half of the seventeenth century nearly a quarter of a million metres of

Table 38. *Textile imports Gdańsk-Cracow, 1589–1655*

Total for period	Cloth						Fine English/Dutch cloth				Kersey					Silk Fabrics		
	Bale	Terling	Warp	Bundle	Horse	Approx. metre equiv.	Bale	Warp	Bundle	Approx. metre equiv.	Bale	Terling	Warp	Bundle	Approx. metre equiv.	Bolt	Bundle	Approx. metre equiv.
1589–1600	168	4	4	4	14	93,184	–	–	–	–	6	1	–	–	3,880	–	–	–
1601–1610	88½	9½	22½	–	6	56,030	2	15	–	1,260	10½	–	16	–	5,360	–	–	–
1611–1620	50	–	1	1	–	24,036	½	10½	–	450	1½	–	150½	2	3,762	–	–	–
1621–1630	–	–	–	–	–	–	–	–	–	–	–	–	50	–	1,000	–	–	–
1631–1640	46	–	2½	–	–	22,130	–	1,194½	–	23,890	6½	–	3,104	–	65,200	46	–	1,426
1641–1650	4	–	6	–	–	2,040	–	112½	–	2,250	–	–	138	–	2,760	48	1	1,504
1651–1655	60½	–	–	45	–	29,760	8	94	3	5,768	–	–	28	–	560	20	–	620
Sub-totals	417	13½	36	50	20	227,180	10½	1,426½	3	33,618	24½	1	3,486½	2	82,522	114	1	3,550

Totals: Bales = 452; Terlings = 14½; Warps = 4,949; Bundles = 56; Horses = 20; Bolts = 114
Approximate metre equivalent = 346,870

Sources: Calculated from Wojewódzkie Archiwum Państwowe w Krakowie, *Księgi celne*, rkps. 2115–69; J. M. Małecki, *Związki handlowe miast polskich z Gdańskiem w XVI i pierwszej połowie XVII wieku*, (P.A.N.) Prace Komisji Nauk Historycznych No. 20, Wrocław-Warszawa-Kraków, 1968, Table XVIII, pp. 236–8.

various cloth types were transported from Gdańsk to Cracow (i.e. 213,098 metres). Further, nearly two-thirds of the metre length is referred to in documents as 'cloth' of no specific type, while kersey is explicitly mentioned for a quarter of the total metre length; moreover, English/Dutch fine cloth ('Falendysz') amounted to a tenth and only 1 per cent was silk fabrics. Supportive evidence of the Gdańsk–Cracow cloth trade besides customs registers also came from shop inventories of certain Cracow textile merchants suspected of customs evasion and subsequent search of their property.[534]

Italy was the other main source of Cracow's western textiles during the seventeenth century. Italian fabrics were fashionable throughout Poland at this time; the city's trading community had several local merchants, mainly of Italian origin, who were involved in importing valuable Italian luxury fabrics. Although the north Italian woollen cloth industry suffered from erratic raw material supplies, their silk weavers were more fortunate. Many countries in the Mediterranean basin produced raw silk, which was collected and dispatched from Messina and Naples to northern Italy. For example, at the beginning of the seventeenth century the production of fine cloth and refined silk fabrics was highly developed in Venice.[535] Obuchowska-Pysiowa has analysed the 1604 Cracow customs registers and shown that such fabrics arrived from Venice by the overland route via Vienna.[536]

The Italians formed a nucleus of wealthy traders in Cracow around 1600, who also held important administrative posts in the city.[537] One of the most successful was Valery Montelupi; he specialised in expensive Italian fabrics, particularly for the royal court, importing his goods mainly via Nuremburg and, to a lesser extent, Krems, Vienna, etc.[538] He, like other Italian merchants in Cracow, had the advantage of continued links with north Italian workshops, unlike Polish and other foreign traders who were discouraged from visiting Italy. Consequently, non-Italian merchants could only travel to intermediary towns (e.g. Nuremburg, Vienna) where they obtained Italian products at local fairs.

Throughout the first half of the seventeenth century Cracow's customs registers provide evidence of continued Italian fabric imports, especially silk; taffeta, a thin glossy silk material, came mainly from Lucca, the chief production centre,[539] as well as a rougher version from Milan.[540] Closely woven silks, like satin, also arrived in Cracow from Florence[541] and Lucca;[542] smaller quantities arrived from Genoa, Naples, Milan and Bologna.[543] Silk damask was exported to Cracow from three main centres – Genoa, Lucca and Venice.[544] Velvet, however, appears to have been less popular on the Polish market.[545] Nevertheless, quantities arrived in Cracow from Genoa, Florence, Venice, Lucca and Naples[546] and, to a lesser extent, Milan and Reggio nell' Emilia.[547] Of the more exotic imported fabrics, cloth of gold – a tissue of gold, silk, and wool threads – should be mentioned. *Drappi d'oro* and its silver equivalent (*drappi d'argento*) were sent to Cracow from workshops in Florence, specially made to order for the royal court.[548] Similarly, brocade came to Cracow from Milan and Florence.[549]

The third quarter of the seventeenth century saw a lull in the intensive import of Italian silks; largely this was due to the uncertain situation existing on the Polish market, caused by the wars suffered in many parts of the country, including Cracow. However, after 1670, more peaceful times led to a rekindling of former contacts in north Italy from whence once again fabrics reached Cracow. Certainly the Italians were still producing considerable quantities of silk cloth for, according to Thornton, 'The demand for silk materials grew as the seventeenth century continued . . . so that there was still sufficient work to keep a fairly large Italian industry active.'[550] Cracow's archival material suggests the existence of two major textile firms, one belonging to Marco Antonio Frederici, the other Paolo Antonio Ricciardetti. They both maintained direct contact with Italian silk manufacturers in Venice, Florence, Lucca and (after 1682) with Bergamo. Goods came overland directly from the warehouses in Verona, via southern Germany (Augsburg) to Leipzig, Wrocław and Cracow, a journey of about six weeks.[551] Usually, three journeys were made annually with cloth arriving in Cracow for the New Year, Easter and the end of September (St Michael's Day).

Cracow's documentary sources also suggest that Italian silk fabrics were of two types – traditional material commonly imported during the first half of the century, and new varieties previously unrecorded in the archives. Amongst the conventional silk cloths were taffetas from Lucca and Florence, as well as Mantua and Pisa,[552] together with satin from Florentine, Venetian, Luccese and Bolognian workshops.[553] A similar location pattern emerges for damask (Lucca, Florence and Venice).[554] Velvet, however, seems to have experienced a changed location emphasis; Venice was now the main importing centre[555] whilst places formerly more significant (Genoa, Lucca, Florence) now sent much less material.[556] On rare occasions velvet from Cremona was imported.[557] More expensive silks (cloth of gold, brocade) previously in such demand by Cracow's royal court, now found few purchasers; the Church was the main buyer for use in making liturgical capes.[558]

Several new types of Italian silk cloth were found in Cracow shops during the latter part of the seventeenth century. Amongst these were lamé,[559] and samite from Venice, first mentioned in 1669.[560] Other cloths included silk gauze from Venice (1681 onwards),[561] and poplin (after 1676).[562] Finally, examples of sagathy came from Milan and Florence, the first appearing in the late 1670s.[563]

Silks were not the only fabrics to be sent from Italy to Cracow; analysis of the 1604 customs registers by Obuchowska-Pysiowa have shown that other fabrics were often more numerous but, as they were much cheaper than silks, their value as transit commodities was lower.[564] Taszycka also has stated that late in the century certain quantities of woollen and cotton textiles arrived in Cracow,[565] but it was silk fabrics that obtained the highest profits in the city's transit trade.

If woollens and luxury silk fabrics predominated in Cracow's cloth imports from West Europe during the seventeenth century, in contrast, linen from central Europe provided much of the cheaper material for the city's transit trade. Linen manufacture was a traditional craft in Silesia and lively trade connections in this cloth existed with Cracow. The textile industry, particularly cloth and flax weaving, became the most important sector in the Silesian economy during the sixteenth and early seventeenth centuries; this was largely due to foreign merchants, including those from Cracow or their agents, entering the Silesian countryside to organise a domestic industry which utilised cheaper rural labour.[566] Industrial growth continued unabated in Silesia during the early years of the seventeenth century, prosperity in the textile centres reflecting the continuous expansion of linen manufacture and exports.[567]

Cracow provided one of the main outlets for Silesian cloth; these came to the city together with west European textiles sent via Wrocław. Disruptions caused by the Thirty Years War (1618–48) seem to have had only a minimal effect on this trade, certainly during the early stages of the conflict. As Jeannin points out, 'on the Continent itself, despite the ravages of warfare, the expansion of trade was often dramatic . . . In 1629 the exports of cloth from Breslau [Wrocław] to Cracow broke all records.'[568] This statement is supported by Wolański's research on the volume of trade between Wrocław and Cracow in 1629, which was at least twice the annual average for the earlier part of the seventeenth century. Similarly, his figures for cloth exports reflect this overall trend (Table 39).

After its peak in 1629 cloth exports to Cracow declined; towards the end of the century they amounted to only a fifth of their level in 1600. According to Inglot 'The Thirty Years War and the counter-reformation policy of the Hapsburgs checked the development of cloth weaving', followed by 'the emigration of Silesian cloth weavers from the regions near the Greater Poland frontier; they settled down in towns on the other side of the frontier'.[569]

Linen manufacture was also a traditional craft in the Czech lands (Bohemia/Moravia); Cracow's imports of both linen and other woollen cloth, from the 1624 customs registers, seem to have come mainly from N. Moravia (Table 40).

These places, however, were mentioned in documents only as exporting centres, but the cloth may have come from a number of workshops spread widely throughout the Czech lands. The main centres trading in cloth were Nový Jičín, Frýdek and Příbor and, to a lesser extent, Karviná, Fulnek and

Table 39. *Silesian cloth exports to Cracow in the seventeenth century*

Year	1600	1609	1619	1629	1639	1644	1654	1664	1674	1684
No. Warp	10,700	8,453	17,789	27,499	9,202	4,815	1,926	3,638	3,103	2,033
Metre equivalent	214,000	169,060	355,780	548,980	184,040	96,300	38,520	72,760	62,060	40,660

Source: Calculated from M. Wolański, *Związki handlowe Śląska z Rzecząpospolitą w XVII wieku z szczególnym uwzględniem Wrocławia*, Wrocław, 1961, p. 149.

Table 40. *Cracow: linen and cloth imports from N. Moravia, 1624*

Place/source	Linen		Other cloth		Fabrics	
	Bolts	Approx. metre equiv.	Bales	Approx. metre equiv.	Warps	Approx. metre equiv.
Lipník	255	7,905	–	–	–	–
Fryšták	494	15,314	–	–	–	–
Frýdek	22	682	–	–	124	2,480
Valašské Meziríči	5	155	–	–	–	–
Fulnek	–	–	–	–	16	320
Nový Jičín	–	–	–	–	1,775	35,500
Karviná	–	–	6	2,880	–	–
Opava	–	–	–	–	8	160
Příbor	–	–	–	–	696	13,920
'from Moravia'	75	2,325	–	–	–	–
Cracow merchants	–	–	–	–	675	13,500
Unknown	–	–	12	5,760	–	–
Sub-totals	851	26,381	18	8,640	3,294	65,880
	Total approximate metre length = 100,901					

Source: Calculated from Wojewódzkie Archiwum Państwowe w Krakowie, *Księgi celne*, rkp. 2142.

Opava. The origin of cloth bought by Cracow merchants is not recorded in the customs notes. Linen came to Cracow mainly from Lipník and Fryšták and was largely in the hands of Moravian merchants; they usually delivered it to Cracow's chief linen buyer, one Martin Pacoszka. Although metre lengths in Table 40 are only approximate they suggest linen totalled over a quarter of Cracow's textile imports from the Czech lands in the 1620s. After 1650 this trade suffered decline, for the Swedish Wars had a catastrophic effect on Cracow's trade relations both with the Czech lands and Silesia.

Polish cloth imports also reached Cracow. The emigration of weavers from Silesia to Greater Poland strengthened textile development in such centres as Poznań, Wschowa and Leszno.[570] The Silesian cloth weavers were welcomed not only by the business-minded landed aristocracy, but also the townspeople of Greater Poland, both seeing their arrival as an economic advantage.[571] Cloths produced by them entered the domestic market, and reached the main urban centres of Little Poland, including Cracow. However, Greater Poland's developing cloth industry indirectly led, through strong competition, to a decline in the textile centres of Little Poland; for example, by the second half of the seventeenth century textile manufacture had all but disappeared in Biecz, formerly so important for Cracow. Much of the cause was related to Biecz's dependence on raw material supplies from northern and north-western parts of Poland, which were increasingly in demand from centres like Poznań, etc. Linen manufacture in Biecz was more fortunate, as it relied on local flax supplies; even so linen exports to Cracow declined during the first half of the seventeenth century (Table 41).

Cracow's textile merchants looked to their traditional markets south of the Carpathian Mountains and, in eastern Poland, from exports during the seventeenth century. Closest links were with East Slovakia in Northern Hungary, and Transylvania. Although Cracow lost some of its former importance in Central European trade during this century, the textile trade maintained its former significance, certainly in the early decades. Cracow's customs sources suggest two East Slovakian towns, Prešov and Levoča, were pre-eminent in the city's cloth exports.

Prešov, located in the middle Torysa valley, lay on the main trade route from Cracow via the R. Poprad en route to Košice. It was a major route centre for connections with Silesia, Moravia and Austria as well as farther south into central Hungary. Marečková's analysis of Prešov's customs registers for the first three decades reveals the significance of Moravian cloth and Polish linen imported via Cracow (Table 42). Clearly, the dominance of Polish linen is most notable, forming two-thirds (67.73 per cent) of total cloth imports, while Moravian was another sixth; all other textiles together formed less than a fifth of the overall amount. Admittedly, the table is only based on surviving archival material covering thirteen random years, but the figures do suggest an average of over 3,300 warps (approx. 66,000 metres) annually were dispatched to Prešov at this time.

Besides the obvious importance of Polish linen, Cracow's customs register for 1631 gives some indication of the main cloth types sent through Prešov to northern Hungary, destined for the markets of Slovakia and Transylvania (Table 43). This table reveals the virtual monopoly enjoyed by merchants from Cracow in supplying some of the more expensive cloths to Prešov around 1630. Moreover, they also seem to have controlled the trade in some cheaper cloths; for example, in 1631 of the 4,550 warps (91,000 metres) of Silesian cloth sent from Cracow, to Prešov, 3,958½ warps (79,170 metres) or 87 per cent were organised by the city's own traders.[572] Several of the merchants mentioned in Table 43 were also members of the city council[573] who monopolised Silesian cloth imports to Cracow; however, merchants from Prešov and other East Slovakian towns, seemed to dominate in other materials – linen, silk, Milan cloth and mohair.[574]

Levoča was the other significant importing centre for Cracow's textiles. Situated 50 km to the west of Prešov on the southern slopes of the Levoča Mountains, its advantageous location at the crossroads of trade routes to

Table 41. *Cracow: linen imports from Biecz, 1589–1650*

Quantity	1589	1593	1594	1595	1597	1600	1603	1604	1610	1611	1612	1615	1619	1624	1629	1644	1650
No. of horses	22	43	33	65	43	9	–	3	3	–	–	2	4	–	–	–	–
Approx. metre equiv.	13,200	25,800	19,800	39,000	25,800	5,400	–	1,800	1,800	–	–	1,200	2,400	–	–	–	–

Source: Calculated from Wojewódzkie Archiwum Państwowe w Krakowie, *Księgi celne*, rkp. 2115, 2117–20, 2125–26, 2132–34, 2137, 2141–43, 2146, 2159, 2164.

Table 42. *Prešov's textile imports through Cracow 1600–1630*

Quantity	Moravian cloth	Kersey	Kęty cloth (L. Poland)	Great Pol. (Międ-zyrzecz)	Bohemian cloth (Jihlava)	Silesian cloth (Zgorzelec)	Falendysz	Polish soft linen	Polish rough linen	Total
No. of warps	6,192	1,592	1,250	3,335	755	613	153	19,415	9,745½	43,050½
Approx. metre equiv.	123,840	31,840	25,000	66,700	15,100	12,260	3,060	388,300	194,910	861,010
%	14.38	3.70	2.90	7.75	1.75	1.43	0.36	45.10	22.63	100.00

Sources: Calculated from Okresní archiv, Prešov, *Registra tricessimy* for the years 1601, 1607–10, 1612, 1614–15, 1618–19, 1623–24, 1629; M. Marečková, 'Prešov v Uhersko-Polských obchodních vztazích počátkem 17 století', *Historický Časopis*, Vol. XXI, No. 3, Bratislava, 1973, pp. 430–1.

Table 43. *Cloth exports from Cracow to N. Hungary, via Prešov, 1631*

Name of Cracow merchant	Type of cloth											
	Kersey		Crepe		Arras cloth		Falendysz		Brussels cloth		Międzyrzecz cloth	
	Warps	Metre equiv.	Warps	Metre equiv.	Warps	Metre equiv.	Warps	Metre equiv.	Warps	Metre equiv.	Warps	Metre equiv.
Cyrus bros.	394	7,880	177	3,540	156	3,120	7½	150	24	480	4	80
M. Domzal	337	6,740					9½	190				
Formankowicz bros.	274	5,480					45½	910				
J. Herman	242	4,840	178	3,560	80	1,600	5½	110			4	80
M. Pacoszka	193	3,860	166	3,320	6	120	16	320	25	500	6	120
S. Januszowicz	109	2,180	90	1,800					14½	290	10	200
Zalaszowski bros.	39	780	68	1,360			2	40				
M. Winkler	74	1,480			74	1,480						
Roikowski			148	2,960								
Others	197½	3,950	63	1,260	83	1,660	23	460				
Totals	1,859½	37,190	890	17,800	399	7,980	109	2,180	63½	1,270	24	480
Total exports to Prešov	2,060	41,200	917	18,340	612	12,240	144	2,880	77	1,540	25	500
% Cracow merchants	90.26		97.05		65.19		75.69		82.46		96.00	

Sources: Calculated from Wojewódzkie Archiwum Państwowe w Krakowie, *Księgi celne*, rkp. 2142; M. Marečková, *Dálkový obchod Prešova v prvýchtřech desetiletích 17 století a Krakovský trh*, Brno, 1971, p. 125.

Table 44. *Levoča: textile imports from Cracow, 1603–1624*

Type of cloth	1603		1604		1605		1606		1607		1619	
	W	M	W	M	W	M	W	M	W	M	W	M
Moravian	870	17,400	1754	35,080	782	15,640	1893	37,860	2290	45,800	5992	119,840
Silesian	717	14,340	990	19,800	50	1,000	352	7,040	1338	26,760	501	10,020
Kersey	305	6,100	252	5,040	19	380	54	1,080	143	2,860	1594	31,880
Totals	1892	37,840	2996	59,920	851	17,020	2299	45,980	3771	75,420	8087	161,740
Ranking		7		5		9		6		4		1

	1620		1621		1622		1623		1624	
	W	M	W	M	W	M	W	M	W	M
Moravian	4190	83,800	3697	73,940	1183	23,660	2	40	0	0
Silesian	347	6,940	330	6,600	9	180	–	–	–	–
Kersey	1231	24,620	711	14,220	410	8,200	710	14,200	638	12,760
Totals	5768	115,360	4738	94,760	1602	32,040	712	14,240	638	12,760
Ranking		2		3		8		10		11

W = No of warps
M = Approx. metre equivalent
Sources: Calculated from Arhiv města Levoče (Okresní arhiv, Spišská Nová Ves), *Registra tricessimy*, Trieda IV A. sign., for 20/1–20/20, pp. 375–80 relevant years; F. Hejl and R. Fisher, 'Obchod východoslovenských měst se zahraničím ve století protihabsburských povstání', *Sborník Prací Filosofické Fakulty*, *Brněnské Univ.* (Řada Hist.), Vol. 31 (C. 29), Brno, 1982, pp. 113–14.

Hungary, Poland and Silesia favoured its development as an international commercial centre. Levoča was a free, royal town and occupied a leading political and economic position in the Spiš region of northern Hungary. It was noted for its linen and cloth woven products. Like Prešov, it figured prominently in Cracow's export trade during the first three decades of the seventeenth century. From surviving archival records, the above materials seem to have been most important in Cracow/Levoča trade. The peak, 1619, also coincided with the second highest year for Cracow's Silesian cloth imports (compare Tables 39 and 44) in which nearly half (45 per cent) found their way onto the Levoča market. The drop in imports between 1604 and 1605 may have been due to the disruption of commerce around Levoča caused by the Bocskay uprising.[575] The noticeable decline in cloth imports in the early 1620s was linked with a cholera epidemic in the Levoča area in 1623.[576] Finally, the earlier momentum of Cracow's cloth trade with Levoča, was somewhat lost in the late 1620s, but the place still remained an active trade centre amongst East Slovakian towns, especially for more expensive cloths from Cracow.[577]

Košice was apparently much less significant for Cracow's textile trade;[578] however, cloth may have arrived in Košice from Prešov, 30 km to the north, but although originally bought on the Cracow market. Pach has stated that 'The merchants of Kassa [Košice] ... purchased in Poland not only Silesian and South German [Wrocław and Nuremburg] cloths, but also English fabrics ("karasia" = kersey, "londis").'[579] Price lists from the Košice tailors guild dated 1627 and 1635, mention 'English coats and skirts';[580] manorial accounts and payment lists also testify to the popularity of English cloths in northern Hungary at this time,[581] whilst the cheaper, lighter kersey cloth, manufactured either in England, Gdańsk or Silesia, proceeded through commercial centres like Cracow to other east European markets.[582]

The second half of the seventeenth century saw a marked decline in Cracow's textile exports to northern Hungary. The main reasons were apparently deteriorating commercial conditions caused by the upheavals of the Thirty Years War, and the growing use of alternative routes. East Slovakian merchants increasingly used the Váh-Hornád valley route connecting their towns with Moravia, and hence direct access to cheaper central European cloth; furthermore the dearer west European textiles could also be purchased in Wrocław. Therefore, the former middle-man role held by Cracow's cloth merchants was being by-passed, spurred on no doubt by the Swedish invasion of Little Poland. Northern Hungary now had to find alternative routes for its cloth supplies. Once the Cracow connection had been severed, the route through the south Carpathian foothills linking the Czech lands and East Slovakia became the main thoroughfare; this, in turn, led to the eclipse of Cracow's dominant role in transit commerce in cloth.

Cracow also sent textiles to Transylvania but, unlike East Slovakia,

Table 45. *Polish textile imports from south-east Europe via Turnu-Roşu 1673–1686*

Cloth type	1673		1682		1683		1684		1685		1686	
	Quantity	Approx. metre equiv.	Quantity	Approx. metre equiv.	Quantity	Approx. metre equiv.	Quantity	Approx. metre equiv.	Quantity	Approx. metre equiv.	Quantity	Approx metre equiv.
A Cotton twill												
1 Ordinary					10 loads	4,700	6 loads 10 pieces	2,995	4 loads	1,880		
2 Thick			2 loads	940	1 load 15 pieces	732½	6 loads	2,820	5 loads	2,350		
3 Coloured							1½ loads	705				
4 Indian	2,075 pieces	36,212½										
5 from Brusa							1½ loads	705				
6 Black/ Green mixture					120 rolls	3,000						
B Flannel			20 rolls	500	13 rolls	325	60 rolls	1,500				
C 'Aba' (coarse woollen cloth)	300 rolls	7,500	13 loads 81 rolls	8,135	280 rolls	7,000	8 loads	3,760	198 rolls	4,950	238 rolls	5,950

D Coarse cloth	6 rolls	150										
E Mohair					11½ loads 10 rolls	5,655	13 rolls	325	11½ loads	5,405		
F Blankets	140 pieces	2,450	130 pieces	2,275	75 pieces	1,312½						
G Cotton					9½ loads	4,465			20 loads	9,400	20 loads	9,400
H Silk	3,948 litres	4,277	10 loads 190 litres	4,905¾	9½ loads 66 litres	4,536½	9 loads	4,230	7 loads	3,290	2 loads	940
I Muslin			8½ bales	323	6 loads	2,820	18 bales 907 rolls	23,359	40½ loads	19,035		
J Linen												
1 Turkish	787 rolls	19,675										
2 Indian	269 rolls	6,725										
K Turkish taffeta	40 rolls	1,000										
L Velvet			3 rolls	75								
Approximate total metreage		78,089½		17,153¾		34,546½		40,399		46,545		16,290

Sources: Calculated from Arhivele Statului, *Sibiu, Zwanzig und Dreissig Rechnungen*, Cutia XXVII Nos. 1–10; L. A. Demény, 'Comerţul de tranzit spre Polonia prin Ţara Românească şi Transilvania (ultimul sfert al secolului al XVII-lea)', *Studii Revista de Istorie*, Vol. 22, No. 3, Bucureşti, 1969, pp. 469–73.

appears to have imported certain oriental cloths in return. Agents for Cracow merchants were active in exporting material to the main urban centres of Transylvania, particularly Oradea, Cluj, Sibiu and Braşov. Kavka's analysis of Cluj customs registers of 1599–1636 reveals a number of Cracow agents sending Czech and Moravian cloth to the town.[583] In 1613, a Polish merchant arrived in Cluj from Cracow with a consignment of textiles including kersey cloth and linen.[584] Cracow's own registers e.g. 1624[585] prove that Cluj merchants themselves came to the city for textiles,[586] whilst on other occasions cloth from Cracow was purchased by Translyvanian merchants in Prešov.[587]

Oriental textiles were brought back on the return journey to Cracow. These came from the Ottoman Empire through Transylvania to Poland; they arrived in Lwów and Cracow even during the troubled years of the later seventeenth century. Archival evidence tends to refute the view that the Turks deliberately hampered Polish trade contacts with south-east Europe. The second half of the seventeenth century has been generally considered to be one of economic decline in Transylvania but analysis of customs registers for Turnu-Roşu suggest intense commercial activity, including textile exports to Poland. (Table 45). The table suggests the importance of certain textiles, silks,[588] cotton and muslin, which travelled via the Olt river and Turnu-Roşu on to Poland at the end of the seventeenth century.[589] It is most likely that some of these more expensive cloths would have reached Cracow's market along with other cheaper varieties of rough cloth, blankets etc., popular amongst the Polish peasantry.

Cracow's other eastern markets were Lwów and Lublin. In the early part of the seventeenth century cloth from Cracow arrived in Lwów for the Moldavian and Near East markets. Obuchowska-Pysiowa's work on the 1604 Cracow customer returns show the popularity of Silesian and Moravian cloth for these south-eastern markets (Table 46). Total metreage reveals a fairly even split between the two cloths, with slightly more Silesian (52.22 per cent) compared with that from Moravia (47.78 per cent). The Cracow/Italian merchant Valery Montelupi was involved in this trade, and mentioned four times in 1604 as sending various silks and other fabrics to Lwów. Some Cracow cloth merchants were known to have branch shops in Lwów during the late 1620s,[590] and in 1666.[591]

The tumultuous events in Moldavia however during the latter half of the seventeenth century, must have taken their toll on the cloth trade. Frequent Turkish, Tartar and Polish incursions on Moldavian territory during the conflicts of 1672–76 and 1693–99, combined with Poland's loss of Podolia and the Turkish capture of Kamieniec Podolski, disrupted commercial links with Lwów, Cracow and the Baltic coast.[592]

Russia was Cracow's other main textile market to the East during the seventeenth century; although the main commercial links were through

Table 46. *Cloth exports from Cracow to Moldavia, via Lwów, 1604*

Type of cloth	Bales	Approx. metre equiv.	Warps	Approx. metre equiv.
Silesian	184	88,320	3,465	69,300
Moravian	–	–	7,212	144,240

Source: H. Obuchowska-Pysiowa, 'Struktura handlu Krakowa z krajami południowo-wschodnimi i z państwem Moskiewskim w 1604 r. '*Roczniki Dziejów Społecznych i Gospodarczych*, Vol. XXXVIII, Warszawa-Poznań, 1977, p. 100.

Table 47. *Cloth exports from Cracow to Russia via Lublin, 1604*

Type of cloth	Bales	Approx. metre equiv.	Warps	Approx. metre equiv.	%
Moravian cloth			6,925	138,500	79.59
Silesian cloth	23	11,040	460	9,200	11.64
Bielsko-Biała linen			763	15,260	8.77
Approximate total metreage = 174,000 metres					

Source: Calculated from H. Obuchowska-Pysiowa, 'Struktura handlu Krakowa z krajami południowo-wschodnimi i z państwem Moskiewskim w 1604 r', *Roczniki Dziejów Społecznych i Gospodarczych*, Vol. XXXVIII, Warszawa-Poznań, 1977, p. 111.

Lublin, other towns were known as intermediaries for Cracow's exports to Moscow, including Jarosław, Vilna and Volhynia (Wołyn). Again Obuchowska-Pysiowa's researches for the early years illustrate the attraction of central European cloth/linen for the Russian market (Table 47).

Unlike the Moldavian market, Russian imports of Moravian cloth seem to have been far more popular than those from Silesia; also here linen formed nearly a tenth of total cloth exports. Certainly, merchants such as Montelupi were active in this trade, transporting cloth from Cracow via Lublin to the Russian market on ten different occasions during 1604.[593] Jeannin has pointed out, with reference to Italian merchants, that 'they also went much farther afield, and in the case of Montelupi even as far as Russia'.[594]

During the 1620s branch shops of Cracow textile merchants existed in Lublin[595] and Vilna,[596] whilst some Lublin merchants, according to the 1624 customs returns, not only came to Cracow, but also went further to Moravia for cloth.[597] Contacts between Cracow and Volhynia also existed; however, high transport costs for the six-day journey tended to limit purchases to the more expensive cloths, with a greater return on profits.[598] Besides cloth, the Russian market was also important for various ready-made goods including expensive hemming material, hosiery products, and various items of ladies' apparel;[599] many originated in Prague, but were organised through Jewish merchants in Cracow. Even at the end of the seventeenth century these and other cloth goods continued to be sent from Cracow to Lwów, Jarosław, Łowicz, Piotrków Trybunalski, Lublin and Warsaw, often through the Italian merchant community in the city.[600]

The eighteenth century was a period of shrinking markets for Cracow's textile merchants. Signs were evident in the previous century; after 1640 a high turnover of cloth of some 30,000 warps (600,000 metres) annually dropped dramatically to only a few thousand warps each year. During the last quarter of the seventeenth century, the annual import of cloth from Silesia, formerly the city's chief supplier, was only about 2,000 warps (40,000 metres)[601] or 6 per cent of pre-1640 years. This was at a time when Silesian textile production was still prosperous, especially linen manufacture for, according to Kisch, 'Silesian linen remained unchallenged in world markets ... the late 17th and early 18th century were the brightest periods of Silesian development.'[602]

Perhaps the answer lay in events beyond the control of Cracow's traders. In south-east Europe the last thirty years of the seventeenth and early eighteenth centuries witnessed the extension of trading activity by Levantine companies, particularly by Greeks, into such towns as Sibiu and Braşov. These merchants of Balkan origin (Greek, Bulgarian, Serbian, etc.)[603] settled in the larger towns for trade purposes. After Transylvania was annexed to the Hapsburg Empire in 1701, privileges granted these Balkan companies were reinforced. The political situation of Moldavia and Wallachia under Turkish rule[604] in the eighteenth century did little to improve commercial relations with former Polish markets. Another blow for Cracow's trade links came in 1714; in that year Peter the Great decided to direct all Russian trade towards St Petersburg, causing commercial stagnation between Cracow and its Ukrainian markets.[605] It was hoped that closer commercial links could be made by Cracow with Silesia, but this again was partially thwarted in 1742, when Silesia became part of the Prussian kingdom of Frederick the Great.[606] Finally Gdańsk, formerly an important supplier of English and Dutch cloth to Cracow, was itself partially eclipsed by increased Swedish trading activity. New outlets for western cloth were being found by Swedish merchants in Russia; according to Unger, 'after 1714 the young St. Petersburg drew 10% of the supply from the Netherlands, and 25% from England; the figures for Danzig were then 10% and 11% respectively. An economic landslide seems to have taken place.'[607] Its impact was also felt in Cracow.

The second half of the eighteenth century for Cracow saw an intensified search for cloth markets nearer home, and greater reliance on Wrocław for

Table 48. *Cloth supplied to Bajer's shop in Cracow, 1752–1755*

Origin	1752/53			1753/54			1754/55		
	E	M	%	E	M	%	E	M	%
Silesia (Wrocław)	9,291	4,645½	60.44	853	426½	10.81	3,516½	1,758½	36.37
Grt. Poland (Leszno)	4,917¾	2,459	32.00	6,484	3,242	82.13	5,817¼	2,908½	60.18
Moravia (Frýdek)	1,168	584	7.56	312	156	3.95	–	–	–
Little Poland (Bielsko)	–	–	–	245	122½	3.11	334	167	3.45
Total	15,376¾	7,688½	100	7,894	3,947	100	9,667¾	4,834	100

E = Ells; M = Approx. metre equivalent
Sources: Calculated from Wojewódzkie Archiwum Państwowe w Krakowie, *Inwentarz-Fascykut aktów Luźnych (1750–1797)*, rkp. 823; M. Kulczykowski, 'Handel Krakowa w latach 1750–1772', *Prace Historyczne*, No. 4 (Historia No. 4), Kraków, 1960, p. 87.

foreign supplies. Unfortunately, concerning imports from Wrocław, Cracow's customs books for this period only note cloth quantities and the merchant involved; Wrocław was a source not only of western textiles, but also of cloth from Greater Poland. Customs data, however, does not allow the different origins to be distinguished. Kulczykowski has found that fifty-seven merchants sent a total of 874 packets of cloth from Wrocław to Cracow in 1750[608] and estimated this as totalling 3,500 warps (i.e. 70,000 metres); for 1755 this came to 1,108 packets, an increase of 4,540 metres over the five years. Other evidence may be found in cloth merchants' ledgers; an inventory from the shop of Jan Bajer for 1752/53[609] shows an annual purchase of 500 warps (10,000 metres: English, Dutch, Belgian, French, Silesian and cloth from Greater Poland) had arrived there from Wrocław; consignments were dispatched originally by merchants in Aachen, Verviers, Leipzig, Wrocław and Opava.[610]

After mid-century increasing amounts of Polish cloth were on sale in the city. This reflected the growing strength of domestic textile production, especially in Greater Poland,[611] accompanied by a decline in imports of top-quality foreign material. Louis, in 1775, noted that 'instead of the foremost Italian goods arriving in Cracow's warehouses ... they were replaced by second-rate Moravian cloth from Opava'.[612] Increasing quantities of domestic cloth were also evident in Bajer's shop during the early 1750s (Table 48). Over half of Bajer's cloth came from Polish workshops, whilst some of the 41 per cent attributed to Wrocław may have been manufactured in Greater Poland; Moravian cloth purchases were insignificant (less than 5 per cent). Kula's research has shown that by the late 1770s Greater

Table 49. *Cloth exports from Cracow to Lwów and Jarosław, 1750–1763*

Year	1750	1751	1752	1753	1754	1755	1756	1757	1758	1759	1760	1761	1762	1763
No. of warps	48	133	330	660	384	270	162	542	420	540	645	564	570	990
Approx. metre equiv.	960	2,660	6,600	13,200	7,680	5,400	3,240	10,840	8,400	10,800	12,900	11,280	11,400	19,800

Sources: Calculated from Wojewódzkie Archiwum Państwowe w Krakowie, *Księgi celne*, rkps. 2216, 2223, 2231, 2232 for relevant years; M. Kulczykowski, 'Handel Krakowa w latach 1750–1772', *Prace Historyczne*, No. 4 (Historia No. 4) Kraków, 1960, p. 84.

Poland supplied four-fifths of Poland's textile needs,[613] although some of the Polish cloth in Cracow had first been sent to Silesia for dyeing, and returned for sale.[614] Jewish merchants were particularly active purchasers of cloth from Little Poland (Bielsko-Biała, Kęty, Staszów, Radomsko)[615] for the Cracow market; this trade was curbed by King Stanisław August in a decree of 1 September 1775, which forbad Jews selling goods in Cracow and its suburbs.[616]

From the city, textiles were exported to the markets and fairs of Eastern Poland. Fairs at Berdyczów and Brody were popular centres[617] but Lwów and Jarosław were the main markets, (Table 49), along with Sądziszów, Tarnopol, Zółkiew, Zamość, Opatów, Dobromil and Wodzisław.[618] The table indicates growing cloth sales to these two important commercial centres during the late 1750s and early 1760s whilst over the fourteen-year period on average 8,500 metres of cloth were sent from Cracow annually. After 1784 Cracow's trade with these two towns declined considerably; the Austrian occupation of Galicia and subsequent high customs tariffs imposed on manufactured goods, meant Cracow lost another of its more prosperous eastern markets.

In spite of threats during the second half of the eighteenth century to the central European linen industry from the growth of cotton manufacture, documentary sources suggest that Cracow merchants were still active in this trade. Certainly parts of Little Poland and, especially, the Carpathian foothill region, were well suited agriculturally for flax growing; here rural industry thrived with several centres (Bielsko, Andrychów, Gorlice, Łańcut, Przeworsk, Sanok, etc.)[619] able to supply Cracow's linen market. Abroad, as with cloth, Cracow's customs books do not give sufficient details to judge the significance of centres like Wrocław; much of the foreign linen was transported by Cracow's own traders and therefore not recorded in customs data. Nevertheless, Kulczykowski has estimated that in 1750 between 3,300 and 3,500 bolts (i.e. 102,000–108,500 metres) of linen were sent from Wrocław to Cracow.[620] Most of the foreign varieties of linen were imported from Silesia, but quantities also came from Saxony (via Leipzig) whilst the most expensive came from Lyons in France.[621] Domestically, linen came from various parts of Poland[622] but most was local, from the Carpathian foothills. Besides rolls of cloth, manufactured linen goods (tablecloths, napkins, towels and drill) were sent to the Cracow market; documentary evidence suggests that Andrychów and its surrounding villages (Wieprz, Roczyny, Sułkowice, Targanice, Rzyki, Inwałd, Zagórnik and Czaniec), provided most linen during the decade following mid-century as seen in Table 50. Over the fourteen-year period about 34,000 metres of linen were sent to Cracow, an average of 2,500 metres annually. If the first and last year are compared, a fourfold increase is noted. Greatest contrast is between 1752 (lowest) and 1762 (highest) – a fourteen-fold difference in the amount;

Table 50. *Linen cloth sent from Andrychów to Cracow, 1750–1763*

Year	1750	1751	1752	1753	1754	1755	1756	1757	1758	1759	1760	1761	1762	1763
No. bolts	35	99	11	52	71	69	68	90½	84½	59½	76	87	152	146
Approx. Metre Equiv.	1085	3069	341	1612	2201	2139	2108	2805½	2619½	1844½	2356	2697	4712	4526

Sources: Wojewódzkie Archiwum Państwowe w Krakowie, *Księgi celne*, rkp. 2216, 2223, 2231, 2232; M. Kulczykowski, *Kraków jako ośrodek towarowy Małopolski zachodniej w drugiej połowie XVIII wieku*, Warszawa 1963, p. 110.

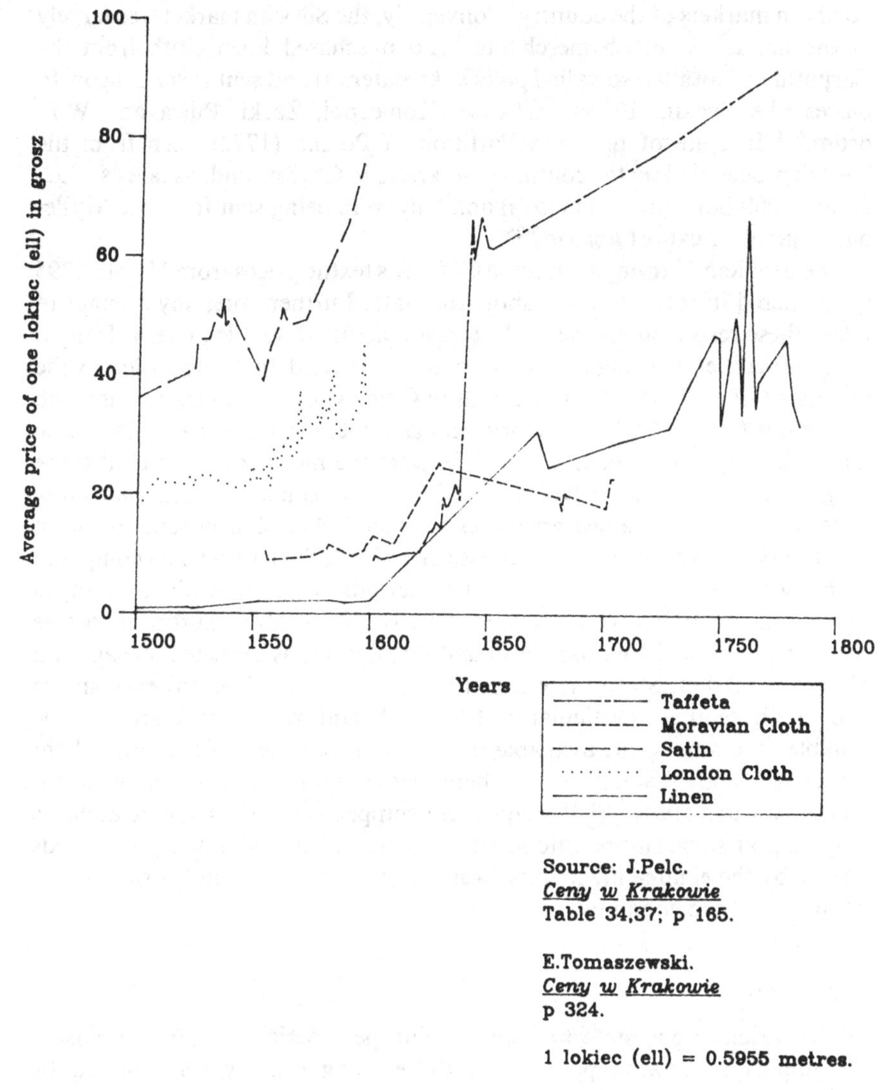

67 Price of certain textiles in Cracow, 1500–1795.

improved trading conditions during the early 1760s are clearly reflected in this change.

Linen exports from Cracow were sent to Holland, Hungary, Moldavia/Wallachia and other parts of the Ottoman Empire during the second half of the eighteenth century,[623] among them the popular products of the Andrychów region. Some producers in Little Poland, however, successfully bypassed Cracow's middle-man role and dealt direct with Gdańsk and the northern markets of the country. Conversely, the Silesian market was largely in the hands of Jewish merchants, who purchased linen cloth from the Carpathian foothills (so-called *podgórski* material) and sent it via Cracow to places like Będzin, Bytom, Gliwice, Koniecpol, Żarki, Pilica and Wolbrom.[624] In spite of the First Partition of Poland (1772) linen from the Austrian-occupied sector continued to arrive in Cracow and, as late as 1792, about 2,000 bolts (62,000 metres) annually were being sent from the Myślenice region for export abroad.[625]

The problem of trying to analyse Cracow's textile prices from 1500 to 1795 is hampered in some cases by short-run data. Furthermore, any attempt to relate these results to an overall European picture of textile prices is fraught with difficulties although some scholars have tried to bring order to the confusion of evidence.[626] Price trends in Cracow for certain textiles are seen in Figures 67 and 68. Their shortcomings aside, some comparisons can be made. Clearly velvet, taffeta and satin were the most expensive cloths; the longer time series for velvet and taffeta show similar trends, indicating considerable and sustained price rises after the 1640s. Shorter series for satin and London cloth do show sharp rises after 1550; the latter, when compared with Lwów and Frankfurt for similar periods reveal that all three series moved in much the same way.[627] Finally, the central European textile materials, such as Moravian cloth and linen, were, as expected, cheaper on Cracow's market. Steep price rises in linen are noticeable, however, in the eighteenth century, particularly after 1750. Unfortunately, there are few reliable data among the available European price series able to throw light on textile trends. Even so 'everywhere there are innumerable piles of documents awaiting study'.[628] Perhaps these snippets from the Cracow archives may help at some future date to clarify broader secular textile price trends which, by the eighteenth century, was the most important industrial branch of the European economy.

(ii) Beverages

While agriculture in sixteenth-century Europe experienced rapid expansion to support a growing population, the ensuing century was marked by stagnation; renewed growth in the eighteenth century was to continue well beyond 1800. Although this periodisation may be imprecise, agricultural fluctuations were to influence those crops utilised for the production of wine, beer and distilled alcoholic beverages. The rise in alcoholism was continuous in Europe during these centuries according to Braudel[629]; certainly the whole of Europe drank wine, but only a part produced it. Beer, brewed from wheat, oats, barley or millet, became established outside the vine-growing regions and was most popular in the northern lands, including Poland. The great innovation in Europe, however, was the appearance of brandy and spirits

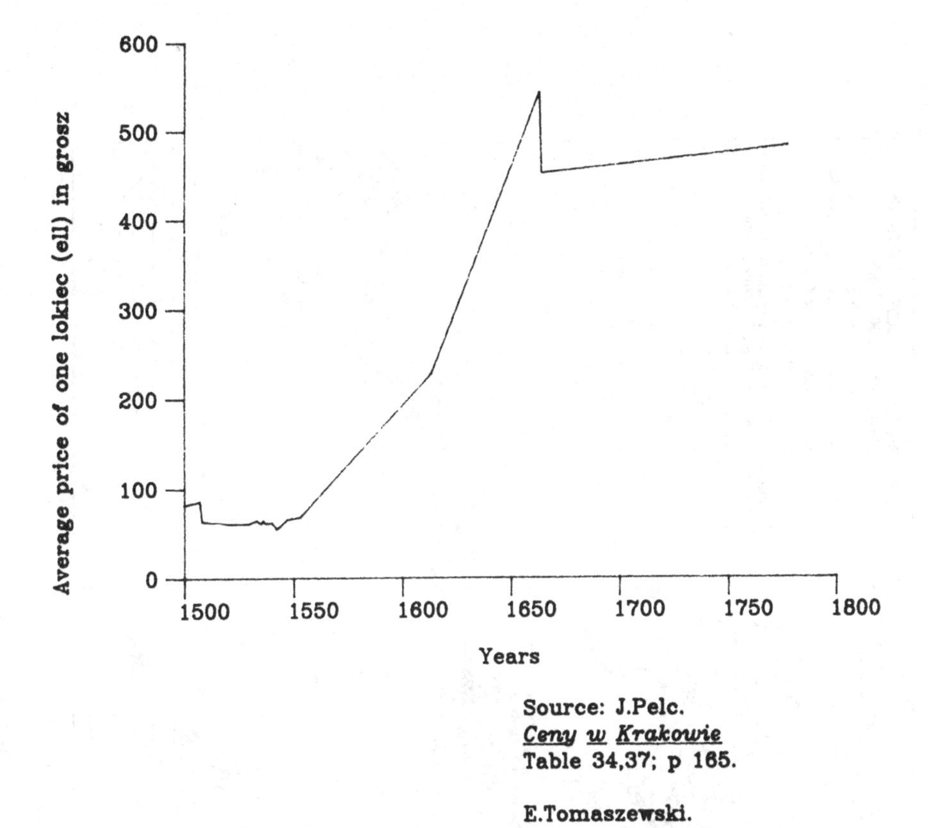

68 Price of velvet in Cracow, 1500–1795.

made from grain, with higher alcohol content, which the sixteenth century created, the seventeenth consolidated and the eighteenth popularised. Documentary evidence from Cracow helps support this general impression of beverage production during the Early Modern Period (Fig. 69). Wine, even from local sources, cost much more than beer, artisans indulging in it only on special occasions; for wealthier people wine was the everyday beverage, but alcohol such as vodka was often seen as harmful in the sixteenth century and only used for medicinal purposes.[630]

Wine was imported into Cracow in large quantities either for local consumption or as transit trade for more northerly destinations. Polish viticulture, according to Morawski[631] had been in decline since the early fifteenth

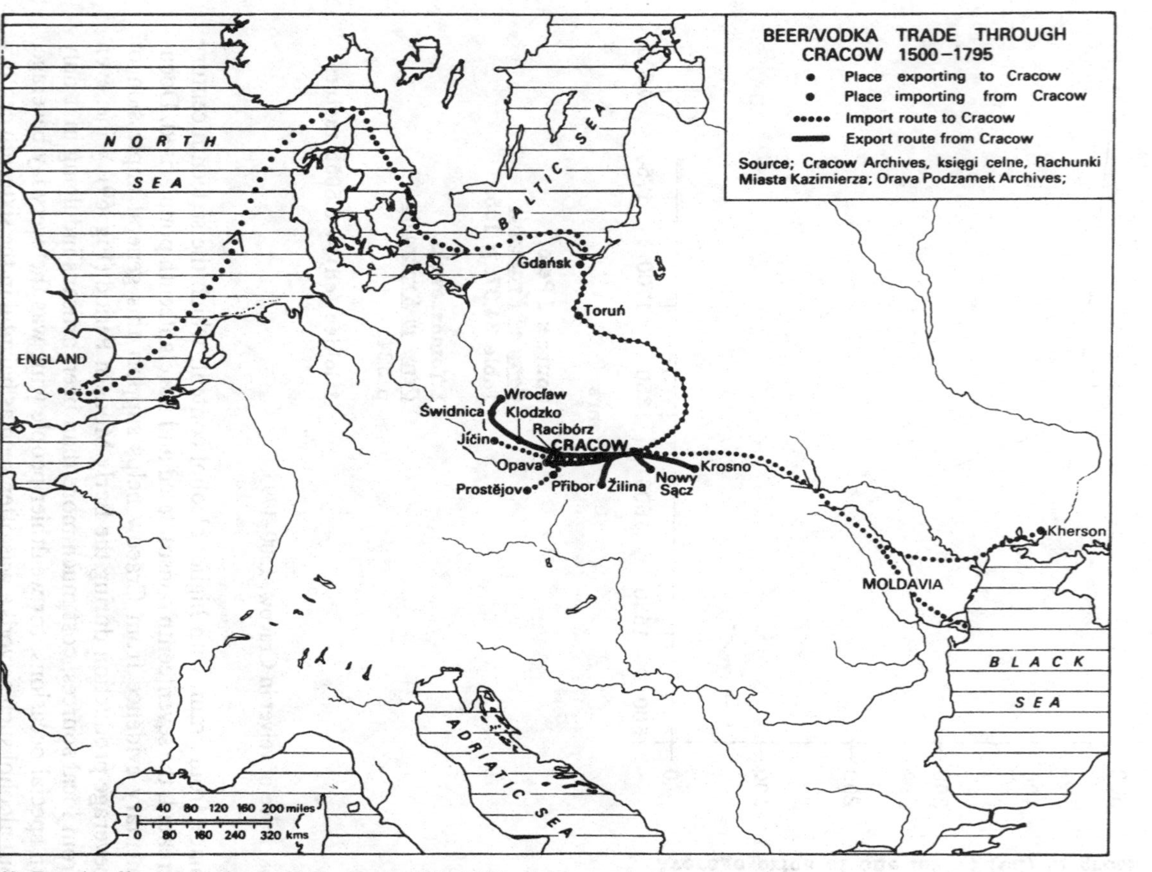
BEER/VODKA TRADE THROUGH CRACOW 1500–1795
Place exporting to Cracow
Place importing from Cracow
Import route to Cracow
Export route from Cracow
Source; Cracow Archives, księgi celne, Rachunki Miasta Kazimierza; Orava Podzamek Archives;
NORTH SEA
BALTIC SEA
ENGLAND
Gdańsk
Toruń
Wrocław
Świdnica
Klodzko
Racibórz
Jičin
CRACOW
Opava
Prostějov
Příbor
Žilina
Nowy Sącz
Krosno
Kherson
MOLDAVIA
BLACK SEA
ADRIATIC SEA
0 40 80 120 160 200 miles
0 80 160 240 320 kms

69 Beer/vodka trade through Cracow, 1500–1795.

century, but demand continued to rise. This was not only for supplying ecclesiastical needs but also increased consumption by the wealthier inhabitants. The deficit led to an intensive growth of foreign wines entering Poland when, according to Rutkowski, 'Wine was imported through the Baltic ports from Spain, France and Portugal, and by land from Hungary and Austria, and from Moldavia and Greece.'[632]

The pattern of Cracow's wine imports reflects this overall national picture, with oriental varieties being sent via Lwów, Hungarian and Moravian vintages arriving from over the Carpathians, and west European types coming up the Vistula river through Gdańsk. One suspects, however, that the city's customs books do not give a complete picture of the size and importance of this trade. Quantities specifically imported for the royal court by Cracow merchants during the sixteenth century were not noted in customs documents. This tends to belie the significance of Hungarian imports at this time.[633] According to Rybarski,[634] Moravian wine was more important than Hungarian in the city's market place during the sixteenth century; Hungarian varieties came to Poland through the Carpathian foothill settlements (Nowy Targ, Biecz, Grzybów, etc.), towns which had their own wine-stapling rights.[635] Conversely Cracow, with its advantageous geographical position, may have found it easier to obtain Hungarian wines via Moravia and Silesia, than by the Carpathian route at this time.

Viticulture played a significant part in the Hungarian economy during the Early Modern Period. Even in the fourteenth and fifteenth centuries wines from Syrmia, between the Danube and the Sava rivers, and the hilly country west of the Danube were already sent abroad; the Great Plain supplied wine which was exported to Silesia and Poland.[636] The north-eastern wine-growing district, extending from Gyöngyös through Eger, Miskolc and Tokaj to Beregszász, began to outrival former areas during the sixteenth century, due to improved technical innovations in viticulture, including prepared terraced soil, new dressing methods and postponement of grape collection till late October, allowing the grapes to dry, when their essence was mixed with ordinary wine to produce the famous aszu of Tokaj.[637] The twenty wine-growing towns of the Tokaj district introduced common regulations for viticulture as early as 1561 with strict production methods. Rapid increase in exports to Poland meant that by the end of the century the hilly country of the Tokaj region became a leading producer in the Hungarian wine trade. Elsewhere, viticulture in territories under Turkish occupation (the Ottomans drank no wine, but levied high taxes on it), not only survived, but were enlarged by new plantations e.g. south of Lake Balaton, whilst the market towns on the Great Plain began to introduce vines on the sandy soils, previously only grown in the hillier areas.[638]

Two towns, Košice (Kassa) and Bardejov (Bartfa), were the main purchasing centres for Polish merchants. Košice, with its excellent geographical location in the Carpathian foothills, controlled the north–south trade routes in this part of Hungary. Ever jealous of its position, municipal laws in the sixteenth century prevented Polish merchants, particularly from Cracow, purchasing goods including wine to the south of the town. In 1522, the Hungarian king Ludwik II (1516–26) forbad merchants transporting wine to Poland if purchased south of Košice, each barrel having to observe the town's stapling laws.[639] Bardejov was an important collecting centre for wines from north-east Hungary. In 1528 the Polish diet (*sejm*) agreed to a joint request from Cracow and Bardejov to act as middlemen for importing wine into Poland;[640] although repealed two years later by the Polish king, Zygmunt I (1506–48),[641] it does suggest Bardejov's importance for Cracow's wine merchants. Information is sparse on transport costs paid by these merchants for Hungarian wines, but surviving documents suggest it was about a third of the total transaction.[642]

One of the major consumers of Magyar wines was the royal court, Cracow. In spite of the king's court accounts being insufficient for detailed estimates of the actual quantities to be made, fragmentary documentation does suggest its importance. One Cracow merchant, Meidel, originally from Eger in Hungary, was active in this trade, and rewarded by the king with citizenship of the Polish capital.[643] Between 1540 and 1548 two Italian merchants in Cracow supplied the king's court with Hungarian wine, receiving presents of woollen cloth and fustian in return.[644] In northern Hungary the Spiš Chamber of Commerce of 1571 recorded sending 600 barrels (163,123 litres) of Hungarian wine to the Polish royal court;[645] several other transactions are recorded for the 1570s.[646]

Besides the royal court, the landed aristocracy and rich urban dwellers were the other important consumers of Hungarian wine. Merchants in Cracow had a steady market for their wine, either in the city itself, or for transit to other parts of Poland and beyond. One of the main wine merchants in Cracow, Sindler, amassed large profits and property based on importing Hungarian vintages, whilst some merchant families such as Gutterer, Dubowski (transported wine to Vilna) and Sapieha (towards the end of the sixteenth century) specialised in Hungarian wine imports.[647] Up to 1571 much of their trade had been aided by no Hungarian customs' levies on wine exports; in that year the Spiš Chamber of Commerce not only imposed duty on such exports, but insisted that merchants possess a valid passport issued through their organisation[648] if they wished to send wine abroad. In spite of these new formalities Polish imports continued unabated and the more detailed customs books from Cracow at the end of the century reflect this; in 1593, over 450,000 litres of Hungarian wine entered the city[649] whilst, in 1594–95, the Spiš Chamber of Commerce alone sent 370,000 litres to Cracow. The years 1599–1600 also had recorded consignments of 11,500 and 23,000 litres sent respectively from Bardejov and Košice to Cracow.[650]

Moravian wines also arrived in Cracow during the sixteenth century. One suspects that some of these vintages may have originated in northern Hungary, but designated as 'Moravian' due to the dispatch route through Moravia and Silesia. Occasionally, the place of origin is mentioned, revealing so-called 'Moravian' vintages which were produced in northern Hungary (e.g. 'Sv. Jure' now a suburb of Bratislava; 'edenburgski' from Sopron). However, in 1591, specific mention was made of Moravian wine (30,000 litres) together with Hungarian varieties,[651] whilst Małecki maintains that small quantities of Austrian wine came through Moravia en route to Cracow.[652]

Wines also arrived overland from Turkey and Transylvania to sixteenth-century Cracow. Greeks, Armenians and Jews were particularly active in this trade operating through Lwów. Oriental varieties, so-named because they arrived from lands under Ottoman control, were in fact mainly of southern European origin. Greek and Cretan vintages were sent to the Black Sea ports (Kiliya, Galaţi)[653] and then overland to Lwów and Cracow. Malmsey and muscatel predominated and were highly prized by members of the royal court. For example, in 1557, a Greek merchant from Suceava sent malmsey wine to Cracow[654] whilst Saxon merchants in Braşov bought quantities of such wines in Galaţi for export to Poland;[655] the Polish king, Stefan Batory (1576–86) ordered malmsey wine from Braşov in 1571[656] whilst, in 1584, a small quantity of this wine was sent from Venice via Lwów to Cracow.[657] Towards the end of the century increasing conflicts between Turkey and Venice discouraged merchants from Crete (under Venetian occupation) from venturing into the Black Sea with supplies[658] and henceforth these wines came to Cracow via western Europe.

The vineyards of Transylvania exported wine to Cracow. The two main collecting centres appear from documentary sources to have been Cluj and Oradea; merchants brought wine, along with honey, wax, skins, etc. for sale in Cracow, where they would purchase textiles and other manufactured goods in exchange.[659] In 1597, for example, a Cluj merchant, Mihail Luther, transported wine and hydromel from Transylvania, via northern Hungary to Melchior Tylis *civis Cracoviensis*;[660] three years later, documents record, Ferenc, a merchant from Oradea, delivered 6 barrels (1,631 litres) of Transylvanian wine to Stanisław Wiatr, *civis Clepardiensis*, i.e. Kleparz, a Cracow suburb.[661]

West European wines came to Cracow either through Nuremburg or Gdańsk. More generally, in Poland, Rhine wines were the most popular, but other vintages entered the country from France, Portugal, Spain and the eastern Mediterranean.[662] Some of these varieties must have reached the royal court for, in a document signed by Stefan Batory in 1580, there is a list of imported wines with origins including 'France especially from Auxois in Burgundy, Italy, Spain particularly Alicante, and the Islands of Madeira and

Canary'.[663] The choice variety of Canary wine was malmsey, the grape having reached the islands from the eastern Mediterranean via southern Spain and provided the basis for Tenerife's viticulture.[664]

Italian merchants in Cracow were particularly active in the wine trade, foremost amongst them the del Pace family. Wine shops in the city owned by Italians contained a variety of vintages, not only from their homeland, but also from Austria, the Rhine valley, France, Spain and Hungary. Although few in number Italian wine merchants possessed some of the best shopping sites in Cracow. A 1595 inventory of wine shops totalled 84, a quarter (21) of which were located in the main square (Rynek). Only nine Italian owners are listed, but six of them had shops in the Rynek, and were amongst the largest in the city.[665] This situation probably resulted from a municipal decree dated 1581. This stated that no foreign merchant could store quantities of wine in his cellar, but must sell it on the city market.[666] Trading privileges were only accorded to foreigners with Cracow citizenship or those who in some way had served the monarch. They were few in number. Of the other merchants some appear to have specialised in certain types, such as Stanisław Węgrzynek (French and Rhenish red and white wines), G. Lizybon (Spanish), J. Francuz (French) and J. Niederland (Hungarian) varieties.[667]

West European wines were usually transported through Nuremburg or Gdańsk to Cracow, but Italian supplies often came direct via Villach, Vienna and Oświęcim.[668] Nuremburg had for many years helped channel eastern and Mediterranean wares northwards. In fact, the city had maintained commercial relations with Poland since the fourteenth century, and Nuremburgers could be found in Łódź, Cracow, Warsaw and Lublin.[669] The decline of Venice and Genoa, paralleled by the rise of Antwerp as the nerve centre of European trade, meant Nuremburg, along with other German towns entered a period of difficult adjustment. This was noticeable in the wine trade, with the arrival of Portuguese caravels and carracks to Antwerp. They contained wines en route for Gdańsk, leaving Nuremburg rather isolated from the main northern avenues of this trade. This was already reflected in Cracow's documents of 1538/1539, when only 8,000 litres of Nuremburg wine entered the city;[670] by 1584 this had dropped to a mere 300 litres of mainly French and Rhenish wines.[671]

What Nuremburg lost from Cracow's wine trade, Gdańsk seems to have gained, particularly as the city took over Lwów's former role as supplier of oriental vintages, towards the end of the sixteenth century. Quantities of west European wines arrived in Cracow from Gdańsk; although sometimes only a few dozen barrels each year, the choice was impressive; Rhine wines, French vintages, Spanish including Peter-see-me, and Alicante and dry 'sec' varieties, malmsey from the Canaries and muscatel from the eastern Mediterranean.[672] Some deliveries came overland from Gdańsk (e.g. 1584, 1,360 litres),[673] others up the Vistula river (e.g. 1573, 1,087 litres).[674] The

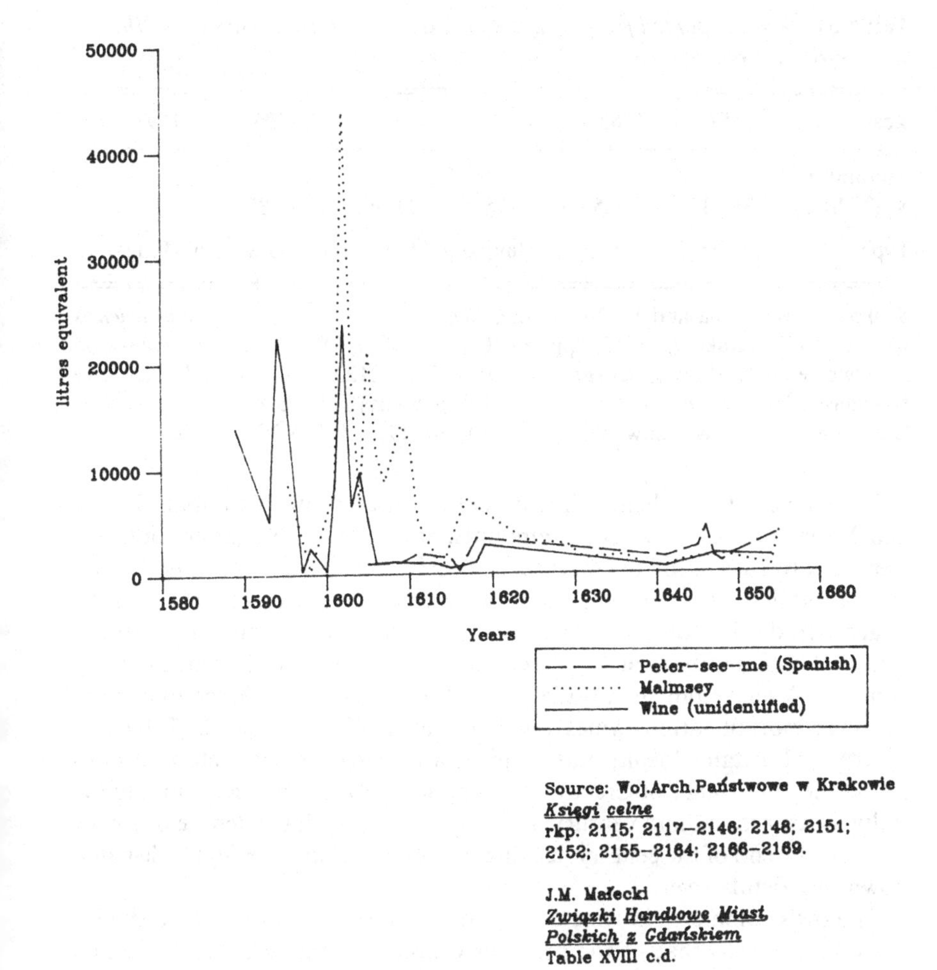

70 Certain wines sent via Gdańsk to Cracow, 1589–1655.

fuller Cracow customs registers from the end of the century reveal the role played by Gdańsk in wine imports (Fig. 70). Considerable variations in quantity are evident over the last decade, but such differences may have been due to the vagaries of the wine trade itself, dependent not only on the annual harvests, but also on demands from west European countries for these vintages. For example, 1589,[675] 1594 and 1595 were obviously years of more plentiful supply on the Cracow market, compared with the leaner times of 1593 and 1597–1600.[676]

Wine exports from Cracow could only be made when the needs of the royal court and sundry wealthy inhabitants had been sufficed. There is very

Table 51. *Wine exported from Cracow to Gdańsk in the second half of the sixteenth century*

Year	1557	1560	1573	1576	1595	1597
Amount in equiv. litres	31,537	6,525	2,175	11,962	1,087	5,166
Type	malmsey	muscatel	Hungarian	Hungarian	Hungarian	Hungarian

Sources: S. Kutrzeba and F. Duda (eds), *Regestra thelonei aquatici Wladislaviensis saeculi XVI*, Kraków, 1915, pp. XLII–LV, 458–9; Wojewódzkie Archiwum Państwowe w Krakowie, *Księgi celne*, rkp. 2117–2120; J. M. Małecki, *Związki handlowe miast polskich z Gdańskiem w XVI i pierwszej połowie XVII wieku* (P.A.N. Odd. w Krakowie) Wrocław-Warszawa-Kraków, 1968, Tables VI c.d., IX.

little evidence of wine being directly sent north-east towards Lublin, Brześć and Vilna from Cracow; small amounts, specially of Hungarian vintages, were sent to Lublin for the markets at Brześć and Vilna[677] from the Carpathian foothill towns, but some of this trade could have been organised by the Cracow trader Dubowski, who exported wine to Vilna during the sixteenth century.[678] Similarly Cracow merchants are known to have sent malmsey wine to Silesia,[679] but such trade was of little significance. More important was the export of various wines down the Vistula river to Gdańsk (Table 51).

Although extant documentation gives a rather irregular picture, it does reflect the importance of Cracow as a supplier of Mediterranean wines to the Baltic coast around mid-century; this role was lost with the demise of Lwów towards the end of the century, leaving only Hungarian varieties for dispatch down the Vistula river.

Cracow's commerce in wine during the seventeenth century had a similar spatial pattern to that of the previous century. Hungary certainly figured large in this trade; Turkish expansion there, in the southern and central parts of the country, tended to disrupt commercial relations with other parts of Central Europe and, as Pieradzka noted, 'Hungary found itself outside the world trade routes'.[680] It was understandable that given this situation, connections between Ottoman-free Hungary (largely Slovakia) and Poland were sure to increase. Hungarian exports to Poland were dominated by the wine trade; for example, at the beginning of the seventeenth century wine exported from the Tokaj region to Poland totalled about 50,000 hectolitres annually,[681] and was mainly in the hands of merchants from Prešov, Košice and Bardejov.[682] In 1601, a Slovakian merchant, Michał Spali, sent 237 barrels (64,500 litres) of wine to Cracow, which was considerable for one journey.[683] On occasions, the Hungarian nobility sent wine directly to Poland, to avoid the urban middleman,[684] but most trade seems to have been

Table 52. *Wine exports from North Hungary to Cracow, 1624, 1629 (in litres equivalent)*

Town of origin	1624	%	1629	%
Prešov	206,670	19.99	35,719	29.13
Bardejov	135,138	13.08	68,240	55.64
Levoča	124,907	12.09	18,668	15.23
Kežmarok	86,238	8.35	–	–
Košice	63,434	6.14	–	–
Banská Bystrica	3,864	0.37	–	–
Cracow merchants (wine origin not given)	413,267	39.98	–	–
TOTAL	1,033,518	100.00	122,627	100.00

Sources: Wojewódzkie Archiwum Państwowe w Krakowie, *Księgi celne*, rkp. 2142; 2143; F. Hejl and R. Fišer, 'Obchod východoslovenských měst se žahraničím ve století protihabsburských povstání', *Sborník Prací Filosofické Fakulty Brněnské University*, Vol. 31 (C.29) (Řada Hist.), Brno, 1982, p. 115.

controlled by merchants either from Cracow itself, or the main exporting towns of Slovakia (Table 52). Prešov,[685] Bardejov[686] and Levoča[687] were the three major exporting centres to Cracow. Collectively their merchants in 1624 sent nearly half a million litres of Hungarian wine to the city. The total may have been higher, but documentary sources do not reveal the origin of purchases by Cracow merchants. Other centres like Kežmarok, Košice and Banská Bystrica were less significant in Cracow's wine trade, with only altogether about 150,000 litres in 1624. The other noticeable factor is the rapid decline during the late 1620s of Slovakian wine in Cracow's trade structure. By 1629, it was only a tenth of the total sent five years earlier. This resulted from agricultural depression experienced throughout Hungary at this time,[688] compounded by problems associated with a war-torn country.[689] Nevertheless, regional differences in wine production were apparent; in 1631 Prešov merchants sent 172,270 litres to Cracow, a five-fold increase over 1629, but Bardejov's traders, with only 26,726 litres, sent less than half that dispatched two years before.[690]

Depression in Hungarian wine-making continued well into the second half of the century; some revival of commercial contacts with Cracow took place (Fig. 71) but not on the scale of the early 1620s. The steep drop in sales of wine to Cracow in 1663–64 corresponded with the general political and military crisis in Hungary at this time, caused by Ottoman/Hapsburg conflicts.[691] By 1669 the situation had improved, and a peak export year to Cracow was recorded with over 300,000 litres. A drop in imports in 1670, coincided with a period of long-term stagnation in the Tokaj region and

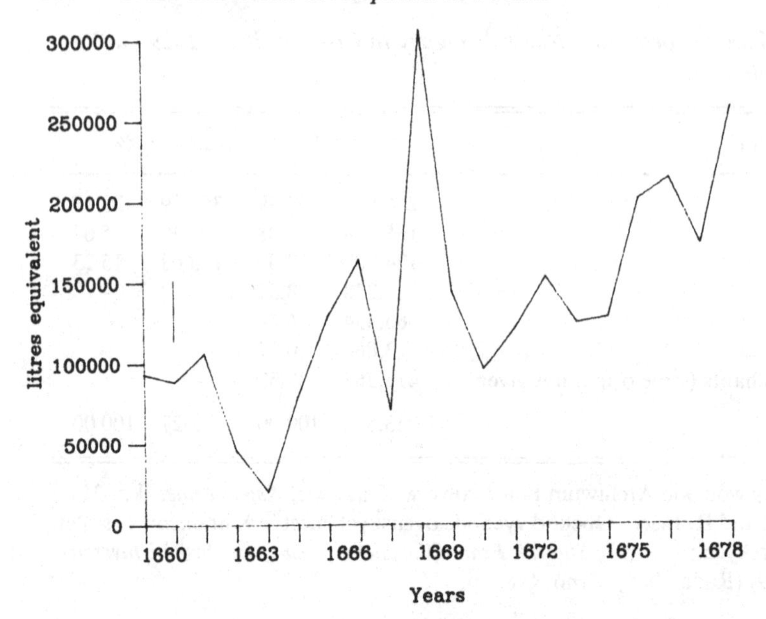

71 Hungarian wine exports to Cracow, 1660–1679.

throughout much of eastern Slovakia. Henceforth, merchants in Cracow had increasingly to rely on western and central Slovakia for their wine imports during the latter years of the century (Table 53). Table 53 emphasises this growing dependence by Cracow on western Slovakia, particularly for wine from Pezinok; this town supplied nearly half Cracow's needs between 1660 and 1679. The earlier significance of eastern Slovakia is seen, with over a third of the exports to the city, dominated particularly by Prešov. Central Slovakia appears to have been of little significance. Unfortunately, over a tenth of Cracow's wine imports in Table 53 have no specified origin given in the documents. The absence of former important centres like Levoča, suggests growing commercial interest elsewhere in places such as Vienna and Wrocław rather than Cracow. A document dated 1673 confirms the fear of some Slovakian wine merchants about the dangerous situation on some routes in Poland, and competition from so-called 'Hungarian merchants' who were in fact inhabitants of Polish towns.[692] Even so, wine from Hungary still arrived in Cracow in 1688 (the last extant customs book for the city in the seventeenth century), but quantities were small and often mentioned together with other merchandise, such as honey, iron and copper.[693]

Polish wine imports involved both domestic and foreign merchants.

Table 53. *Hungarian wine exports to Cracow, 1660–1679*

Eastern Slovakia			Central Slovakia			Western Slovakia			Remainder (origin unknown)	
Town	Litres equiv.	% of total	Town	Litres equiv.	% of total	Town	Litres equiv.	% of total	Litres equiv.	% of total
Prešov	666,086	24.54	Banská Bystrica	125,061	4.60	Pezinok	1,124,734	41.46	318,090	11.74
Kežmarok	184,872	6.82				Sv. Jure (Bratislava)	137,839	5.08		
Košice	156,598	5.76								
Sub-totals	1,007,556	37.13		125,061	4.60		1,262,573	46.53	318,090	11.74

Total in litres equivalent = 2,713,280 litres

Sources: Wojewódzkie Archiwum Państwowe w Krakowie, *Księgi celne*, rkp. 2251–2270; F. Hejl and R. Fišer, 'Obchod východoslovenských měst se zahraničím ve století protihabsburských povstání', *Sborník Prací Filosofické Fakulty Brněnské University*, Vol. 31, (C.29) (Řada Hist.), Brno, 1982, p. 118.

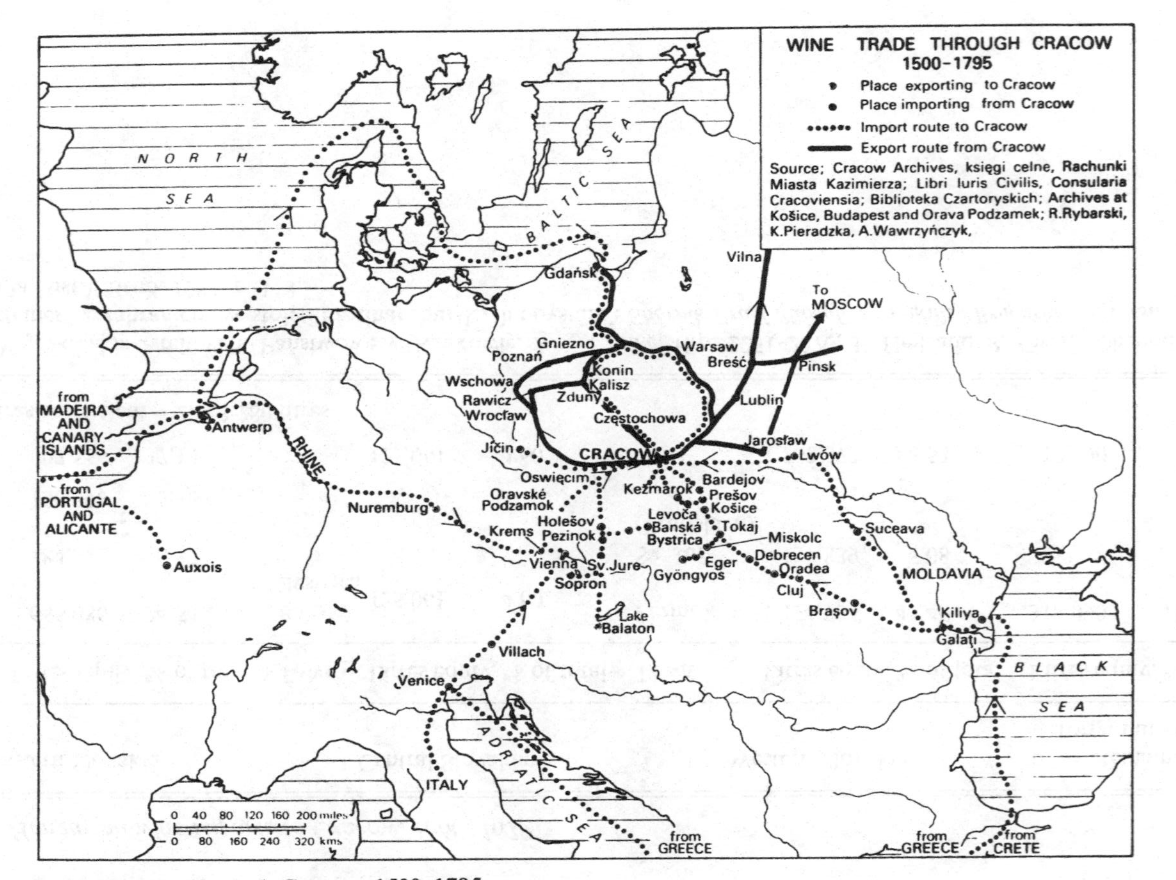

WINE TRADE THROUGH CRACOW
1500-1795
Place exporting to Cracow
Place importing from Cracow
Import route to Cracow
Export route from Cracow
Source; Cracow Archives, księgi celne, Rachunki Miasta Kazimierza; Libri Iuris Civilis, Consularia Cracoviensia; Biblioteka Czartoryskich; Archives at Košice, Budapest and Orava Podzamek; R.Rybarski, K.Pieradzka, A.Wawrzyńczyk,
NORTH SEA
BALTIC SEA
BLACK SEA
ADRIATIC SEA
from MADEIRA AND CANARY ISLANDS
from PORTUGAL AND ALICANTE
Antwerp
Auxois
RHINE
Nuremburg
Gdańsk
Vilna
To MOSCOW
Gniezno
Poznań
Warsaw
Brześć
Pinsk
Konin
Kalisz
Wschowa
Rawicz
Zduny
Wrocław
Częstochowa
Lublin
Jarosław
Jičín
CRACOW
Lwów
Oswięcim
Bardejov
Kežmarok
Prešov
Oravské Podzamok
Košice
Levoča
Holešov
Banská Bystrica
Tokaj
Krems
Pezinok
Miskolc
Suceava
Debrecen
Vienna
Sv. Jure
Eger
Oradea
Gyöngyos
MOLDAVIA
Sopron
Cluj
Brașov
Kiliya
Lake Balaton
Galați
Villach
Venice
ITALY
from GREECE
from GREECE
from CRETE
0 40 80 120 160 200 miles
0 80 160 240 320 kms

72 Wine trade through Cracow, 1500–1795.

Usually Polish merchants were members of the richer urban hierarchy, or nobility. They used their own employees to transport wine from purchasing centres to destinations like Cracow, and some of the wine merchants had their own special trade guilds.[694] Among the more active merchants mentioned in Cracow customs books was Prince Ostrowski (Chancellor of Cracow);[695] foreign traders included those from smaller Carpathian centres (Dukla, Lubovl'e, Sabinov) and from northern Hungary (Prešov, Bardejov, Levoča, Košice, Prievidza, Kežmarok and Tokaj itself). Nearer the Silesian border much of the trade was controlled by Jewish merchants with contacts amongst the Cracow Jewry. Moravian wine often reached the city through their organisation,[696] along with west Slovakian and 'Hungarian' vintages.[697] Towards the end of the seventeenth century (1688) small quantities of these wines were still being recorded in the customs ledgers.[698] From Oradea merchants occasionally dispatched Transylvanian wines via Debrecen for the Polish market.[699]

The west European wine trade was particularly active in Cracow during the early years of the seventeenth century. Most vintages came via Gdańsk (Fig. 72), with 1602 a peak importing year (43,910 litres). Demand remained high in Cracow during the first decade not only for more ordinary wines, but also for the prized malmsey variety, a favourite of the royal court. Bogucka has shown the popularity of Spanish wine imports ('Sek', Bastard, Alicante) to Gdańsk during the early years of the century[700] while Cracow customs books give details of wine from the Canary Islands,[701] Alicante,[702] the sweet 'sek' variety[703] as well as Peter-see-me. Sweet Italian wine, popular with ladies of the court,[704] arrived in Cracow at this time[705] from Venice, by way of Krems and Vienna. The transference of the capital to Warsaw (*circa* 1610) drastically affected Cracow's wine trade (Fig. 70). After 1610 wine imports through Gdańsk for Cracow remained at a low key, further complicated by disturbances during the Thirty Years War and the traumas of the Swedish invasions.

Consignments of wine also went down the Vistula river from Cracow towards Warsaw and Gdańsk; south Carpathian vintages were in demand after the royal court moved to Warsaw as seen in Table 54. The immediate years following this move found continued demand for Hungarian wine from Cracow merchants; in fact 35 per cent of the total in Table 54 was specifically designated as 'Hungarian' although it is probable that this wine was part of many other consignments. Twenty-six merchants from Little Poland, according to Obuchowska-Pysiowa, were involved in sending wine down the Vistula River seven of whom were from Cracow.[706]

Northern Hungary remained Cracow's main source for wines during the eighteenth seventeenth. The towns of Kežmarok (Késmárk) and Prešov (Eperjes) provided the greater part of this profitable trade with Poland in which some Hungarian families (e.g. Pulszky) acquired considerable

Table 54. *Wine exported from Cracow down the Vistula River, 1611–1623*

Year	Cracow Merchant	No. of Barrels	Litre Equivalent
1611	B. Węgrzyn	100	27,187
1611–1612	J. Jachman	61 (Hungarian)	16,584
1611	Z. Hipolit	56	15,225
1612	J. Krongowski	30 (Hungarian)	8,156
1611	S. Sidłowski	21	5,709
1611	J. Benkowicz	10	2,719
1623	J. Zawisza	10 (Hungarian)	2,719
Total		288	78,299

Source: Calculated from H. Obuchowska-Pysiowa, *Handel wiślany w pierwszej połowie XVII wieku* (Ossolineum), Wrocław-Warszawa-Kraków, 1964, Table 17, p. 143.

wealth.[707] Archival information from Oravský Podzámok in west Slovakia confirms wine movement up the Orava valley to Poland in 1720.[708] After 1740 fear from the Turks in Hungary ceased and, by mid-century, Cracow's merchants were playing a leading role in wine imports from south of the Carpathians. Hungarian wine was recognised throughout the Hapsburg Empire as extremely palatable and cheap; if it was allowed to enter Austrian markets demand for Lower Austrian vintages would have rapidly disappeared; as a result energetic measures were taken to prohibit Hungarian wine imports to Austria. Therefore the main markets open for Magyar wines were in Poland and northern Europe; thus Cracow wine merchants were able not only to supply the needs of the nobility and church, but increased demand from town dwellers, including craftsmen artisans.[709] Cracow's city merchants role in this trade is seen in Table 55. Over two-thirds of Hungarian wine imports were controlled by Cracow merchants around mid-century; a further fifth came from Slovakian traders (Levoča, Kežmarok), or from Lubovle and Nowy Targ in the Carpathian mountains.

Large quantities of Hungarian wine continued to reach Cracow during the last quarter of the eighteenth century. It formed the chief source of income for towns in the north-west Highlands of Hungary (Slovakia), and the Little Alföld.[710] Grosmann has estimated that, in 1775, about 40 per cent of Poland's imported Hungarian wine from these regions went through Cracow.[711] The Wieliczka toll-bridge returns suggest similar amounts to those noted in the 1760s (Table 55) were being sent in transit through Cracow, or for local consumption (Table 56). There appears to have been some change in wine destinations over the ten-year period. In 1775 nearly all imported vintages were for local consumption; by 1785 it was about half and half.

Table 55. *Import of Hungarian wine to Cracow, 1750–1763*

	1750			1755			1760			1763		
	No. of barrels	Litre equiv.	%	No. of barrels	Litre equiv.	%	No. of barrels	Litre equiv.	%	No. of barrels	Litre equiv.	%
Total no. of barrels	2,973	808,275	100	3,852	1,047,251	100	3,806	1,034,745	100	4,441	1,207,384	100
of which:												
Cracow merchants	2,075	564,134	69.8	2,542	691,099	66.0	2,604	707,955	68.4	3,019	820,782	67.9
other merchants	478	129,955	16.1	449	122,070	11.7	391	106,302	10.3	877	238,432	19.7
nobility	186	50,568	6.2	413	112,283	10.7	485	131,858	12.7	204½	55,598	4.6
the Church	234	63,618	7.9	448	121,799	11.6	326	88,630	8.6	340½	92,575	7.8

Sources: Calculated from Wojewódzkie Archiwum Państwowe w Krakowie, *Księgi celne*, rkp. 2216, 2223, 2231, 2232; M. Kulczykowski, 'Handel Krakowa w latach 1750–1772', *Prace Historyczne*, Vol. 26, No. 4, Kraków, 1960.

Table 56. *Hungarian wine through Cracow 1775–1785*

	1775			1780			1785		
	No. of barrels	Litre equiv.	%	No. of barrels	Litre equiv.	%	No. of barrels	Litre equiv.	%
For city & suburbs	3,345	909,412	87.5	2,968	806,915	69.0	1,805	490,729	52.6
In transit	481	130,770	12.5	1,331	361,862	31.0	1,626	442,064	47.4
Total	3,826	1,040,182	100	4,299	1,168,777	100	3,431	932,797	100

Sources:. Calculated from Wojewódzkie Archiwum Państwowe w Krakowie, *Rachunki miasta Kazimierza*, rkp. 660, 665, 670; *Księgi celne*, rkp. 2235, 2238; M. Kulczykowski, *Kraków jako ośrodek towarowy Małopolski zachodniej w drugiej połowie XVIII wieku* (P.W.N.), Warszawa, 1963, p. 59.

After 1785 Cracow's wine trade declined considerably, partly due to the Partitions. Moreover, towards the close of the eighteenth century, there was a general complaint in Hungary that the great demand for their wines abroad had ceased. This particularly affected the Tokaj (or Hegyalja) district, together with Sopron. It was not the result of falling Polish demand, but the Prussian occupation of Silesia that was considered by the Hungarians as sealing the fate of their once profitable wine trade.[712]

Other wines besides Hungarian arrived in Cracow, but throughout the century their importance was dwarfed by the products from the Magyar vineyards. Wines from Moldavia,[713] France[714] and Italy, malmsey from Crete, and vintages from Austria and Moravia[715] entered the city cellars, but were mainly for local consumption.

From Cracow, Hungarian wine was also in transit to other parts of the country. Greater Poland, with purchasing centres in Poznań, Warsaw, Wschowa, Częstochowa, Sławków, Kalisz, Gniezno, Rawica, Zduny and Konin, was a major consumer as seen in Table 57. Small quantities of Hungarian wine were also sent eastwards from Cracow to Galicia[716] e.g. Jarosław, attracting customers there from as far away as Moscow.[717]

Wine prices in Cracow for the period 1500–1795 are portrayed in Figure 73. More generally, in Europe, everything points to an increase in the consumption of wine between the fifteenth and eighteenth centuries. Braudel and Spooner maintain that 'A first increase in the sixteenth century involves the towns',[718] where, 'the towns are drinking increased quantities and not paying overmuch attention to the quality of the wine'.[719] The second advance came in the eighteenth century, wine 'this time affected the countryside', when 'the peasants, who traditionally lived soberly ... turned to drinking more and more wine'.[720] Poland throughout much of this time was in that part of northern Europe devoid of the vine; again quoting Braudel

Table 57. *Hungarian wine exports from Cracow to Greater Poland, 1750–1785*

Year	1750	1755	1760	1763	1775	1780	1785
No. of barrels	42	136½	1,620½	1,862½	439	1,274	1,613
Litre equiv.	11,419	37,111	440,569	506,362	119,352	346,365	438,530

Sources: Calculated from Wojewódzkie Archiwum Państwowe w Krakowie, *Księgi celne*, rkp. 2216, 2223, 2231, 2232; Ibid. *Rachunki miasta Kazimierza*, rkp, 660, 665, 670.

and Spooner 'we have to distinguish further between regions: those on the one hand which were old clients for the wines of the south, and those on the other hand for whom wine was still a luxury, a display of wealth, something relatively new. We have only to think of the Polish nobleman who insisted on drinking wine so as not to be confused with his beer-swilling peasants.'[721] New wine varieties in the sixteenth century often coincided with festive occasions and 'In Kraków the new wines from Moravia and from Hungary arrived every year towards September, and were not allowed to be sold without guarantee of their origin. That is why the consuls ordered branches of greenery to deck the fronts of inns serving Moravian wine, and a shock of straw for those where Hungarian wine was on sale.'[722]

Sixteenth-century wine prices in Cracow appear to have been fairly stable,[723] rising gradually towards 1600. The main exceptions were malmsey and muscatel which peaked in the early 1570s. This may have been related to commercial disruption in the Mediterranean basin caused by the Ottoman conflict with the Holy League, which culminated in the decisive Battle of Lepanto in 1573.[724] Prices for these wines practically doubled during these critical years, returning to a more stable, if higher pattern during the last two decades of the century.

This stability continued into the seventeenth century, but the Thirty Years War seems to have caused a recession in the wine trade. Wine prices appear to have followed the general fluctuations experienced by other goods in the city, compounded by the effects of the Swedish invasions, and complete ruin of Cracow by 1655. During the second half of the seventeenth century all wine prices suffered inflationary tendencies with more expensive varieties, like malmsey, reaching new heights towards 1700.

In spite of considerable oscillations during the first half of the eighteenth century the overall trend of wine prices was downward, reaching a trough in the 1740s, followed by increases over the next thirty years; a low occurred around 1775 and a drop in 1781 to a price similar to a century before. This price drop was short lived, with a dramatic rise the following year, characteristic of the unstable price conditions found with many products in Cracow during the years leading up to the final Partition.

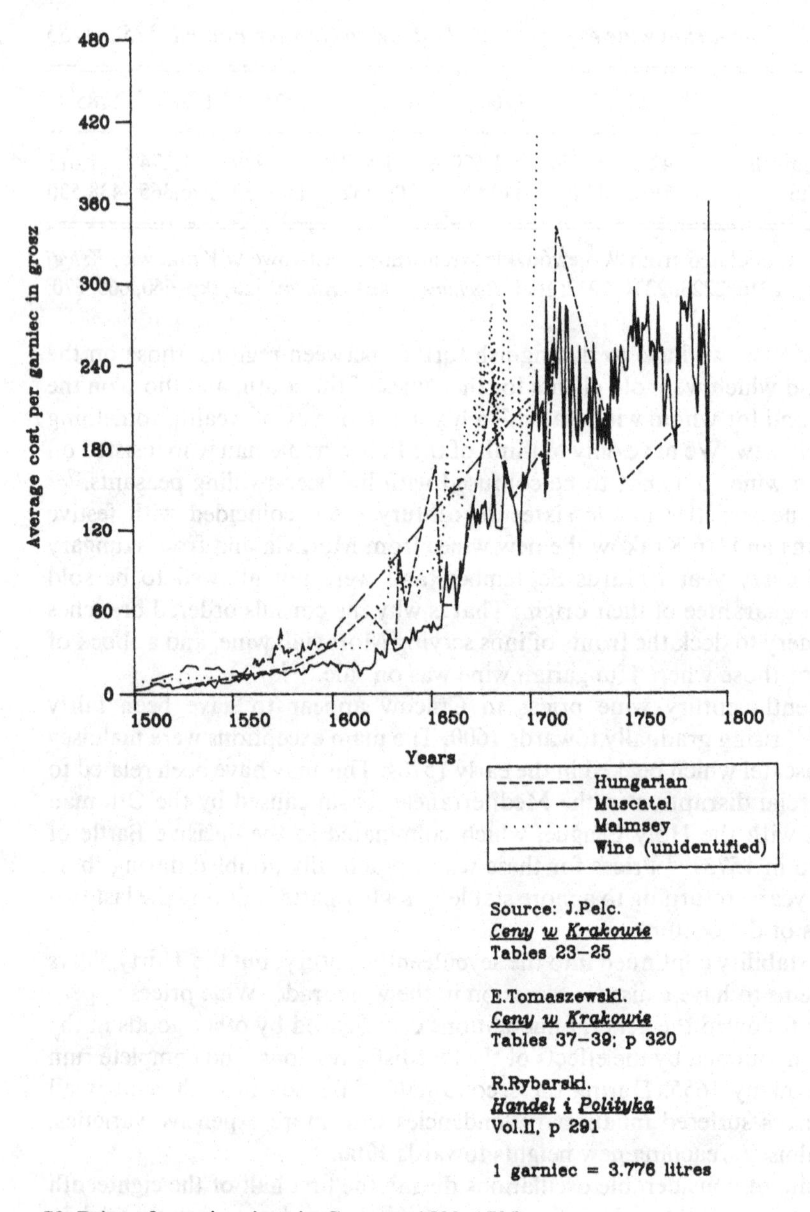

73 Price of certain wines in Cracow, 1500–1795.

As already intimated beer was a beverage of some significance in the city. Beer was the ordinary man's drink in northern Europe, and enormous quantities of it were consumed, for nourishment as well as pleasure, because it was a rich brew of wholesome ingredients. This ancient beverage was

Table 58. *Świdnica beer exports to Cracow 1517–1538*

Year	1517–1519	1524	1531	1538
Number of barrels	185	85	111	62
Litre equivalent	50,296	23,109	30,178	16,856

Source: Calculated from *Prawa, przywileje i statuta miasta Krakowa (1507–1795)*, F. Piekosiński (ed.), Kraków 1885, Vol. I, pp. 950, 965, 969, 983.

brewed in most areas of Europe where wine was not made, and formed the main drink from Brittany to Poland.[725] Any grain could be used for brewing, though oats were the commonest, and malted barley the preferred cereal. The use of hops for flavouring purposes was first introduced in the sixteenth century for they were widely grown in Germany, Bohemia,[726] the Low Countries and France.[727] Most beer was a weak home-brew, which according to R. J. Forbes was 'a harmless and healthy drink in an age when water was commonly impure'.[728] Quality varied from place to place and in Poland cities had several varieties; Braudel and Spooner state that 'At Warsaw there were prices of five sorts; at Lwów, two; at Kraków, six.'[729]

Beer does not appear to have played an important role in Cracow's sixteenth-century trade; if quantities were imported from abroad, however, they were free of customs duty provided the beverage was 'for domestic use' (*pro uso domestico*).[730] Cracow's beer imports came from two main sources, namely Silesia and Gdańsk. Several Silesian breweries were recorded in documents as sending beer to the Polish capital, especially Świdnica, Table 58. Compared with the wine trade, these are small amounts; moreover, according to Rybarski, consumption of foreign beer in Cracow did not greatly increase throughout the century.[731] Towards the end of the century several Silesian breweries were listed as beer exporters to Cracow (Table 59). The importance of Opava is noticeable. Matějek's analysis of Cracow's archival sources during the last decade of the sixteenth century,[732] also suggests its significance.

During the reign of Zygmunt August (1548–72) an increasing amount of beer brewed in Gdańsk, according to Kutrzeba,[733] was imported into Poland. Small consignments reached Cracow on several occasions near the end of the century.[734]

During the early part of the seventeenth century documentary evidence suggests the continuing dominance of Opava beer in Cracow. In 1624 a total of 52 barrels (14,137 litres) was recorded,[735] mainly delivered by Opava merchants, but those from Racibórz, and Cracow itself (e.g. Pavel Bonkowiecki)[736] are also mentioned. Later in the century, beer arrived from Krosno,[737] whilst from Nowy Sącz the trade was dominated by Jewish merchants.[738]

Table 59. *Silesian beer sent to Cracow, 1584, 1591/2*

Town brewery	1584	1591/1592
Świdnica	1½ barrels (408 lit. eq.)	5½ barrels, 8 horses (4,215 lit. eq.)
Wrocław	6½ barrels (1,767 lit. eq.)	14½ barrels (3,942 lit. eq.)
Opava	3 barrels, 8 horses (10,336 lit. eq.)	23½ barrels, 9 horses (9,449 lit. eq.)
Kłodzko	–	7 barrels (1,903 lit. eq.)
Unknown origin	8 barrels (2,175 lit. eq.)	55 barrels, 29 horses (24,813 lit. eq.)
Total	19 barrels, 28 horses (14,686 lit. eq.)	105½ barrels, 46 horses (44,322 lit. eq.)

Sources: Calculated from R. Rybarski, *Handel i polityka handlowa Polski w XVI stuleciu*, Warszawa, 1958, Vol. II, p. 198; Wojewódzkie Archiwum Państwowe w Krakowie, *Księgi celne*, rkp. 2116.

The market for beer in eighteenth-century Cracow was largely local, demand being mainly covered by the city brewery. In 1780 it alone produced 2,400 barrels (652,500 litres).[739] Regarding foreign brews it became fashionable in Cracow to drink English beer, imported through Gdańsk and Toruń.[740] Žilina in central Slovakia[741] also sent supplies of their noted beer.

Between the sixteenth and eighteenth centuries (Fig. 74) beer prices in Cracow reveal a similar pattern to wine (Fig. 73). Fairly stable prices were recorded between 1500 and 1620, followed by pronounced leaps in the 1620s and 1650s. Unfortunately, in the eighteenth century fewer beer data points exist than wine. Nevertheless comparison shows both beverages reached a higher price plateau then; even so considerable variations are noticeable for select years and certain periods, e.g. drops in prices during the 1770s and 1780s.

Besides wine and beer, vodka was part of Cracow's trading network during the Early Modern Period. In fact, northern Europe was noted for its high consumption of spirits and brandies and, even more so, of liquors distilled from grain; this arose with improved use of the still in the sixteenth century which, by the eighteenth, saw greatly expanded production. While Amsterdam was the main centre for liquor (both grain spirits and brandy), Poland in the early part of the sixteenth century[742] became noted for its own particular brandy – *gorzalka.* Gorzałka (the term 'vodka' is of later use), like brandy or other distilled liquor was, in areas lacking cheap transport, a desirable product since its greater value allowed freight charges to be made over longer distances.

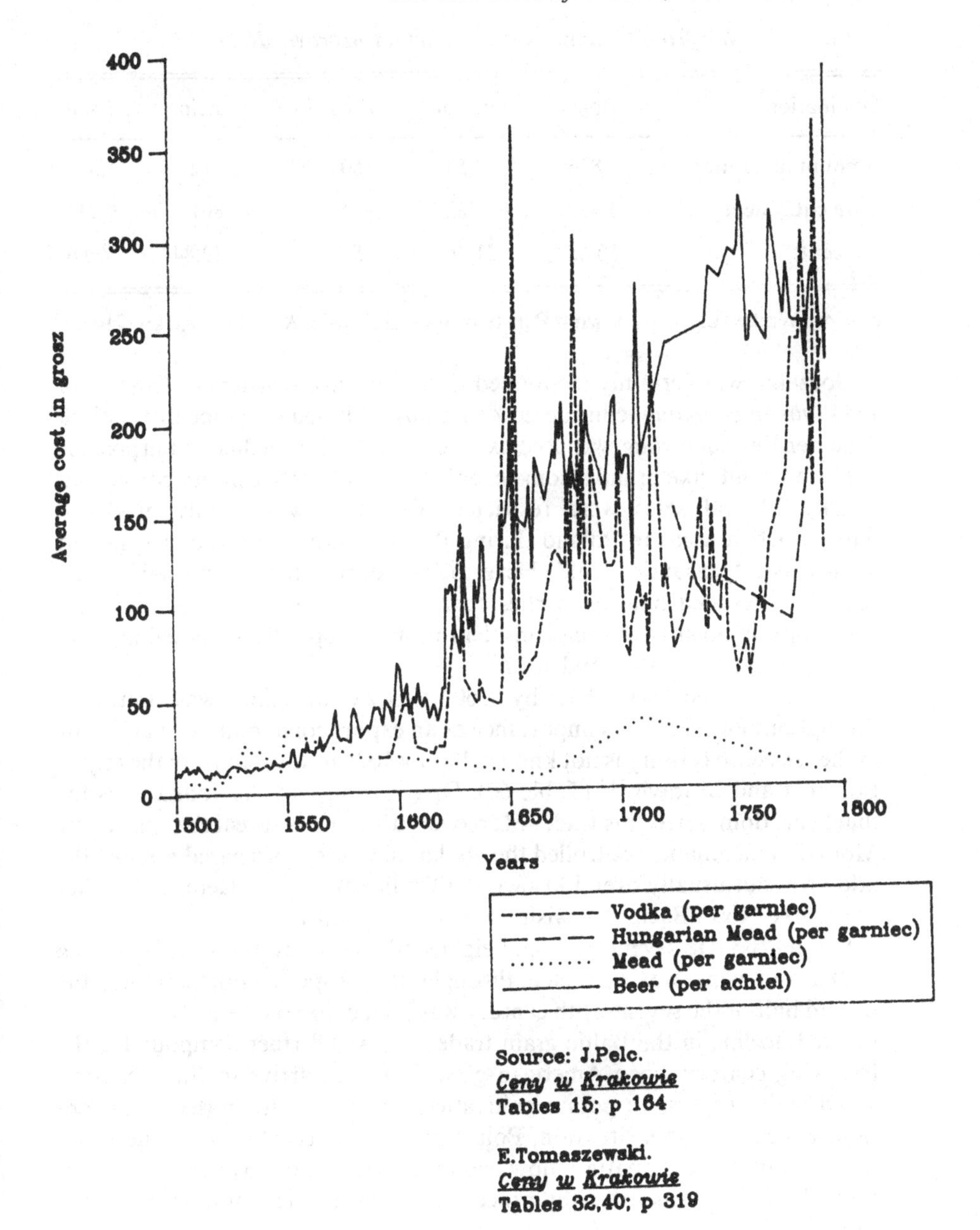

74 Prices of certain beverages in Cracow, 1500–1795.

Table 60. *Vodka from Cracow sent to Bohemia/Moravia 1624*

Destination	Opava	Přibor	Prostějov	Jičín	Total
Amount in garniec	836	353	250	212	1,651
Litre equivalent	3,157	1,332	944	801	6,234
Per cent	50.65	21.36	15.15	12.84	100.0

Source: Wojewódzkie Archiwum Państwowe w Krakowie, *Księgi celne*, rkp. 2142.

Gorzałka was certainly mentioned in the corkage registers of Cracow in 1544 (*vinum crematum* considered *destillato*),[743] based on grape distillation. Whether this *aqua vitae* was used as a beverage or for medicinal purposes is not known but, like brandy, some people in the sixteenth century considered it harmful and sales were restricted. Gorzałka was mainly produced outside urban areas in Poland during the sixteenth century, and regarded mainly as a 'peasant's drink'. Nevertheless, during the second half of the century[744] its medicinal properties began to be recognised. Contemporary consumption must have been considerable as it is specifically noted in royal decrees dated 1564, 1565 and 1577.[745]

Kuchowicz believes that by 1600 vodka drinking was universal throughout Poland.[746] Its importance as an export commodity from Cracow in the sixteenth century is not known, but by 1624 it was sent from the city to Bohemia and Moravia[747] (Table 60). Opava, with half the vodka sales for that year, dominated this trade. Moreover, documents reveal that just a few Moravian merchants controlled the vodka sales; they purchased not just the odd litre, but usually over 100 garniec (378 litres)/year.[748] Here, as in other towns like Nowy Sącz,[749] Jewish traders were important.

During the late seventeenth and eighteenth centuries, the art of distilling vodka and wine was perfected throughout Europe.[750] Furthermore, the second half of the seventeenth century was a time, for Western Europe,[751] of marked decline in the Baltic grain trade; this was further compounded the following century when American grain began to arrive in Europe, combined with progress in potato cultivation, which revolutionised consumption. Faced with this situation, Polish producers accustomed to the traditional grain economy pattern now had to look for alternatives. A way out of these difficulties was found by processing grain for alcohol in larger quantities.

In Cracow, the vodka market, like that of beer, was local in character with consumption coming from the nobility and city people. Most supplies came from the surrounding villages of Cracow (Biskupice, Smolensk, Wesoła, Wygoda, Strzelin; Garbary, Zwierzyniec and Zawisła), documentary

Table 61. *Vodka supplies to Cracow from surrounding villages, 1756–1765*

Year	1756	1758	1759	1760	1761	1762	1763	1764	1765
Amount in garniec	14,536	10,908	14,069	16,759	19,034	20,561	25,705	29,565	28,657
Litre eq.	54,888	41,189	53,124	63,282	71,872	77,638	97,062	111,637	108,209

Sources: Calculated from Wojewódzkie Archiwum Państwowe w Krakowie, *Regestra Percepta od wódek i win różnych do miasta wprowadzonych do miasta*, rkp. 2243, 2245, 2247, 2248 (1756–1765); M. Kulczykowski, *Kraków jako ośrodek towarowy Małopolski zachodniej w drugiej połowie XVIII wieku* (P.W.N.), Warszawa, 1963, p. 60.

evidence giving some idea of the volume of the vodka trade (Table 61). Vodka consumption in the city doubled over this period, perhaps reflecting use of grain made by the surrounding landed estates, resulting from declining export opportunities. The continued importance of local vodka supplies for the Cracow market is suggested by toll-bridge data for 1775 and 1785.[752]

The city also imported vodka from abroad. In 1758 two separate consignments of Russian vodka, totalling 1,100 garniec (4,153½ litres) and organized by Jewish merchants,[753] arrived in Cracow, together with small quantities of the famous Gdańsk vodka.[754]

There is limited evidence of vodka exports from Cracow, but the eighteenth century experienced a formidable trade with the Ottoman Empire in this beverage, particularly Moldavia.[755] During the later decades Polish vodka and other liquors were mentioned in the lists of merchandise arriving at Turkish ports, especially Kherson.[756] It is possible, but unproven, that some of these alcoholic beverages came from Cracow.

Only in the 1590s were vodka prices first quoted in Cracow's documents, coinciding with the time it became a firmly established drink in the city. Early peak prices relate to the unstable economic and political situation around 1630 (Thirty Years War), and Swedish invasions of 1650s. Vodka prices suffered distinctive oscillations during the period 1680–1720, but by mid-century they dropped to a level found a hundred years before; after 1760 increased demand meant prices rose sharply in 1784 and 1790, reaching levels noted in 1654 and 1681 (Fig. 74).

Mead was another popular beverage. Unfortunately sixteenth-century documents did not always differentiate between honey and mead but, according to Małecki, mead was sent to Cracow in small quantities.[757] This probably came from villages in the more wooded parts of the Carpathian foothills, or from Bohemia.[758]

More information is forthcoming from the eighteenth century. Quantities were imported from Levoča (E. Slovakia),[759] and referred to in documents as 'Hungarian mead'. Compared with more locally produced varieties its price (Fig. 74) was much higher, attributed by Tomaszewski to the superior flavour of the Hungarian drink.[760] Little is known of the other beverages, but Bogucka found that certain fruit juices (particularly lemon) were imported from the Iberian peninsula by Gdańsk merchants at the beginning of the seventeenth century;[761] it is probable that some citrus fruit drinks along with other goods were dispatched up the Vistula river to the old Polish capital.

(iii) Handicraft products

During the three centuries from the sixteenth to the late eighteenth, craftsmanship in many European towns evolved a more sophisticated production style than previously, partly to cater for increased extravagance by the land-owning nobility and greater wealth of urban citizens. Most goods were still made without the help of machines, or an organised factory system,[762] but urban artisans developed specialised skills which had existed since the Middle Ages in the more advanced parts of Europe. In Nuremburg, probably the most advanced handicraft centre in Europe, goldsmiths, cutlers, furriers, belt-makers, various kinds of wool/cloth-makers, armourers, swordsmiths, scythesmiths, pewterers and mirror-makers had been there since the thirteenth century.[763] In Poland and its neighbouring countries evidence of well-developed handicraft industries was present already in the sixteenth century,[764] particularly in towns like Toruń.[765]

In Cracow too, there was a long tradition of handicraft production, largely due to the needs of the royal court: Kleparz in the Later Middle Ages had developed as the city's main craft centre,[766] whilst handicrafts formed one of the city's largest guilds.[767] Important pictorial and written information from the beginning of the sixteenth century on Cracow's handicraft trades is the codex of Baltazar Behem (1505). Over twenty various crafts are mentioned including those of local importance (bakers, tailors, hatters, potters, tanners, bell founders, carpenters, shoe, needle, soap and purse makers, coopers and barber-surgeons); also those specialising in equine needs (harness makers, saddlers, wheel and cartwrights, and smiths); weaponry (bow and arblast makers) and luxury articles (goldsmiths and artistic painters).[768] Franaszek has recently shown, with examples from Cracow and its region, that later in the century the main state income besides trade was from handicrafts.[769]

It is now pertinent to enquire how much this activity generated trade in Cracow either through raw material imports, and products not readily available, or surplus craft exports (Fig. 75). Some products have been previously discussed e.g., textiles and skins, but of the remainder perhaps the most important were connected with the metal industry. Metal ware production, a widely esteemed Cracow craft, was unlike that in many other cities, as it did not rely solely on the immediate vicinity for main markets. Products were designed to suit international taste, with a wide variety of choice. For example, in Cracow, one could purchase all sizes of bells – from thimble bells to huge church bells – candelabra, table lamps, lanterns, mortars, flat irons, spigots, thimbles, scale weights, hinges, doorknobs, locks, buttons and buckles, trumpets and horns, curtain rods, nails and screws, scales, warming-pans, scissors and tools of every description; weapons ranging from swords and poignards to small cannons, as well as many other items, were produced. As the century progressed, however, some Cracow craft guilds could not compete with Western merchandise, partly due to inferior quality, and partly the effect of heavy taxation. Thus, foreign merchants, who could afford such financial impositions, increasingly

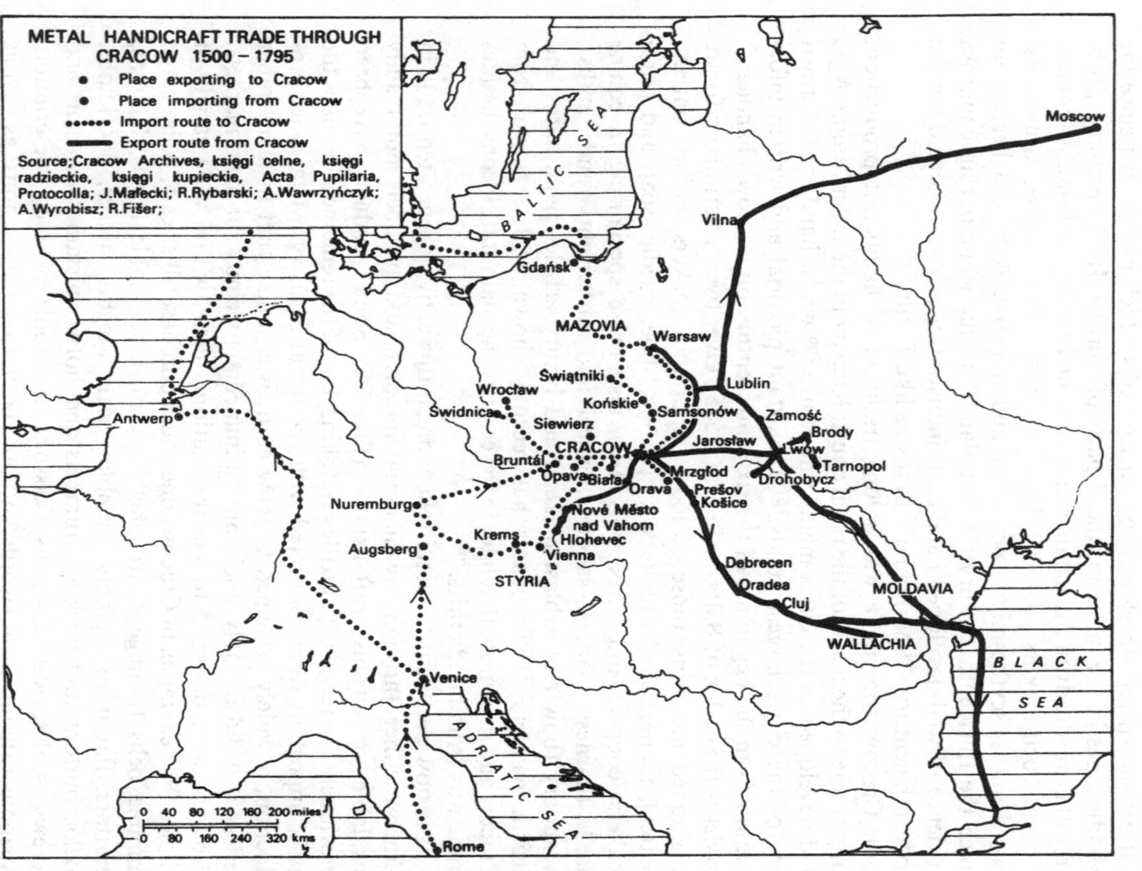

75 Metal handicraft trade through Cracow, 1500–1795.

imported certain craft goods into Cracow,[770] but evidence suggests that for some commodities, local artisans could be competitive.

Recent analysis by Wyrobisz[771] of craft structure in certain Polish towns has revealed that at the end of the sixteenth century the metal sector in Cracow was, after textiles, leather and food, the most important group. Supplies of iron ore from Moravia and Slovakia,[772] together with local deliveries from the Holy Cross Mountains, were utilised near Cracow for iron smelting,[773] and provided metal craftsmen with much-needed source material. Steel imports from Austria, Silesia and Moravia[774] were worked by the city's cutlers for manufacturing swords, knives and other sharp instruments;[775] over sixty cutlers received the coveted citizenship of Cracow during the sixteenth century,[776] emphasising their importance for the city's prosperity. Gold- and silversmiths also enjoyed privileged status,[777] some involved in making jewellery from pearls and other precious stones.[778]

Metal goods also came to Cracow for local use or as transit trade. The main imported iron goods were scythes, sickles, nails, needles, semi-manufactured sheet metal and wire. Scythes, indispensable for harvesting crops like cereals, were the most popular item, in demand both locally and as a part of a larger international transit trade. Some scythes arrived in Cracow from Styria (Austria),[779] larger consignments came from Opava[780] and other towns in Silesia.[781] Opava was also Cracow's main supply centre for wire towards the end of the sixteenth century,[782] whilst Wrocław[783] was an important source for sheet metal. From Slovakia came various iron/copper metal products (pots, iron rods, religious crosses and cannon balls),[784] whilst metal hoes were sent in exchange for salt.[785] Cracow's local region also imported quantities of nails,[786] their weight presumably being a restrictive factor, but demand was high in the city, mainly for building purposes.

Metal-goods exports from Cracow in the sixteenth century concentrated on two main markets: south over the Carpathians to Hungary, and north-eastwards to Russia. Growth in the importance of Cracow's cutlers' guild in the sixteenth century[787] was reflected in its product demand, both long distance and for local consumption. Cracow knives had a considerable reputation, particularly in Hungary, with exports to places like Debrecen, Cluj[788] and Oradea,[789] together with nearer towns like Košice.[790] Some metal objects e.g., scythes and sickles, were sent in their thousands to Hungary,[791] but were made mainly in Nowy Sącz (noted for its famous forges) rather than Cracow.[792] Spades and scythes also went via Wallachia to Turkey.[793] North-eastwards the two main markets were Vilna[794] and Moscow,[795] for both knives and scythes; some were re-exported to Siberia in exchange for furs.[796] In fact, Cracow's cutlers specially produced a highly decorative lady's knife, the so-called 'Muscovian'.[797] Sheet metal also left Cracow for Lublin.[798]

Paper manufacture and printing steadily increased in importance from the sixteenth to eighteenth centuries, aided by the expansion of book production and spread of literacy. Paper, usually made from reprocessed rags, replaced parchment as the main conveyor of the written word, except for legal and state documents. Paper mills appeared in most large European cities, where an ample supply of rags and demand for paper existed; closely associated with this was the growth of printing which utilised most of the paper made. According to Braudel, 'Paris saw its first printed book in 1471, Potiers in 1479, Venice in 1470, Louvain in 1479 and Cracow in 1474',[799] which places the latter amongst the European forerunners of the printing craft. By 1500, a total of 236 cities in Europe had their own printing presses.[800]

At the end of the fifteenth century, the first paper mill near Cracow was located at Prądnik Czerwony, whilst many more were established in the environs of the city during the following century (Tenczynek, Krzeszowice, Podchełmie and Balice in the Rudawa valley; Czajowice, Grembienice, Prądnik Biały and Prądnik Czerwony in the Prądnik valley; and Wilczkowice, Młodziejowice and Mogiła in the Dłubnia valley) along Vistula river tributaries.[801] Other paper mills some 40–65 km from Cracow (around Chrzanów, Olkusz and Jędrzejów) also supplied the Polish capital,[802] so that the city was not only self-sufficient, but had an export surplus at this time. According to Ptaśnik, the Cracow region was the largest paper manufacturing centre in Poland (72,000 reams in 1581[803]); much smaller quantities came from Poznań, Lublin, Warsaw, Vilna, Toruń etc.[804]

Two by-products of Cracow's paper industry were the manufacture of playing cards and printing. Playing cards, of Eastern origin, were already known in Europe by the fourteenth century, but not in common use till *circa* 1400.[805] In Cracow, the playing-card makers' guild frequently entered the city statutes during the second half of the sixteenth century.[806] Printing, however, was of greater importance in Cracow's handicraft structure, as it played both significant cultural and commercial roles in Cracow's development. Kasper Hochfeder, a printer from Metz, set up a typographical workshop in the city centre (św. Anna st. No. 3) in 1503 under the auspices of a rich Cracow wine merchant, Jan Haller; in 1510 another workshop was established by a Bavarian, Florian Ungler.[807] Others followed, producing not only books in various languages (Latin, Greek, Hebrew, Polish and Hungarian) but also almanacs and school textbooks;[808] in 1526 the first geographical map was published in Polish by Bernard Wapowski.[809] Cracow's dominance of Poland's printing trade during the sixteenth century is seen from the number of official workshops; according to Baranowski, of the 52 in Poland, 21 (40 per cent) were in Cracow, a total of 16 were in Poznań, Lwów, Vilna and Lublin (30 per cent) and the remaining 15 were scattered throughout the country.[810]

Cracow's longer distance paper trade and its derivatives was mainly export based, although small quantities were imported into the city. During the early part of the century paper was sent from Opava and Nysa to Cracow,[811] whilst further late-sixteenth-century examples again reveal paper imports from Silesia.[812] The main paper export market was Hungary with consignments sent to Bardejov, Prešov, Košice and routes farther south.[813] Examples from the end of the century show it was transported to the north-east, with markets in Vilna and Minsk.[814]

Book production, stimulated not only by local demand from the city's university and various religious authorities, also had longer distance commercial appeal.[815] Trade in books was particularly lively in sixteenth-century Hungary, spurred on by increasing literacy and the impact of the Reformation. This trade was significant in northern Hungary, where polemical religious literature, psalters and grammatical texts were in demand;[816] as the whole of northern Hungary contained no printing workshops during the first half-century[817] Cracow enjoyed a virtual monopoly there. Books from Cracow printed in Hebrew also had a wide market, especially in Western Europe. The first printed Hebrew text in Cracow was by Jan Halicz in 1535; from then on trade in Hebraic literature gathered momentum, especially in the latter part of the century with consignments sent to Germany (Nuremburg), Austria, Silesia (Wrocław), Moravia and Hungary.[818] Izaak ben Aron Prostice, a Jewish printer in Kazimierz, was given royal consent in 1586 to have monopoly rights for printing the Talmud.[819]

Other products in the sixteenth century with narrower trade implications for Cracow included ceramics and glass. These goods, like textiles and leather, had a wide range of appeal, from the high quality of Venetian glass and Meissen porcelain, to roughly formed earthenware and poorly made vitrium. Cracow's neighbourhood contained ceramic clay beds,[820] the principal location factor for pottery making, together with access to a nearby market.[821] Problems of weight and/or fragility restricted the development of longer distance trade, although the distinctive Cracovian pottery style[822] may have travelled to other major market centres in Poland and beyond. Unfortunately these were often included in the 'unnamed' category for goods mentioned in city documents. Glass products seem to have had less commercial importance for Cracow,[823] often purchased locally or from the Sandomierz region mainly for limited consumption. The demand for window glass had grown during the sixteenth century, as well as for domestic and ornamental glassware of which some, on rare occasions, left Cracow for Gdańsk.[824]

Towards the end of the sixteenth century a crisis appears to have developed in Poland's handicraft industries, and continued into the early part of the next century. It is thought to have been caused by contemporary problems associated with the general level of Polish industry. Limited progress in handicraft production has been blamed by Wyrobisz on the lack of readily available investment funding, shortages of wood fuel, and failure to find suitable alternatives such as coal. Moreover, unlike some West European countries, Poland failed to appreciate the benefits of mass production methods, particularly in the handicraft sector.[825]

Imports played an increasing role in the decline of Poland's domestic handicraft production; transport of metal goods, particularly iron, in the Baltic was dominated by the Dutch and Swedes. The latter were important exporters of commodities like guns, whilst the former were major carriers of metal manufacture from western Europe.[826] Land routes, also played a significant role in Poland's imports at this time according to Rusiński, amongst which metal goods (scythes, knives, etc.) were significant.[827] Cracow, as an important trade centre on the north–south commercial axis, intensified its transit role in such products especially for lands beyond the southern border. Some intimation of Cracow's importance for metal goods is seen from a shop inventory for 1650.[828] Amongst iron goods for sale were battle- and pole-axes, hatchets, saws, screwdrivers, crochet-hooks, drills and spoons, all included under the heading 'goods from Nuremburg'; brass objects included holy-water basins, Jewish lamps, stirrups, store-lids, pipes, capstans, ink stands, compasses and thimbles. Also mentioned were silver-framed mirrors and spectacles from Augsburg, Venetian tin-plate, pewter brushes, combs of various materials, knives, scissors and weights; musical instruments (from Germany, Vienna, and lutes from Rome), along with painted metal-boxes, fishing rods, elegant French iron work and vases, were often erroneously categorised as 'Nuremburg goods', although they may have arrived from Italy via Augsburg, or via Gdańsk from Holland. However, most probably came to Cracow via Wrocław from Nuremburg, hence the popular collective name for these products.

Various Silesian towns were the sources of numerous metal manufactures for Cracow; iron works and forges were scattered throughout Silesia, but the Bytom-Tarnogórze region became the main iron-smelting centre, whilst Kowary played an important part in metal exports.[829] Wrocław was the leading Silesian centre for arts/crafts and exports to Poland, and included such metal products as scythes, knives, saws, sheet metal, wire and arms; miscellaneous articles included candlesticks, hammers, screws, pipes, padlocks, keys, bells, awls, horse-combs, etc. Guldon and Stępkowski have stated recently, with reference to the early seventeenth century, that 'Toruń and Cracow were the main markets for Silesian metal products, of which scythes, sickles, knives and gunsmiths implements were the main items.'[830] Certainly during the 1620s, Cracow was a transit centre for Silesian metal products to Slovakia (e.g. Levoča).[831] Northern Moravia also supplied Cracow with scythes,[832] steel, cutlery and wire from handicraft centres like Opava and Bruntál.[833]

During the first half of the seventeenth century export markets for

Table 62. *Scythes and knives sent from Cracow to Levoča, 1603–1607*

	1603	1604	1605	1606	1607	Total
No. of Scythes	3,400	5,250	1,300	600	4,100	14,650
No. of Knives	15,800	18,400	–	–	5,600	39,800

Sources: Wojewódzkie Archiwum Państwowe w Krakowie, *Księgi celne*, rkp. 2123–2127; Okresní archiv, Spišská Nová Ves, *Trieda IV A*, signatury 20/3–20/7.

Table 63. *Scythes and knives sent from Cracow to Levoča, 1620–1624*

	1620	1621	1622	1623	1624	Total
No. of Scythes	2,900	10	3,920	–	100	6,930
No. of Knives	–	–	9,300	–	–	9,300

Sources: Wojewódzkie Archiwum Państwowe w Krakowie, *Księgi celne*, rkp. 2138–2142; Okresní archiv, Spišská Nová Ves, *Trieda IV A* signatury 20/20–20/24.

Cracow's transit trade in metal goods were mainly southward (Slovakia, central Hungary) and eastward to the Ukraine. In Slovakia, Levoča (Table 62) and Prešov[834] were the main markets and, to a lesser extent, Hlohovec, Nové Město nad Váhom and the Orava valley.[835]

Such detailed information suggests that, on average, nearly 3,000 scythes and 8,000 knives left Cracow annually for this Slovakian town alone. Obviously, a considerable proportion was dispatched to other markets, but it shows that despite a depression in the Polish handicraft industry, transit trade was still a profitable exercise. Comparison with the early 1620s reveals a considerable decline in this trade (Table 63) when direct contact between Slovakian markets and Silesian manufactures reduced Cracow's transit role.

Less than half the number of scythes and a quarter of the knives were sent from Cracow for this period compared with the first decade (Table 62). Obuchowska-Pysiowa's work has shown in the 1604 customs returns for Cracow that some trade in metal goods was also evident; she found four merchants sending metal goods to Lwów and stations south/east;[836] Moldavia and Wallachia were ready markets for all types of metal agricultural implements. Similarly for Cracow, Muscovy was an important market for Cracow's transit metalware trade. The key point on this route was Lublin; merchants like Jan Topszel from Cracow in the early seventeenth century imported Silesian metal hooks for dispatch to Lublin;[837] steel from Styria also came via Silesia (Opava), and Cracow for Lublin and the Russian market.

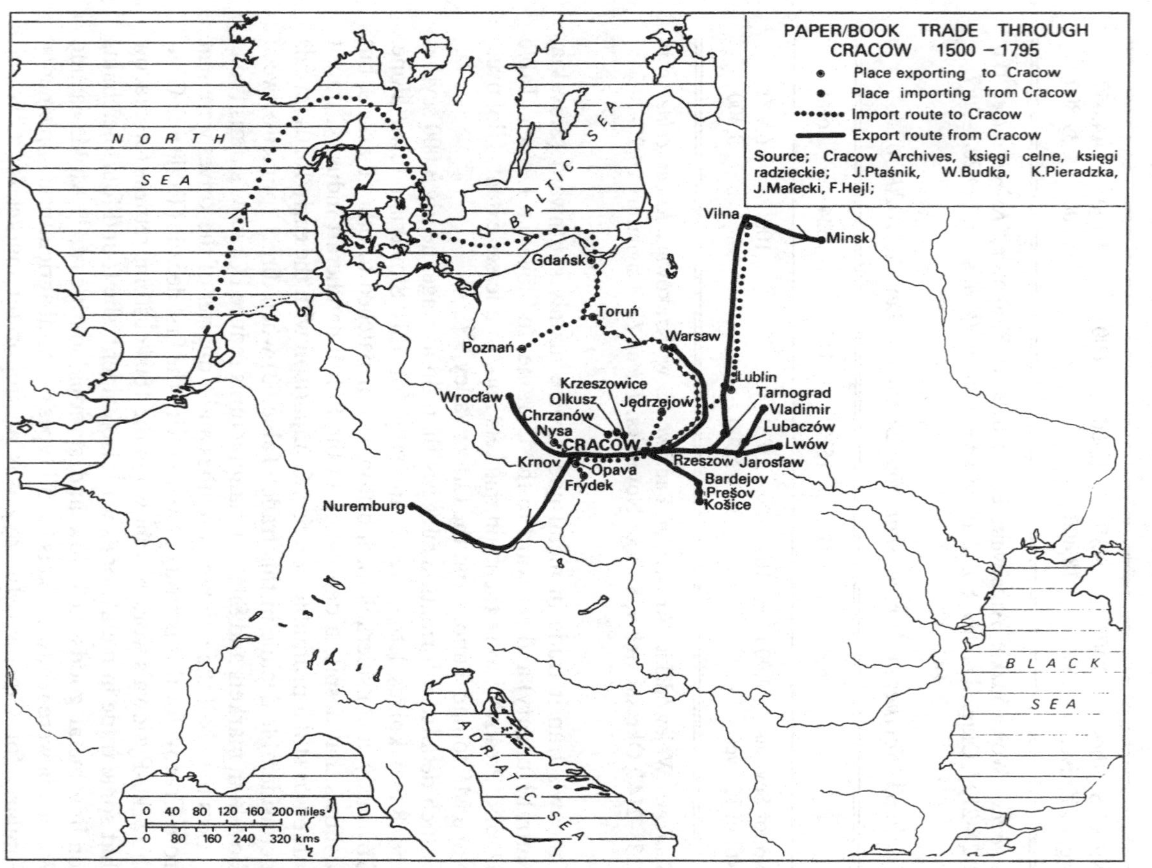
PAPER/BOOK TRADE THROUGH CRACOW 1500 – 1795
Place exporting to Cracow
Place importing from Cracow
Import route to Cracow
Export route from Cracow
Source; Cracow Archives, księgi celne, księgi radzieckie; J.Ptaśnik, W.Budka, K.Pieradzka, J.Małecki, F.Hejl;
NORTH SEA
BALTIC SEA
BLACK SEA
ADRIATIC SEA
Vilna
Minsk
Gdańsk
Toruń
Poznań
Warsaw
Lublin
Krzeszowice
Olkusz
Jędrzejów
Tarnograd
Vladimir
Lubaczów
Lwów
Wrocław
Chrzanów
Nysa
CRACOW
Krnov
Opava
Frydek
Rzeszow
Jarosław
Bardejov
Prešov
Košice
Nuremburg
0 40 80 120 160 200 miles
0 80 160 240 320 kms

76 Paper/book trade through Cracow, 1500–1795.

As with other goods much of Cracow's role in this transit trade was lost through the Swedish invasion of Poland in the 1650s. Previous trade links to former markets in metal products from Cracow had not revived, even by the 1670s, and was further complicated by the decline of forging at Nowy Sącz.[838] In fact, by 1670 places like Bardejov in Slovakia were producing their own metal goods, and the following decade was to witness mass export of scythes and other tools to Polish market centres.[839]

Cracow's book and paper trade continued into the seventeenth century (Fig. 76). Locally produced paper, together, occasionally with supplies from Silesia, formed the basic raw material for the book trade. However, paper from the city was sometimes exported over the Carpathians to markets in northern Moravia (Krnov),[840] Hungary (Slovakia) and Transylvania.[841] Book printing in Cracow failed to flourish as it had the previous century, largely due to a brake on cultural development attributed to long periods of warfare (Thirty Years War) in Europe and, although Poland was on the margins of this conflict, later the Cossack/Russian and Swedish invasions were to affect cultural development in the country. Furthermore, the transfer of the king's court from Cracow to Warsaw at the turn of the century meant the city's printers lost many privileges they formerly had for printing court materials and other notabilia connected with a capital whose functions controlled the political and social arena in Poland.[842] Nevertheless, it could be argued that even by mid-century Cracow's printers were still decidedly superior to their Warsaw counterparts, both in print quantity and quality.[843] Books were also imported to Cracow from Gdańsk, Warsaw and Lublin, a particularly flourishing trade after 1650.[844]

Cracow was a net importer of glass. Morano, the Italian glass-making centre, developed during the seventeenth century,[845] and glassware became popular amongst Cracow's wealthier citizens; for example, Obuchowska-Pysiowa found evidence for 1604 of mirrors, Venetian glass and decorated caskets arriving via Prague, whilst through Vienna two boxes of Venetian glass came together with two cases via Opava;[846] window-panes were imported from Bohemia (Lipník).[847] One gains the impression that most glass products were for local consumption, but it is probable some were sent further on, especially to Hungary. On occasions (e.g. early seventeenth century) Hungary was detached by the advancing Turkish army from other transit centres supplying western goods.

Eighteenth-century Poland suffered from a general lack of state policy on industrialisation;[848] state-owned estates contained no industrial establishments except those geared to the needs of the royal court. What manufacturing took place in urban centres was often achieved through partnership with local feudal lords, who controlled raw materials and labour. There were no state subsidies for manufacturing, so entrepreneurs were often dependent on the landed nobility for capital investment; disagreement between urban manufacturer and landed estate-owner inevitably led to the eclipse of the whole enterprise. In the Kielce region, the Bishops of Cracow owned the largest iron works in the country, founded around mid-century.[849] It was the only mass-commodity producer in Poland yet supplied mainly semi-manufactured products (iron bars) rather than finished goods. Such facts help explain why Cracow could continue to be a transit centre for foreign metal goods on the longer international trade routes due to the lack of finished goods of Polish origin.

Increased demand for metal goods during the second half of the eighteenth century meant Cracow's artisans were largely dependent on foreign iron supplies; Silesia was the major supplier, although occasionally much-prized Swedish iron arrived in the city from Warsaw.[850] Silesian exports declined after 1760, due to the Seven Years War (1756–63) and meant Cracow was increasingly dependent on domestic supplies from Polish metallurgical works at Końskie, Mrzygłód, Siewierz, Samsonów[851] and from Mazovia.[852]

Local production appears to have been insufficient to satisfy demand. Metal goods were imported into Cracow by merchants from Opava and Krems with products made in Styria and Vienna.[853] A large variety of iron goods arrived in Cracow from such sources; these included parts for ploughs, sickles, hatchets, saws, chisels, files, two-handled carpenters' blades, pliers, drills, knives, flowerpots, braziers, spits, scissors, hammers and many other instruments referred to in documents as 'fancy iron goods'. Certain metal goods such as scythes, sheet metal and wire were in great demand farther east. From Cracow they were sent to Jarosław (often by the Vistula/San rivers), Brody, Lwów, Lublin, Zamość, and even northward on occasions to Warsaw.

The largest market was for scythes. Several Cracow merchant families together with foreign traders were involved; according to Kulczykowski there were twice as many towns receiving scythes from Cracow in 1763 than thirteen years previously.[854] Whilst demand for scythes throughout Little Poland was largely in Cracow hands, supplies were also sent to more distant Polish towns, probably for further transit (Table 64). Over this same period, Cracow's scythe imports rose from 91,687 (1750) to 255,000 (1763), a near trebling over the thirteen-year period.[855] The popularity of Jarosław as a market centre may have been due to cheaper river transport and the presence there of a warehouse for Cracow merchants to store scythes and other metal goods. Some decline in Cracow's scythe trade was affected by the establishment, after the First Partition of Poland, of a scythe-making factory at Podgórze under Austrian control on the opposite river bank to Cracow. In 1780, only 43,800 scythes were imported into Cracow from Austria[856] and the heyday of the city's trade in this product seemed to be over.

Table 64. *Exports of scythes from Cracow to selected markets 1750–1763*

Importing centre	1750	1755	1760	1763
Jarosław	24,000	19,200	64,200	86,730
Drohobycz	2,000	1,550	800	0
Brody	800	7,800	15,250	21,000
Lublin	300	0	0	0
Lwów	400	700	0	3,750
Tarnopol	0	0	700	0
Gródek	0	0	870	0
Warsaw	0	200	2,100	0

Sources: Wojewódzkie Archiwum Państwowe w Krakowie, *Księgi celne*, rkp. 2216, 2223, 2231, 2232, M. Kulczykowski, *Kraków jako ośrodek towarowy Małopolski zachodniej w drugiej połowie XVIII wieku*, Warszawa, 1963 (P.W.N.), pp. 120–1.

Metal-producing centres within Poland were also important suppliers. For example, Świątniki near Łódź supplied the Cracow guild of swordmakers[857] whilst Sułkowice and Biała in the Cracow region were important producers of nails,[858] a commodity in constant demand for the city. By the second half of the eighteenth century the once significant trade in Hungarian copper had almost disappeared, but small quantities were still sent to Cracow for the city's boilermakers.[859] Similarly lead, which had once been such an important commercial commodity, played no role in the city's trade, but limited supplies arrived for local needs.[860] Meanwhile, other metal handicraft centres were developing, particularly in Greater Poland;[861] places like Poznań[862] were successfully competing in some of Cracow's former market area, reducing the latter's importance as a transit centre for metalware goods.

Cracow's paper industry suffered considerably from the Swedish incursions during the second half of the seventeenth century; local paper mills at Mogiła, Młodziejowice and Prądnik were destroyed by the Swedes,[863] whilst others at Balice, Podchełmie, Żuradnie and Siedlec suffered decline. This meant greater reliance on imported paper mainly from Silesia (Table 65). As much as three-quarters of Cracow's paper imports came from Silesia, mainly via Opava and Frýdek; paper exports, unlike metal goods, do not seem to have been discontinued as a result of the Seven Years War. Domestic supplies from various Polish paperworks (Polcza, Krzeszowice, Okleśna) were minimal in 1750, but by 1763 had reached a third of the Silesian total. From Cracow, paper exports went to Warsaw, Lwów, Tarnogród, Jarosław, Lubaczów, Rzeszów and Vladimir[864] (Fig. 76).

Table 65. *Cracow: paper imports 1750–1763 (in bales)*

Source	1750	1755	1760	1763	Total	%
Silesia	183	112	141	308	744	74.14
Poland	8	86	71½	94	259½	25.85

Sources: Wojewódzkie Archiwum Państwowe w Krakowie, *Księgi celne*, rkp. 2216, 2223, 2231, 2232; M. Kulczykowski, *Kraków jako ośrodek towarowy Małopolski zachodniej w drugiej połowie XVIII wieku*, Warszawa (P.W.N.), 1963, pp. 116–17.

Table 66. *Cracow: glass imports 1750–1763 (no. of carts)*

Source	1750	1755	1760	1763	Total	%
Silesia	16	18	19	13	66	45.5
Poland	7	8	40	24	79	54.5

Sources: Wojewódzkie Archiwum Państwowe w Krakowie, *Księgi celne*, rkp. 2216, 2223, 2231, 2232; M. Kulczykowski, *Kraków jako ośrodek towarowy Małopolski zachodniej w drugiej połowie XVIII wieku* (P.W.N.), Warszawa, 1963, p. 119.

After Cracow's so-called 'Golden Years of Printing' in the sixteenth century came to an end,[865] there appears to have been no major revival of the industry until the beginning of King Stanisław August Poniatowski's reign in 1764. There had been developments in journalism,[866] and books were still produced[867] but this had been accompanied by a relative decline of cultural interest in Poland during the period of rule by the Saxon kings (August II Mocny, 1697–1733; and August III, 1733–63). A new cultural epoch emerged with Poniatowski's reign, when intellectual pursuit and reasoning received fresh impetus and, at last, the Age of Reason was accepted. In 1773 the first government department for public education in Europe was established. An education commission was also founded to reorganise Cracow and Vilna universities, as well as the supervision of nearly 100 grammar schools. Book printing and the publication of magazines/journals as a result flourished, and the presses in Cracow once more rolled at full speed, but were now geared mainly to a national, rather than international, market.

As with paper, Cracow's glass imports came mainly from Silesia and the wide variety – sheet glass, panes, tiles, bottles, 'small and white glass' – suggests a considerable assortment of products. Initially, some of the glass probably came from Bohemia, imported into Silesia and hence to Cracow. Domestic Polish glass was also purchased for the Cracow market, largely

Table 67. *Cracow: export of pottery from Cracow to Galicia, 1775–1785*

Year	1775	1780	1785
Number of carts	11	60	154

Sources: Wojewódzkie Archiwum Państwowe w Krakowie, *Rachunki miasta Kazimierza*, rkp. 600, 665, 670; *Księgi celne* rkp. 2235, 2238; M. Kulczykowski, *Kraków jako ośrodek towarowy Małopolski zachodniej w drugiej połowie XVIII wieku* (P.W.N.), Warszawa, 1963, p. 128.

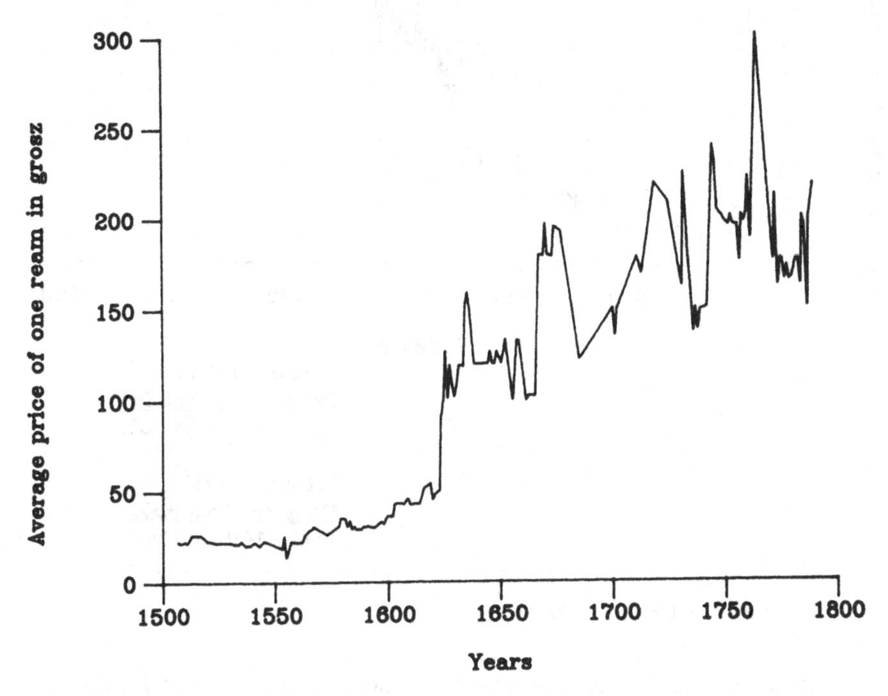

Source: J.Pelc.
Ceny w Krakowie
Table 59; p 67–68.

E.Tomaszewski.
Ceny w Krakowie
Table 60; p 76–77,
Tables 56,57; p 74–75.

77 Price of paper in Cracow, 1507–1791.

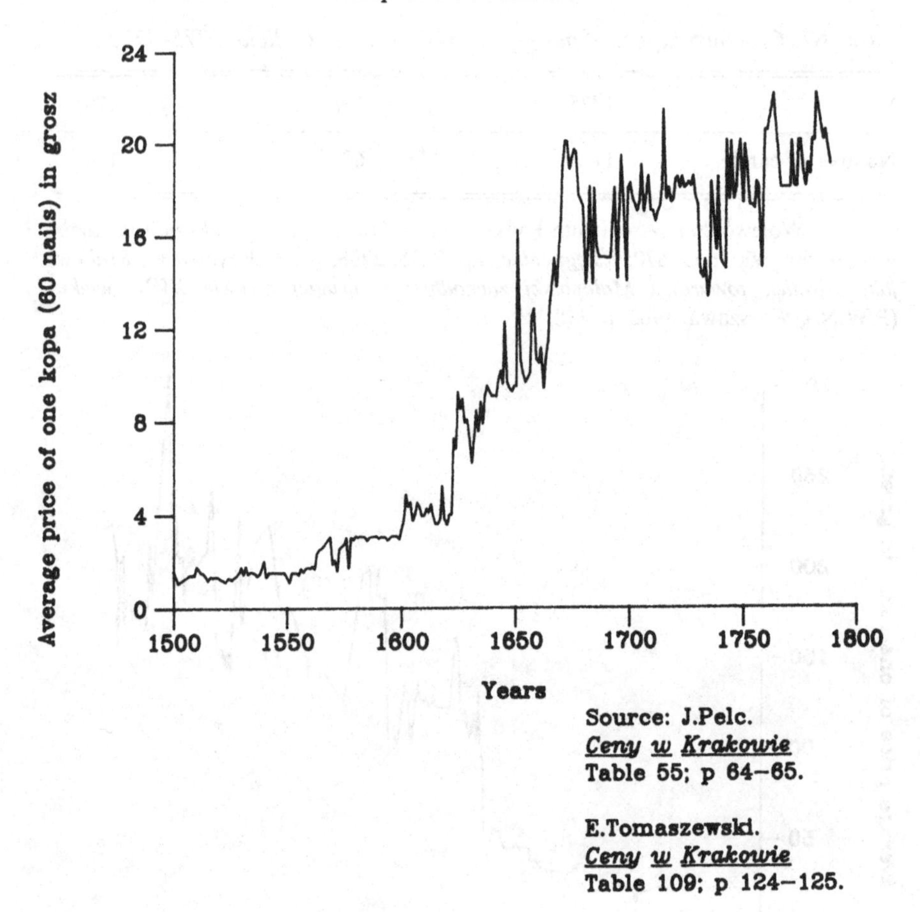

78 Price of nails in Cracow, 1500–1795.

from near Olkusz (Rodaki) but also from more local sources at Bieńczyce, Pądzichów, Retoryki and Biskupie[868] (Table 66). Like paper, Polish glass imports to Cracow grew over the thirteen-year period, providing on average over half the demand between 1750 and 1763; by the latter year two-thirds of the city's glass imports were of Polish origin. For comparison, Poznań's glass imports for these years were 10, 2, 7 and 6 carts respectively.[869]

The ceramic industry also saw lively movement towards the end of the eighteenth century, especially in various pots and pottery objects. The weight factor meant transporting them as far as possible by river; ceramic goods arrived in Cracow from the Vistula's lower reaches and then transferred to carts for further dispatch to Galicia and beyond. This trade's growth is seen in Table 67.

Over the ten-year period a fourteenfold increase was recorded; between 1780 and 1785 there was a one and a half times increase. This fact perhaps reflects growing demand for cheaper pottery in Galicia, one of the poorer parts of Poland at that time.

Lengthy price time-series for handicraft goods were difficult to find for Cracow, not only due to the complexity of products manufactured, but also the lack of useful continuous data over the three-century period. Fortunately information both for paper and nails (Figs. 77, 78) were available and provides some insight into price changes between 1500 and 1795. For both products the sixteenth century was characterised by steady prices until the 1560s; then a new price plateau was reached which continued up to *circa* 1620. The violent devaluation of the grosz suffered during the early 1620s, initiated a series of repercussions which continued for the rest of the century.[870] This phenomenon was experienced by most other goods in Cracow at that time. The period 1660–80 experienced price increases resultant most probably from the aftermath of the Swedish invasions, whilst the eighteenth century had pronounced periods of price oscillation. Most notable of these came in the 1760s, possibly generated by the effects of the Seven Years War in Europe. Certainly Kuklińska noted dynamic price changes for certain goods in Cracow for the period 1750–63[871] including paper, whilst the unstable political conditions during the last quarter of the eighteenth century in Poland are reflected in varied price movements both for paper and nails at this time.

Conclusion

Commercial relations during the period 1500–1795 saw Cracow trying to adapt herself to the changing conditions of the time, and was particularly discernible between the sixteenth and eighteenth centuries. The Turkish occupation of south-east Europe meant a decline in former connections with the Black Sea ports, together with increased competition from Balkan traders. Such events somewhat impeded the flow of more expensive articles to Cracow, like spices, although alternative routes were found via Spain and Holland. As the sixteenth century progressed a gradual reorientation of Cracow's trade took place with greater emphasis henceforth on cheaper raw materials from the central and eastern parts of the European continent. Manufactured goods were also affected, particularly textiles; competition from Greeks, Bulgarians, Armenians and other local merchants in the Balkan markets led to greater concentration on markets in northern Hungary (Slovakia), Transylvania, Silesia, Moravia, Lithuania and Muscovy.

By 1600, Cracow was past its peak as a great emporium, for it no longer helped control the trade turnover of the entire country. Moreover, it did not supervise total exports and imports, accumulate and sell agricultural products, conduct retail and wholesale commerce both in luxury articles and in all basic commodities at a national level. According to Wyrobisz 'In pre-partition Poland only Gdańsk played this role.'[872] Yet Cracow was more than just a large centre which collected the agricultural products of a sizeable region, sold them and engaged in the trade of luxury and basic articles at a provincial level.[873] It has been seen how Cracow had a much wider trading sphere in the sixteenth century in spite of the loss of earlier political significance whilst its reputation as an international route and commercial centre remained for certain products.

The sixteenth and early seventeenth centuries were to see more intensive trade links with Silesia, Bohemia, and northern Hungary/Transylvania; two-way traffic enabled the purchase of raw materials such as minerals and processed goods like wine, in exchange for more expensive west European and cheaper domestic textiles, and the products of metal handicraft. In this trade Cracow played an important transit role. Even the growing menace of Ottoman incursions failed to deter commercial links with that part of Hungary free from Turkish control. Later in the seventeenth century, the Swedish invasion, Thirty Years War, and disputes between Turkey and Poland were to disturb Cracow's trade pattern, but certain commodities, like textiles, continued to flow through the city en route for destinations in Ottoman-held lands.

These two centuries also witnessed a strengthening of contacts with Italy, and a wider choice of goods entering the city from Spain, France, England and the Low Countries. These were predominantly luxury articles (textiles), varied wines, and the more sophisticated products of the West European artisan workshop. For such commodities there was always a ready market in Eastern Poland, Lithuania and Russia. In the opposite direction, raw materials, such as skins, wax, honey and livestock, entered Cracow en route for the markets of Germany and, to a lesser extent, the Baltic Coast. Finally, there was a need to import products for Cracow itself and the surrounding region of Little Poland; there was always a market here for foodstuffs (e.g. cereals, fish), beverages and building materials. By 1700, therefore, Cracow still satisfied certain pre-conditions necessary for international commodity movement in textiles, wax and livestock but, at the same time, developed an increasing regional role in commercial exchange. In this way Cracow slowly adapted to the changing commercial milieu of the European scene.

The eighteenth century was to see a greater intensification of the middleman role at the regional/national level rather than the wider international sphere; indications suggest, however, that Cracow was still more than a provincial trading centre. At the regional level, Cracow's influence extended from the Silesian border to Lwów, and northward as far as the middle Vistula; the southern boundary coincided with the Carpathian Mountains – a natural barrier, whilst to the north there was some trading overlap with cities like Warsaw and Poznań. The First Partition of Poland (1772) divided Cracow's regional market into two, with the city located on the border between them. At once this limited Cracow's former nodal position in local trade. Later partitions were to prove even more depressing, the city situated between Austrian-controlled Galicia and the rest of Poland. At the international level trade contacts also diminished, but some links remained with Silesia, lands now under Austrian and Russian control and even southward to areas of diminishing Ottoman control. Unfortunately for Cracow the halcyon days of intensive long-distance traffic were long since gone. Cracow had lost a great trading emporium, and found herself after 1795 a city under Austrian rule which in turn contributed to her impending decline.

8
Cracow: a final appraisal

This book has attempted to examine the role of commerce and trade links in the development of Cracow from its origins to 1795. The approach has been that of an historical geographer, who has strived to reconstruct the geography of past times through the utilisation of historical data combined with a geographical spatial methodology. This has inevitably involved the use of maps in the cartographical presentation of results. The map is foremost a tool, but is also an expression of thinking in terms of areal distribution. Throughout the work emphasis has been placed on the spatial impact of commercial distribution to try and discover the territorial extent of Cracow's influence on European trade over time, a similar strategy to that adopted by the author in an earlier study of the city-state of Dubrovnik (Ragusa).[1]

The period chosen for this review has stretched from the city's earliest trade contacts through to the end of the eighteenth century, when Cracow came under Austrian control in 1795 as a result of the Third Partition of Poland. Cracow, blessed with an advantageous natural position on the crossroads of several routeways, and a safe haven of refuge, went through the embryonic stages of commercial development between the ninth and thirteenth centuries. By the twelfth century, the city already contained a much-frequented market, and the flowering of its commercial prosperity was to steer Cracow towards European recognition as a significant emporium during the ensuing centuries. This lasted until the eighteenth century, by which time much of its former trading glory had disappeared, whilst the various partitions of Poland during the latter decades disrupted the smooth functioning of even her regional trade.

Significant stages of development

The granting of a Municipal Charter under Magdeburg Law in 1257 was the central factor in the city's general economic development. It came at an important and decisive phase of Cracow's early development and was proof by the mid-thirteenth century of its acceptance as a major urban centre in southern Poland. Favoured by a beneficial geographical position, the city enjoyed the early impetus of commercial contact through its location as a trade route node on roads leading to it from the north, south, east and west of Europe. Three important factors were to dominate Cracow's political situation in the Later Middle Ages; close ties with Hungary and Lithuania, the rise and fall of the Teutonic Order along the Baltic coast, and the growth of Ottoman domination in southern Europe. Such components played their effective role in trade contacts, for there were strong links between politics and commerce.

Analysis of Cracow's commercial development up to 1500 has justified the conclusion that the great bulk of trade was conducted between the more underdeveloped regions to the south and east, such as Slovakia and Moldavia, and the more developed parts of western and southern Europe. Cracow's political contacts with these parts of Europe during the Later Middle Ages were largely dependent on the broader backcloth of wider events.

Whilst Poland's dynastic unions with Bohemia (1293–1306) and Hungary (1370–85) had a relatively short existence, the link with Lithuania through the Jagiellonian House (1385–1572) was of fundamental importance for Polish history and of benefit for Cracow's development. Not least of the successes achieved by this Polish–Lithuanian union was the defeat of the Teutonic State at the Battle of Grünwald (1410), which assured future access to the Baltic coast.[2] On a more continental scale, however, Christian Europe was having little success against the advancing tide of the Infidel, in the form of the Ottoman Empire, which was eventually to restrict Cracow's links with south-east Europe.[3]

The political events of the Early Modern Period after 1500 were also to pose new problems for the city. The discovery of new sea routes, Poland's involvement in the European Wars, continued Ottoman expansion and the territorial ambitions of Sweden, Prussia, Russia, and Austria eventually led to a reduction in the size of Poland. This, in turn, meant a shrinkage in Cracow's marketing sphere. Commercially, the Early Modern Period saw Cracow trying to adapt to changing conditions in the commercial world, especially from the sixteenth century onwards.

Alternative routes for the distribution of expensive commodities in Europe such as spices, meant that the city's merchants had to reorientate their trading activities; greater emphasis was now placed on cheaper raw materials from central and eastern Europe, whilst competition arose, e.g. from Balkan merchants, in those parts of the continent for the selling of imported manufactured goods, like textiles, from western Europe. By 1600, it was becoming obvious that Cracow had passed its peak as a significant European emporium, a role increasingly being taken over by Gdańsk, at the expense of Cracow. Nevertheless, the latter still maintained its significance as a transit trade centre for some commodities on the international market; wax, honey, livestock and its products continued to go through the city from eastern/south-eastern Europe to Germany, whilst some luxury goods from Italy, Spain, France, England and the Low Countries, were sent via Cracow to northern and eastern European markets.

By the eighteenth century, Cracow was predominantly a regional/national rather than an international trade centre. The city's merchants intensified their middleman role at both the regional (largely foodstuffs) and national level, although trade links were maintained with Silesia and some of the former lands of the Ottoman Empire.

Cracow's trading sphere

From the evidence presented in this work, it has been possible to make some attempt to judge Cracow's commercial sphere of influence through links over the European continent. It would probably be unfair to acclaim the city as one of the greatest emporia in Europe during the Later Middle Ages and Early Modern Period; this would place it alongside top-ranked commercial centres like Venice, Nuremburg, Bruges and Antwerp, which would be too complimentary. However, it can be said that Cracow certainly justified selection amongst the second-ranked group of Europe's commercial centres, together with places like Dubrovnik, Augsburg and Leipzig.

Analysis of the commodities handled by Cracow's merchants and the markets they served has allowed the mapping of places connected to the city through trade. In keeping with the two major periods under discussion, namely 1390–1500 and 1500–1795, it has been possible to locate those centres mentioned in Cracow documents and in other archives, and place them against a European background. Furthermore, by means of simple schematic presentations, it is hoped to present Cracow's market within the framework of East–West European trade during the two epochs and illustrate graphically the circulation of major commodities entering/leaving its emporium.

(a) Cracow's trade in the Later Middle Ages

Whilst emphasis throughout this work has been on Cracow's international transit trade, it should not be forgotten that it was also an important local and regional centre for Little Poland during the Later Middle Ages, which had a population estimated on the basis of Peter's Pence at between 250,000–300,000 inhabitants in the first half of the fourteenth century.[4] Much of the local and regional trade probably never entered the confines of contemporary recorded data, due to its limited value or restricted commercial significance; agricultural produce and livestock must have featured largely in this trade to support the 12,000–15,000 residents of the capital at this time.

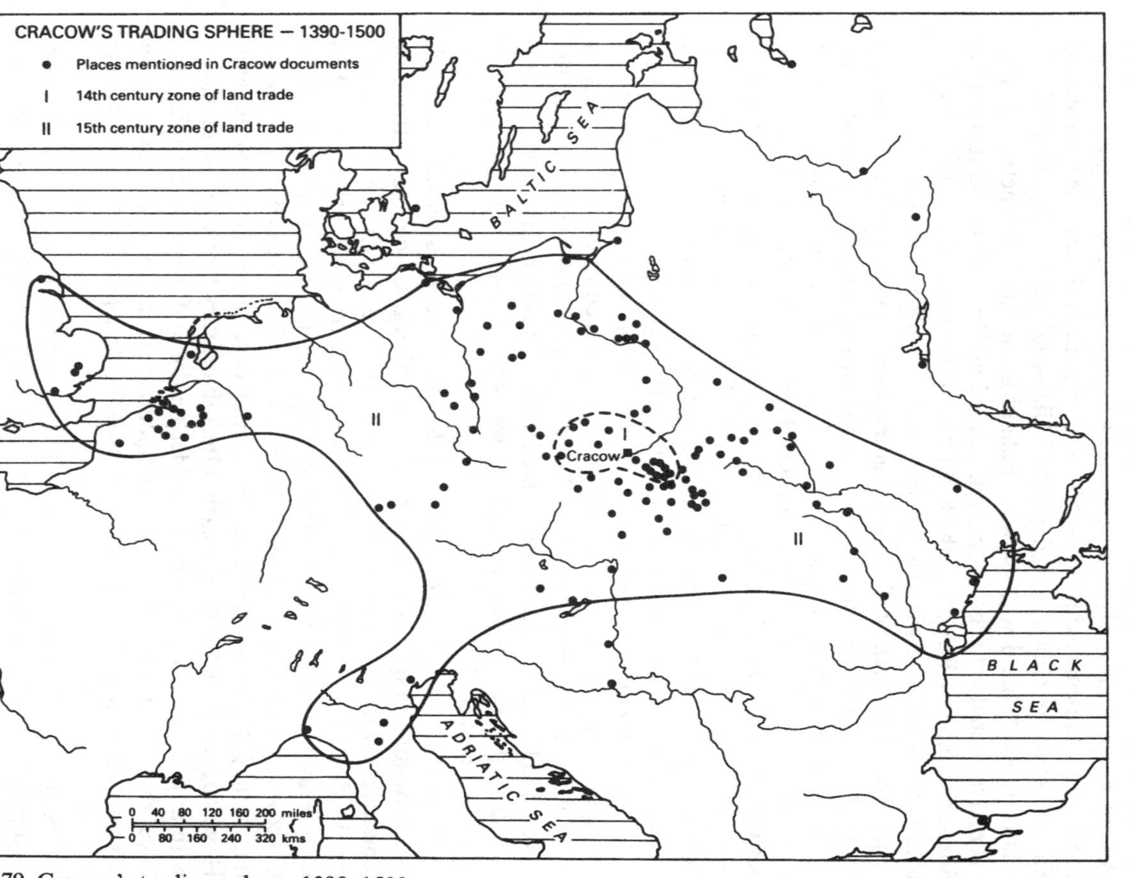

79 Cracow's trading sphere, 1390–1500.

As recently stated elsewhere 'the trade of Cracow, and of many smaller towns in Little Poland, was based primarily on close contacts with Hungary. In spite of this, however, it would be an exaggeration to describe, as is traditional in Polish studies, the long-distance trade of the towns of Little Poland as an exclusively transit trade.'[5] Mapped evidence of places mentioned in documents from 1390–1500 (Fig. 79) reflect these close ties with the Slovak areas of Hungary and part of Silesia (Zone I) in contrast to the relatively weak bond with parts of Poland immediately to the north. This may have been due to the strong links forged by Cracow merchants with the Silesian capital of Wrocław which permitted them to engage in commodity transactions between western Europe and the Black Sea region.

The restricted area of Cracow's trading sphere in the fourteenth century (Zone I) is in great contrast to that of the ensuing century (Zone II). The city's commercial links spread from the towns and cities of Eastern England to the shores of the Black Sea and heralded a period of intense economic activity for the Polish capital. The exploitation of natural resources, especially minerals, encouraged the general economic expansion of Little Poland coupled with the extension of a functional pattern of long-distance trade routes across east-central Europe. Moreover, they linked up with ports in the Lower Don and Volga valleys, and Black Sea coast which formed termini for the major caravan trade routes, connecting Europe with Persia and China. The foresight of the Polish king Kazimierz Wielki (Casimir the Great, 1333–70), in conquering the lands connecting Poland with south-west Russia in the mid-fourteenth century, was still of immense importance for Cracow a century later.

Westwards, the long-distance trade of Cracow stretched through Wrocław and Prague to Regensburg, Nuremburg, Cologne and the Low Countries, with its major port of Bruges. This was reached either on this overland route or northward by land and then from Gdańsk by sea. Ties with Flanders were strong even during the fourteenth century when Cracow merchants and burghers were entrusted on occasions with carrying money to Bruges collected from papal levies in southern Poland. However, it was not until the early fifteenth century that Cracow managed with any success to penetrate the wider markets of north-western Europe, once the threat of commercial competition from the Prussian city of Toruń had receded, and Prussian merchants were prohibited from trading in Russia.

The papal connection also stimulated links with Italy. Even in the thirteenth century Italians had been in contact with Little Poland for papal tax collection purposes, but it was along trade routes from the Black Sea that merchants from Florence, Venice and Genoa visited southern Poland, some taking permanent residence in places like Cracow. By the fifteenth century their influence was clearly felt in Little Poland, not only through trading

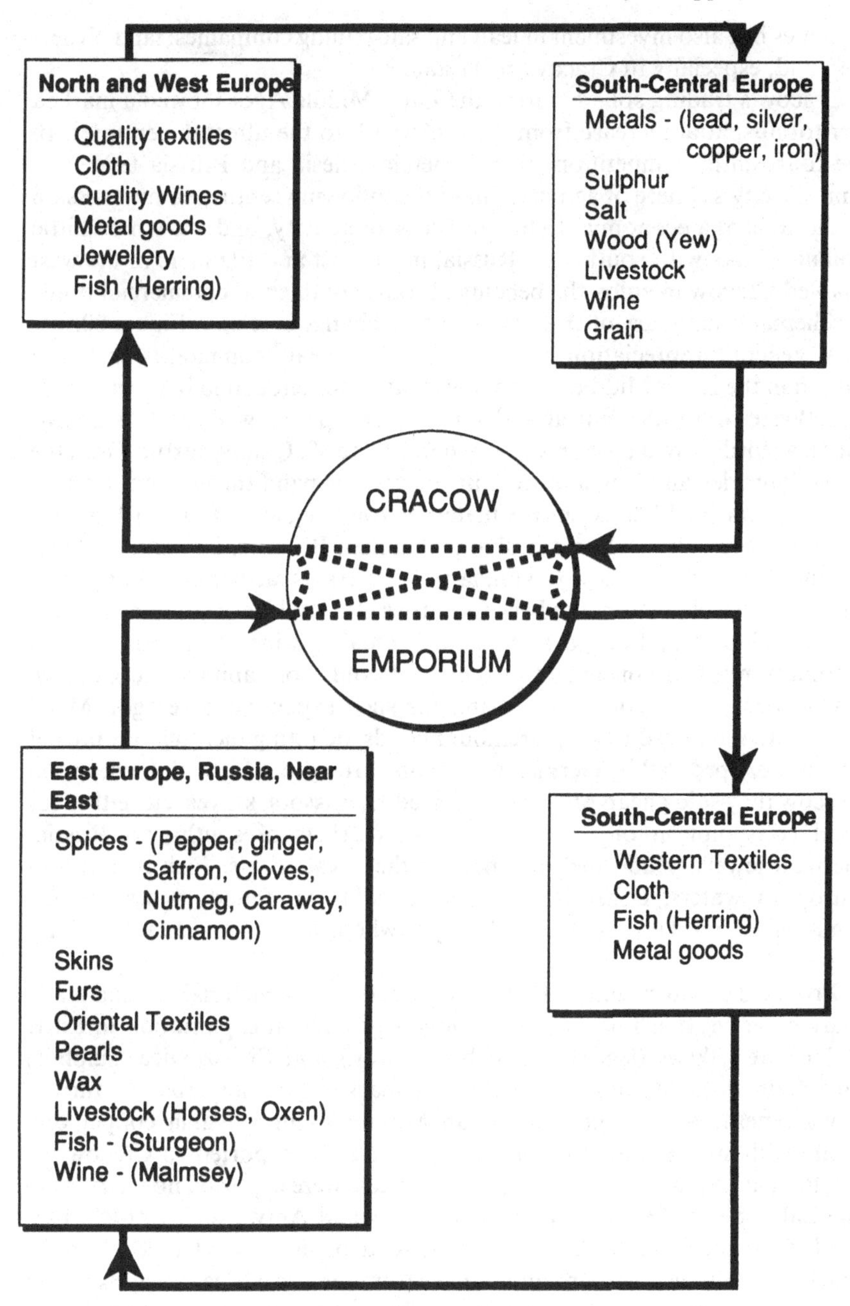

80 Main commodities circulating between East–West Europe via Cracow in the Later Middle Ages.

activities but also investment in lead and salt-mining companies, land ownership and, especially in Cracow, real estate.[6]

Cracow's trading sphere during the Later Middle Ages showed a marked territorial/spatial increase from the fourteenth to the fifteenth centuries. In the fourteenth, competition from Bohemia, Silesia and Prussia tended to limit the city's sphere of activity, but in the following century some recession in the Bohemia economy, decline in Prussian activity, and growth of Little Poland's links with south-west Russia, in the east and Flanders in the west enabled Cracow to enjoy the benefits of trans-continental commercial trade. A schematic summary of this commodity exchange is seen in Figure 80.

A general appreciation of Cracow's European commodity-exchange system in the Later Middle Ages shows that goods circulated between North and Western, South-Central and Eastern Europe (as well as Russia/Near East), according to a pattern suggested in Figure 80. Quality textiles from the Low Countries and England were in constant demand throughout much of Europe, and by 1400 expensive material from England (Kersey) Flanders (Ghent, Ypres, Bruges), and Brabant, (Malines, Brussels, Louvain etc.) was finding its way via Cracow to Hungary and markets farther east. Lengths of less expensive cloth from the Low Countries also could easily find markets in south and central Europe at this time. Quality wines from France and Germany reached consumers in the royal courts of capitals like Cracow, where there was a constant demand for such expensive beverages. Metal wares, often referred to as 'Nuremburg goods' denoting their main source of origin, were peddled by German merchants throughout Little Poland, and in Cracow they sold cheap products like needles, scissors, knives, etc. either for local consumption or the more distant markets of south-west Russia. German jewellery adorned members of the royal courts. Fish from north European waters, either in dried or salted form, found its way to the land-locked markets of central Europe, where herrings were particularly popular.

Products of south-central Europe, particularly raw materials, found ready markets in western Europe. The products of mineral exploitation in Little Poland at Olkusz (lead), Częstochowa (iron) and Swoszowice (sulphur) and their close proximity to Cracow's market gave impetus for further development. Across the Carpathian Mountains in Slovakia copper ore, along with smaller amounts of silver, was also transported to Cracow for smelting and resale. From the city these metals were exported northwards to the Baltic Sea for trans-shipment to Bruges and Antwerp. Livestock, particularly oxen from Moldavia, were herded on hoof via Cracow for the markets of Silesia and Germany; timber from the Carpathians was exported to Flanders, whilst the famous yews from these mountains found sure markets in England for long- and cross-bow archery manufacture.

More local/regional trade was found with salt, wine and grain. Salt from

Cracow's nearby mines at Wieliczka and Bochnia went via the city to outlets in Hungary, Silesia and northern Poland; wine from Hungary had little international significance until the sixteenth century, whilst grain (wheat, barley, rye, oats) was mainly for local consumption.

The lands to the east of Poland provided various raw materials and oriental products. Foremost amongst them were the spices obtained in south-west Russia via the caravan routes from the Near East. Pepper, saffron, ginger, cloves, nutmeg, cinnamon and caraway seeds were invaluable for improving the central and west European cuisine. Oriental silks, cottons, and pearls also appealed to the wealthier west European consumers, whilst the taste for malmsey wine from Crete found favour in Cracow's royal court.

The plains and forests of Eastern Europe and Russia provided Cracow merchants with a variety of products. Skins and leather goods from livestock, especially cattle and sheep, vied in importance with expensive furs from such animals as lynx, marten and fox. The forests of north-east Europe and Russia were noted for their quality beeswax, whilst tallow was obtained from cattle. Herds from Moldavia and Podolia reached Cracow via the cattle fairs e.g. in Lublin, and Transylvanian horses came via Lwów. The Black Sea was a source of sturgeon for the Polish royal court.

Western textiles proved the greatest attraction for Eastern Europe and Russia, Cracow merchants usually sending them through Lwów. From here they were exported, along with varieties of cheaper cloth, to amongst others, the distant markets of Moldavia and Multenia. Metal goods and fish (herring) also found their way to these eastern markets.

By the end of the fifteenth century, therefore, Cracow's traders were providing a valuable middleman's service. They supplied the more economically advanced countries of western Europe, particularly the Low Countries, with much-needed raw materials like minerals; at the same time, in the opposite direction, western quality goods were sold to the less-developed regions of east and south-east Europe.

(b) Cracow's trade in the Early Modern Period

In Europe, the sixteenth and first half of the seventeenth century experienced a pronounced development of overland trade resulting from increased demand and hence higher volume, accompanied by closer ties between contiguous regions, provinces and states. This greater volume of trade meant increased circulation of goods at state and regional level within east-central Europe, as well as continuing bonds between western Europe and lands farther east. The sixteenth century was a time of flux, when overland transcontinental trade mixed increasingly with maritime commercial expansion. The following century, and particularly after 1650, maritime trade began to excel at the expense of transcontinental commerce; this lost ground was to be felt in the Cracow market place, together with commercial damage suffered from Ottoman control of former trading outlets in southern Europe.

Yet another factor was the growing disquiet within Poland itself. As early as the third decade of the seventeenth century there were signs of a deterioration in the Polish economy, largely resulting from a decline in the Baltic grain trade, and a slump in internal commerce and urban growth. For Cracow, the loss of its capital city status to Warsaw early in the century, and decline in central European copper mining in the face of Swedish competition, were also critical events. Further blows were suffered in the 1650s and 1660s; wars led to commercial disruption of the Vistula river trade, particularly resulting from the second Swedish–Polish war of 1655–60; again Cracow, recovering from a visitation of the 'black death' in 1652, was devastated three years later when the Swedes captured and looted the city, an incident that was to repeat itself in 1702.

The eighteenth century fared no better for Cracow. The collapse of Poland's military establishment evident in the Great Northern War (1700–21) signalled the growing power and expansion of Russia at the expense of Polish territory. Poland's armies were of little significance in disputes featured in the Wars of the Polish (1733–35), and Austrian (1740–48) Successions, or the Seven Years War (1756–63), yet these conflicts were partially fought out on Poland's terrain. Russian influence over Poland's eastern territory grew and, during the reign of Stanislaw August Poniatowski (1764–1795), he suffered the ignominy of three partitions, the last committing Cracow to Austrian control.

With the advent of the Early Modern Period, mapped evidence of places mentioned in documents between 1500 and 1795 (Fig. 81) reveals how the city's trading sphere became even more extended during the sixteenth and seventeenth centuries (Zone III). Commercial contacts were established and commodities sent via Cracow as far as London in the west, Moscow in the north-east, Naples in the south-west and the Danube delta in the south-east. Thus, in spite of a change in emphasis from predominantly expensive goods to their cheaper unprocessed counterparts, the city's commercial sphere of influence was greater than it had ever been in its history. Cracow's importance as a long-distance transit trading centre can therefore be appreciated from this extensive distribution of markets having trading links with this Polish city. Perhaps the greatest change from earlier periods was by extension eastward into Russian markets, stretching through Belorussia and Lithuania right to the heart of the old Muscovy capital. Southward penetration into the Balkan peninsula seems to have been somewhat restricted, largely due to competition from Greek, Levantine, Ragusan (from Dubrovnik) and other merchants operating in this area.

In a westward direction, the long-distant trade extended through

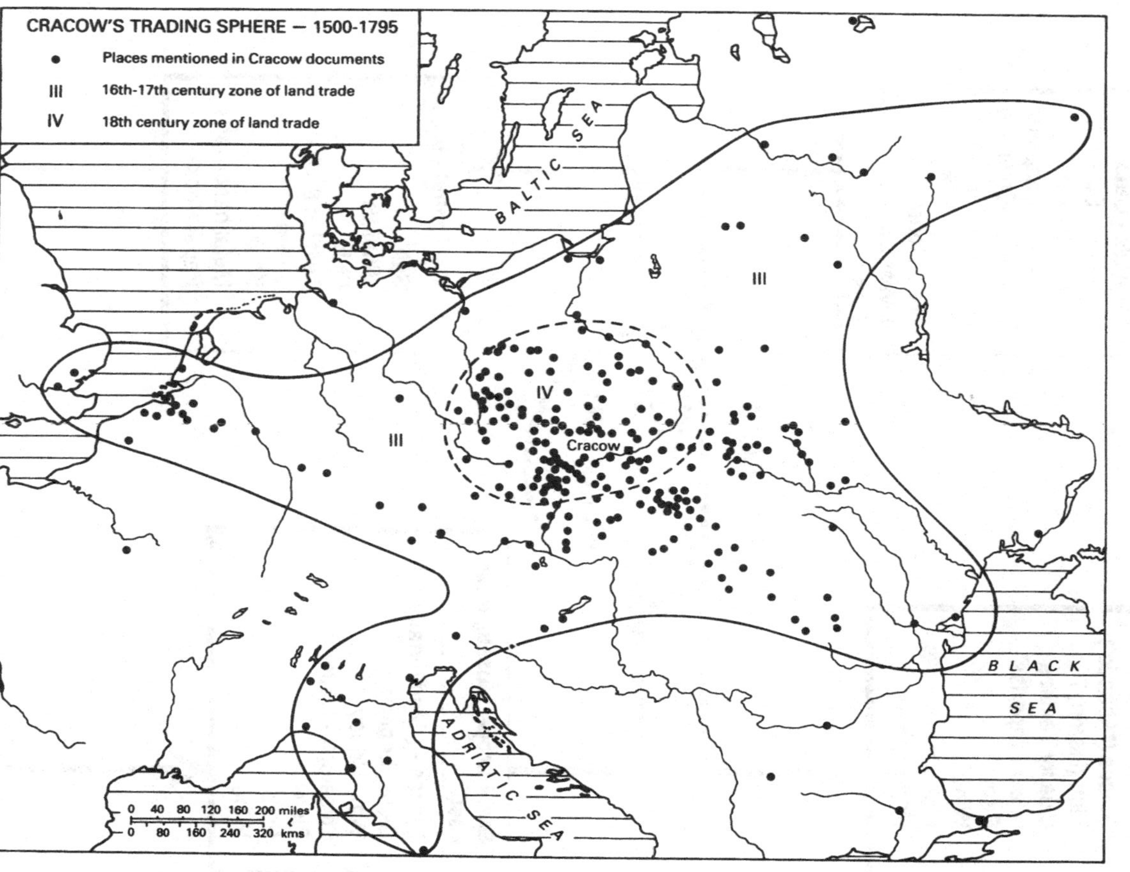

81 Cracow's trading sphere, 1500–1795.

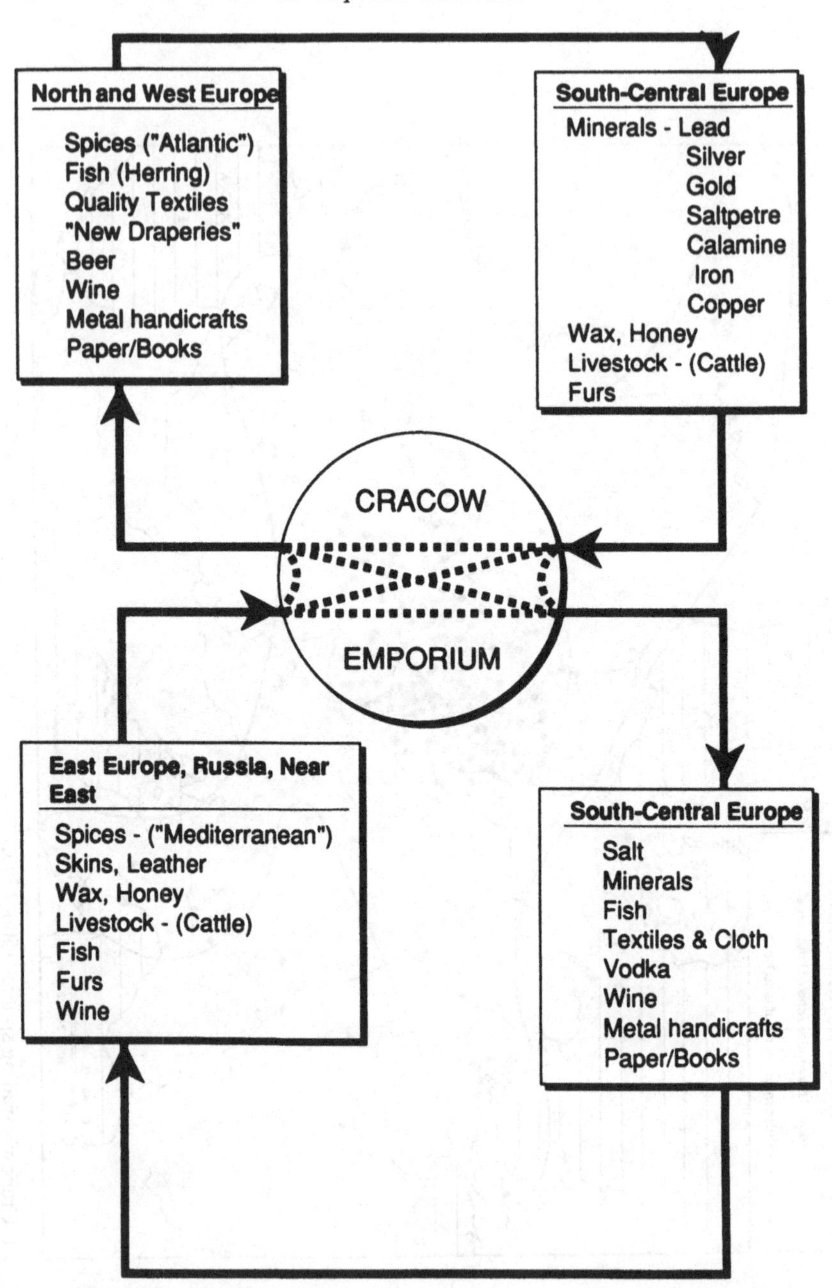

82 Main commodities circulating between East–West Europe via Cracow in the Early Modern Period.

Germany to a cluster of markets on the North Sea coast, including Bruges and the rising centres of Antwerp and Amsterdam, together with London and Colchester in England. South-west links with northern Italy remained strong, especially Venice – the queen of the Adriatic ports. Finally, it should not be forgotten that Cracow continued as a major commercial centre for its own immediate hinterland and most of Little Poland.

As the seventeenth century progressed, inevitably the toll of war and changing economic conditions over the European continent meant some contraction of this trading sphere during the eighteenth century (Zone IV). Much greater emphasis was now being placed on the local, regional and national markets, rather than the international scene. Central and southern Poland dominated the trade pattern, although links with Silesia, Moravia and Slovakia continued beyond the Polish borders. As the eighteenth century wore on, Little Poland tended to overshadow other areas in Cracow's marketing sphere, although there was still some longer distance trade in certain commodities, e.g. alcoholic beverages, with areas to the south over the Carpathian Mountains. For Cracow, northern Poland was noticeably an area of paucity in markets, largely due to the dominance of merchants from Gdańsk and Szczecin who controlled the organisation of commerce in this area.

The pattern of Cracow's European commodity exchange system in the Early Modern Period is seen in Figure 82. While the pattern in the variety of goods exchanged is somewhat similar to the Later Middle Ages, there was a growth in the volume of trade during the sixteenth and first half of the seventeenth century, due to greater demand. This escalation in commercial relations was noticeable not only between western and east-central Europe, but also between the differing states and areas of central and eastern parts of the continent. Some commodities bought in eastern Europe, such as furs and skins, were acquired for local consumption whilst en route to western markets. On the other hand, domestic cloth production expanded in central Europe not only to satisfy local/regional needs, but with a surplus for markets farther east. There was also a growth in the number of fairs, not only at local, but also regional and international levels.

Commodities sent from western Europe were dominated by textiles; these consisted of the quality cloth, mixed stuff, linen, silk, etc., as before, as well as ready-made products but on a much smaller scale. There was also a shift away from the more traditional older heavy-quality woollen products, to the so-called 'New Draperies' which consisted of a variety of lighter articles such as bays and fustians. Technically, these contained less wool, and fulling was unnecessary and demanded less skilled labour, resulting in a cheaper product with lower transport costs. Spices, such as pepper, were always in demand; however, in the sixteenth century both the 'Mediterranean' (i.e. spices transported via the Levant) and 'Atlantic' (conveyed by sea via

Lisbon) competed for predominance; in the seventeenth century the 'Atlantic' route came into its own centred on England and the Low Countries.[7] Spices then continued their voyage via the Baltic to Gdańsk and down the Vistula River to Cracow. Quality wines and beer also came this way, as did fish, mainly herring from the northerly waters of the Norwegian coast. Other products included a wide variety of metal handicrafts, particularly knives together with other industrial products such as glass, paper, books and dyes.[8]

In the opposite direction products from south-central Europe found their way via Cracow to the west European markets. Whilst Cracow was never a significant grain market, precious metals such as silver, copper and lead were important. Of these, copper proved the major driving force for trade during the sixteenth century especially under the entrepreneurial expertise of Jakob Fugger and his family. Demand for copper, from Antwerp in particular, came not only for coinage but also countless other manufactured articles. Fugger's choice of Cracow as a major smelting and transit centre for Hungarian copper en route to western Europe was to have a significant effect on the city's trading fortunes. Cracow was also situated on the main cattle route from eastern Europe to markets in the west, particularly centres in west and central Germany, e.g. Frankfurt, all dependent on east European meat supplies. Furs, skins and leather goods, particularly from Russian sources, were in constant demand in western markets, eventually arriving via Cracow and other Polish market centres at international fairs like Leipzig and Frankfurt. The demand for wax and honey followed similar routes.

South-central Europe also had commodities to offer the eastern markets. Textiles of varying types remained the major article of trade; besides west European kinds there was now a growing domestic production in places like Bohemia, Silesia, Moravia and Greater Poland, all of which entered the Cracow emporium. This helped to supply the Russian, Hungarian, Moldavian and Wallachian markets where domestic cloth production was lacking. Demand for wine from southern Europe, and especially from Hungary, could be found in both the eastern and western markets of Europe, whilst Polish vodka had a more traditional outlet to the east and south. Salt from the Cracow district was distributed farther eastward although most was sold within Poland. Some of the various industrial goods imported into the city from western Europe were placed, after local needs had been satisfied, in transit for destinations as far away as Moscow.

The fairs in Poland also helped distribute goods from Russia, Lithuania, Belorussia and the Ukraine towards western markets. Cracow along with centres like Lublin, Wrocław and Poznań, figured largely in this trade, particularly for types of skins, leather goods, wax, honey, furs, fish and wine. The transport of livestock from Moldavia, Wallachia and the western Ukraine arrived in Cracow from Lwów, from where it followed the well-trodden path to Wrocław and thence to Germany. 'Mediterranean' spices from the Levant also came from the Black Sea up to Cracow via Lwów, but in ever-decreasing amounts due to Ottoman incursions in the region, and Cracow had to augment its supplies more and more from Venice and the Low Countries.

This state of affairs became increasingly disrupted towards the end of the seventeenth and eighteenth century, when the exchange system seemed to break down due to external factors. Swedish incursions into Poland upset the network of international fairs and the trade centres like Cracow that depended upon them; the supremacy of maritime trade over transcontinental links, largely through lower transport costs, favoured ports like Gdańsk at the expense of places like Cracow deep inland. Alternative sources of meat from Denmark for the German market impeded Cracow's position on the cattle route from the east; the traditional trade in furs, skins and leather goods from Russia began to be threatened by imports from the New World, e.g. Canadian furs, coupled with better livestock husbandry in western Europe providing improved qualities of skins and leather. Precious metals from America (silver) and northern Europe (Swedish copper) led to an even deeper recession in Cracow's commerce, forcing the city to pay even greater attention to its markets in the immediate hinterland and surrounding region.

The role of prices

The wealth of material in Cracow's archives has also allowed some interpretation of price changes over the period under review. Here they have been utilised mainly as a means of giving some idea of the various commodities sold on the Cracow market and their local significance, and not as an indicator for wider interpretation on a European scale. Whilst prices and wages within Cracow acted as a sort of temperature gauge for economic conditions within the city, they did not act in a vacuum and were subject to factors outside the city's control. Just like other commodities, labour had its price, which also reflected the subtlety of supply and demand. As Pounds has stated, 'Medieval economic opinion, as a general rule, regarded the just price and the just wage as those which were arrived at as a result of open competition in the market.'[9]

Under medieval conditions, controlling wages and prices was often bedevilled with difficulties, and efforts to stabilize them were usually restricted to times of disaster linked to bad harvests or the visitation of the plague and other epidemics. More reliable data on wages in Cracow can be seen in longer runs in the Early Modern Period. For example, here those for carpenters and bricklayers (Fig. 83) are illustrated for the critical period from the mid-sixteenth to mid-seventeenth centuries. Average daily wages for these two skilled employees showed a similar pattern to that experienced by price changes over this period, namely rapid increase noted after 1610,

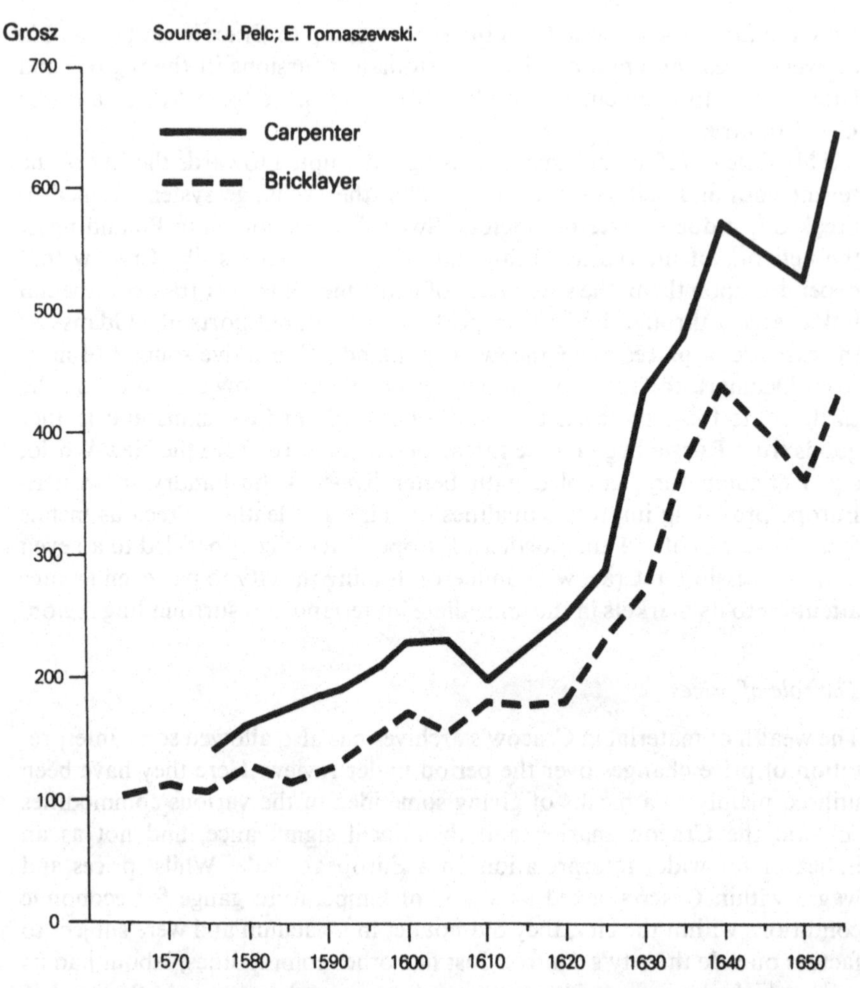

83 Average daily wage of certain construction workers in Cracow, 1565–1651.

when inflation in the city, as elsewhere in Poland, became a dramatic event in the everyday lives of Cracow's residents. Nevertheless, as Knoll points out, the dispersed and incomplete nature of the data on costs and income suggest caution in their interpretation.[10]

In spite of such obvious weaknesses, some useful runs on prices of various commodities for sale in the city have been included at the end of each commodity section discussed in the various chapters. Whilst by no means complete, they do give some information on the trade value of particular merchandise over time; certainly when dealing with more inelastic goods such as foodstuffs and other agricultural products, price fluctuations can usually be regarded as a useful measure for the adequacy of supply. Manufactured and luxury goods were more elastic in demand and could easily be dispensed with, particularly if their prices rose too sharply, leading to fewer sales and the rapid restoration of more realistic price levels. Overall, most of the prices analysed on the Cracow market tended to follow the general changes within Europe, although often characterised by a 'delayed action' compared with price differences in the more westerly parts of the continent.

The basic monetary unit in the city was the 'grosz', the Polonised name for the silver 'grossus', originally introduced into Poland during the reign of Kazimierz Wielki (Casimir the Great 1333–70), as the 'grossus Cracoviensis'. From the fourteenth to the sixteenth centuries, they were the most predominant and sought after silver coins and included all kinds of 'grosze', often minted in Hungary, Bohemia, Germany and Pomerania, and were easily recognisable in international trade.[11] Throughout the fourteenth century the 'grosz' proved a stable currency but the ensuing century saw its value decline, as did prices in general during the second half of that century. Research done on the history of prices in Cracow in the Later Middle Ages by Pelc, tends to suggest that price levels for manufactured goods remained high during the second half of the fourteenth century, but fell by a fifth in the following century.[12] This may be partly explained by the large imports of foreign merchandise into Cracow which may have exerted some influence on the drop in prices, but more research is needed on this problem before such issues can be resolved. Overall, however, the picture is one of a stable economy in Cracow during the Later Middle Ages.

This situation was to contrast sharply with the price eruption associated with the upward turn during the later years of the fifteenth century, which was to become the price revolution of the sixteenth century, and ascribed to the inflow of silver from the New World. Some of this change must, however, be attributed to the rising output from the mines of central Europe, although it should be remembered that not all goods responded to the same degree, with agricultural commodities experiencing sharper rises than many other goods. The violent devaluation of the 'grosz' suffered in Poland during the early 1620s helped initiate a series of repercussions which continued for the rest of the century, and was experienced by most goods in Cracow at that time. The price rises during the period 1660–80 were most probably the outcome of the Swedish invasions and their aftermath, whilst the eighteenth century had pronounced periods of oscillation. Most notable of these came in the 1760s, and was possibly induced by the effects of the Seven Years War in Europe. Finally, the unstable political conditions in Poland during the last quarter of the eighteenth century were reflected in the varied price movements suffered by most types of merchandise.

Cracow – a great trading city?

In conclusion, one may well consider whether Cracow was a great commercial emporium, but this depends on the attitude one adopts towards its evaluation. It should be stated that Cracow's trading successes were dependent upon, and a reaction to the economic and political conditions prevailing in central and eastern Europe at certain times. Inevitably, Cracow's future was closely connected with Poland's own destiny; for example, Topolski has intimated, with reference to the period of the sixteenth–eighteenth centuries, 'two essential elements may be isolated from among the causes of economic decline in Poland: the destruction due to wars, and the system of farming in the larger properties based on self-labour and on the export of cereals'.[13] Similarly, Małowist has commented on the problem of inequality of European economic development in the Later Middle Ages and concludes that, 'From the thirteenth century there is to be observed in eastern Europe a dynamic development, but one which is not equal in all regions . . . The weakness of the towns was an important factor', and from the fifteenth century onwards, 'Urban industry, which was undeveloped, struggled against the competition of foreign products and the difficulties caused by the smallness of interior markets.'[14]

Cracow had to operate within this framework and its limitations. Its early development was generated by a favourable geographical location which facilitated the growth of trade at the junction of major trade routes. It became the main urban centre in Little Poland, which entered a phase of quickening development in the thirteenth century and furthered by its choice as the state capital yet more enhancement came in the fourteenth century with the accelerated pace of commercial growth in Europe. Its location as a node on two major long-distance trade routes was paramount in this advancement; the east–west road operated from the vast Russian plain through Ruthenia via Cracow to Wrocław in Silesia and on to southern Germany; the north–south road stretched from the Baltic through Cracow down to Hungary and the northern Balkans. This route became increasingly important after the mid-fourteenth century following the demise of the Teutonic Knights in 1466 at the termination of the Thirteen Years War. From now on the River Vistula flowed through Polish territory from Cracow via Toruń to Gdańsk, enabling at a national level increased trade between Little Poland and Greater Poland (especially Poznań), as well as more international links with western Europe, and Lithuania, factors which could only strengthen and intensify Cracow's commercial significance.

During the fifteenth century its trading importance was heightened still further thanks to its advantageous geographical situation. Route control was the key factor. Cracow lay on the main road from Wrocław to Ruthenia, and by forbidding foreign merchants access through the city, its merchants could more or less monopolise merchandise being carried in an easterly direction. If it so wished, the Cracow authorities could also compel foreign merchants travelling either eastward from Silesia, or westward from Ruthenia to Greater Poland to sojourn in the city and, through stapling rights, force them to sell their goods in the city, all adding fiscal gains to the municipal coffers. Finally, Cracow's crucial position on the route from Hungary via the Vistula river, Baltic coast and hence to the Low Countries for Slovak copper supplies proved vital. The Cracow city council's masterstroke came with its insistence on the warehousing and refining within the Polish capital, of all Hungarian copper ore travelling westward to the Baltic. This gave Cracow a monopolistic position in a prized commodity of international trade, which was later enlarged to include all Hungarian products transported through the city. The financial spin-off from all this international trade through customs duty, taxes and rented accommodation helped build up considerable local fortunes, benefit the city's infrastructure, and spread the word abroad of Cracow's commercial significance.

Cracow's days as a noteworthy European emporium began to wane towards the end of the fifteenth century, and by 1600 it was clear that some deterioration had taken place in its European trading role. The new conditions operating during the Early Modern Period in western Europe, and characterised by a shift of commercial emphasis from the Mediterranean to the North and Baltic Seas, meant Cracow lost its dominant position as a centre of international commercial exchange. The Hanseatic League ceased to be the great trading and political bond between the Bay of Finland and the British Isles, a change which was also felt in Cracow. The city had to reorientate its commercial undertakings under these changed conditions with new trading contacts in the European mart. This it managed to do, except that now its merchants faced much stiffer competition from other transit trade centres on the European circuit.

In spite of losing its capital city status to Warsaw early in the seventeenth century, Cracow was still part of the Polish–Lithuanian Commonwealth, a major intermediary between east and west Europe, during its 'golden' period under the Jagiellonian dynasty. This time, however, Cracow's geographical position was very peripheral to the centre of activity along the Vilna–Warsaw axis. The end of the sixteenth and beginning of the seventeenth century saw this commonwealth experience some recession (the so-called 'silver' period), when its economic and political situation worsened. By the mid-seventeenth century this state had embarked on its 'iron' period, troubled by conflicts with Moscow, Sweden and Turkey whilst, closer to home, Cracow was subjected to the incursions of the Swedish wars. Polish territory further succumbed to various European conflicts in the eighteenth century, including the western extension of Russian territorial advances, all at a time when Cracow could no longer be considered as anything but a local and regional trading centre.

Given all these difficulties, judging whether Cracow was a great trading emporium also demands the need for further comparison. Wyrobisz, in his attempt to classify the functional categories of Polish urban centres in the sixteenth–eighteenth centuries, states that, 'The peak of the hierarchy of the trade towns was composed of the great emporia through which was concentrated the trade turnover of the entire country, and which supervised total exports and imports, accumulated and sold agricultural products, conducted retail and wholesale trade both in luxury articles and in all basic commodities. In pre-Partition Poland only Gdańsk played this role.'[15] Cracow was relegated by Wyrobisz to a lower rank of urban centre, along with Poznań, Warsaw, Vilna, Lwów etc., where their main trading function was similar to that in the previous category, but at a provincial level.

In isolation it is difficult to estimate Cracow's full impact on European trade. This requires many detailed further studies and attempts at generalisations on other cities of similar commercial standing. For example, the trading spheres of Nuremburg, Hamburg, Augsburg or Leipzig could be studied spatially in the historical perspective. Results could then be compared, and the real geographical effects of such topics as the New World discoveries and the impact they had on the changing trade routes of central European and Baltic commerce could then be more accurately measured. This sort of work has been hampered by too few basic studies of this type in the field of urban historical geography. Even in urban history, as Buchkovitch pointed out some years ago 'the Polish historians suffer from being a bit too far ahead of some of their colleagues in other countries. Particularly in the case of Germany, where in most cases urban history stops in 1500, work comparable to that done in Poland does not exist.'[16] Even so he complains that, 'Polish historians have not in all cases taken advantage of what does exist. Often the result is that they combine an enormous erudition with a certain narrowness in the circle of ideas, a phenomenon that is partly inevitable in view of the state of the urban history in Central Europe'.[17]

To ask the question what makes a city great one often lacks the ability to say exactly why one urban centre may be chosen in preference to another. The whole city phenomenon takes on irrational qualities when one considers in depth the values of urban versus rural life. Yet, despite everything, the truly great city is the material for legends and stories and is a place with an eradicable fascination.[18] Certainly, Cracow is a city of legends and appreciation from afar, as witnessed by its elevation to top priority for conservation by UNESCO. It was Poland's leading city up to the early seventeenth century when its mantle of power as the country's capital was transferred permanently to Warsaw. Cracow also contained the basic amenities that made life tolerable, not only for the wealthier élite, but also for more ordinary residents in the city. It was a place of worship and centre of religious activity, through the cathedral on Wawel Hill, and its extensive bishopric in the surrounding region; it provided, therefore, some spiritual attraction for those who migrated to the city for permanent residence. The city's university proved a magnet for students not only from Poland but also from abroad, who wished to further their interests in culture, science and education; not until the second quarter of the sixteenth century did it begin to decline in importance, a trend that was to continue during the ensuing century and part of the next. Finally, from its varied and interesting history one can see that Cracow was never a dull city to reside in; a constant flow of visitors from both home and abroad, the attraction of the Royal Court, and the display of numerous goods on its market stalls must have been appealing. It was also a city of culture, especially after the influx of Italians into its midst, who brought with them new ideas and influences from a country that was then one of Europe's most advanced areas of civilisation.

Given all these advantages and indicators on the necessary demands of the great city, the question begs itself as to whether Cracow was a great *trading* city. The ideas put forward here can only be regarded as tentative. Work on markets and trade in all parts of the European continent is gradually being extended and our knowledge slowly becoming more complete, but there still remains a need for more detailed field studies and comparative work in other parts and places of this land mass before more profound judgements can be made. There is obviously a need for a more conceptual general framework to be developed, which would consider the complexities of such matters as market institutions, time–distance factors and other significant influences. Thus, without such a framework the study of these individual cities and their commercial importance are in danger of being a series of mere descriptive studies. Here attempts have been made not just to present the bare historical facts about the city, but to place it under a geographical microscope which has taken account of its trading activity through historical time, and of which some of the remnants may still be seen in the city today. Certainly, Cracow was and still is a great city, but perhaps during the Middle Ages its political and administrative significance as the country's capital tended to overshadow its celebrated commercial prominence. After it lost its function as Poland's capital city the peak years of its mercantile development also diminished. Cracow's former commercial glory suggests it was a 'child of its time', which like many other events in history experienced a particular combination of favourable factors. According to F. Scott Fitzgerald, until people are known by numbers alone, the great city will continue to exist.[19] He was referring of course to Manhattan. He may well have been speaking of London, Paris or even Cracow, where former commercial glory can still be seen through the architectural wealth and spacious market place of the contemporary city.

Streszczenie
Handel a rozwój miast w Polsce: geografia gospodarcza Krakowa od początków miasta do roku 1795

F. W. CARTER

Praca F. W. Cartera poświęcona jest badaniom roli handlu i stosunków handlowych w rozwoju Krakowa od jego początków aż do roku 1795. W całej pracy położono nacisk na analizę znaczenia przestrzennego rozmieszczenia handlu w celu określenia geograficznego zakresu wpływów na rozwój miasta na przestrzeni wieków. Okres, któremu poświęcono badania sięga najdawniejszych kontaktów handlowych miasta, aż do końca XVIII wieku, kiedy to na skutek trzeciego rozbioru Polski w 1795 roku, Kraków znalazł się w zaborze austriackim.

W pracy oparto się na obszernych materiałach zródłowych. Podstawowymi zródłami były dostępne dokumenty, przede wszystkim bogate materiały archiwalne Krakowa, jak również inne źródła polskie. Aby rozszerzyć pole widzenia przebadano archiwa znajdujące się w innych krajach: szczególnie byłej Czechosłowacji i Rumunii. Innym rodzajem źródeł były materiały opublikowane, począwszy od wydawnictw polskich XIX i XX wieku, a skończywszy na publikacjach innych krajów europejskich, tak nawet odległych od Polski jak Hiszpania, Portugalia, Bułgaria i kraje skandynawskie; lub tak bliskich Polski jak Niemcy, Wegry, Rumunia, była Czechosłowacja i Związek Radziecki.

W pracy zastosowano przede wszystkim metodologię badań geograficznych. Badania przestrzennych rozmieszczeń głównych towarów, które były przedmiotem eksportu i importu Krakowa, zostały zilustrowane w postaci map, gdzie wykorzystano najrozmaitsze źródła kartograficzne. Mapy te, w większości wypadków ilustrując pozycję Krakowa na tle kontynentu europejskiego, obrazują strefy handlowe dla poszczególnych towarów na przestrzeni określonego czasu i w ścisłym ich powiązaniu z miejscem pochodzenia towaru i miejscem jego handlowego przeznaczenia. Podjęto się również zadania, by przedstawić ceny towarów w Krakowie w sposób analityczny na wykresach, wykorzystując do tego dostępne statystyki na temat miar, wielkości, wag, ilości i jakości, pod warunkiem że mogły być one obliczone na podstawie oryginalnych źródeł.

Pierwsze trzy rozdziały przedstawiają w ogólnym zarysie tło głównego tematu. Rozdział I omawia główną problematykę i podaje zarys najważniejszych elementów badań, jak również podejścia metodologicznego. Rozdział II dokonuje przeglądu materiałów źródłowych oraz publikacji z zakresu geografii historycznej Krakowa, studiów związanych z tematem, jak również różnorodnych źródeł w zakresie dokumentacji i kartografii. Rozdział III zajmuje się najwcześniejszym rozwojem grodu w kontekście jego położenia geograficznego, roli politycznej i wczesnych stosunków handlowych.

Następne dwa rozdziały (IV i V) analizują okres od 1257 roku (data nadania Krakowowi dokumentu lokacyjnego) aż do roku 1500 – daty traktowanej jako punkt przełomowy (benchmark), oznaczający schyłek późnego średniowiecza (Later Middle Ages) i początek nowej ery (Early Modern Period). Tło polityczne miasta zarysowane jest (Rozdział IV) w konfiguracji układów z państwami, które utrzymywały stosunki handlowe z Krakowem. Rozdział V poddaje szczegółowym badaniom przepływ handlu europejskiego przez Kraków. W takim kontekście praca bada rolę organizacji handlowych, sieci dróg lądowych, szlaków rzecznych, kosztów transportu oraz rolę czynnika czasu.

Cenne surowce (metale, przyprawy, sól) zilustrowane są na mapach i poddane analizie. To samo dotyczy tańszych towarów (skóra, wosk, produkty pochodzenia bydlęcego, ryby, drewno, zboża), jak i wyrobów z manufaktur (tkaniny, wino i piwo). Przebadane zostały ceny towarów, zanalizowano rolę traktatów handlowych oraz znaczenie kupców zagranicznych.

Takie samo metodologiczne podejście do zagadnienia zastosowano w następnych dwóch rozdziałach (VI i VII), które zajmują się okresem od 1500 do 1795 roku. Rozdział VI poddaje ocenie zmienne losy polityczne zachodniej, wschodniej, jak również południowo-wschodniej Europy, uwzględniając pozycję Krakowa w kontekście wojen toczących się w tym czasie na terenie Europy. Podstawowym dla polskiej sceny politycznej tego okresu jest przeniesienie stolicy Polski z Krakowa do Warsszawy. Wydarzenia analizowane są po kolei aż do 'Finis Polonia' pod koniec XVIII wieku. Zmienia się także scenariusz handlowy (Rozdział VII), następuje upadek rynku na drogie surowce, rośnie rynek na towary tańsze (jak na przykład: skóry, futra, wosk, miód i bydło). Zdobywają znaczenie wyroby produkowane przez manufaktury: tkaniny, różnorodne napoje i towary ręcznie robione, takie jak gwoździe, broń, kosy, papier, druk, szkło i ceramika.

Rozdział ostatni poświęcony jest ocenie stref handlowych Krakowa na przestrzeni obejmującej prawie pięć i pół wieku. Każde centrum handlowe, dla każdego odpowiedniego okresu i bez względu na towar, zilustrowane jest w postaci mapy. Ośrodki te, przeżywając zarówno rozwój jak i zmiany, na przestrzeni czasu ułożyły się w cztery zasadnicze strefy wpływów handlowych. Najbliższa Krakowowi i Małopolsce (Strefa I) rozszerzyła się po rok 1500 aż do Anglii, Bałtyku, Morza Czarnego i Włoch (Strefa II). Podobny wzorzec reprezentuje Strefa III, ale sięga ona na wschód do Moskwy. Zasadniczym restrykcjom poddana jest Strefa IV, która obrazuje silne powiązania w obrębie środkowej i południowej Polski w XVIII wieku. Analizując czynnik czasu stwierdzono że Kraków, mimo tego iż nigdy nie był tak pierwszorzędnym emporium jak Wenecja czy Antwerpia, śmiało może być zaliczony w poczet emporiów drugorzędnych, których wpływy na rozwój handlu europejskiego były bardzo znaczne.

Praca F. W. Cartera zawiera ponadto: Streszczenie Podziękowania, Spis treści, Spis ilustracji i tablic oraz Bibliografię.

Appendix

Important dates in the history of Cracow up to 1795

200,000 BC – Traces of human settlement discovered in Ojców caves near Cracow, dating from the Old Stone Age.

50,000 BC – Remains of human occupation and flint-tool workshops discovered on Wawel hill (known as the 'Wawel industry').

1,700 BC – During the Bronze Age, stone dies for casting implements, ornaments etc. found around Nowa Huta and Cracow. New settlements established and proof of ceramic and metal manufacture.

600–800 AD – Period of greater stability and reconstruction in the aftermath of devastation caused by earlier population migrations. There was a gradual transition from tribal and family bonds to early feudal forms of social integration. Krakus Mound is dated from this period. The Early Middle Ages saw the rise of the Vislane state in the ninth century, and the castle settlement and fortified suburbs on Wawel hill date from this era.

965 – Ibrahim ibn Jakub, a Jewish merchant from Spain, recorded a town called 'Krakwa' on a journey through Slavonic territory.

1000 – Founding of the Bishopric of Cracow.

1020 – King Bolesław Chrobry (the Brave, 992–1025), commences construction of Wawel Cathedral.

1038 – Cracow becomes capital of the country.

1125 – First recorded destruction by fire in Cracow.

1146–1173 – Prince Bolesław Kędzierzawy (the Curly), a Mazovian Prince, rules over Cracow. He is the first of the Piast dynasty to be buried in Cracow.

1200 – The first recorded earthquake in Cracow.

1207 – Wincenty Kadłubek (*c.* 1150–1223) author of the chronicle *Chronica Polonorum*, became Bishop of Cracow.

1228 – Testimony of municipal organisation in Cracow, from document referring to Piotr, a city magistrate.

1241 –	First Tartar invasion of Poland and devastation of Cracow. The city burghers garrisoned themselves in St Andrew's Church.
1244 –	The Holy Ghost Order of Hospitallers settled in Cracow. Their hospital building remained intact until the end of the nineteenth century.
1257 –	King Bolesław Wstydliwy (the Shy, 1243–79), granted Cracow its Municipal Charter.
1259 –	The Second Tartar invasion.
1286 –	Fortification walls are begun to be constructed around Cracow.
1287 –	Another Tartar invasion of Cracow is thwarted.
1300 –	Wawel Castle undergoes expansion.
1320 –	Coronation in Cracow of King Władysław Łokietek (the Short, 1306–33).
1335 –	Creation of Cracow's suburb, Kazimierz.
1364 –	King Kazimierz Wielki (the Great, 1333–70), decrees an act for the establishment of Cracow Academy.
1364 –	Consecration of the new Gothic Cathedral on Wawel hill.
1364 –	Congress of Monarchs takes place in Cracow, and the famous royal banquet at Wierzynek.
1366 –	Foundation of the town suburb of Kleparz.
1393 –	A water-supply system is established in Cracow.
1400 –	King Władysław Jagiełło (1386–1434) rejuvenates the Cracow Academy.
1411 –	Fifty-one Teutonic banners, seized during the Battle of Grünwald, are exhibited in Wawel Castle.
1473 –	The first printed work in Poland was made in Cracow (*Calendarium Cracoviense*, an astronomical calendar for 1474). It was made by the printer Kaspar Straube.
1475 –	The first book printed in Poland (*Explanatio in psalterium* written by Jan de Turrecrematus) was produced in Cracow.
1477–1489 –	The altar of St Mary's Church was carved by Wit Stwosz (Veit Stoss).
1491–1495 –	The famous astronomer Nicolaus Copernicus studied in Cracow.
1493 –	A description and personal impression of Cracow is given in Hartmann Schedel's book *Opus de historiis aetatum mundi . . .*, published in Nuremberg.
1500 –	Commencement of the Renaissance Castle construction on Wawel hill.
1505 –	The appointment of a city doctor was instituted in Cracow.
1513 –	Florian Ungler's printing shop in Cracow produced the first book in Polish (*Raj duszny* – Soul's Paradise), written by Biernat of Lublin.
1520 –	Poland's largest bell (the Sigismund) was placed in the bell tower of Wawel Cathedral.
1525 –	The hereditary Duke of Prussia, Albrecht Hohenzollern, pays

homage to King Zygmunt I Stary (the Old, 1506–48), in Cracow's Market Square.

1526 – The Polish king makes the first order for sixteen tapestries from Antwerp to adorn the walls of Wawel Castle.

1533 – The building of Wawel Cathedral's Sigismund Chapel is finished.

1543 – The following famous works appeared in Cracow: *Krótka rozprawa między trzema osobami Panem, Wójtem a Plebanem* (A Short Dialogue between the Lord, Bailiff and Parish Priest) by Mikolaj Rej; *O karze za mężobójstwo* (Penalty for Genocide) by Andrzej Frycz Modrzewski; and *Fidelis subditus*, by Stanisław Orzechowski.

1549 – The murder of a scholar in Cracow by church servants led to a student revolt in the city; failure to punish the guilty parties led to students leaving the city.

1555 – Establishment of the first Evangelical Congregation.

1556 – Reconstruction commenced of the Cloth Hall (Sukiennice) after the original was destroyed by fire.

1565 – King Zygmunt August (1548–72) offers the Gun Cock Fraternity a silver cockerel which was a masterpiece of the Polish silversmith's work.

1574 – The first surge of the Counter Reformation in which the Evangelical Congregation came under attack.

1587 – Cracow is besieged by Maximillian Hapsburg and his army.

1588 – Cracow's first lay secondary school is opened.

1593 – The dome of Sigismund Chapel in Wawel Cathedral is gilded.

1609 – King Zygmunt III Waza (1587–1632), his family and court leave Wawel Castle on 25 May and, in spite of no declaration of a formal act, the country's capital moves to Warsaw.

1621 – On 27 November, Prince Władysław rides triumphantly through Cracow after his defeat of the Ottoman Turks at Chocim.

1630 – The Cracow Academy's Library now contained 20,000 volumes.

1652 – Cracow's inhabitants suffered a visitation of the Black Death plague, in which 24,000 people died.

1655 – Sweden invades Poland, Cracow is captured, looted and subjected to paying heavy financial tributes.

1661 – Poland's first periodical (*Merkuriusz Polski*) appears in Cracow, edited and published by Jan Aleksander Gorczyn.

1673 – The Piotrowczyk printing press was purchased by Cracow University, forerunner of today's Cracow University Press.

1702 – The second occasion when Swedish troops entered the city; negligence on the part of her soldiers led to a serious fire in Wawel Castle.

1711 – Units of the Russian army entered the city and occupied Wawel Castle for five years.

1734 – The last coronation of a Polish King (August III, 1733–63) took place in Wawel Cathedral on 17 January.

1745 – Cracow Academy purchased the first telescope for the Chair of Astronomy.

1777 – The emissary for the National Education Commission (Hugo Kołłątaj) arrived in Cracow and introduced educational reforms for the university.

1794 – Cracow was involved in the Insurrection against the armies of Russia and Prussia, under Tadeusz Kościuszko (1746–1816), which failed after eight months of conflict.

1795 – Under the Third Partition of Poland, Cracow becomes part of Austria.

Source: J. Adamczewski, *In Cracow* (Interpress Publications), Warsaw, 1973, pp. 44–50.

Notes

1 Introduction: Cracow in context

1 R. Jamka, 'Początki głównych miast wczesnośrednowiecznych w Polsce południowej w świetle badań archeologicznych' (Część II: Przemyśl, Lublin, Sandomierz, Wiślica i Opole) *Prace Archeologiczne*, No. 15, Kraków, 1973, 183 pp.

2 W. Hensel, 'The Origins of Western and Eastern European Slav Towns', Ch. 21 in M. W. Barley (ed.) *European Towns, Their Archaeology and Early History* (Academic Press) London-New York-San Francisco, 1977, p. 375.

3 T. Lalik, 'Recherches sur les origines des villes en Pologne', *Acta Poloniae Historica*, Vol. I, Warsaw, 1959, p. 121.

4 T. Lalik, 'Markte des 12 Jahrhunderts in Polen' *Ergon*, Vol. III, 1962, p. 366.

5 W. Hensel and L. Leciejewicz, 'En Pologne médiévale: L'Archéologie au service de l'Histoire I: Villes et Campagnes', *Annales: Economies-Sociéties-Civilisations*, Vol. XVII, 1962, No. 2, p. 215; A. Gieysztor, 'Les Origines de la Ville Slave', *Settimane di Studi sull'alto medioevo*, Spoleto, 1959, p. 299.

6 K. Maleczyński, 'Najstarsze targi w Polsce i stosunek ich do miast przed kolonizacją na prawie niemieckiem', *Studya nad Historyą Prawa Polskiego*, Vol. X, No. 1, 1926, p. 20.

7 *Kodeks dyplomatyczny klasztoru tynieckiego* W. Kętrzyński and S. Smolka (eds) Lwów, 1875, No. 1 (year 1123/5); *Kodeks dyplomatyczny Śląska*, K. Maleczyński (ed.) Wrocław, 1857, Vol. I, No. 55 (1175).

8 T. Ładogórski, *Studia nad zaludnieniem Polski XIV wieku* (P.A.N., Instytut Historii), Wrocław, 1958, 229 pp.

9 P. W. Knoll, 'The Urban Development of Medieval Poland, with particular reference to Kraków', Ch. 2 in *The Urban Society of Eastern Europe in Premodern Times*, B. Krekić (ed.) University of California Press, Berkeley and Los Angeles, 1987, pp. 103, 120–21. See also J. Ptaśnik, 'Towns in Medieval Poland' Ch. 2 in *Polish Civilization: Essays and Studies*, M. Giergielewicz (ed.), New York University Press, New York, 1979, pp. 25–50.

10 N. Davies, *Heart of Europe: A Short History of Poland* (Oxford University Press), Oxford and New York, 1986, pp. 306–11.

11 J. Topolski, 'Economic Decline in Poland from the Sixteenth to the Eighteenth

Centuries' Ch. 6 in *Essays in European Economic History 1500–1800*, P. Earle (ed.) Clarendon Press, Oxford 1974, pp. 127–42.

12 P. Bairoch, *Cities and Economic Development (From the Dawn of History to the Present)* (Trans. by C. Braider), University of Chicago Press, Chicago, 1988, p. 153.

13 A. Davies, 'A Study in City Morphology and Historical Geography', *Geography*, Vol. XVIII, 1933, p. 30.

14 A. Buttimer, *The Wake of Erasmus: Saints, Scholars and Studia in Mediaeval Norden*, Lund Studies in Geography (Ser. B. Human Geography, No. 54), Lund 1989, p. 17.

15 L. Hajdukiewicz and M. Karaś, *The Jagiellonian University: Traditions, The Present, The Future*, Cracow, 1978, p. 17.

16 S. Leszczycki and B. Modelska-Strzelecka, 'Six Centuries of Geography at the Jagiellonian University in Cracow', *Geographia Polonica*, Vol. XI, Warsaw, 1967, pp. 5–28.

17 J. Z. Łoziński and A. Miłobędzki, *Guide to Architecture in Poland* (Polonia Pub. House), Warsaw, 1967, pp. 10–35; 115–27.

18 Anon, 'What Makes a City Great?' *Time Magazine* (14/11/1969), pp. 47–8.

19 P. M. Hohenburg and L. H. Lees, *The Making of Urban Europe, 1000–1950* (Harvard University Press, Cambridge, Massachusetts and London, 1985, p. 38.

20 P. Claval, 'Chronique de géographie économique XV: les économistes et la ville', *Revue Géographique de l'Est*, Vol. XXI, No. 3, Lyon, 1981, pp. 213–25.

21 J. Brutzkus, 'Trade with Eastern Europe, 800–1200' *Economic History Review*, Vol. XIII, London, 1943, p. 34.

22 For example see T. R. Slater, 'Medieval and Renaissance Urban Morphogenesis in Eastern Poland', *Journal of Historical Geography*, Vol. XV, No. 3, 1989, pp. 239–59.

23 P. Bushkovitch, 'Polish Urban History: Sixteenth and Seventeenth Centuries', *Polish Review*, Vol. 18, No. 3, New York, 1973, p. 89.

24 D. Denecke and G. Shaw (eds), *Urban historical geography: recent progress in Britain and Germany* (Cambridge University Press), 1988, Cambridge, 232 pp.

25 A. H. Clark, 'Geographical Change: A Theme for Economic History, *Journal of Economic History*, Vol. XX, 1960, p. 607.

26 W. L. Blackwell, 'Geography, History, and the City in Europe and Russia', *Journal of Urban History*, Vol. VI, No. 3, 1980, pp. 339–56.

27 H. C. Darby, *An Historical Geography of England before AD 1800* (University Press), Cambridge, 1936; D. S. Whittlesey, 'Sequent Occupance', *Annals of the Association of American Geographers*, Vol. XIX, 1929, pp. 162–5.

28 A. H. Clark, 'Geographical Change', p. 611.

2: Source materials and published literature

1 Norton, W., *Historical Analysis in Geography* (Longman) London, 1984, p. 24.

2 Guelke, L., and Lai, P. C., 'Computer Cartography in Historical Geographical Research' *Canadian Geographer*, Vol. XXVII, 1983, pp. 207–22, Pourez, C., Poy, R., and Martin, F., 'The Linkage of Census Data Problems and Procedures' *Journal of Interdisciplinary History*, Vol. XIV, 1983, p. 129–52; Doherty, J. C.,

and Gibson, A. J. S., 'Computer-assisted Data Handling in Historical Geography' *Area*, Vol. XV, 1983, pp. 257–60; Dennis, R. J., 'Historical Geography; Theory and Progress', *Progress in Human Geography*, Vol. VIII, No. 4, 1984, pp. 536–43.

3 Baker, A. R. H., *Historical Geography and Geographical Change* (Aspects of Geography Series) (Macmillan Education), London, 1975, p. 1.

4 Krawczyk, A., 'Present-day Aims and Methods of Historical Geography', *Prace Geograficzne*, Vol. LXI, Kraków, 1984, p. 81.

5 Arnold, S., *Geografia historyczna Polski* (P.W.N.), Warszawa, 1951, 112 pp. Buczek, K., 'O teorii badań historyczno-osadniczych' *Kwartalnik Historyczny*, Vol. LXV, pp. 65–86: Dobrowolska, M., 'Przedmiot i metoda geografii historycznej' *Przegląd Geograficzny*, Vol. XXV, No. 1, pp. 57–77: 1953. Kotarski, S., *Geografia historyczna, Wiadomości wstępne*, Warszawa, 1951, 65 pp.: Labuda, G., 'Uwagi o przedmiocie i metodzie geografii historycznej' *Przegląd Geograficzny*, Vol. XXV, No. 1, 1953, pp. 5–56: Semkowicz, W., 'O potrzebie i metodzie badań nad krajobrazem pierwotnym' in *Pamiętnik IV Zjazdu Historyków Polskich w Poznaniu*, Lwów, 1925, pp. 1–8. Ibid., 'Potrzeby w zakresie nauk pomocniczych historii' *Nauka Polska*, Vol. I, Warszawa, 1946, pp. 287–312: Szymański, J., *Nauki pomocnicze historii* (P.W.N.), Warszawa, 1976, pp. 231.

6 Krawczyk, A., 'Present-day Aims', p. 82.

7 Blackwell, W. L., 'Geography, History, and the City in Europe and Russia' *Journal of Urban History*, Vol. VI, No. 3, 1980, p. 342.

8 Wynot, E. D., 'Urban History in Poland: a Critical Appraisal', *Journal of Urban History*, Vol. VI, No. 1, 1979, pp. 31–79.

9 Ibid. p. 31.

10 Ibid. p. 55.

11 Davies, A., 'A Study in City Morphology and Historical Geography', *Geography*, Vol. XVIII, Manchester, 1933, pp. 25–37.

12 Innocenti, P., *La Città di Cracovia*: (Origini, aspetti e funzioni dell' organismo urbano. Suo ruol nella struttura geografico-economica della Republica Popolare Polacca), Firenze, 1973, pp. 78–98.

13 Bromek, K., 'Zarys rozwoju historycznego i terytorialnego Krakowa' *Folia Geographica* (Series Geographica-Oeconomica), Vol. VIII, Kraków, 1975, pp. 15–35; Z. Górka, 'Akcja odnowy Krakowa a użytkowanie przestrzeni centrum miasta' *Prace Geograficzne*, Vol. LVI, Kraków, 1983, pp. 7–20.

14 Carter, F. W., *Dubrovnik (Ragusa) – A Classic City-State* (Seminar Press), London/New York, 1972, 10 pp.

15 Carter, F. W., *An Industrial Geography of Prague: 1848–1921*, University Microfilms International, Ann Arbor/London, 1982, 568 pp.; Bater, J. H., *St Petersburg: Industrialization and Change* (Edward Arnold), London 1976, 469 pp.

16 Bastié, J., *La Croissance de la Banlieue Parisienne*, Paris, 1964, 624 pp.

17 Kutrzeba, S., 'Handel Krakowa w wiekach średnich na tle stosunków handlowych Polski', *Rozprawy Wydziału Historyczno–Filozoficznego Akademii Umiejętności*, Vol. XLIV, Kraków, 1903, pp. 1–196; Kutrzeba, S., 'Handel Polski ze Wschodem w wiekach średnich', *Przegląd Polski*, Vol. II, Kraków, 1903, pp. 189–219, 462–96; Vol. III, pp. 512–37; Vol. IV, pp. 115–45; Kutrzeba, S., and J. Ptaśnik, 'Dzieje handlu i kupiectwa krakowskiego' *Rocznik Krakowski*, Vol. XIV, Kraków, 1910, p. 183. See also, Vetulani, A., and

Wyrostek, L., 'Bibliografia prac Prof. Dr. S. Kutrzeby' (w 1897–1937), *Studia Historyczne ku czci Stanisława Kutrzeby*, Kraków, 1938, Vol. I, pp. IX–XXXIII. Dąbrowski, J., 'Stanisław Kutrzeba', *Nowa Polska*, Vol. VI, Kraków, 1946, pp. 185–92.

18 Małecki, J., *Studia nad rynkiem regionalnym Krakowa w XVI wieku* (P.W.N.) Warszawa, 1963, 251 pp.

19 Kulczykowski, M. 'Handel Krakowa w latach 1750–1772' *Prace Historyczne*, No. 4 (Historia No. 4) Kraków, 1960, pp. 79–105; Kulczykowski, M. *Kraków jako ośrodek towarowy Małopolski zachodniej w drugiej połowie XVIII wieku* (P.W.N.), Warszawa, 1963, 173 pp.

20 Pieradzka, K., 'Handel Krakowa z Węgrami w XVI w., *Biblioteka Krakowska*, No. 87 (Towarzystwo Miłośników Hist. i Zabytków m. Krakowa) Kraków, 1935, 280 pp.; M. Małowist, 'Le développement des rapports économiques entre la Flandre, La Pologne et les pays limitrophes du XIII au XIVs'. *Revue Belge de Phil. et d'Histoire*, Vol. X, Brussels, 1931, pp. 1013–65; M. Małowist, 'The Baltic and the Black Sea in Medieval Trade' *Baltic and Scandinavian Countries*, Vol. II, No. 1 (5), Gdynia, 1937, pp. 36–42; M. Małowist, 'Polish-Flemish trade in the Middle Ages' *Baltic and Scandinavian Countries*, Vol. IV, No. 1 (8), Gdynia, 1938, pp. 1–9; J. Topolski, 'Rola Gniezna w handlu europejskim od XV do XVIII wieku', *Studia i Materiały do Dziejów Wielkopolski i Pomorza*, Vol. VII, No. 2, Poznań, 1962, pp. 6–78; Z. Guldon and L. Stępkowski, *Z Dziejów handlu Rzeczypospolitej w XVI–XVIII wieku: Studia i Materiały* (Wyższa Szkoła Pedagogiczna im. Jana Kochanowskiego), Kielce, 1980, 326 pp.

21 Bieniarzówna, J., and J. M. Małecki, *Kraków w wiekach XVI–XVIII* (Dzieje Krakowa, Vol. II) (Wyd. Lit.) Kraków, 1984, 666 pp.

22 Wyrozumska, B. *Drogi w ziemi krakowskiej do końca XVI wieku* (P.A.N.) Wrocław-Warszawa-Kraków-Gdańsk, 1977, 106 pp. Wyrozumska, B. *Lustracja dróg województwa krakowskiego z roku 1570* (P.A.N.) Wrocław-Warszawa-Kraków-Gdańsk, 1971, 114 pp.

23 Falniowska-Gradowska, A. (ed.), *Lustracja województwa krakowskiego 1765 (Część I.)* (P.A.N.), Warszawa-Kraków, 1973, 355 pp.

24 Müller, J., 'Der Umfang und die Hauptrouten des Nürnberger Handelsgbietes im Mittelalter' *Vierteljahrschrift für Social und Wirtschaftsgeschichte* Vol. VI Nuremberg, 1908, pp. 1–38.

25 Fišer, R., 'Obchodní styky Levoče se zahraničím v 17 století' *Československý Časopis Historický*. Vol. XXVI, No. 6, Praha, 1978, pp. 844–67; F. Hejl, 'Český obchod na Krakovském trhu po Bílé Hoře' *Sborník Prace Filosofické Fakulty Brněnské University* Vol. X (řada historická č.8) Brno. 1961, pp. 228–250: O. R. Halaga, 'Krakovské právo skladu a sloboda obchodu v Uhersko-Pol'ski-Pruskej politike pred Grunwaldom' *Slovanské Štúdie* Vol. XVI, Bratislava, 1975, pp. 163–91.

26 Dan, M. P., 'Din relaţiile comerciale ale Transilvaniei cu Polonia la sfîrşitul secolului al XVI-lea. Produse textile importate de clujeni de la Cracovia'. *Omagui lui P. Constantinescu – Iaşi cu prilejul implinirii a 70 de abi*' Bucureşti (ed. Acad R P R) 1965, pp. 289–94; Dan, M. P., 'Le Commerce de la Transilvanie avec la Pologne au XVIe siècle', *Revue Roumaine d'Histoire* Vol. VIII, No. 3, Bucureşti, 1969, pp. 621–34; 'Relaţiile Commerciale dintre Oradea şi Cracovia la sfîrşitul secolului al XVI-lea'. *Crisia*, No. 3, Oradea, 1973, pp. 167–81.

27 Pounds, N. J. G., *An Historical Geography of Europe 1500–1840*, Cambridge University Press, Cambridge, London, New York, Melbourne, Sydney, 1979, preface p. xv.
28 *Tabeller over Skibsfart og Varetransport gennem Øresund 1497–1660*, N. E. Bang (ed.), Del. 1, 2, København, 1906–1933 N. E. Bang, *1661–1783 og gennem Storebaelt 1701–1748*, Del. 1, 2, København, 1930–1953; *Svensk handelsstatist 1637–1737*, E. F. Heckscher (ed.), Stockholm, 1938; *Tamozhennye knigi moskovskogo gosudarstva XVII veka*, A. I. Jakovleva (ed.), 3 Vols., Moskva, 1950–1951; *Regestra thelonei aquatici Wladislaviensis saeculi XVI*, S. Kutrzeba and F. Duda, Kraków, 1915; R. Rybarski, *Handel i polityka* (Ref. 46).
29 Cobb, R., *A Sense of Place*, London, 1975, p. 47.
30 Harley, J. B., 'Historical Geography and Its Evidence: Reflections on Modelling Sources', in *Period and Place (Research Methods in Historical Geography)* A. R. H. Baker and M. Billings (eds), Cambridge University Press, Cambridge, London, New York, Melbourne, Sydney, 1982, p. 261.
31 *Ibid.*, p. 263.
32 Baker, A. R. H., On Ideology and Historical Geography in *Period and Place (Research Methods in Historical Geography)*, p. 238.
33 Wojewódzkie Archiwum Państwowe w Gdańsku, dział 300, 19, Gdańsk; for Elbląg there are a total of 33 registers, one of which is in the Czartoryski Biblioteka (Oddział Muzeum Narodowego) in Cracow, the rest in Gdańsk Archives. For the English reader their value may be appreciated from A. Zins, *England and the Baltic in the Elizabethan Era* (transl. H. C. Stevens, Manchester University Press/Rowman & Littlefield), Totowa, N.J. 1972, 347 pp.
34 Wojewódzkie Archiwum Państwowe w Krakowie, *Regestra thelonei civitatis Cracoviensis*, containing 136 hand-written volumes (syg. 2115–2250) beginning on 12 June 1589 and ending 20 September 1792.
35 Małecki, J., 'Księgi celne krakowskie i ich znaczenie jako źródła do historii handlu', *Sprawozdania z posiedzeń komisji*, styczeń-czerwiec 1960, P.A.N. oddział w Krakowie), Kraków, 1961, p. 49. For a more lengthy discussion see Małecki, J., 'Krakowskie księgi celne i problem ich wydania' *Kwartalnik Historii Kultury Materialnej*, Vol. IX, No. 2, Warszawa, 1961, pp. 251–73.
36 Dudik, B., *Archive im Königreiche Galizien und Lodomiren* (Archiv für österreichische Geschichte Bd. XXXIX), Wien, 1868, pp. 19–34; 83–7; K. Kaczmarczyk, 'Das historische Archiv der Stadt Krakau: Seine Geschichte. Bestände und wissenschaftliche Erforschung', *Mitteilungen, des k.k. Archivrates*, Bd. i, Wien, 1913, pp. 155–200.
37 Piekosiński, F., *Codex diplomaticus civitatis Cracoviensis (1257–1506)*, Vol. I, Kraków, 1879, Vols. II–IV, Kraków, 1882.
38 Piekosiński, F., and J. Szujski, *Najstarsze księgi i rachunki miasta Krakowa* (Libri antiquissimi civitatis Cracoviensis (1300–1400), Kraków, 1878.
39 Daenell, E. R., 'Polen und die Hansa um die Wende des XIV Jahrhundert', *Deutsche Zeitschrift für Geschichtswissenschaft*, N.F. 2, Vierteljarshefte, Freiburg, 1897–8, p. 330.
40 Kutrzeba, S., 'Finanse Krakowa w wiekach średnich', *Rocznik Krakowski*, Vol. III, Kraków, 1900, pp. 27–152.
41 Friedberg, M., 'Kancelaria miasta Kazimierza pod Krakowem, 1335–1802',

Archeion, Vol. XXXVI, Warszawa, 1962, p. 137; Friedberg, M., *Inwentarz Archiwum Miasta Kazimierza pod Krakowem, 1335–1802*, Warszawa, 1966, pp. 31–8.

42 Kutrzeba, S. and Ptaśnik, J., 'Dzieje handlu i kupiectwa "krakowskiego" *Rocznik Krakowski* Vol. XIV, Kraków, 1912, p. 138.

43 Małecki, J. M., 'Księgi celne krakowskie', pp. 47–9; Małecki, J. M., 'Krakowskie księgi celne i problem ich wydania' *Kwartalnik. Historia Kultury Materialnej*, Vol. IX, Warszawa, 1961, pp. 251–73; F. Hejl, 'Městký archiv v Krakově a jeho význam pro studium česko-polských a slovenské (uhersko) – polských vztahů od konce XV. do XVIII stol. *Sborník Matice Moravské*, Vol. 79, Brno. 1960, pp. 288–90.

44 Friedberg, M., 'Kancelaria Miasta Kazimierz'; pp. 137–9.

45 Filipowski, J. (ed.), *Katalog Archiwum Aktów Dawnych Miasta Krakowa*, Vol. II (Kraków: Rękopisy Nr. 1–3568), Kraków, 1915, pp. xi–xxvi.

46 Anon., 'Straty archiwów i bibliotek w zakresie rękopiśmiennych zródeł historycznych', Vol. I., *Archiwum Główne Akt Dawnych*, Warszawa, 1957, pp. 88–90; S. Kutrzeba and F. Duda (eds.), *Regestrathelonei aquatici Wladislaviensis saeculi XVI*. Kraków, 1915, 364 pp.; R. Rybarski, *Handel i polityka handlowa Polski w XVI stuleciu*, Vol. I (Rozwój handlu i polityka handlowa) Vol. II (Tablice i materiały statystyczne). Poznań, 1928–9, Reprinted Warsaw, 1958.

47 Wojewódzkie Archiwum Państwowe Gdańsku, *dział* 300,19: Archiwum miasta Elbląg, *sygn*. R.H. 43.

48 Hoszowski, S., 'Polski eksport wiślany w 1784 roku', *Kwartalnik Historyczny*, Vol. 53, Warszawa, 1956, Nos. 4–5, pp. 64–80.

49 Wojewódzkie Archiwum Państwowe, Poznań, *rękopis* G 3, 1206.

50 Małecki, J. M., 'Krakowskie księgi celne' pp. 254–5.

51 Domanovszky, S., *A harmincasvám eredete*, Budapest, 1916, 254 pp.

52 Státní Ustrední Archiv v Bratislavě, *Archive rodiny Révay*, Reskripty, fasc. 2. No. 10 (dated 1529); Ibid., *Regesta super incasatione tricesimae Solnensis ab anno 1529 usque 1537*.

53 Janáček, J., 'České soukenictvi v 16 století,' *Československý Časopis Historický*, Vol. IV, Prague 1956, pp. 553–5.

54 Archiv mesta Trenčina, *Rationes* z r. 1530–1536. OA v Trenčine.

55 Fejérpataky, L., *Magyarországi várasok régi számadáskönyvei*, Budapest, 1886, 341 pp.; B. Iványi, *Bártfa szabad királyi város levéltara* Vol. I, 1319–1501, Budapest, 1910, 230 pp.; Archiv mesta Bardejova, No. 433/424 (1444); No. 522/515 (1448); No. 523/516 (1448–1452) No. 687/680 (1452–4). See also L. Deák, 'Bardejovský obchod a Bardejovská obchodná cesta v prvej polovici 15 storočia', *Sborník Filozofskej Fakulty Univerzity Komenského*, Vol. XOV, Bratislava, 1963, pp. 107–9.

56 M. Marečková, Majetková struktura Bardějovských obchodníků v prvé polovině 17 století, *Sbornik Prací Filozofické Fakulty Brněnské Univerzity* (Řada Historická) Vol. XXVII–VIII, C 25/26, Brno, 1978–1979, pp. 131–2.

57 Horváth, P., 'Obchodné styke Levoče s Pol'skom v druhej polovici XVI storočia' *Historické Štúdie*, I, Bratislava, 1955, pp. 105–6.

58 Granasztói, G., 'La ville de Kassa dans le commerce Hungro-Polonais au XVIe siècle', in *La Pologne et la Hongrie aud XVIe–XVIIIe siècles*, V. Zimányi (ed.) (Akadémiai Kiadó), Budapest, 1981, p. 57.

59 Anon, *Quellen zur Geschichte der Stadt Kronstadt*, Vols. I–III, Braşov, pp. 186–96.
60 Archivele Statului Cluj, *fondul Arhiva oraşului Cluj*, Socotelile Oraşului Cluj, 12 v. VII–VIII; 13 a VI, XVI, XX–XXIII; 13 v. II–IX; 14 a I–XI, XXII, XXV; 15 a III–XII; 15 v. XIII–XXII; 16 VII–XVIII; 18 v. IV; 19 VII, XI; 20 I, II.
61 Panova, S., 'Zur Frage der Handelsbeziehungen zwischen den bulgarischen und den rumänischen Landen im XVII Jahrhundert', *Études Balkaniques*, No. 1, Sofia, 1975, pp. 103–14.
62 Dan, M., 'Negustorii clujeni la Cracovia în ultimul deceniu al secolului al XVI-lea', *Acta Museum Napocensis*, Vol. VIII, Cluj, 1971, pp. 205–17.
63 Dan, M. and S. Golderberg, 'Marchands Balkaniques et Levantins dans le commerce de la Transylvanie aux XVIe et XVIIe siècles' in *Actes du Premier Congrès International des Études Balkaniques et Sud-Est Européenes*, Vol. III, Sofia 1969, pp. 641–8.
64 Demény, L. A., 'Le Commerce de la Transylvanie avec les régions du sud du Danube effectué par la douane de Turnu Roşu en 1685', *Revue Roumaine d'Histoire*, Vol. VII, No. 5, Bucureşti, 1968, pp. 761–77.
65 *Katalog Dokumentów Tureckich (Dokumenty do dziejów Polski i krajów ościennych w latach 1455–1672)*, Z. Abrahamowicz (ed.) (Pan. Wyd. Nauk.), Warszawa, 1959, 362 pp. *Catalogul documentalor turceşti*, M. Guboglu (ed.), Vol. I, Bucureşti, 1960, Vol. II, Bucureşti, 1965; *Documente privind istoria României*, I. Bogdan (ed.), Serie B. secolului al XVI-lea, Vol. II, Burcureşti, 1951; *Documente privitóre la Istoria Romanilor*, E. de Hurmuzaki (ed.), 19 vols. Bucureşti, 1876–1922, especially Vol. XV.
66 Wawrzyńczyk, A., *Studia z dziejów handlu Polski z Wielkim Księstwem Litewskim i Rosją w XVI* w., Warszawa, 1956, pp. 41–2; J. Perenyi 'Villes hongroises sous la domination ottomane aud XVIe et XVIIe siècles, Les chefs-lieux de l'administration ottomane' *Studia Balcanica*, Vol. III, Sofia, 1970, p. 28; R. Klier, 'Der schlesische und polnische transithandel durch Böhmen nach Nürnberg in den Jahren 1540 bis 1576', *Mitteilungen des Vereins für Geschichte der Stadt Nürnberg*, Vol. LIII, Nuremburg, 1965, p. 195; W. S. Unger, 'Trade through the Sound in the Seventeenth and Eighteenth Centuries', *Economic History Review* (2nd Series) Vol. 12, No. 2, London, 1959, p. 206.
67 Krassowski, B., 'Poland on Old Maps', *Poland*, No. 4 (284), Warszawa, 1978, pp. 23–6.
68 Kałkowski, T., *Tysiąc lat monety polskiej* (3rd edition) (Wyd. Lit.), Kraków, 1981, 490 pp; J. A. Szwagrzyk, *Pieniądz na ziemiach polskich X–XX w.*, Wrocław, 1973, 360 pp.
69 Szelągowski, A., *Pieniądz i przewrót cen w XVI i XVII w. Polsce*, Lwów, 1902, 305 pp.
70 Hoszowski, S., 'The Revolution of Prices in Poland in the 16th and 17th Centuries', *Acta Poloniae Historica*, Vol. II, Warsaw, 1959, p. 7.
71 Pelc, J., *Ceny w Krakowie w latach 1369–1600*, Lwów, 1933, 182 pp. E. Tomaszewski, *Ceny w Krakowie w latach 1601–1795*, Lwów, 1934, 350 pp.
72 Nef., J. U., 'Silver Production in Central Europe, 1450–1618', *Journal of Political Economy*, Vol. XLIX, Chicago, 1941, pp. 575–91.
73 Pounds, N. J. G., *An Economic History of Medieval Europe* (Longman), London,

New York, 1984, pp. 477–81; F. Braudel and F. C. Spooner, 'Prices in Europe from 1450 to 1750', Ch. VII in *The Cambridge Economic History of Europe*, Vol. IV, Cambridge, 1967, p. 381.

74 Reddaway, W. F., *et al.* (eds.), *The Cambridge History of Poland from the Origins to Sobieski (to 1696)* (Octagon Books), New York, 1978, p. 29; J. Rutkowski, *Historia Gospodarcza Polski*, Warszawa, 1953, p. 133; S. Hoszowski, 'L'Europe centrale devant la révolution des prix aux XVIe et XVIIe siècles' *Annales, Economiques, Sociétés, Civilisations*, Vol. XVI, No. 1, Paris, 1961, p. 448.

75 Trevor-Roper, H. R., 'The General Crisis of the 17th Century', *Past and Present*, No. 16, London, 1959, p. 38.

76 Bogucka, M., 'The Monetary Crisis of the XVIIth Century and Its Social and Psychological Consequences in Poland', *Journal of European Economic History*, Vol. IV, No. 1, Rome 1975, p. 141.

77 Kuklińska, 'Les rôles joués par les marches intérieurs et extérieurs dans le développement du commerce polonais au XVIIIe siècle', *Studia Historia Oeconomicae*, Vol. III, Poznań, 1976, p. 100.

3 Cracow's early development

1 F. Braudel, *Capitalism and Material Life, 1400–1800*, London 1977, p. 387 (Fontana/Collins).

2 L. Leciejewicz, 'Wczesnośredniowieczne przemiany socjotopograficzne osad miejskich u Słowian zachodnich w świetle archeologii', Ch. 2 in *Miasta doby feudalnej w Europie środkowo-wschodniej*, A. Gieysztor and T. Rosłanowski (eds.) (Pań.Wyd.Nauk), Warszawa-Poznań-Toruń, 1976, pp. 35–66.

3 F. Rörig, *The Medieval Town* (B. T. Batsford), London, 1967, p. 181.

4 R. Jamka, '*Początki głównych miast wczesnośredniowiecznych w Polsce południowej w świetle badań archeologicznych*' Część I: Kraków, Zeszyty Naukowe U. J., CCLXXII, Prace Archeologiczne, zesz. 13, Kraków, 1971, 166 pp.; R. Jamka, *Kraków w pradziejach* (Biblioteka Archeologiczna, Vol. XVI), Wrocław-Warszawa-Kraków, 1963.

5 W. Abraham, 'Początek biskupstwa i kapituły katedralnej w Krakowie' *Rocznik Krakowski*, Vol. IV, 1901, pp. 177–200, Kraków.

6 F. Braudel, *Capitalism and Material Life*, p. 387.

7 Further discussion may be found in: J. Kondracky, 'Types of Natural Landscapes in Poland', *Przeglad Geograficzny*, Vol. XXXII, Warsaw 1960, Supplement pp. 29–39; M. Klimaszewski, 'Podział morfologiczny południowej Polski', *Czasopismo geograficzne*, Vol. XVII, Wrocław, 1946, No. 3–4, pp. 138–82; M. Tyczyńska, 'Rzeźba terytorium miasta Krakowa', *Folia Geographica*, Series Geographica-Physica, Vol. VIII, Kraków-Warszawa, 1974, pp. 19–43; A. Pierzchała, 'Rzeźba obszaru miasta Krakowa', *Czasopismo geograficzne*, Vol. XXXI, Wrocław, 1960, No. 1, pp. 27–46.

8 S. Dżułyński, 'Tektonika południowej części Wyżyny Krakowskiej', *Akta Geologiczne Polski*, Vol. III, Warszawa, 1953, pp. 325–440; R. Gradziński, *Przewodnik geologiczny po okolicach Krakowa* (Wydawnictwa Geologiczne), Warszawa, 1972, 335 pp.; R. Gradziński, 'Budowa geologiczna terytorium Krakowa', *Folia Geographica* Series Geographica-Physica, Vol. VIII, Kraków-Warszawa, 1974, pp. 11–17.

9 K. Bąkowski, 'Dawne kierunki rzek pod Krakowem', *Rocznik Krakowski*, Vol. V, Kraków, 1902, pp. 138–72; I. Kmietkowicz-Drathowa, 'Zagadnienia starorzeczy Wisły na obszarze Krakowa', *Sprawozdania z Posiedzeń Komisji Oddziału PAN w Krakowie* (I–VI/1965), P. W. N., Kraków, 1965; M. Tobiasz, 'Historyczny rozwój sieci wodnej Krakowa i jej wpływ na urbanistykę miasta', *Zeszyty Naukowe Politechniki Krakowskiej; Architektura*, No. 2, Kraków, 1958, pp. 25–40; Ibid., *Pierwsze Wieki Krakowa* (W.O.I.T.), Kraków, 1974, pp. 15–26.

10 K. Bromek, 'Zarys rozwoju historycznego i terytorialnego Krakowa', *Folia Geographica*, Series Geographica-Oeconomica, Vol. VIII, Kraków-Warszawa, 1975, pp. 14–17.

11 T. Komornicki, 'Gleby terytorium miasta Krakowa', *Folia Geographica*, Series Geographica-Physica, Vol. VIII, Kraków-Warszawa, 1974, pp. 145–51.

12 K. Mamakowa, 'Późnoglacjalna i wczesnoholoceńska flora z terenu Krakowa', *Acta Paleobotanica*, Vol. 11, No. 1, pp. 3–12, Kraków 1970; J. Kornaś and A. Medwecka-Kornaś, 'Szata roślinna Krakowa', *Folia Geographica*, Series Geographica-Physica, Vol. VIII, Kraków-Warszawa, 1974, pp. 153–69.

13 Further discussion may be found in J. Mitkowski, 'Dawne warunki geograficzne jako podłoże, na którym rozwinął się zespół osad krakowskich', Ch. 2 in *Kraków: Studia nad rozwojem miasta*, J. Dąbrowski (ed.) (Wyd. Lit.) Kraków, 1957, pp. 39–64.

14 S. Zakrzewski (ed.), *Descriptio civitatum ad septentrionalem plagem Danubii*, Lwów, 1917; H. Łowmiański, 'O identyfikacji nazw Geografa Bawarskiego' *Studia Zródłoznawcze*, Vol. III, Warszawa, 1958, pp. 1–6; J. Natanson-Leski, *Rozwój terytorialny Polski, od czasów najdawniejszych do okresu przebudowy państwa w latach 1565–1572* (P.W.N.) Warszawa, 1964, pp. 11–12.

15 J. F. N. Bradley, *Czechoslovakia: A Short History* (University Press), Edinburgh, 1971, pp. 6–7; T. Wasilewski, 'Wiślańska dynastia i jej zachlumskie państwo w IX–X w.', *Pamiętnik Słowiański*, Vol. XV, Warszawa, 1965, pp. 23–6.

16 P. Goudy, 'Racial Origins', Ch. 1 in *The Cambridge History of Poland from the Origins to Sobieski (to 1696)*, W. F. Reddaway et al. (eds.), Octagon Books, New York, 1978, p. 12.

17 J. F. N. Bradley, *Czechoslovakia*, p. 7.

18 S. Arnold and M. Zychowski, *Outline History of Poland: From the Beginning of the State to the Present Time* (Polonia Pub. House), Warsaw, 1965, 2nd edn. p. 11.

19 R. Holtzmann 'Böhmen und Polen im 10 Jh.' *Zeitschrift der Vereins für Geschichte Schlesiens*, Vol. LII, Breslau, 1918, pp. 25–32.

20 S. Kętrzyński, 'The introduction of Christianity and the Early Kings of Poland', Ch. 2 in *The Cambridge History of Poland*, p. 21.

21 H. Łowmiański, 'Dynastia Piastów we wczesnym średniowieczu', *Początki państwa polskiego* Vol. I (Organizacja polityczna), Poznań, 1962, pp. 111–17; J. Natanson-Leski, p. 21.

22 *Monumenta Poloniae Historica*, Nova Seria, Vol. I, Kraków-Warszawa, 1946, p. 48.

23 J. Natanson-Leski, p. 11.

24 K. Buczek, 'Pierwsze biskupstwa polskie', *Kwartalnik Historyczny*, Vol. LII, 1938, Kraków, pp. 169–72.

25 M. Z. Jedlicki (ed.), *Kronika Thietmara*, Poznań, 1953, pp. 224–5; Słownik starożytności słowiańskich (Lexicon Antiquitatum Slavicarum), Vol. I, Wrocław-Warszawa-Kraków, 1961, p. 312. For further discussion see A. Gieysztor, 'Les Territoires de la Pologne aux IXe et Xe siècles', *Godišnjak Društva Istoričara Bosne i Hercegovine*, Vol. XV, Sarajevo, 1966, pp. 7–24.
26 G. Slocombe, *A History of Poland: Earliest Times to 1939* (T. Nelson & Sons), London, New York, 1939, p. 28.
27 S. Kętrzyński, 'The Introduction of Christianity', p. 21.
28 M. Klimaszewski, 'Warunki geograficzne rozwoju Małopolski', Ch. 1 in '*Kraków i Małopolska przez dzieje: Studia i szkice profesorów Uniwersytetu Jagiellońskiego*, C. Bobińska (ed.), Wyd. Lit. Kraków, 1970, pp. 11–31.
29 D. Borawska, '*Kryzys monarchii wczesnopiastowskiej w latach trzydziestych XI w.*' Warszawa, 1964, 245 pp.
30 J. F. N. Bradley, *Czechoslovakia*, p. 13; G. Slocombe, *A History of Poland*, p. 34.
31 S. Kętrzyński, 'Kazimierz Odnowiciel (1034–1058)', *Rozprawy A. U. Wydziału Historyczno-Filozoficznego*, Vol. II, No. 3, Kraków, 1899, pp. 366–80.
32 W. Dziewulski, '*Postępy chrystianizacji i proces likwidacji pogaństwa w Polsce wczesnofeudalnej*', Wrocław, 1964, 285 pp.; J. Dowiat, 'Dyskusja nad kryzysem państwa piastowskiego w pierwszej połowie XI wieku', *Kwartalnik Historyczny*, Kraków, 1964, pp. 753–9; S. Kętrzyński, 'The Introduction of Christianity', p. 38; Archbishop Aaron died in 1059.
33 M. Frančić, *Kalendarz dziejów Krakowa* (Wyd. Lit.), Kraków, 1964, p. 11; J. Adamczewski, *In Cracow* (Interpress Pub.), Warsaw, 1973, p. 13, also states, 'In 1038 . . . Cracow was made the capital of Poland.'
34 J. Garbacik, 'Przeniesienie stolicy Polski do Krakowa w XI w.', Ch. 6 in *Kraków i Małopolska*, pp. 127–41; J. Natanson-Leski, p. 33.
35 S. Kętrzyński, 'The Introduction of Christianity', p. 39.
36 J. M. Małecki, 'Kiedy i dlaczego Kraków przestał być stolicą Polski', *Rocznik Krakowski*, Vol. XLIV, Kraków, 1973, pp. 21–36.
37 S. Kętrzyński, 'The Introduction of Christianity', p. 41; G. Slocombe, *A History of Poland*, pp. 38–9.
38 For further discussion see T. Wojciechowski, *Szkice historyczne jedenastego wieku* (3rd edn.), Warszawa, 1951, 300 pp. Z. Wojciechowski, 'La condition des nobles et le problème de la féodalité en Pologne au Moyen-Âge', *Revue historique de droit français et étranger*, Vols. XV–XVI, Paris, pp. 936–7, 651–700.
39 *Monumenta Poloniae Historica*, Vol. II, pp. 832, 874; Vol. III, pp. 150, 154–5, Warszawa, 1960–1.
40 W. Łuszczkiewicz, 'Architektura romańska kościoła św. Andrzeja w Krakowie', *Sprawozdania Komisji do Badania Historii Sztuki*, A.U., Vol. VII, Kraków 1902, pp. 10–20.
41 G. Slocombe, *A History of Poland*, p. 40; M. Frančić, *Kalendarz*, p. 11.
42 Based on information recorded in *Rocznik kapituły krakowskiej*, written in 1266/67, one of the richest sources of information on Cracow during the Middle Ages. See M. Frančić, *Kalendarz*, pp. 12, 20.
43 *Miasta Polskie w Tysiącleciu*, Vol. I, Warszawa, 1965, pp. 613–17.
44 M. Frančić, *Kalendarz*, p. 14.
45 R. Grodecki, *Dzieje polityczne Śląska. Historia Śląska*, Kraków, 1933,

Vol. I, p. 208; A. Gieysztor, et al., *History of Poland* (PWN), Warszawa, 1968, pp. 110–12.

46 S. Trawkowski, 'Die Rolle der Deutschen Dorfkolonisation und des Deutschen Rechtes in Polen im 13 Jahrhundert' in *Die Deutsche Ostsiedlung des Mittelalters als-Problem der Europäischen Geschichte*, H. Beumann (ed.), Sigmaringen, 1975, pp. 349–68.

47 A. Bruce Boserll, 'Territorial Division and the Mongol Invasions, 1202–1300', Ch. V in *The Cambridge History of Poland*, p. 97.

48 J. Długosz, *Historia Polonica libri XII* (A. Przeździecki edt.), Vol. II, Kraków, 1865, p. 377; E. Power, 'The Opening of the Land Routes to Cathay', Ch. VII in *Travel and Travellers of the Middle Ages*, A. P. Newton (ed.), London (Kegan Paul), 1930, pp. 126–7; B. Grekow and A. Jakubowski, *Złota Orda i jej upadek* (translated from Russian), Warszawa, 1953, 243 pp.; B. Spuler, *Die Goldene Horde. Die Mongolen in Russland, 1223–1502*, Leipzig, 1943, 348 pp.; H. Cordier, L'Invasion mongole au Moyen-Âge et ses conséquences', *Mélanges d'histoire et de géographie orientales*, Vol. II, Paris, 1920, pp. 10–31.

49 K. Czechowicz, 'Najazdy Tatarów na Polskę w XIII w.', *Teki Historyczne* (Polskie Towarzystwo Historyczne w Wielkiej Britanii), Vol. XVI, London, 1969–71, pp. 55–78; S. Krakowski, *Polska w walce z najazdami tatarskimi w XIII wieku*, Warszawa, 1958, 241 pp.; H. Morel, 'Les Champagnes mongoles au XIIIe siècle', *Revue militaire française*, Nos. 12–13, Paris, 1922, pp. 20–41; G. Strakosch-Grassmann, *Der Einfall der Mongolen in Mittel-europa in den Jahren 1241 und 1242*, Innsbruck, 1893, 352 pp.; W. Zatorski, 'Pierwszy najazd mongolski na Polskę w latach 1240–1241', *Przegląd Historyczno-Wojskowy*, Vol. IX, Warszawa, 1937, pp. 16–35.

50 E. P. Kelly, *The Trumpeter of Kraków: A Tale of the XV Century* (McMillan) New York, 1928, 208 pp.

51 A. Gieysztor, et al., p. 110.

52 J. Milan, 'Napad Tatarów na Polskę za Leszka Czarnego w roku 1287', *Sprawozdania gimnazjum w Stanisławowie za rok 1905–1906*, Stanisławów, 1906, pp. 25–48; K. Czechowicz, 'Najazdy Tatarów', p. 55 maintains that between the first expedition in 1241 to the last in 1698, 164 major invasions are known to have taken place, but there may have been many more, which went unrecorded.

53 M. Małowist, 'The Problem of the Inequality of Economic Development in Europe in the Later Middle Ages', *Economic History Review*, Vol. XIX, No. 1, 1966, London, p. 15. See also M. Keen, *The Pelican History of Medieval Europe* (Penguin Books), Harmondsworth, 1978, p. 84.

54 See, for example, G. Labuda, 'Villes de droit polonais', in *Les origines des villes polonaises*, P. Francastel (ed.), École Pratique des Hautes Études – Sorbonne, Congrès et Colloques, Vol. II, Paris (Mouton), 1960, pp. 52–71; T. Lalik, 'Recherches sur les origines des villes en Pologne', *Acta Poloniae Historica*, Vol. II, Warszawa, 1959, pp. 101–31.

55 G. Sjoberg, *The Preindustrial City Past and Present*, New York/London, 1965, p. 27.

56 Ibid., p. 31.

57 The remarkable information of Ibrāhīm ibn Ya'qūb has come down to us through the Spanish-Arab Abū Obaid 'Abdallāh al-Bekrī (+1094) in his book,

Kitâb ab-masâlik mal-mamâlik. See A. A. Kunik and V. Rosen 'Izvestiya Al-Bekri i drugikh avtorov o Rusi i "Slavyanakh"', *Zapiski Imperatorskoy Akademià Nauk v Sankt Peterburge*, St Petersburg, 1878, Vol. XXXII, p. 46; F. Westberg, 'Ibrahim-ibn-Jacobs Reiseberichte', *Mémoires de l'Académie des Sciences de St. Petersbourg*, Series 8, Vol. III, No. 4, St Petersburg, 1898; G. Jacob, 'Arabische Berichte von Gesandten an germanische Fürstenhöfe aus dem 10 Jahrhundert', *Quellenzur deutschen Volkskunde*, Vol. I, Leipzig, 1927.

58 J. Widajewicz, *Studia nad relacją o Słowianach Ibrahima ibn Jakuba*, Kraków, 1946, p. 70; R. Jakimowicz, 'Kilka uwag nad relacją o Słowianach Ibrahima ibn Jakuba', *Slavia Antiqua*, Vol. I, Poznań, 1948, pp. 444, 451. For further discussion see G. Labuda, 'Ibrahim ibn Jakub', *Roczniki Historyczne*, Vol. XVI, Warszawa, 1947, pp. 100–83; L. Koczy 'Relacja o Słowianach Ibrahima ibn Jakuba', *Teki Historyczne*, Vol. III, Warszawa, 1949, No. 1/2, pp. 7–27; A. Miquel, *La Géographie humaine du monde musulman jusqu'au milieu du 11e siècle* (Mouton), Paris, 1973, pp. 316–19.

59 J. Brutzkus, 'Trade with Eastern Europe, 800–1200', *Economic History Review*, Vol. XIII, London, 1943, p. 34; Z. Hilczerówna, 'Przyczynki do handlu Polski z Rusią Kijowską', *Przegląd Archeologiczny*, Warszawa, Vol. IX, No. 1, 1950, pp. 8–23.

60 A. Bielowski (ed.), *Monumenta Poloniae Historica*, Vol. II, Lwów, 1872, p. 394.

61 H. Kapiszewski, 'Droga z Panonii do Polski w roku 966. Przyczynek do dziejów przełęczy dukielskiej', *Acta Archeologica Carpathica*, Vol. II, Nos. 1–2, Kraków, 1960, pp. 107–21; O. R. Halaga, 'Portae Poloniae – w kwestii szlaków karpackich we eczesnym średniowieczu', *Acta Archeologica Carpathica*, Vol. VII, 1965, Nos. 1–2, Kraków, pp. 9–15.

62 A. Zbierski, 'Ośrodek handlowy i portowy', Ch. IV, in *Historia Gdańska* (Tom I do roku 1454), E. Chieślak (ed.), Wyd. Morskie, Gdańsk, 1978, pp. 195–7; K. Śląski, 'Lądowe szlaki handlowe Pomorza w XI–XIII w.' *Zapiski Historyczne*, No. 3, Toruń, 1969, pp. 29–44; K. Śląski, 'North-Western Slavs in Baltic Sea Trade from the VIIIth to the XIIIth Century', *Journal of European Economic History*, Vol. 8, No. 1, Rome, 1979, pp. 83–108.

63 T. Lewicki, 'Ze studiów nad handlem Polski i innych ziem zachodniosłowiańskich z krajami arabskimi w IX–XI w', *Biuletyn Numizmatyczny*, No. 2, Warszawa, 1954, pp. 1–25.

64 A. Szelągowski, '*Najstarsze drogi z Polski na Wschód w okresie Bizantyńsko-Arabskim*', Kraków, 1909, 145 pp.

65 T. Lalik, p. 118.

66 A. Jaubert, *Kitāb nuzhat al-muštāk fi'ḫtirāk al-āfāk*. *Géographie d'Edrisi* (Traduite de l'arabe en français d'après deux manuscrits de la Bibliothèque du Roi et accompagnée de notes), Vol. I, Paris, 1836, p. 375. See also T. Lewicki, *Polska i kraje sąsiednie w świetle 'Księgi Rogera' geografa arabskiego z XII w. Al-Idrīsī'ego*, Part I (P.A.U.), Kraków, 1945, pp. 113–15.

67 F. Kupfer and T. Lewicki, *Żródla hebrajskie do dziejów Słowian i niektórych innych ludów środkowej i wschodniej Europy*, Wrocław-Warszawa, 1956, p. 43.

68 H. Münch, 'Kraków do roku 1257 włącznie', *Kwartalnik Architektury i Urbanistyki*, Vol. III, No. 1, Warszawa, 1958, p. 14.

69 Gallus Anonymus, *Kronika polska*, 4th edn Wrocław, 1975, p. 152. J. Mitkowski,

'Nationality Problems and Patterns in Medieval Polish Towns: The Example of Cracow', *Prace Historyczne*, No. 59, Warszawa-Kraków, 1978, Vol. 32.

70 J. Mitkowski, 'Nationality Problems', p. 32.

71 M. Borowiejska-Birkenmajerowa, *Kształt średniowiecznego Krakowa* (Wyd. Lit.), Kraków, 1975, pp. 85–9.

72 K. Radwański, 'Budowle drewniane odkryte pod poziomami romańskimi kościoła św. Wojciecha w Krakowie', *Materiały Archeologiczne*, Vol. XI, Kraków, 1970, pp. 7–23.

73 A. Vetulani, 'Krakowska biblioteka katedralna w świetle swego inwentarza z r. 1100', *Slavia Antiqua*, Vol. IV, Warszawa, 1953, pp. 163–92. On the coinage of Władysław Herman, see M. Gumowski, *Corpus nummorum Poloniae*, Vol. I, Kraków, 1939, p. 232; R. Grodecki, 'Polityka mennicza Książąt polskich w okresie piastowskim', *Wiadomości Numizmatyczno-Archeologiczne*, Warszawa, 1921, p. 47.

74 L. Lepszy, p. 15; L. Lepszy, 'Przemysł artystyczny i handel', *Rocznik Krakowski*, Vol. VI, Kraków, 1904, pp. 263–4.

75 K. Radwański, 'Budowle drewniane, pp. 279–376.

76 K. Radwański, *Kraków przedlokacyjny: Rozwój przestrzenny*, Kraków, 1975, pp. 218–91; K. Radwański, 'Wyniki badań archeologicznych prowadzonych na terenie wczesnośredniowiecznej osady "Okół"w Krakowie', *Biuletyn Krakowski*, Vol. I, Kraków, 1959, pp. 5–89.

77 K. Radwański, Kraków przedlokacyjny, pp. 279–376.

78 K. Pieradzka, 'Kraków w relacjach cudzoziemców X–XVII wieku', *Rocznik Krakowski*, Vol. XXVIII, Kraków, 1937, p. 186.

79 C. Verlinden, '*L'Esclavage dans l'Europe Médiévale*', Vol. I (Péninsule Ibérique-France), Bruges, 1955, pp. 211–25; T. Lewicki, 'Osadnictwo słowiańskie i niewolnicy słowiańscy w krajach muzułmańskich według średniowiecznych pisarzy arabskich', *Przegląd Historyczny*, Vol. XLIII, No. 3/4, Warszawa, 1952, pp. 473–91.

80 M. Małowist, 'The Baltic and the Black Sea in Medieval Trade', *Baltic and Scandinavian Countries*, Vol. III, No. 1(5), Gdynia, 1937, p. 37; F. Balodis, 'Latviešu tarptañiške sakari ap 1000 y.kr', *Latvija Vēstures Institut Žurnals*, No. 9, Riga, 1934, pp. 9–10; K. Śląski, 'Stosunki krajów skandynawskich z południowo-wschodnim wybrzeżem Bałtyku od VI do XII w.', *Przegląd Zachodni*, No. 5/8, Poznań, 1952, pp. 30–45.

81 M. N. Tikhomirov, *Drevnerusskiye goroda*, Moscow (2nd edn), 1956; B. A. Rybakov, *Remeslo drevney Rusi*, Moscow-Leningrad, 1948; F. Graus, *Dějiny venkovského lidu v Čechách v době předhusitské*, Vol. I, Prague, 1953; E. Mühle, *Die städtischen Handelszentren der nordwestlichen Ruś* (F. Steiner Verlag), Stuttgart, 1991, 371 pp.

82 S. Suchdolski, 'A propos de l'intensité de l'échange local sur les territoires polonais au X–XIe siècles', *Wiadomości numizmatyczne*, Vol. XXI, No. 1, Warszawa, 1977, pp. 1–11; R. Kiersnowski, 'Coin Finds and the Problem of Money in Early Medieval Poland', *Wiadomości Numizmatyczne*, Vol. V (supplement), Warszawa, 1961, 22 pp; J. Sztetyłło, 'Czeski i morawski pieniądz: pozakruszcowy wczesnego średniowiecza', *Kwartalnik Historii Kultury Materialnej*, Vol. XI, Nos. 3–4, Warszawa, 1963, pp. 13–21.

83 A. Gorlicki, 'Miocene Salt Deposits in Poland', *Northern Ohio Geological Society, IV Symposium on Salt*, Vol. I, Cleveland, 1974, pp. 23–30.

84 A. Gaweł, 'Budowa geologiczna złoża solnego Wieliczki', *Prace Instytutu Geologicznego*, Vol. XXX, No. 3, Warszawa, 1962, pp. 305–31; J. Poborski, 'Złoże solne Bochni na tle geologicznym okolicy', *Biuletyn Państwowego Instytutu Geologicznego*, No. 78, Warszawa, 1952, pp. 5–12.

85 A. Jodłowski, *Ekspoloatacja soli na terenie Małopolski w pradziejach i we wczesnym średniowieczu*, Kraków-Wieliczka, 1971, p. 51.

86 Ibid., p. 127. See also ibid., *Technika produkcji soli na terenie Europy w pradziejach i we wczesnym średniowieczu*, Kraków-Wieliczka, 1976, 257 pp.

87 Z. Kozłowska-Budkowa, *Repertorium polskich dokumentow doby piastowskiej*, No. 1, Kraków, 1937, pp. 29–33; H. Łabęcki, 'Najdawniejsze dzieje salin krakowskich aż do żupnictwa Jana Bonera czyli do r. 1515', *Biblioteka Warszawska*, No. 2, Warszawa, 1856, pp. 266–9.

88 *Monumenta Poloniae Historica*, Vol. II, 1872, p. 805, stating 'Sal durum in Bochnia est repertum, quod nunquam ante fuit'. See also A. Jodłowski, 'Bochnia wczesnośredniowieczna', *Studia Historyczne*, Vol. XV, No. 4, Kraków, 1972, pp. 501–35.

89 W. Bogusławski, *Dzieje Słowiańszczyzny północno-zachodniej*, Vol. II, Warszawa, 1889, p. 564.

90 F. Skibiński, 'Handel solny we wczesnym średniowieczu polskim', *Księga pamiątkowa M. Handelsmana*, Warszawa, 1929, p. 455.

91 K. Śląski, North-Western Slavs, p. 104.

92 O. R. Halaga, 'Polská sůl na Slovensku v Středověku', *Tisíc let česko-polské vzájemnosti*', Opava, 1966, p. 232.

93 D. Csánki, *Magyarország történelmi földrajza a Hunyadiák korában I*, Budapest, 1890, p. 362; V. Šmilauer, *Vodopis starêho Slovenska*, Praha-Bratislava, 1932, p. 421; G. Györffy, *Az Árpádkori Magyarország történeti földrajza*, Budapest, 1963, p. 803–4; J. Stanislav, *Slovenský juh v stredoveku*, Vol. II, Martin, 1948, p. 460.

94 V. Černý, 'Polská sůl na Oravě', *Roczniki Dziejów Społecznych Gospodarczych*, Vol. III, Lwów, 1934, p. 143; O. R. Halaga, p. 240; K. Wutke, 'Die Versorgung Schlesiens mit Salz während des Mittelalters' *Zeitschrift des Vereins für Geschichte und Alterhum Schlesiens*, Vol. XXVII, Breslau, 1893, pp. 257–8, 261; A. Aristov, *Promyshlennost' drevniey Rusi*, St Petersburg, 1866, pp. 69–70; *Kodeks dyplomatyczny Małopolski* (Codex diplomaticus Poloniae Minoris) F. Piekosiński (ed.), Vol. II, Kraków, 1880, No. 375 (for the year 1198); S. Ciszewski, 'Sól', *Wisła*, Vol. XXI, No. 1, Warszawa, 1922, p. 16; H. Burchard, A. Keckowa, L. Leciejewica, 'Die Salzgewinnung auf polonischen Boden im Alterum und in frühen Mittelalter', *Kwartalnik Historii Kultury Materialniej*, Vol. XIV (*Ergon*, Vol. V), Warszawa, 1966, pp. 745–60.

95 K. Tymieniecki, 'Organizacja rzemiosła wczesnośredniowiecznego a geneza miast polskich', *Studia wczesnośredniowieczne*, Vol. III, Warszawa-Wrocław, 1955, pp. 33–40.

96 J. Mitkowski, p. 32.

97 A. Gieysztor et al., *History of Poland*, p. 89.

98 S. Smolka, *Mieszko Stary i jego wiek*, Warszawa, 1881, Vol. 53, finds the same in later times.

99 H. Łowmiański, 'Economic Problems of the Early Feudal Polish State', *Acta Poloniae Historica*, Vol. III, Warszawa, 1960, pp. 22–3.

100 Z. Wojciechowski, *Państwo polskie w wiekach średnich. Dzieje ustroju*, Wyd. II, Poznań, 1948, p. 80; J. Bardach, *Historia państwa i prawa Polski do roku* Vol. I, Warszawa, 1957, p. 141; K. Buczek, 'Publiczne posługi transportowe i komunikacyjne w Polsce średniowiecznej', *Kwartalnik Historii Kultury Materialnej*, Vol. XV, No. 2, Warszawa, 1967, pp. 255–99.

101 *Kodeks dyplomatyczny Małopolski*, Vol. II, No. 451 – a document from King Bolesław Wstydliwy (The Shy) 1243–79, dated 1256 mentions 'nullum powoz neque currus'.

102 T. Wąsowiczówna, 'Research on the Medieval Road System in Poland', *Archaeologia Polona*, Vol. II, Warszawa, 1959, p. 130; A. Gieysztor, 'Local Markets and Foreign Exchange in Central and East Europe before 1200', *Kwartalnik Historii Kultury Materialnej*, Vol. XIV (*Ergon*, Vol. V), Warszawa, 1966, p. 769.

103 A. Bochnak, 'Sztuka Krakowa w Tysiącleciu', Ch. 7 in *Kraków i Małopolska przez dzieje*, p. 145.

104 *Monumenta Poloniae Historica*, Vol. II, p. 257; J. Z. Jakubowski (ed.), *Literatura Polska od średniowiecza do pozytywizmu* (P.W.N.), Warszawa, 1975, p. 25.

105 A. Gieysztor et al., *History of Poland*, p. 91.

106 J. Adamczewski, *In Cracow* (Interpress) 1973, p. 21, which states 'The oldest parochial school (in Cracow) was established in 1223, founded by Bishop Iwo Odrowąż in the St. Mary's Church parish.'

107 Gallus Anonymus, *Cronicae et gesta ducum sive principum Polonorum*. See also *Monumenta Poloniae Historica*, Nova Series, Vol. II, Kraków-Warszawa, 1948, pp. 75, 83; J. Z. Jakubowski, pp. 22–4.

108 A. Gieysztor et al. *History of Poland*, p. 93.

109 L. Lepszy, *Cracow: The Royal Capital*, p. 15.

110 J. Muczkowski, 'Historia rzeźby', *Rocznik Krakowski*, Vol. VI, Kraków, 1904, pp. 157–98.

111 W. Ostrowski, 'History of the Urban Development and Planning', Ch. 1, in *City and Regional Planning of Poland*, J. C. Fisher (ed.), Cornell Univ. Press, Ithaca, New York, 1966, p. 11. For further comparison see R. Marsina, 'Pour l'Histoire des villes en Slovaquie au Moyen Âge', *Studia Historica Slovaca*, Vol. VIII, pp. 21–75, Bratislava.

112 Some German historians are inclined to regard Polish medieval towns as a result of German municipal colonization in the east, or even as east German towns. Towns, however, existed in Poland long before German colonists came, and the urban centres contained numerous nationalities as well as Poles. For further discussion see, F. Rörig 'Die deutsche Stadt in der deutschen geschichte' in *Die Stadt des Mittelalters*, C. Haase (ed.), Vol. I, Darmstadt, 1969, pp. 7–33; K. Buczek, *Targi i miasta na prawie polskim (Okres wczesnośredniowieczny)*, P.A.N. Odd. w Krakowie (Prace Komisiji Nauk Historyczanych No. 11) Wrocław-Warszawa-Kraków, 1964, 139 pp.

113 M. Bukowski and M. Zlat, *Ratusz wrocławski*, Wrocław 1958, pp. 11–14; W. Kalinowski, *Zarys historii miast w Polsce do połowy XIX wieku*, Toruń, 1966, pp. 10–11; H. Münch, 'Początki średniowiecznego układu miejskiego

w Polsce ze szczególnym uwzględnieniem Śląska', *Kwartalnik Architektury i Urbanistyki*, Vol. V, No. 3, Warszawa, 1960, pp. 357, 361, 362, 364.

114 K. Bromek, p. 19.

115 *Zbiór dyplomów klasztoru mogilskiego*, E. Janota and F. Piekosiński (eds.), Kraków, 1865, pp. 7, 9–10.

116 A. Gieysztor et al., *History of Poland*, p. 100.

117 R. Grodecki, *Podziały i zjednoczenie państwa polskiego* (Teksty źródłowe z 18) Kraków, 1924, p. 16.

118 S. Zachorowski, 'Kraków biskupi', *Rocznik Krakowski*, Vol. VIII, Kraków, 1906, pp. 103–26; H. Münch, 'Kraków do roku' p. 27; H. Münch, 'Podstawowe problemy urbanistyczne przedlokacyjnego Krakowa', *Teka Komisji Urabnistyki i Architektury*, Vol. II, Kraków, 1968, p. 183; J. Jamroz, 'Układ przestrzenny miasta Krakowa sprzed lokacji 1257 r. stan dotychczasowych badań nad miastem', *Biuletyn Krakowski*, Vol. II, Kraków, 1960, pp. 10–25.

119 *Zbiór dyplomów klasztoru*, p. 17.

120 K. Bąkowski, 'Historia Krakowa w zarysie', *Biblioteka Krakowska*, No. 6, Kraków, 1898, p. 15.

121 *Kodeks dyplomatyczny Wielkopolski* (Codex diplomaticus Majoris Poloniae), I. Zakrzewski and F. Piekosiński, Vol. I, No. 321, Poznań, 1877; A. Kłodziński, *Przywileje lokacyjne Krakowa i Poznania* (Biblioteka Zródeł Historycznych No. 6), Poznań, 1947, pp. 24–7.

122 *Osiemnaście wieków Kalisza*, A. Gieysztor and K. Dąbrowski (eds.), Vol. I, Kalisz, 1960, pp. 13–15.

123 A. Kłodziński, pp. 17–20; *Kodeks dyplomatyczny miasta Krakowa 1257–1560* (Monumenta medii aevi historica t.V), Vol. I, Kraków, 1879, No. 1.

124 W. Ostrowski, p. 13. See also K. Dziewoński, 'The Plan of Cracow: Its Origin, Design and Evolution', *The Town Planning Review*, Vol. XIX, Liverpool, 1943–7, pp. 29–30.

125 A. Kłodziński, p. 17.

126 D. E. Friedlein, *O magistratach miast polskich a w szczególności miasta Krakowa*, Kraków, 1845, p. 288; Z. Kaczmarczyk, 'Początki miast polskich. Zagadnienia prawne', *Czasopismo proawno-historyczne*, Vol. XIII, Warszawa, 1961, pp. 9–45; A. Roykowska-Płachcińska, *Gmina miejska w poczatkach XIII w. w Polsce*, Warszawa, 1962, pp. 143–50; G. Labuda, pp. 52–71.

127 S. Estreicher, *Kraków i Magdeburg w przywileju fundacyjnym krakowskim* (Księga ku uczczeniu Bolesława Ulanowskiego), Kraków, 1911, pp. 10–12.

128 M. Borowiejska-Birkenmajer, 'Problem pierwszej lokacji i wielka lokacja Krakowa z r. 1257 w świetle ostatnich badań', *Teka Komisji Urbanistyki i Architektury*, Vol. VIII, Kraków 1974, pp. 19–36; M. Tobiasz 'Założenia przestrzenne przedlokacyjnego Krakowa', *Teka Komisji Urbanistyki i Architektury*, Vol. VIII, Kraków, 1974, pp. 37–50.

129 The name comes from Polish 'gród' (monarchical dwelling), the road leading to the royal residence, 'platea Castrensis'. See J. Mitkowski, 'Nationality Problems', p. 33.

130 T. Ruszczyńska and A. Sławska, *Poznań*, Warszawa, 1953, pp. 25–7.

131 K. Pieradzka, 'Rozkwit średniowiecznego Kraków w XIV i XV wieku', Ch. 6 in *Kraków: Studia nad rozwojem miasta*, p. 154.

132 S. Świszczowski, 'Mury miejskie Krakowa', *Ochrona Zabytków*, Vol. VIII, Warszawa, 1955, p. 162.
133 A. Kłodziński, pp. 18–19.
134 Ibid.

4 The political situation of Cracow 1257–1500

1 For further discussion see D. Hay, *Europe in the Fourteenth and Fifteenth Centuries*, Ch. 9 'The Central Monarchies' (Longman), 1976, London, pp. 212–41.
2 See P. W. Knoll, *The Rise of the Polish Monarchy: Piast Poland in East Central Europe, 1320–1370* (Univ. of Chicago Press), Chicago and London, 1972, pp. 2–3.
3 G. Slocombe, *A History of Poland: Earliest Times to 1939* (T. Nelson & Sons), London/New York, 1939, pp. 58–9. Further discussion may also be found in O. Halecki, *A History of Poland* (J. M. Dent & Sons), New York/London, 1961, Ch. 5, 'The Reconstruction of the Kingdom', pp. 41–51.
4 A. Gieysztor, et al, *History of Poland* (P.W.N.), Warszawa, 1968, pp. 112–13.
5 Further discussion may be found in E. Ennen, *Frühgeschichte der Europäischen Stadt* (L. Rohrscheid Verlag), Bonn, 1953, 345 pp.; Ibid., 'Zur typologie des Stadt-Land-Verhältnisses im Mittelalter', *Studium Generale*, Vol. XVI, 1963, pp. 445–56; H. Planitz, *Die deutsche Stadt im Mittelalter* (Böhlau-Verlag), Cologne, 1965, 365 pp.
6 N. J. G. Pounds, *An Economic History of Medieval Europe* (Longman), 1974, London/New York, p. 248.
7 J. Garbacik, 'Przeniesienie stolicy Polski do Krakowa w XI w.', Ch. 6 in *Kraków i Małopolska przez dzieje: Studia i szkice profesorów Uniwersytetu Jagiellońskiego*, C. Bobińska (ed.), Wyd. Lit., Kraków, 1970, pp. 127–41.
8 L. Lepszy, *Cracow: The Royal Capital of Ancient Poland; Its History and Antiquities* (T. F. Unwin Publ.), London, Leipzig, 1912, p. 23.
9 O. Balzer, *Królestwo Polskie 1295–1370*, 3 Vols. Lwów, 1919–20, Vol. I, pp. 172–97.
10 J. Baszkiewicz, *Powstanie zjednoczonego państwa polskiego (na przełomie XII i XIV w.)*, Warszawa, 1954, pp. 440–5; D. Borawska, *Z dziejów jednej lengendy*, Warszawa, 1950, p. 50.
11 'Łokietek' means 'one ell high' – a nickname given to this ruler on account of his small stature.
12 P. W. Knoll, *The Rise of the Polish Monarchy*, p. 16.
13 It was noted in Rocznik Kapitulny Krakowski, '1265 Castrum edificatur in Cracovia super totum montem cum lignis', *Monumenta Poloniae Historica* (Nova Series), Vol. II, Kraków-Warszawa, 1946, p. 808.
14 M. Frančić, *Kalendarz dziejów Krakowa* (Wyd. Lit.) Kraków, 1964, p. 20.
15 P. W. Knoll, *The Rise of the Polish Monarchy*, p. 16.
16 K. Bąkowski, *Historya Krakowa w zarysie* (Wyd. 'Czasu'), Kraków, 1898, p. 22.
17 J. Garlicki, J. Kossowski and L. Ludwikowski, *A Guide to Cracow* (Sport i Turystka Wyd.), Warszawa, 1969, p. 24; S. Świszczowski, 'Gródek i mury miejskie miedzy Gródkiem a Wawelem', *Rocznik Krakowski*, Vol. XXXII, Kraków, 1950, pp. 1–41.

18 K. Bąkowski, *Historia Krakowa*, p. 23.
19 S. Zachorowski, 'Wiek XIII i panowanie Władysława Łokietka', *Encyklopedya Polska*, Vol. V, pt. 1, Warszawa-Kraków-Lublin-Łódź, 1920, pp. 134–309.
20 J. Garlicki, J. Kossowski and L. Ludwikowski, *A Guide to Cracow*, p. 25.
21 K. Bąkowski, *Historia miasta Kazimierza pod Krakowem do XVI w.*, Kraków 1903, 303 pp.; W. Konieczna, 'Początki Kazimierza (do r. 1419) (Studia nad przedmieściami Krakowa)', *Biblioteka Krakowska*, No. 94, Kraków, 1938, pp. 7–90.
22 J. Dzikówna, 'Kleparz do 1528 roku', *Biblioteka Krakowska*, No. 74, Kraków 1932, pp. 26–30; S. Witek, *Z dziejów Kleparza* (W 600 rocznicę nadania praw miejskich przez króla Kazimierza Wielkiego (Wyd. Art-Graf) Kraków 1968, 93 pp.
23 J. Dużyk and S. Salmonowicz, *Cracovie et son Université* (Wyd. Art.-Graf.) Kraków, 1966, pp. 8–90; L. Hajdukiewicz and M. Karaś, *The Jagiellonian University (Traditions, The Present, The Future)*, Cracow, 1975, 108 pp.; W. M. Bartel, 'L'Université de Cracovie jusqu'à l'année 1500', *Ius Romanum Medii Aevi*, Pars V, 8 bis, Mediolani (Typis Giuffrè), 1981, pp. 1–44.
24 R. Kiersnowski, *Wstęp do numizmatyki polskiej wieków średnich*, Warszawa, 1964, pp. 128–31.
25 R. Grodecki, *Kongres Krakowski w roku 1364* (Gebethner & Wolff), Warszawa, 1939.
26 P. W. Knoll, *The Rise of the Polish Monarchy*, p. 216.
27 M. Frančić, *Kalendarz*, p. 34.
28 W. Łuszczkiewicz, 'Sukiennice krakowskie', *Biblioteka Krakowska*, No. 11, 1899 Kraków, 204 pp.; J. Dużyk, *Sukiennice* (Kraków Dawniej i Dziś No. 4) Kraków, 1952, 23 pp.
29 L. Lepszy, 'Pergameniści i papiernicy krakowscy w ubiegłych wiekach i ich wyroby', *Rocznik Krakowski*, Vol. IV, Kraków 1901, pp. 233–48.
30 J. Ptaśnik, 'Towns in Medieval Poland', Ch. 2 in *Polish Civilization: Essays and Studies*, M. Giergielewicz (ed.) (New York Univ. Press), New York, 1979, p. 37; See also, F. Giedroyć, *Wodociągi i kanały miejskie*, Warszawa, 1910, pp. 4–5; J. Rajman, 'Początki "przemysłowego" Krakowa' *Kraków*, No. 2, 1987, pp. 53–4.
31 J. Długosz, *Historia Polonicae libri IV* (ed. A. Przeździecki), Kraków, 1869, pp. 35–8; P. Garou, 'Tannenberg: Jagellon contre les chevaliers Teutoniques', *Histoire Magazine*, No. 17, Paris, 1981, pp. 14–16.
32 L. Lepszy, *Cracow: The Royal Capital*, p. 30.
33 Ibid.
34 Ibid. According to M. Frančić, *Kalandarz*, p. 41 the first printing in Cracow was in 1473 by Kaspar Straub; see also A. Gieysztor, et al., *History of Poland*, p. 159.
35 L. Lepszy, *Cracow: The Royal Capital*, pp. 30–1; The first printer permanently living in Crakow was Kasper Hochfeder employed by the merchant Jan Haller. See K. Bąkowski, p. 43; M. Frančić, *Kalendarz* p. 47; J. S. Bandtkie, *Historya drukarń krakowskich, od zaprawadzenie druków do tego miasta aż do czasów naszych, wiadomością o wynalezieniu sztuki drukarskiej poprzedzone*, Warszawa, 1974, 504 pp.
36 J. Bogdanowski, 'Barbakan krakowski jako "dzieło kluczowe" miasta', *Teka komisji urbanistyki i architektury*, Vol. IX, Kraków, 1975, pp. 5–24.

37 K. Bromek, 'Zarys rozwoju historycznego i terytorialnego Krakowa', *Folio Geographica* (series Geographica-Oeconomica) Vol. VIII, Kraków, 1975, p. 23; M. Mastalerzówna, 'Przedmieście Stradom w wiekach średnich', *Biblioteka Krakowska*, No. 94, Kraków, 1938, pp. 91–4.
38 K. Pieradzka, 'Garbary przedmieście Krakowa' (1363–1587), *Biblioteka Krakowska*, No. 71, Kraków, 1939, 154 pp.
39 F. Kopera, 'Wit Stwosz w Krakowie', *Rocznik Krakowski*, Vol. X, Kraków, 1907, pp. 1–121; T. Szydłowski, 'Ze studiów nad Stwoszem i sztuką jego czasów, *Rocznik Krakowski*, Vol. XXVI, Kraków, 1935, pp. 1–72.
40 A. Gieysztor et al., p. 152; See also S. Gierszewski, *Obywatele miast Polski przedrozbiorowej* (P.W.N.), Warszawa, 1973, Ch. 1, pp. 15–30; T. Ładogórski *Studia nad zaludnieniem Polski XIV wieku* (P.A.N. Inst. Hist.) Wrocław, 1958, 229 pp.
41 H. Pirenne, *Economic and Social History of Medieval Europe* (Routledge & Kegan Paul), London (7th edn), 1961, p. 213.
42 R. Grodecki, *Wybór źródeł do dziejów społeczno-gospodarczych Polski feudalnej*, Kraków, 1959, pp. 39–40.
43 K. Stachowska 'Prawo składu w Polsce do 1565 r.', *Sprawozdania z czynności i posiedzeń Polskiej Akademii Umiejętności*, Vol. LI, Nov. 1950, No. 9, Kraków, 1951, pp. 586–93; see also S. Lewicki, *Prawo Składu w Polsce*, Lwów, 1910.
44 A. Jelicz, *W średniowiecznym Krakowie (wiek XII–XV)* (P.I.W.), Warszawa, 1966, pp. 101–2.
45 M. Bobrzyński, 'Bunt wójta krakowskiego Alberta z roku 1311', *Biblioteka Warszawska*, Vol. XXXVII, No. 3, Warszawa, 1877, pp. 329–48; E. Długopolski, 'Bunt wójta Alberta' *Rocznik Krakowski*, Vol. VII, Kraków 1905, pp. 135–86; A. Kłodziński, 'Jeden czy dwa bunty wójta Alberta', *Sprawozdanie Akademii Umiejętności: Wydział Historyczno-Filozoficzny*, Vol. XLIII, Kraków, 1938, pp. 47–51; A. Kłodzinski, 'Z dziejów pierwszego krakowskiego buntu wójta Alberta', *Zapiski Towarszystwa Naukowego w Toruniu*, Vol. XIV, Toruń, 1948, pp. 45–56.
46 K. Dziwik, 'Rozwój przestrzenny miasta Nowego Sącza od XIII do XIX wieku na tle stosunkow gospodarczych', *Rocznik Sądecki*, Vol. V, Nowy Sącz, 1962, pp. 165–73.
47 S. Lewicki, pp. 135–8; K. Dziwik, p. 172.
48 K. Myśliński, 'Rola miast Polski środkowej w handlu z krajami na południe od Karpat do końca XV wieku', *Prace Historyczne*, Vol. LVI, Warszawa-Kraków, 1977, pp. 22–4; S. Kutrzeba and J. Ptaśnik, Dzieje handlu i kupiectwa krakowskiego', *Rocznik Krakowski*, Vol. XIV, Kraków, 1912, p. 21; K. Myśliński, 'Lublin a handel Wrocławia z Rusią w XIV i XV wieku' *Rocznik Lubelski*, Vol. III, Lublin, 1960, pp. 5–36, L. Białkowski, 'Lublin na starych szlakach handlowych', *Pamiętnik Lubelski*, Vol. III, Lublin, 1938, pp. 288–93.
49 Ł. Charewiczowa, 'Handel Lwowa z Mołdawią i Multanami w wiekach średnich', *Kwartalnik Historyczny* Vol. XXXVIII, Nos. 1–2, Lwów, 1924, pp. 37–67; see also L. Charewiczowa, *Handel średniowiecznego Lwowa* (Studja nad Historiją Kultury w Polsce, Vol. I), Lwów, 1925, pp. 61–82; *Kodeks dyplomatyczny miasta Krakowa 1257–1506* (F. Piekosiński ed.), Vol. I, Kraków 1879, no. 104; S. Kutrzeba, 'Handel Krakowa w wiekach średnich na tle stosun-

ków handlowych Polski' *Rozprawy Akademii Umiejętności: Wydział Historyczno-Filozoficzny*, Vol. XLIV, Kraków, 1902, pp. 106–15.

50 M. Frančić, Kalendarz, p. 36; K. Pieradzka, 'Rozkwit średniowiecznego Krakowa w XIV i XV wieku', Ch. 6 in *Kraków: Studia nad rozwojem miasta* (J. Dąbrowski ed.) (W. Lit.) Kraków, 1957, p. 182.

51 J. Mitkowski, 'Nationality Problems and Patterns in Medieval Polish Towns: the Example of Cracow', *Prace Historyczne*, Vol. LIX, Warszawa-Kraków, 1978, p. 38.

52 Ibid.

53 For a more general discussion see M. Bałaban, *Historia Żydów w Krakowie na Kazimierzu 1304–1868*, Vols. I–II, Kraków, 1931–6; A. Gieysztor et al., pp. 154–5.

54 A. Jelicz, pp. 41–5.

55 M. Frančić, *Kalendarz*, pp. 33–4.

56 Ibid.

57 S. Kutrzeba, 'Ludność i majątek Kazimierza w końcu XIV stulecia' *Rocznik Krakowski*, Vol. III, Kraków 1900, pp.183–201; K. Pieradzka, pp. 182–3; S. Świszczowski, 'Założenie i rozwój miasta Kazimierza' *Biuletyn Krakowski*, Vol. III, Kraków, 1962, pp. 10–25.

58 H. Samsonowicz, 'Warszawa w handlu średniowiecznym', *Warszawa Średniowieczna* No. 2 (Inst. Hist. P.A.N.), Warszawa, 1975, pp. 19, 22, 24, 28; N. Krebs, 'Krakau und Warschau als Spiegelbilder polnischer Geschichte' *Zeitschrift der Gesellschaft für Erdkunde*, Nos. 1/2, Berlin, 1940, pp. 45–62; W. A. Wagner 'Handel dawnego Jarosławia do połowy XVII wieku', *Prace Historyczne* (Wydane ku uczeniu 50-lecia akademickiego koła historyków uniwersytetu Jana Kazimierza we Lwowie 1878–1928), Lwów, 1929, pp. 122–47; L. Koczy, 'Handel Litwy przed połową XVII wieku', *Pamiętnik VI zjazdu historyków polskich w Wilnie*, Lwów, 1935, Vol. I, pp. 272–8.

59 S. Piekarczyk, *Studia z dziejów miast polskich w XIII i XIv w.*, Warszawa, 1955; H. Münch, 'Melioratio terrae nostrae. Szkice z dziejów miast Polski średniowiecznej', *Kwartalnik Architektury i Urbanistyki*, Vol. X, No. 2, Warszawa, 1965, pp. 27–38.

60 S. Weymann, *Cła i drogi handlowe w Polsce Piastowskiej* (Poznańskie Towarzystwo Przyjaciół Nauk, Prace Komisji Historycznej, Vol. XIII, No. 1), Poznań, 1938, 144 pp.; S. Kutrzeba, 'Taryfy celne i polityka celna w Polsce od XIII do XIV w.', *Ekonomista*, Vol. II, Warszawa, 1902, pp. 33–48, 187–211.

61 H. Samsonowicz, 'Przemiany osi drożnych w Polsce późnego średniowiecza' *Przegląd Historyczny*, Vol. LXIV, No. 4, Warszawa, 1973, pp. 697–716; T. Wąsowicz, 'Sandomierska sieć drożna w wiekach średnich', *Studia Sandomierskie*, Vol. I, Sandomierz, 1967, pp. 113–20.

62 A. Żaki, 'Rola Krakowa w postępach rozwoju miast Małopolskich', *Sprawozdania z Posiedzeń Komisji Naukowych Oddział* (P.A.N.), Vol. XII, No. 1, Kraków, 1962, pp. 29–31; W. M. Bartel, 'Stadt und Staat in Polen im 14 Jahrhundert', in *Stadt und Stadtherr in 14 Jahrhundert: Entwicklungen und Funktionen* (herausgegeben von Wilhelm Rausch), Österreichischen Arbeitskreises für Stadtgeschichtsforschung, Linz/Donau, 1972, pp. 129–64.

63 D. Hay, p. 216.

64 F. J. Jeckel, *Pohlens Handelsgeschicte*, Vol. I, Vienna/Trieste, 1809, p. 165.
65 J. Dąbrowski, 'Z czasów Łokietka. Studia nad stosunkami polsko-węgierskimi w XIV', *Rozprawy Akademii Umiejętności: Wydział Historyczno-Filozoficzny*, Series II, Vol. XXXIV, Kraków 1916, pp. 278–326; Ibid., 'Lokietek Ulászló és Magyarország 1300–1315 -ben' *Történeti Szemle*, Vol. X, Budapest, 1921, pp. 76–84; E. Kovács, 'Az Anjouk Lengyel politikája' Ch. II in *Magyarok és Lengyelek a Történelem Sobrában* (Gondolat), Budapest, 1973 (Karoly Robert és Lokietek Ulászló), pp. 34–42.
66 F. Piekosiński and J. Szujski (eds), *Najstarsze księgi i rachunki miasta Krakowa od r. 1300 dod 1400*, Vol. I, Kraków 1878, p. 4; in which one Jakób Węgier, who died in Cracow in 1301, left 'his family and considerable belongings and other non-movable items in the city where he and his commerce had long been'.
67 W. Felczak, *Historia Węgier* (Zak. Narod. im Oss. Wyd.), Wrocław/Warszawa/Kraków, 1966, p. 60.
68 F. Piekosiński (ed.) *Kodeks dyplomatyczny miasta Krakowa 1257–1506*, Vol. I, Kraków, 1879, No. 23 dated 7 December, 1339 in Wyszegrad.
69 Ibid. See also J. Dąbrowski 'Kraków a Węgry w wiekach średnich,' *Rocznik Krakowski*, Vol. XIII, Kraków, 1911, pp. 189–250; The problem of the Angevin succession has been well documented in J. Dąbrowski, *Ostatnie lata Ludwika Wielkiego 1370–1382*, Kraków, 1913, whilst an alternative viewpoint is offered by O. Halecki, 'O genezie i znaczeniu rządów andegaweńskich w Polsce' *Kwartalnik Historyczny*, Vol. XXXV, Kraków, 1921, pp. 31–68; This was answered in J. Dąbrowski, 'Polityka andegaweńska Kazimierza Wielkiego' *Kwartalnik Historyczny*, Vol. XXXVI, Kraków 1922, pp. 11–40; see also J. Dąbrowski 'Dzieje Polski od r. 1333 do r. 1506', Vol. II of R. Grodecki, S. Zachorowski and J. Dąbrowski, *Dzieje Polski średniowiecznej*, Kraków, 1926.
70 O. Halecki, 'Kazimierz Wielki', *Historia Polski* (do roku 1764), Vol. I, H. Łowmiański (ed.), Warszawa, 1958, pp. 339–40.
71 F. Piekosiński (ed.), *Kodeks*, Vol. I, No. 32, dated 7 Dec. 1358.
72 Ibid., document dated 15 July 1368.
73 *Codex diplomaticus Hungarie ecclesiasticus et civilis*, G. Fejér (ed.), Vol. IX, Pt. 3, Buda, 1840, pp. 241–3; 'quod omnes mercatores extranei, de regnis scilicet Ruscie, Polonie et provinciis eorumdem in regnum nostrum Hungarie venientes pretextu vendicionis suarum rerum et mercium, ultra ipsam civitatem Cassensem'; R. Marsina, 'Pour l'histoire des villes en Slovaque au moyen âge' *Studia Historica Slovaca*, Vol. VIII, Bratislava, 1975, pp. 21–72; O. R. Halaga 'Spojenia slovenských miest s Pol'skom a Rusou do 16 stor.', *Historické Stúdie*, Vol. XI, Bratislava, 1966, pp. 139–42; Ibid., 'Počiatky dial'kového obchodu cez stredné karpaty a kősického práva skladu', *Historica Carpatica*, Vol. IV, Bratislava, 1973, pp. 12–30.
74 A. Divéky, *Felsö–Magyarország kereskedelmi összekötetése Lengyelországgal fölog a XVI-XVII században*, Budapest, 1905, pp. 7, 11.
75 S. Kutrzeba, 'Handel Krakowa w wiekach średnich', p. 32.
76 *Kodeks dyplomatyczny*, Vol. I, No. 57; document dated 11 November 1380.
77 E. Mályusz, *Zsigmondkori oklevéltár*, Vol. I, Budapest, 1951, p. 4620; G. Fejér, Vol. X, Pt. 2, p. 504; for further discussion see J. Bartl, 'Political and Social Situation in Slovakia at the Turning Point of the 14th and 15th Centuries and the

Reign of Sigismund of Luxemburg' *Studia Historica Slovaka*, Vol. IX. Bratislava, 1979, pp. 41–84; O. Halaga, 'Krakovské právo skladu a sloboda obchodu v Uhorsko-Pol'sko-Pruskej politike pred Grunwaldem', *Slovanské Stúdie*, Vol. XVI, Bratislava, 1975, pp. 163–91.

78 L. Fejér-Pataky, *Magyarországi városok régi számadáskönyvei*, Budapest, 1885, index; W. Semkowicz, 'Akt zastawa XVI miast spiskich Polsce w 1412' *Wierchy* (Rocznik poświęcony górom i góralszczyźnie), Vol. VIII, Warszawa, 1930, pp. 152–8.

79 *Kodeks dyplomatyczny*, Vol. I, No. 137: document dated 10 October 1440.

80 *Ibid.*, No. 183: document dated 26 October 1473.

81 M. Dogiel, *Codex Diplomaticus Regni Polonia et Magni Ducatus Lituaniae in quo pacta, foedera tractus pacis*, Vol. I, Wilno, 1758, p. 87.

82 J. F. N. Bradley, *Czechoslovakia: A Short History* (University Press), Edinburgh, 1971, p. 21.

83 J. Fiedler, 'Böhmens Herrschaft in Polen. Ein urkundlicher Beitrag', *Archiv für Kunde österreichischer Geschichtsquellen*, Vol. XIV, Berlin, 1885, pp. 182–5; J. Dąbrowski, 'Dzieje polityczne Śląska w latach 1290–1402', Vol. I, pp. 334–6 in *Historia Śląska od najdawniejszych czasów do roku 1400*, S. Kutrzeba et al. (ed.), 3 Vols., Kraków, 1933.

84 *Kodeks dyplomatyczny katedry krakowskiej św. Wacława*, F. Piekosiński (ed.), 2 Vols., Vol. I, No. 94, 95, Kraków, 1874.

85 Chronicon aule regiae, in *Fontes rerum Bohemicarum*, F. Palacký and J. Emler (eds.), 7 Vols., Vol. IV, Prague, 1890, p. 82.

86 J. Długosz, *Historia Polonicae*, Vol. III, p. 21.

87 R. Grodecki, 'Pojawienie się groszy czeskich w Polsce' *Wiadomości Numizmatyczno-Archeologiczne*, Vol. XVLLL, Kraków, 1936, p. 77.

88 E. Długopolski, *Władysław Łokietek na tle swoich czasów*, Wrocław, 1951, p. 46.

89 P. W. Knoll, p. 27.

90 *Kodeks dyplomatyczny Polski*, L. Rzyszczewski, A. Muczkowski and J. Bartoszewicz (eds.), 3 Vols. Warszawa, 1847, 1858, Vol. III, No. 67; see also S. Morawski, *Sądecczyzna*, Kraków, 1863, Vol. I, pp. 177–9; J. Syganski, *Nowy Sącz jego dzieje i pamiątki dziejowe*, Nowy Sącz, 1892, pp. 14–16; R. Kesserling, *Neu-Sandez und das Neu Sandezer Land. Ihre deutsche Vergangenheit und Aufbauarbeit (1230–1940)*, Krakau, 1941, pp. 62–3; A. Rutkowska-Plachcinska, *Sądecczyzna w XIII i XIV wieku*, Przemiany gospodarcze i społeczne, Wrocław – Warszawa-Kraków, 1961, pp. 126–30; K. Dziwik, 'Majętność ziemska Nowego Sącza w wiekach średnich,' *Rocznik Sądecki*, Vol. IV, Nowy Sącz, 1960, pp. 53–5.

91 *Kodeks dyplomatyczny Małopolski*, F. Piekosiński (ed.), 4 Vols., Kraków, 1876–1905, Vol. II, No. 536, 'magis aptus propter necessitatem defensionis terre nostrae Cracoviensis'.

92 See for example, D. Długopolski, 'Bunt wójta Alberta', pp. 140, 169; *Kodeks dyplomatyczny Polski*, Vol. III, No. 75; *Monumenta Poloniae Historica*, Vol. II (A. Bielowski ed.), Lwów, 1873, No. 557; J. Szujski and F. Piekosiński, *Stary Kraków*, Kraków 1901, pp. 40, 44, 62; for events surrounding the dispute leading to the agreement between Cracow and Nowy Sącz, 30 May 1329, see *Kodeks*

dyplomatyczny miasta Krakowa, Vol. I, No. 16; *Monumenta Poloniae Historica*, Vol. I, No. 178.

93 Treaty signed between King Ludwik Węgierski and Karel IV, in which Cracow merchants could go to Prague for six years under a free trade agreement. Unfortunately the Hussite Wars prevented full utilization of this pact. *Kodeks dyplomatyczny miasta Kraków*, Vol. I, No. 52.

94 S. Kutrzeba, Handel Krakowa w wiekach, p. 87; for further discussion see J. Cělakowsky, *Privilegia Královskich měst venkovských*, Prague 1892, No. 303, p. 445; F. L. Hübsch, *Versuch einer Geschichte des böhmischen Handels*, Prague 1849, pp. 204, 251, 265.

95 J. F. N. Bradley, p. 50.

96 S. Kutrzeba and J. Ptaśnik, 'Dzieje handlu i kupiectwa krakowskiego' *Rocznik Krakowski*, Vol. XIV, Kraków 1912, pp. 88–90; evidence of the first Czechs in Cracow date from 1440.

97 J. Heřman 'Obchod pražských židů s Polskem do roku 1541', *Tisíc let česko-polské vzájemnosti*, Vol. I, Opava, 1966, pp. 197–203.

98 S. Krakowski, *Polska w walce z najazdami tatarskimi w XIII wieku*, Warszawa, 1956, pp. 146–8.

99 J. Długosz, *Historia Polonicae*, Vol. IV, pp. 112–14.

100 H. Wendt, *Schlesien und der Orient. Ein geschichtlicher Rückblick*, Breslau, 1916, pp. 52, 55, 191, 209; see also M. Morelowski, 'Rozwój urbanistyczny Wrocławia średniowiecznegp', Ch. 2 in *Wrocław-Rozwój Urbanistyczny*, K. Maleczyński, M. Morelowski and A. Ptaszycka (eds.) (Budow. i Arch.), Warszawa, 1956, pp. 11–79.

101 The original document of 1242 has been lost but the reconfirmation of the municipal charter in 1261 is extant. See K. Maleczyński and J. Reiter, *Teksty źródłowe do historii Wrocławia*, Pt. I, Wrocław, 1951, pp. 22–3.

102 J. Korn, *Breslauer Urkundenbuch*, Breslau, 1870, No. 43. For further discussion see S. Kutrzeba, Handel Krakowa w wiekach średnich, pp. 72–3; L. Koczy, *Związki handlowe Wrocławia z Polską do końca XVI wieku* (Polski Śląsk: Odczyty i Rozprawy No. 18), 34 pp.

103 K. Maleczyński and J. Reiter, pp. 57–8.

104 O. Balzer, *Królestwo Polskie 1295–1370*, 3 Vols., Lwów, 1919–20, Vol. I, p. 272; S. Piekarczyk, p. 124; *Polski Słownik Biograficzny*, W. Konopczyński, K. Lepszy and E. Rostworowski (eds.), 13 Vols. Kraków-Wrocław, 1935–1968, Vol. IX, pp. 405–8.

105 E. Długopolski, *Władysław Łokietek*, pp. 1–4.

106 P. W. Knoll, p. 17.

107 J. F. N. Bradley, p. 30; Włodarski, 'Polityka Jana Luksemburczyka wobec Polski za czasów Władysława Łokietka', *Archiwun Towarzystwa Naukowego we Lwowie*, Vol. XI, No. 3, Lwów, 1933, pp. 17–60.

108 *Kodeks dyplomatyczny miasta Krakowa, op. cit.*, Vol. I, No. 41 and 43 (dated 1372); S. Estreicher (ed.), *Najstarszy zbiór przywilejów i wilkierzy miasta Krakowa*. Kraków, 1936, No. 8, p. 9; *Kodeks dyplomatyczny miasta Krakowa, op. cit.*, Vol. I, No. 63, p. 76 (dated 1387).

109 T. Puymaigre 'Une campagne de Jean de Luxembourg, roi de Bohème', *Revue des questions historiques*, Vol. XLII, Paris, 1887, pp. 168–80.

110 P. W. Knoll, p. 59. See also J. Dąbrowski, 'Dzieje polityczne Śląska w latach 1290–1402', in *Historija Śląska od najdawniejszych czasów do roku 1400*, S. Kutrzeba (ed.), 3 Vols., Kraków, 1933–9, Vol. I, pp. 327–40.
111 J. F. N. Bradley, p. 31; O. Bauer, 'Poznánky k mírovým smlouvám česko-polskyḿ z roku 1335', *Sborńik prací věnovanych prof. dru. Gustavu Friedrichovi k šedestým narozeninám 1871–1931*, Prague, 1931, pp. 9–22.
112 F. L. Hübsch, p. 107; *Kodeks dyplomatyczny miasta Krakowa*, Vol. I, no. 4.
113 J. Korn, no. 189, p. 169.
114 J. Cělakowsky, *op. cit.*, No. 303, p. 445; J. Korn *op. cit.*, No. 189, p. 169 (document dated 27/III/1349); A. Mosbach, *Przyczynki do dziejów Polski z archiwum miasta Wrocławia*, Poznań, 1860, p. 79 (Document dated 24/II/1352).
115 J. Korn, *op. cit.*, No. 189, p. 170 – in 1353 the king closed trade routes to Ruthenia for non-Polish merchants 'Se terram Russie propriis suis hominibus expugnasse, et quod illa via solum suis hominibus et mercatoribus patere deberet'.
116 *Kodeks dyplomatyczny miasta Krakowa, op. cit.*, Vol. I, No. 29, p. 354 (Document dated 1354).
117 L. Koczy, *Związki handlowe ... op. cit.*, p. 17; see also S. Kalfas-Piotrowska, 'Stosunki handlowe Śląsko-Polskie za Kazimierza W.', *Roczniki Twa. Przyj. Nauk na Śląsku*, Vol. 5, Wrocław, 1936, pp. 1–20.
118 *Kodeks dyplomatyczny miasta Krakowa, op. cit.*, Vol. I, No. 41 and 43 (dated 1372); S. Estreicher (ed.), *Najstarszy zbiór przywilejów i wilkierzy miasta Krakowa*, Kraków, 1936, No. 8, p. 9; *Kodeks dyplomatyczny miasta Krakowa, op. cit.*, Vol. I, No. 63, p. 76 (dated 1387).
119 *Kodeks dyplomatyczny miasta Krakowa, op. cit.*, Vol. I, No. 42, 43, and 63; A. Mosbach, *op. cit.*, p. 82; M. Rauprich, 'Breslaus Handelslage im Ausgange des Mittelalters', *Zeitschrift der Vereins für Geschichte und Altertums Schlesiens*, Vol. 26, Breslau, 1892, pp. 14–17.
120 A. Mosbach, *op. cit.*, p. 79; W. Długoborski, J. Gierowski and K. Maleczyński, *Dzieje Wrocławia do roku 1870*, Warszawa, 1958, p. 132.
121 A. Mosbach, *op. cit.*, p. 82; the full text of King Władysław Jagiełło's decree is given in Index Actorum Saeculi XV, in A. Lewicki (ed.), *Monumenta Medii Aevi Res Gestas Poloniae Illustrantia: Pomniki dziejowe wieków średnich do objaśnienia rzeczy polskich służące*, 19 Vols., Kraków (1874–1927), Vol. I, No. 370, Kraków, 1888.
122 K. Myśliński, 'Lublin a handel Wrocławia z Rusią w XIV i XV wieku', *Rocznik Lubelski* Vol. III, Lublin, 1960, p. 11.
123 S. Kutrzeba, Handel Krakowa, *op. cit.*, pp. 100–1; M. Rauprich, *op. cit.*, p. 17; Ibid., 'Der Streit um die Breslauer Niederlage 1490–1515' *Zeitschrift der Vereins für Geschichte und Altertums Schlesiens*, Vol. 27, Breslau, 1893, pp. 101–16.
124 K. Myśliński, 'Lublin a handel ...' *op. cit.*, p. 11.
125 J. Waręzak, 'Polska polityka handlowo-celna względem Śląska i Wrocławia za Zygmunta Starego' *Ekonomista*, Vol. 30, No. 2, Warszawa, 1930, pp. 48–55.
126 A. Mosbach, *op. cit.*, p. 120.
127 M. Z. Jedlicki, 'German Settlement in Poland and the Rise of the Teutonic Order' in *The Cambridge History of Poland*, W. F. Reddaway et al., 2 Vols.,

Cambridge, 1941–1950, Vol. I, pp. 125–47; S. Trakowski, 'Die Rolle der Deutschen Dorfkolonisation und des Deutschen Rechtes in Polen im 13 Jahrhundert' in *Die Deutsche Ostsiedlung des Mittelalters. Als problem der Europäischen Geschichte*, H. Beumann (ed.), Sigmaringen, 1975, pp. 349–68; B. Zientara, 'Nationality Conflicts in the German-Slavic Borderland in the 13th–14th centuries and Their Social Scope', *Acta Poloniae Historica*, Vol. XXII, Warszawa, 1970, pp. 207–25; M. Friedberg, *Kultura polska a niemiecka* (Wyd. Instyt. Zach.), Poznań, 2 Vols, 1946, Vol. I, pp. 266–327; see also F. L. Carsten, *The Origins of Prussia*, Oxford 1954.

128 E. Długopolski, 'Bunt wójta Alberta, *op. cit.*, pp. 135–86; see also, J. Ptaśnik, *Miasta i mieszczaństwo w dawnej Polsce* (2nd edn), Warszawa, 1949, pp. 311–57; W. Kuhn, 'Die Deutsch Rechliche Siedlung in Klein Polen', in *Die Deutsche Ostsiedlung des Mittelalters. Als Problem der Europäischen Geschichte*, H. Beaumann (ed.), Sigmaringen, 1975, pp. 369–415.

129 A. Gieysztor, et al. *History of Poland, op. cit.*, p. 113.

130 S. Arnold and M. Żychowski, *Outline History of Poland: From the Beginning of the State to the Present Time*, 2nd edn (Polonia Publ. Hse.), Warsaw, 1965, p. 30.

131 W. Friedrich, *Der Deutsche Ritterorden und die Kurie in den Jahren 1300–1330*, Königsberg 1912, pp. 36–49, 83–86; E. Keyser, 'Der burgerliche Grundbesitz der Rechstadt Danzig im 14 Jahrhundert', *Zeitschrift des Westpreussischen Geschichtsvereins*, Vol. LVIII, Danzig, 1918, pp. 45–6.

132 J. Długosz, *Historia Polonicae*, Vol. III, Kraków, 1876, pp. 144–5.

133 M. Małuszyński, 'Zabór Pomorza przez Krzyżaków (1308–1309)' *Rocznik Gdański*, Vol. VII–VIII, Gdańsk, 1933–1934, pp. 44–80; The exact border is delineated in E. Sandow, 'Die Polnisch-pommerellische Grenze 1309–1454', *Beihefte zum Jahrbuch der Albertus-Universität Königsberg Pr.*, Vol. VI, Kitzingen/Main, 1954, pp. 4–9.

134 K. Maleczyński, *Polska a Pomorze Zachodnie w walce z Niemcami w wieku XIV i XV*, Gdańsk-Bydgoszcz-Szczecin, 1947, pp. 26–7; S. Nowogrodzki, 'Pomorze zachodnie a Polska w latach 1323–1270' *Rocznik Gdański*, Vol. XI–X, Gdańsk, 1935–36, pp. 23–7.

135 J. Szujski, 'Warunki traktatu kaliskiego r. 1343' *Opowiadania i roztrząsania historyczne*, Warszawa, 1882, pp. 52–66.

136 By 1300 the Lübeck merchants had their own banking house in Elbąg. See E. Carstenn, *Geschichte der Hansestadt Elbing*, Danzig, 1937, 210 pp.

137 In 1339 two Cracow merchants were involved in a legal dispute in Warsaw against the Teutonic Knights. It appears that their boat returning from Flanders with goods, was arrested by the Knights at Świecino in 1309. 'veniebat de Flandria cum aliis mercatoribus de Cracovia et habebant naves suas in flumine Visla, que circumdat castrum Swecze, quod est in Pomerania . . ., ipse et alii mercatores et mercimonia eorum erant arestata . . .'. The other witness confirmed that 'veniebat de Flandria cum mercimoniis suis et navibus fuit in flumine Visla . . . arestatus . . . et alii eciam mercatores, qui cum eo erant . . .' *Lites ac res gestae inter Polonos Ordinemque Cruciferorum, editio altera*, I. Zakrzewski and J. Karwasińka (eds.), 3 Vols Poznań/Warszawa (1890–1935), Vol. I, Poznań, 1890, pp. 385–6; see also R. Grodecki, 'Znaczenie handlowe Wisly w epoce piastowskiej' *Studia historyczne, ku czci S. Kutrzeby*, Vol. II, Kraków 1937, p. 301.

138 H. Dragendorf, 'Hansische Findlinge im Ratsarchiv zu Rostock', *Hansische Geschichteblätter*, Vol. VIII, Danzig, 1902, pp. 13–24; J. Voigt, *Geschichte Preussens von den ältesten Zeiten bis zum Untergange der Herrschaft des Deutschen Ordens*, 9 Vols, Königsberg (1827–1839), Vol. V, p. 120.

139 K. Stachowska, *op. cit.*, p. 588.

140 *Hanserecesse: Die Recesse und andere Acten der Hansetage von 1256–1430*, K. Koppmann (ed.), 8 Vols, Leipzig (1870–1897), Vol. III, 1875, No. 422, 9, p. 436; Document dated 4/IV/1389).

141 Ibid., Vol. V, No. 118, 5, p. 80; (Document dated 3/III/1403 – 'Item haben die heren von Thorun vor unsern herer, den homeister, gebrocht, wy das dy Crakower und andere geste boben der Wyssle keyne nyderloge czu Thorun halden, sunder ire gut dorch dys land czur zeewort furen und brengen; dergelich dy geste dy czur zeewort her in dys land kommen, das sy keyne nyderloge in den obirsteten halden, sunder mit erern gute voren und keren um lande, wor sy wellen; und das hat unser here homeistr befoln eynem itzlichen in syme rate doremme czu sprachin, czum neste tage widr unczubrengen'.

142 D. Hay, *op. cit.*, p. 204.

143 M. Dogiel, *op. cit.*, Vol. IV, p. 112. (Document dated 1422.)

144 W. Kowalenko 'Polska żegluga na Wiśle i Bałtyku w XIV i XV w.', *Roczniki Historyczne*, Vol. XVII, Poznań, 1948, pp. 376–7.

145 H. Oesterreich, 'Die Handelsbeziehungen der Stadt Thorn zu Polen (1232–1577)', *Zeitschrift des Westpreussichen Geschichtsvereins*, Vol. XXVIII, Danzig, 1890, p. 44, for example in 1428; Toruń again started to exert her staple laws in 1440; see *Codex epistolaris nec non diplomaticus saeculi XV (Monumenta Medii Aevi Historica res gestas Poloniae illustrantia)*, 3 Vols, A. Sokołowski, J. Szujski and A. Lewicki (eds.), Kraków, 1876–1894, Vol. II, No. 200.

146 For further details see E. Cieślak and C. Bierna, *Dzieje Gdańska*, Gdańsk 1969, pp. 94–111; E. Cieślak (ed.), *Historia Gdańska* (Wyd. Morskie), Gdańsk, 1978, Vol. I (do roku 1454), pp. 397–416, 507–529; V. Lauffer 'Danzigs Schiffs- und Warenverkehr am Ende des XV Jahrhunderts' *Zeitschrift des Westpreussischen Geschichtsvereins*, Vol. XXXII, Danzig, 1894, pp. 10–20.

147 D. Hay *op. cit.*, p. 238; *Wielkopolska w świetle źródeł historycznych* (Wybór tekstów źródłowych), Poznań, 1958, pp. 28–30.

148 A. Gieysztor, et al., *op. cit.*, p. 143; M. Oehler, *Der Krieg zwischen dem Deutschen Orden und Polen-Litauen*, Elbing, 1910, 330 pp.

149 M. Biskup, 'Z problematyki handlu polsko-gdańskiego drugiej połowy XV wieku' *Przegląd Historyczny*, No. 2–3, 1954, Warszawa, pp. 391–3.

150 J. Rukowski, *Historia gospodarcza Polski*, Warszawa, 1953, pp. 60–3.

151 M. Małowist, 'Le développement des rapports économiques entre la Flandre, la Pologne et les pays limitrophes du XIII au XIV s.' *Revue Belge de Phil. et d'Histoire*, Vol. 10, Brussels, 1931, pp. 1013–65.

152 B. Grekow and A. Jakubowski, *Złota Orda i jej upadek* (translated from Russian), Warszawa, 1953, 243 pp.; B. Spuler, *Die Goldene Horde. Die Mongolen in Russland, 1223–1502*, Leipzig, 1943, 348 pp.

153 H. Cordier, 'L'invasion mongole au Moyen – Âge et ses conséquences', *Mélanges d'histoire et de géographie orientales*, Vol. II, Paris, 1920, pp. 10–31; K. Czechowicz, 'Najazdy Tatarów na Polskę w XIII w.', *Teki Historyczne*

(Polskie Towarzystwo Historyczne w Wielkiej Brytanii), Vol. XVI, London, 1969–1971, pp. 55–78; S. Krakowski, *Polska w walce z najazdami tatarskimi w XIII wieku*, Warszawa, 1958, 241 pp.; H. Morel, 'Les campagnes Mongoles au XIII – e siècle', *Revue militaire française*, No. 12–13, Paris, 1922, pp. 20–41.

154 G. Strakosch-Grassman, *Der Finfall der Mongloen in Mittel-europa in den Jahren 1241 und 1242*, Innsbruck, 1893, 352 pp.; W. Zatorski, 'Pierwszy najazd mongolski na Polskę w latach 1240–1241', *Przegląd Historyczno-Wojskowy*, Vol. IX, Warszawa, 1937, pp. 16–35.

155 J. Milan, 'Napad Tatarów na Polskę za Leszka Czarnego w roku 1287', *Sprawozdania gimnazjum w Stanisławowie za rok 1905–1906*, Stanisławów, 1960, pp. 25–48.

156 K. Czechowicz, *op. cit.*, p. 55 maintains that between the first expedition in 1241 to the last in 1698, 164 major invasions are known to have taken place, but there may have been many more, which went unrecorded.

157 D. Hay, *op. cit.*, p. 262.

158 S. Kutrzeba and J. Ptaśnik *op. cit.*, p. 7.

159 P. W. Knoll, *op. cit.*, p. 123.

160 *Monumenta Poloniae Historica, op. cit.*, Vol. II, Lwów, 1872, p. 860; B. Antonowicz, *Monografji po istorji i jugozapadnoj Rossiji*, Vol. I, Kiev, 1885, p. 74; A. Prochaska, 'Podole lennem Korony 1352–1430', *Rozprawy Akademii Umiejętnosci: Wydział Historyczno-Filozoficzny*, Vol. XXXII, Kraków, 1895, pp. 256–79; P. W. Knoll, *op. cit.*, pp. 123–51.

161 S. Estreicher (ed.) *op. cit.*, No. 3, p. 3, in which Kazimierz Wielki allowed non-Polish merchants to go with merchandise to Ruthenia 'ultra Cracoviam, Wisliciam, Sandomiriam et ad ulterim ad Lublin civitatem iter suum non protendent ... Iterato in Cracoviam et deinde ad propria securiter reversuri'. Thus foreign merchants going from Ruthenia to Poland 'hii transeant per Lublin, Sandomiriam, per Wisliciam in Cracoviam, ibique non ultra res suas vendant seu deponant'. Merchants could not now by-pass Cracow via Wieliczka and the return journey must be 'Iterato in Cracoviam ... securiter reversuri'.

162 Ibid., No. 8, p. 9. Granted by Queen Elżbieta (Elizabeth) in 1371 with reference to Polish-Ruthenian trade 'Omnibus mercatoribus extraneis, ut puta boemis, Moravis, Slesicis ac ceteris quibuscunque ... eundem versus Russiae partes transitum et quaelibet loca Russia pro comparandis mercibus seu mercimoniis ... nostro regio edicto perpetue prohibentes. Sed mercatores iidem extranei ... cum eorum rebus et mercibus ... ad nostram civitatem Cracoviam et non ulterim ... transire poteurent'.

163 *Kodeks dyplomatyczny miasta Krakowa*, *op. cit.*, Vol. I, No. 56.

164 O. Halecki, *op. cit.*, p. 69.

165 V. O. Klyuchevskiy, *Kurs russkoy istorii*, Vol. I, Moscow, 1938, pp. 10–20; D. J. B. Shaw, 'Urbanism and Economic Development in a Pre-industrial Context: the Case of Southern Russia', *Journal of Historical Geography*, Vol. 3, No. 2, London/New York, 1977, p. 107.

166 W. Heyd, *Geschichte des Levanthandels*, Leipzig, 1885, Vol. II, pp. 182; 190–1; 228; G. I. Bratianu, *Recherches sur le Commerce Génois dans la Mer Noire au XIIIe siècle*, Paris, 1929, 354 pp.; J. Lelewel, *Gilbert de Lannoy i jego prodroże* (Rozbiory dzieł obejmujacych albo dzieje albo rzeczy polskie różnymi czasy),

Poznań, 1844, pp. 353–5. See also, S. Kutrzeba, *Handel polski ze Wschodem w wiekach średnich*, Kraków 1902, p. 62; M. Małowist, 'Bałtyk i Morze Czarne w handlu średniowiecznym', *Janitar*, Vol. I, No. 2, Gdynia 1937, pp. 69–74; Ibid., 'The Baltic and the Black Sea in Medieval Trade, *Baltic and Scandinavian Countries*, Vol. III, No. 5, Gdynia, 1937, pp. 36–42.

167 *Akta grodzkie i ziemskie*, E. Kałużniacki (ed.), Vol. VII, Lwów, 1878, No. 76, p. 148; *Kodeks dyplomatyczny miasta Krakowa*, *op. cit.*, Vol. I, No. 102; A. Prochaska, *Materiały archiwalne wzięte z metryki litewskiej*, Lwów, 1890, No. 18; V. A. Ulianickij, *Materaly dlia istorii vzairmych otnošenij Rossii, Polši, Moldavii, Valachii i Turcii*, Moskva, 1887, No. 20; J. W. Niemann, 'Der Handel der Stadt Lemberg im Mittelalter', *Die Burg. Vierteljahreschrift-des Instituts für Deutsche Ostarbeit*, Vol. 2, Krakau, 1941, pp. 69–92; S. Lewicki, *Lemberg's Stapelrecht*, Lwów, 1909, p. 34.

168 S. Kutrzeba, Handel Krakowa w wiekach średnich, *op. cit.*, pp. 108–15; J. Nistor, *Die auswärtigen Handelsbeziehungen der Moldau, im XIV, XV und XVI Jahrhundert*, Gotha, 1911, p. 19; K. Myśliński, 'Rola miast Polski ...' *op. cit.*, pp. 24–8.

169 K. Kantak, J. Szablowski and J. Żarnecki, 'Kościół i klasztor oo. bernardynów w Krakowie', *Biblioteka Krakowska*, Vol. 96, Kraków, 1938, pp. 72–5.

170 W. Felczak, *op. cit.*, pp. 84–92.

171 B. Stachoń, *Polityka Polski wobec Turcji i akcji antytureckiej w XV wieku do utraty Killii i Białogrodu (1484)*, Lwów, 1930, pp. 11–18; 177; J. Reychman, *Historia Turcji*, Wrocław-Warszawa-Kraków-Gdańsk, 1973, p. 51.

172 F. von Kraelitz, 'Osmanische Urkunden in türkischer Sprache aus der zweiten Hälfte des 15 Jahrhunderts', *Mémoires de l'Academia de Vienne* (sect. hist.) Vienna, 1907, 24 pp.; P. P. Panaitescu, 'La route commercial de Pologne à la Mer Noire au Moyen Âge', *Revista Istorica Română*, Vol. 3, No. 2–3, Bucharest, 1934, pp. 18–19; G. C. Brockelman, *Histoire des peuples et des États Islamiques depuis les origines jusqu'à nos jours*, Paris, 1949, p. 240; *Akta grodzkie i ziemskie*, *op. cit.*, Vol. II, No. 3; L. Demeny, 'Comersul de tranzit spre Polonia prin Stara Romeneasca şi Transilvania, *Studii Revista de istoria*, Vol. XXII, No. 3, Bucharest, 1969, pp. 80–2. C. Georgieva 'La rapports de commerce entre l'Empire Ottoman et la Pologne et les terres Bulgares en XVIe siècle', *Bulgarian Historical Review*, Vol. 6, No. 3, Sofia, 1978, p. 41.

173 K. Kosman, *Historia Białorusi* (Ossolineum), Wrocław-Warszawa-Kraków-Gdańsk, 1979, pp. 72–93.

174 *Percepta et exposita civitatis Leopoldiensis od 1460–1518 r*, (Archiwum miast Lwowa), Lwów, 1884, pp. 461, 462, 476, 504, 554, 704; M. Dubiecki, 'Kaffa osada genuenska i jej stosunek do Polski w XV wieku' *Przegląd Powszechny*, Vol. XII, Kraków, 1886, pp. 56–64; 216–27.

175 D. J. B. Shaw, *op. cit.*, p. 107.

176 K. Czechowicz, *op. cit.*, p. 56.

177 A. Gieysztor, et al., *op. cit.*, p. 144; O. Halecki, *op. cit.*, pp. 102–3.

178 E. Rykaczewski (ed.), *Inventorium ... Privilegiorum, Litterarum, Diplomatum, Scripturarum et Monumentorum quaecunque in Archivo Regni in Arce Cracoviensi continentur per commissanos a Sacra Regia Majestate et Republica ... Deputatos Confectum A.D. 1682 ... Cura Bibliothecae Polonicae Editum*, Lutetiae, Parisio-

rum, Berolini et Posnaniae, 1862, p. 143; J. Nistor, *op. cit.*, p. 42; Archiwum Biblioteki Czartoryskich Kraków; *Miscellana Turcica*, Vol. IV, No. 611, pp. 35–8.
179 M. Kosman, *op. cit.*, pp. 100–1.
180 S. Arnold and M. Żychowski, *op. cit.*, pp. 37–8.
181 O. Halecki, *op. cit.*, p. 42.
182 N. J. G. Pounds, *op. cit.*, p. 373.
183 Ibid., p. 376.
184 D. Hay, *op. cit.*, p. 123; D. Kirby, *Northern Europe in the Early Modern Period: The Baltic World 1492–1772* (Longman), London/New York, 1990, pp. 4–12.

5: European trade through Cracow, 1257–1500

1 E. Ennen, *Die europäische Stadt des Mittelalters* (Vanderhoeck & Ruprecht), Göttingen, 1975, 306 pp.; W. Rausche, *Die Städte Mitteleuropas im 12 und 13 Jahrhundert*, Linz, 1963, 251 pp.; T. Rosłanowski, Comparative Sociotopography on the Example of Early-Medieval Towns in Central Europe, *Acta Poloniae Historica*, Vol. 34, Warsaw, 1976, pp. 7–27.
2 M. Małowist, 'Croissance et regression en Europe XIVe–XVIIe siècle' *Cahiers des Annales*, Vol. 34, Paris, 1972, pp. 10–23.
3 A. Lewis, *The Sea and Medieval Civilisation*, London (Variorum), 1978, 324 pp.
4 S. Kutrzeba, 'Handel Polski ze Wschodem w wiekach średnich', *Przegląd Polski*, Kraków, 1903, Vol. IV(2), Rocz. XXXVIII, pp. 189–219; Rocz. XXXVIII (3), pp. 512–537; Rocz. XXXVIII (4), pp. 115–145; A. H. Lybyer, 'The Influence of the Rise of the Ottoman Turks upon the Routes of Oriental Trade' *English Historical Review*, Vol. XXX, London, 1915, pp. 577–588.
5 R. S. Lopez, *The Commercial Revolution of the Middle Ages, 950–1350* (Cambridge Univ. Press), London, 1976, 180 pp.; R. T. Rupp, 'The Unmaking of the Mediterranean Trade Hegemony: International Trade Rivalry and the Commercial Revolution', *Journal of Economic History*, Vol. 35, No. 3, London, 1975, pp. 499–525.
6 G. Below, 'Grosshändler und Kleinhändler im Deutschen Mittelalter', *Jahrbücher für Nationaloekonomie und Statistik*, III Series, Vol. 20, Berlin, 1900, pp. 25–34; F. Keutgen, 'Der Grosshandel im Mittelalter', *Hansische Geschichtsblätter*, Vol. IX, Leipzig, 1901, pp. 22–34.
7 E. M. Carus-Wilson, *Medieval Merchant Venturers*, London, 1954, 230 pp.; B. N. Nelson, 'The Usurer and the Merchant Prince, *Journal of Economic History*, London, 1947 supplement; M. M. Postan, 'Credit in Medieval Trade' in *Essays in Economic History*, E. M. Carus-Wilson (ed.), London, 1966, pp. 61–87; G. Richards, *Florentine Merchants in the Age of the Medici*, Cambridge, Mass., 1932, 354 pp.; H. Samsonowicz, *Untersuchungen ubedr das Danziger Burgerkapital in der zweiten Halfte des 15 Jahrhunderts*, Weimar, 1969, 168 pp.
8 S. L. Thrupp, 'Medieval Industry, 1000–1500', Vol. I, Ch. 7, *The Fontana Economic History of Europe*, C. M. Cipolla (gen. ed.), London and Glasgow, 1971, p. 39; see also S. Herbst, *Toruńskie cechy rzemieślnicze*, Toruń, 1933, 231 pp.; T. Tasiński, 'Rzemiosła kowalskie średniowiecznego Torunia', *Kwartalnik Historii Kultury Materialnej*, Vol. 33, No. 2, Warsaw, 1975, pp. 225–235; M. Mollat, 'Guerre de

course et piratie à la fin du Moyen âge: aspects économique et sociaux. Position et problèmes', *Hansische Geschichtblätter*, Vol. 90, Leipzig, 1972, pp. 1–14; G. Pfeifier, '*Das Breslauer Patriziat im Mittelalter*' (Scientia-Verlag), Aalen, 1973, 412 pp.

9 D. Hay, *Europe in the Fourteenth and Fifteenth Centuries* (Longman), London, 1976, p. 359.

10 H. Samsonowicz, *Rzemioslo wiejskie w Polsce XIV–XVII w*, Warszawa, 1954, 158 pp.; Ibid., 'Le "suburbium" en Pologne, vers la fin du Moyen âge. L'importance économique et sociale des faubourgs a- XIVe–XVe siècles', *Studia Historiae Oeconomicae*, Vol. 13, Warsaw, 1978, pp. 73–82; J. Žemlička, 'Přemyslovská hradská centra a počatký měst v Čechách', *Československý Časopis Historický*, Vol. 26, No. 4, Prague, 1978, pp. 559–586; B. Zientara, 'Z dziejów organizacji rynku w średniowieczu', *Przegląd Historyczny*, Vol. 64, No. 4, Warszawa, 1973, pp. 681–696. Ibid., 'Socio-economic and Spatial Transformation of Polish Towns during the Period of Location', *Acta Poloniae Historica*, Vol. 34, Warszawa, 1976, pp. 57–83.

11 J. C. Russell, 'The Metropolitan City Region in the Middle Ages', *Journal of Regional Science*, Vol. 2, London, 1960, pp. 55–70; T. Lalik, 'Struktura miasta późnośredniowiecznego. Jej geneza i zmiany', *Kwartalnik Historyczny*, Vol. 82, No. 4, Warszawa, 1975, pp. 776–794; Ibid., 'Funkcje miast i miasteczek w Polsce późniejszego średniowiecza, *Kwartalnik Historii Kultury Materialnej*, Vol. 23, No. 4, Warszawa, 1975, pp. 551–565; Ibid., 'Geneza sieci miasteczek w Polsce średniowiecznej', in *Miasta doby feudalnej w Europie środkowo-wschodniej*, Warszawa, 1976, pp. 113–136; Ibid., 'La genèse du réseau urbain en Pologne médiévale', *Acta Poloniae Historica*, Vol. 34, Warszawa, 1976, pp. 97–120; Ibid., 'Les fonctions des villes en Pologne au bas Moyen âge', *Acta Poloniae Historica*, Vol. 37, Warszawa, 1978, pp. 5–28.

12 I. Kieniewicz, 'Droga morska do Indii i handel korzenny w latach 1498–1522', *Przegląd Historyczny*, Vol. 55, Warszawa, 1964, pp. 573–603.

13 M. Małowist, 'W sprawie badań nad historią rzemiosła miejskiego w średniowiecznej Polsce', *Rocznik Dziejów Społeczno Gospodarczych*, Vol. III, Warszawa, 1951, pp. 16–19; H. Dąbrowski, 'Rozwój gospodarki rolnej w Polsce od XII do połowy XIV wieku, *Studia z dziejów gospodarstwa wiejskiego*, Vol. I, Warszawa, 1962, pp. 10–21.

14 Cracow had two main fairs: i) Spring Fair which began around St. Jacob's day at the beginning of May; *Najstarsze księgi i rachunki miasta Krakowa od r. 1300 do 1400*, F. Piekosiński and J. Szujski (eds.), Kraków, 1878, Vol. I, No. 137; it began on Holy Cross day, May 3rd and ended on 14th May, *Kodeks dyplomatyczny miasta Krakowa, 1257–1506*, Vol. I, Kraków, 1879, No. 310 10; ii) Autumn Fair, which was around St. Michael's day, usually at the end of September. It probably lasted the same length as the Spring Fair (i.e. 10 + two half days), but there is lack of documentary proof; there was also a third minor fair at the time of St. Vitus (*Kodeks dyplomatyczny op. cit.*, Vol. II, Kraków, 1882, No. 310, 11) which began on June 11th and lasted till 17th June (i.e. 5 + two half days) and was known as the Small Fair or 'der kleine jarmark" (*Najstarsze księgi op. cit.* Vol. I, No. 109). It was obviously less important than the other two particularly as it came only a month after the Spring Fair. The fairs in Cracow were particularly attractive for all merchants were freed from paying customs duty (*Kodeks*

dyplomatyczny ... op. cit. Vol. II, No. 310, "und dize czwene yormargkte sind frey den gestem, ydoch den quartczoist man der stat allerzeit pflychtig".

15 P. Huvelin, *Essaie historique sur le droit des marchés et des foires*, Paris 1897, 254 pp.; H. Ammann, 'Die Zurzacher Messen im Mittelalter', *Taschenbuch der Historischen Gesellschaft des Kantons Aargau für das Jahr 1923*, Aargau (H. R. Sauerländer), 1923, pp. 1–154; E. Coornaert, 'Caractères et mouvements des foires internationales au moyen âge et au XVIe siècle', in *Studi in Onore di Armando Sapori* (Instituto Editoriale Cisalpini), Vol. I, Milan, 1957, pp. 355–371; R. D. Face, 'Techniques of Business in the Trade between the Fairs of Champagne and the South of Europe in the Twelfth and Thirteenth Centuries', *Economic History Review*, Vol. 10, London, 1957–58, pp. 427–438; M. Mitterauer, 'La continuité des foires et la naissance des villes', *Annales: Économiques, Sociétés, Civilisations*, Vol. 28, No. 3, Paris, 1973, pp. 711–734.

16 H. Pirenne, *Economic and Social History of Medieval Europe* (Routledge & Kegan Paul), London, 1961, p. 169.

17 C. Yver, *Le commerce et les marchands dans l'Italie méridionale au XIIIe et au XIVe siècle* (Bibliothèque des Ecoles Françaises d'Athènes et de Rome, No. 88), Paris 1903, 335 pp.; G. I. Britianu, *Recherches sur le commerce génois dans la Mer Noire au XIIIe siècle* (Paul Geuthner), Paris, 1929, 213 pp.; R. Doehaerd, *Les Relations commerciales entre Gênes, la Belgique et l'Outremont*, 3 Vols. (Institut Historique Belge de Rome), 1941, 254 pp.; F. C. Lane, *Andrea Barbarigo: Merchant of Venice 1418–1449* (Studies in Historical and Political Science Series,s 62, No. 1, Johns Hopkins University Studies) Baltimore 1944, 273 pp.; Ibid., 'Venetian Shipping during the Commercial Revolution', *American History Review*, Vol. XXXVIII, New York, 1932–33, pp. 219–239; R. S. Lopez and A. W. Raymond, *Medieval Trade in the Mediterranean World* (Oxford Univ. Press), 1955, 365 pp.; I. Origo, *The Merchant of Prato*, Knopf, 1957, 156 pp.; H. Simonsfeld, *Der Fondaco dei Tedeschi in Venedig und die Deutsch-Venetianischen Handelsbeziehungen*, 2 vols., Stuttgart, 1887, 438 pp.; G. Luzzato, 'Small and great merchants in the Italian Cities of the Renaissance', in *Enterprise and Secular Change*, F. C. Lane & J. C. Riemersma (eds.), Homewood, Illinois (Irwin), 1953, pp. 41–52; D. S. Chambers, *The Imperial Age of Venice, 1380–1580* (Thames and Hudson), 1970, 423 pp.; C. M. Cipolla, 'The Trends in Italian Economic History in the Later Middle Ages', *Economic History Review*, Vol. I, London, 1949–1950, pp. 181–4; A. Sapori, 'The Culture of the Medieval Italian Merchant', in *Enterprise and Secular Change*, *op. cit.*, pp. 53–65; R. L. Reynolds, 'Genoese Trade in the Late Twelfth Century, Particularly in Cloth from the Fairs of Champagne', *Journal of Economic Business History*, Vol. III, New York, 1931, pp. 362–381; E. Ashtor, 'The Venetian Supremacy in Levantine Trade: Monopoly on Pre-Colonialism', *Journal of European Economic History*, Vol. 3, No. 1, Rome, 1974, pp. 5–53; C. Giurescu, 'Les Génois au Bas-Danube aux XIIIe et XIVe siècles', in *Colocviul Românо-Italian*, Bucharest, 1977, pp. 47–61; R. S. Lopez, 'L'importance de la mer Noire dans l'histoire de Gènes', in *Colocviul Româno-Italian*, Bucharest, 1977, pp. 13–33; M. N. Pelekidis, 'Venise et la Mer Noire du XIe au XVe siècle, in *Venezia e il Levante fino al secolo XV*, A. Pertusi (ed.), Florence, 1973, Vol. I, pp. 541–582; G. Richards, *Florentine Merchants in the Age of the Medici*, Cambridge, Massachusetts, 1932, 198 pp.

18 See for example, A. P. Usher, *The Early History of Deposit Banking in Mediterranean Europe* (Harvard University Press) New York, 1943, 361 pp. See especially pp. 269–300.

19 S. Kutrzeba, 'Handel Krakowa . . . ', *op. cit.*, p. 155.

20 R. de Roover, *Money, Banking and Credit in Mediaeval Bruges*, Cambridge, Massachusetts, 1948, 409 pp.

21 For example, see R. Farkas, *Kassa Árumegállító-Joga*, Kassa, 1893, 102 pp.; A. Felsó, *Magyarország Kereskedelmi útjai a Középkorban*, Budapest, 1908, 254 pp.; G. Kerekes, *Kassa Polgársága, Ipara és Kereskedése a Középkor Végén*, Budapest, 1913, 145 pp.; S. Kutrzeba (ed.), 'Akta odnoszące się do stosunków handlowych Polski z Węgrami z lat 1354–1505, *Archiwum Komisji Historycznej*, Vol. IX, Kraków, 1902, pp. 407–485; R. Marsina, 'Miesto a trh na Slovensku do konca 13 stor.', *Historický Časopis*, Vol. 26, No. 1, Bratislava, 1978, pp. 77–95.

22 Ł. Charewiczowa, *Handel średniowiecznego Lwowa* (Studja nad historiją, kultury w Polsce, Vol. I) (Wyd. Zak. Narod. Imiena Ossolinskich), Lwów, 1925, pp. 6–30.

23 S. Kalfas-Piotrowska, 'Stosunki handlowe Śląsko-Polskie za Kazimierza W.', *Roczniki Towarzystwa Przyjaciół Nauk na Śląsku*, Vol. 5, 1936, pp. 20–31; B. Mendle, 'Breslau zu Beginn des 15 Jahrhunderts. Eine Statistische Studie nach dem Steuerbuch von 1403', *Zeitschrift des Vereins für Geschichte Schlesiens*, Vol. 63, Breslau, 1929, pp. 154–185.

24 Wojewódzkie Archiwum Państwowe w Krakowie, i) *Ksiegi Ławnicze, Krakowskie*, p. 213 (1454). In 1454 'a furman, M. Szymoch from Cracow, received 550 ells of rough linen cloth from a merchant named Zarogowski, which he transported to Toruń and delivered to one H. von Staden'. ii) Ibid., *Consularia Cracoviensia* Vol. III, p. 30. In 1451, a furman delivered cloth to Toruń for J. Sweidniczer.

25 Wojewódzkie Archiwum Państwowe w Krakowie, i) *Consularia Cracoviensia*, Vol. I, p. 190. In 1403 a commission agent took goods to Bohemia from Cracow, but due to errors he made in the transaction he lost his quarter of the profit made. ii) Ibid., *Najstarsze księgi . . . op cit.*, Vol. II, pp. 137–138. A bill is preserved for a commission agent who took goods to Flanders for a Cracow merchant 'Super . . . expensis et perculo . . . '. iii) Ibid. *Consularia Cracoviensia*, Vol. III, p. 143 (1456). A commission agent had two separate contracts with a merchant in Cracow and a foreign merchant in Košice through a furman, and then had merchandise from Košice delivered to Cracow on the return journey. iv) Ibid. *Consularia Cracoviensia*, Vol. I, p. 231 (1406) gives a similar example to iii).

26 See for example 'Family Partnerships and Joint Ventures in the Venetian Republic', in *Enterprise and Secular Change*, *op. cit.*, pp. 86–98.

27 Ł. Charewiczowa, 'Handel Lwowa z Mołdawią i Multanami w wiekach średnich, *Kwartalnik Historyczny*, Vol. XXXVIII, No. 1–2, Lwów, 1924, pp. 37–67.

28 J. Baader, 'Nürnberg Handel im Mittelalter', *Jahresbericht des Historische Vereins von Mittelfranken*, Vol. 34, Unsbach, 1871, pp. 13–25, Vol. 35, 1872, pp. 41–60; C. Nordmann, *Nürnberger Grosshändler im spätmittelalterlichen Lübeck*, Nuremberg, 1933, 278 pp.; F. Schnelbögl, 'Die wirtschaftliche Bedeutung

ihre Landgebietes für die reichstadt Nürnberg' *Beiträge zue Wirtschaftgeschichte Nürnbergs*, Nuremberg, 1967, Vol. I, pp. 261–317.

29 L. Koczy, 'Związki handlowe Wrocławia z Polską do końca XVI wieku', Katowice, 1936, 34 pp.

30 J. Ptaśnik, Przedsiębiorstwa kopalniane krakowian i nawiązanie stosunków z Fuggerami w początku XVI wieku', *Biblioteka Krakowska*, No. 21, Kraków, 1902, pp. 67–88.

31 Wojewódzkie Archiwum Państwowe w *Krakowie, Najstarsze księgi op. cit.*, Vol. II, p. 139. Document dated 1396 refers to a partnership formed in Cracow for trade with Flanders. Two merchants (M. Hambrow and M. Lewnink), involved, the latter accompanying the merchandise to Flanders himself.

32 Ibid., i) *Consularia Cracoviensia*, Vol. I, p. 211 (1405). In 1405 a partnership was formed in which one partner provided capital totalling 150 grzywień and the other 242 grzywień. Profits were to be divided equally. ii) Ibid., same year, refers to a partnership formed where profits were to be divided on a 5:3 basis.

33 A. Doren, *Untersuchungen zur Geschichte der Kaufmannsgilden des Mittelalters*, Leipzig, 1893, 239 pp.

34 A list of city councillors and law benchers is given in F. Piekosiński (ed., *Kodeks dyplomatyczny . . . op. cit.* Vol. I, pp. xx–xxxix and pp. xlvi–lix.

35 Z. A. Helcel, *Starodawne prawa polskiego pomniki*, Kraków, 1856, Vol. I, p. 226.

36 Z. Bocheński, Krakowski cech mieczników, *Biblioteka Krakowska*, No. 92, Kraków, 1937, 90 pp.; A. Bochnak and J. Pagaczewski, *Polskie rzemiosło artystyczne wieków średnich*, Kraków, 1959, 154 pp.; A. Chimel, 'Dawne wyroby nożowników krakowskich i znaki na nich', *Rocznik Krakowski*, Vol. II, Kraków, 1899, pp. 89–108; X. Chotkowski, *Rzemiosła i cechy krakowskie w XV wieku*, Kraków 1891, 134 pp.; L. Lepszy, 'Cech złotniczy w Krakowie', *Rocznik Krakowski*, Vol. I, Kraków, 1898, pp. 135–268; J. Mitkowski, 'Statut cechu garbarzy we wsi Skotniki pod Krakowem i wiadomość o archiwum cechowym', *Prace z dziejów Polski feudalnej ofiarowane R. Grodeckiemu*, Warszawa, 1960, pp. 10–21; Z. Pazdro, *Uczniowie i towarzysze cechów krakowskich od połowy wieku* XVII, Lwów, 1900, 139 pp.; W. Stesłowicz, 'Cechy krakowskie', *Kwartalnik Historyczny*, Vol. VI, Kraków, 1892, pp. 277, 393.

37 *Najstarsze księgi i rachunki miasta Krakowa od r. 1300 do 1400*, F. Piekosiński and J. Szujski, Vol. II, Kraków, 1878, p. 398 (1396–1398) refers to 'vom deme gemeyne Kaufmanne czu Cracow, czum Czanse und von Dobschicz' in connection with the 'Kumpanie der Kaufmann' and 'der gemeyne Kaufmann'.

38 Wojewódzkie Archiwum Państwowe w Krakowie, *Consularia Cracoviensia*, Vol. I, p. 361 (25/II/1410), 'Seniores mercatorum electi per dominos consuls anno quo supra' and gives the names of six merchants, viz. M. Gemelich, W. Marcin, J. Schiler, P. Kaldherburg, Jerzy Morsztyn, P. Homan.

39 Ibid., *Consularia Cracoviensia*, Vol. III, p. 227 (Libri iuiris civilis), 1459 gives four merchants names, viz. H. Parchwicz, M. Belse, T. Ul, H. Wierzynek. See also S. Krzyżanowski, 'Morsztynowie w XV wieku', *Rocznik Krakowski*, Vol. I, Kraków, 1898, pp. 326–358.

40 O. Gönnenwein, 'Das Stapel – und Niederschlagsrecht', *Quellen und Darstellungen zur Hanisischen Geschichte*, Vol. II, Halle, 1939, pp. 14–28; in Świdnica in Poland this lasted up to six weeks, see K. Stachowska, 'Prawo Składu w Polsce

do 1565 r.' *Sprawozdania z czynności i posiedzeń Polskiej Akademii Umiejętności* Vol. LI, No. 9 (Nov. 1950), Kraków, 1951, pp. 486–593.

41 Ibid. See also K. F. Klöden, *Beiträge zur Geschichte des oderhandels*, Berlin, 1845–1852, pp. 25–29.

42 A. Jelicz, Życie codzienne w średniowiecznym Krakowie (wiek XIII–XV) (Państwowy Instytut Wydawniczy), Warszawa, 1965, p. 100, staple rights only referring to copper. From thereon foreign merchants must sell their copper to Cracow merchants only. *Kodeks dyplomatyczny ... op. cit.*, Vol. I, No. 4.

43 S. Kutrzeba and J. Ptaśnik, Dzieje handlu i kupiectwa krakowskiego', *Rocznik Krakowski*, Vol. XIV, Kraków, 1912, p. 10.

44 *Kodeks dyplomatyczny op. cit.*, Vol. I, No. 29.

45 *Hansisches Urkundenbuch*, K. Kunze, C. V. Runstedt and W. Stein (eds.) (Publ. K. Hohlbaum), Munchen-Halle-Leipzig-Weimar, Vols. I–IX, 1876–1958, Vol. III, nos. 532–533.

46 *Codex Epistolaris nec non Diplomaticus seculi XV*, A. Lewicki (ed.), Kraków, 1885, Vol. II, No. 200 refers to complaints by foreign merchants from Toruń and Gdańsk in 1430 about Cracow's rigid staple laws.

47 Wojewódzkie Archiwum Państwowe w Krakowie: *Consularia Cracoviensia*, Vol. III, p. 158, mentions the case in 1457 of a Nuremburg merchant who illegally sold cloth and silk material in Cracow.

48 *Kodeks dyplomatyczny op. cit.*, Vol. I, No. 192.

49 A. Jelicz, *op. cit.*, pp. 111–113.

50 For example salt was sold by the 'bałwan' regulated at 6 cetners in 1395 and 8 cetners in 1400 (*Najstarsze księgi ... op. cit.* Vol. II, pp. 131, 214). Lead had various weights (pondus, gwicht, woge) according to place of origin – from Olkusz (*Consularia Cracoviensia* Vol. II, p. 63; *Ksiegi Ławnicze ... op. cit.* (for 1447) p. 5; *Consularia Cracoviensia*, Vol. IV, pp. 108, 322; Ibid., Vol. V, pp. 47, 175); from Trzebin (*Consularia Cracoviensia*, Vol. V, p. 37); and for Cracow (*Consularia Cracoviensia*, Vol. IV, pp. 172, 244; Ibid., Vol. V, p. 164). Liquid measure was usually by barrel which had different sizes (fundry, kufy and drailing) whilst barrels also possessed various names (fass, vas, barylki and achteke), but relations between them are not known. Cloth measures also greatly differed according to place of origin.

51 R. de Roover, L'Evolution de la Lettre de Change: XIVe–XVIIe siècles', *Ecole Pratique des Hautes Etudes*, 6e section (Affaires et Gens d'Affaires), Paris, 1953, pp. 10–24.

52 Wojewódzkie Archiwum Państwowe w Krakowie, *Consularia Cracoviensia*, Vol. II, p. 247; Ibid., Vol. IV, p. 317.

53 Ibid., Vol. IV, p. 32 (for 1484); one such book is 'regestr Henryk Smet', for the year 1402–1403 (mentioned in A. Pawiński, 'Notatki kupca krakowskiego w podróży do Flandrii', *Biblioteka Warszawa*, Vol. 3, Warszawa, 1872, pp. 53–73). For more general discussion see M. M. Postan 'Credit in Medieval Trade' in *Essays in Economic History, op. cit.*.

54 For example, at Horze in Bohemia a Cracow merchant lost his goods (*Consularia Cracoviensia*, Vol. IV, p. 128) and in Venice a similar problem arose over the purchase of 4,300 Bohemian knives for a Cracow merchant (Ibid., Vol. V, p. 159.

55 D. Hay, *op. cit.*, p. 369.

56 See, for example, J. Hubert, 'Les routes du moyen âge', in *Les routes de France depuis les origines jusqu'à nos jours, Colloques: Cahiers de Civilisation*, Paris, 1959, pp. 25–56; G. Schreiber, 'Mittelalterliche Alpenpässe und ihre Hospitalkultur', *Miscellanea Giovanni Galbiati* (Vol. III, Fontes Ambrosaiani), Vol. XXVII, Milan, 1951, pp. 335–352.

57 R. S. Lopez, 'The Evolution of Land Transport in the Middle Ages', *Past and Present*, No. 9, London, 1956, pp. 17–29, which refers to the 'fluvialisation' of land transport during the early Middle Ages as roads and bridges built by the Romans fell into disrepair.

58 G. A. Sterzel and G. A. Tzschoppe, *Urkundensammlung zur Geschichte des Ursprungs der Städte und der Einführung und Verbreitung deutscher Kolonister und Rechte in Schlesien und Ober-Lausitz*, Hamburg, 1832, No. 54.

59 Evidence from a document signed by King Bolesław Wstydliwy, which mentions 'nullum powoz neque currus', *Kodeks dyplomatyczny Małopolski*, F. Piekosiński (ed.), Kraków, Vols. I–IV, 1878–1905), Vol. II, No. 451; Ibid., No. 516, 544, reference is made to carts and carters in Miechów, near Cracow, 'podwoda a podwodowe'. In 1354, King Kaziemierz Wielki confirmed a statute of 1256 which mentions 'a . . . vecturis . . . a przewód, podwodowe . . . stan et sep'. Ibid., Vol. III, No. 704. Special reference to oxen transport is made in J. Bardach, *Historia państwa i prawa Polski do roku 1795*, Vol. I, Warszawa, 1957, p. 141.

60 K. Buczek, 'Publiczne posługi transportowe i komunikacyjne w Polsce średniowiecznej', *Kwartalnik Historii Kultury Materialnej*, Vol. 15, No. 2, Warszawa, 1967, p. 277.

61 Z. Wojciechowski, *Państwo polskie w wiekach średnich. Dzieje ustroju* (II wyd.), Poznań, 1948, p. 80; Ibid., 'Ustroj polityczny Śląska do 1327/9 r. 'in *Historia Śląska od najdawniejszych czasów do roku 1400*, Vol. I, Kraków, 1933, p. 638. E. Janota, *Monografia opactwa cystersów w Mogile*, Part II (Zbiór dyplomów klasztoru mogilskiego), No. 18, Kraków, 1867 (for the year 1243).

62 V. Dragoun, 'Dějiny poštovního zřízeni', introduction to: F. Fischer, *Stručné dějiny ústředí československých poštovních a telegrafních zřízenců, 1873–1923*, Praha, 1923; see also P. Čtvrtník, *Cesta, Pošty Dějinami (Silnicí, Železnicí, Letadlem)* (Nak, Dop. a Spoju), Praha, 1977, pp. 28–29; 'Les transport au Moyen âge', *Actes du VIIe Congrès des mediévistes, juin 1976*, Rennes, 1978 (Linarmor), 342 pp.

63 *Kodeks dyplomatyczny miasta Krakowa, op. cit.*, Vol. I, No. 4 (dated 1306) and No. 16; see also B. Wyrozumska, *Drogi w ziemi krakowskiej do końca XVI wieku* (P. A. N. Ossolineum) (Prace Komisji Nauk Historycznych No. 41), Wrocław-Warszawa-Kraków-Gdańsk, 1977, pp. 28–35.

64 *Hansiches Urkundenbuch op. cit.*, Vol. III, Nos. 147, 559 (which gives a list of Polish custom houses in 1350–1360), and 631. For more general discussion see S. Weymannm, 'Cła i drogi handlowe w Polsce piastowskiej' (Poznańskie Towarzystwo Przyaciół Nauk) *Prace Komisji Historycznej*, Vol. XIII, No. 1, Poznán, 1938, 144 pp.; Ibid., 'Ze studiów nad zagadnieniem dróg w Wielkopolsce od X do XVIII wieku', *Przegląd Zachodni*, Vol. IX, Nos. 6–8 (Studia Poznańskie ku uczczeniu 1000-lecia miast i 700-lecia samorządu miejskiego) (połowa X w 1253–1953), Poznań, 1953, pp. 194–253.

65 *Hansisches Urkundenbuch, op. cit.*, Vol. IV, Nos. 810, 818, 830. See also

M. Małowist, 'Le développement des rapports économiques entre la Flandre, la Pologne et les pays limitrophes du XIIe au XIV siècle', *Revue Belge de Philologie et d'Histoire*, Vol. X, Brussels, pp. 1013–1065; S. Lewicki, 'Drogi handlowe w Polsce w wiekach średnich', *Sprawozdania z czynności i posiedzeń Akademji Umiejętności*, Vol. X, Kraków, 1906, pp. 31–48.

66 *Kodeks dyplomatyczny miasta Krakowa, op. cit.*, Vol. I, No. 156 (document dated 1453).

67 Ibid.

68 See K. Ślaski, 'Lądowe szlaki handlowe Pomorza w XI–XIII w', *Zapiski Historyczne*, No. 3, Toruń, 1969, pp. 29–44.

69 Wojewódzkie Archiwum Państwowe w Krakowie, *Acta Castrensia Cracoviensis* (Akta Grodzkie i Ziemskie) Vol. IX, No. 14 (dated 1405) refers to a customs station at Czchów; *Kodeks dyplomatyczny miasta Krakowa, op. cit.*, Vol. III, No. 310 (dated 1432).

70 Ibid., Vol. I, No. 16 (1329), and No. 57 (1380).

71 *Kodeks dyplomatyczny Małopolski, op. cit.*, Vol. I, No. 173; *Kodeks dyplomatyczny miasta Krakowa, op. cit.*, Vol. I, No. 22 (dated 1338).

72 Wojewódzkie Archiwum Państwowe w Krakowie, *Acta Castrensia Cracoviensis*, Vol. IX, No. 44.

73 *Kodeks dyplomatyczny miasta Krakowa, op. cit.*, Vol. I, No. 15 (dated 1324).

74 E. Janota, *Bardyów, historyczno-topograficzny opis miasta i okolica*, Kraków, 1862, p. 151, No. 74 (document dated 1363); p. 154, No. 79 (1388); p. 155, No. 80 (1424); p. 155, No. 81 (1434) and p. 156, No. 82 (1450). See also *Acta Castrensia Cracoviensis, op. cit.*, Vol. IX, No. 32 (dated 1434).

75 *Hansisches Urkundenbuch op. cit.*, Vol. III, No. 631.

76 Ibid., No. 559; see also T. Wąsowiczówna, 'Sandomierska sieć drożna w wiekach średnich', *Studia Sandomierskie*, Vol. I, Sandomierz, 1967, pp. 113–120.

77 *Hansisches Urkundenbuch op. cit.*, Vol. III, No. 559 and 631 refers to tributary route from Brześć to Kowal, Gostynin, Łowicz, Rawa and Radom.

78 A. Diószegi, *A Magyarországon keresztülvezetö kereskedelmi útak az Árpádhazi királyok idejében* (Publ. Stief Jenö és Társa), Kolozsvar, 1909, 52 pp.; J. Galicz, 'Przełęcz Jabłonkowska', *Rocznik Towarzystwa Tatrzańskiego* '*Beskid Śląski*', Vol. 3, Cieszyn, 1933, pp. 10–24; A. Gardony, *op. cit.*; O. R. Halaga, 'Portae Poloniae – w kwestii szlaków karpackich we wczesnym średniowieczu', *Acta Archeologica Carpathica*, Vol. VII, No. 1–2, Kraków, 1965, pp. 9–15; Ibid., 'Počiatky dial'kového obchodu cez stredné karpaty a kosického prava skladu', *Historica Carpatica*, Vol. 4, Bratislava, 1973, pp. 12–24; H. Kapiszewski, 'Droga z Panonii do Polski w roku 966: Przyczynek do dziejów przełęczy Dukielskiej', *Acta Archeologia Carpathica*, Vol. II, No. 1–2, Kraków, 1960, pp. 107–121; K. Myśliński, 'Rola miast Polski środkowej w handlu z krajami na południe od Karpat do końca XV wieku', *Prace Historyczne*, Vol. 56, Warszawa-Kraków, 1977, pp. 19–32.

79 F. Kiryk and R. Kołodziejczyk (eds.), *Dzieje Olkusza i regionu olkuskiego*, 2 Vols. (P.W.N.) Warszawa, 1978, 576 pp., 574 pp.

80 *Kodeks dyplomatyczny miasta Krakowa, op. cit.*, Vol. II, No. 310; A. Klose, *Von Breslau, Dokumentirte Geschichte und Beschreibung*, Vol. II, Pt.2, Breslau, 1881, p. 358. Evidence of a customs house at Będzin is documented for 1390, *Naj-*

starsze księgi op. cit., Vol. II, p. 290, and is again confirmed for 1432, *Kodeks dyplomatyczny miasta Krakowa, op. cit.*, Vol. II, No. 310; in 1452 Wrocław merchants confirmed that the oldest route to Cracow went via Opole and Toszek; A. Klose *op. cit.*, p. 358.

81 K. Wuttke, 'Die Versorgung Schlesiens mit Salz während des Mittelalters', *Zeitschrift für Geschichte und Alterthum Schlesiens*, Vol. 27, Breslau, 1893, p. 284.

82 *Kodeks dyplomatyczny miasta Krakowa, op. cit.*, Vol. I, No. 516; in 1435 Cracow merchants are recorded as having paid customs duty along this route, A. Mosbach, *Przyczynki do dziejów Polski z archiwum miasta Wrocławia*, Poznań, 1860, p. 96.

83 K. Raczyński (ed.), *Codex diplomaticus Maioris Poloniae*, Poznań, 1840, No. 126.

84 L. Białkowski, 'Lublin na starych szlakach handlowych', *Pamiętnik Lubleski*, Vol. 3, Lublin, 1938, pp. 288–293; K. Myślinški, 'Lublin a handel Wrocławia z Rusią w XIV i XV wieku', *Rocznik Lubelski*, Vol. III, Lublin 1960, pp. 5–36; A. Mosbach, *op. cit.*, 110, refers to Lublin fair in 1466 and 1476.

85 K. Raczyński, *op. cit.*, No. 126 (document dated 1455). Lwów was therefore connected via Wrocław with Frankfurt/Oder and the towns of Germany and Flanders, K. F. Klöden, *op. cit.*, Vol. II, p. 74.

86 Wojewódzkie Archiwum Państwowe w Krakowie, *Acta Castrensia Cracoviensis*, *op. cit.*, Vol. IV, No. 49 (dated 1419); Ibid., Vol. VII, No. 54 (dated 1454).

87 Ibid., Vol. V, No. 41.

88 Ł. Charewiczowa, *Handel średniowiecznego Lwowa, op. cit.*, pp. 31–52.

89 Wojewódzkie Archiwum Państwowe w Krakowie, *Acta Castrensia Cracoviensis*, *op. cit.*,Vol. V, No. 41.

90 *Kodeks dyplomatyczny miasta Krakowa, op. cit.*, Vol. I, No. 33; *Hansisches Urkundenbuch, op. cit.*, Vol. III, No. 559.

91 Ibid., See also A. Wagner, 'Handel dawnego Jarosławia do połowy XVII w.' *Prace Historyczne* (Wydane ku uczczeniu 50-lecia akademickiego koła historyków uniwersytetu Jana Kazimierza we Lwowie, 1878–1928), Lwów, 1929, pp. 122–147; *Akta Grodzkie i Ziemskie*, Vol. XIX, Lwów 1906, No. 452, p. 87; *Matricularum Regni Poloniae Summaria* (excussis codicibus qui in chartophylacio maximo Varsoviensi asservatur), T. Wierzbowski (ed.), Pars IV, Warszawa, 1910, No. 7447, p. 428; M. Hruszewskyj, *Istorja Ukrainej-Rusy*, Vol. VI, Lwów, 1907, p. 43.

92 A. Kunysz, 'Kontakty handlowe Przemyśla we wczesnym średniowieczu', *Rocznik Przemyski*, Vol. 15–16, 1975, Przemyśl, 1976, pp. 5–21.

93 *Hansisches Urkundenbuch, op. cit.*, Vol. III, No. 559.

94 *Kodeks dyplomatyczny miasta Krakowa, op. cit.*, Vol. I, No. 128 (dated 1430), No. 151 (1415), No. 200 (1493).

96 Ibid. See also J. Garbacik (ed.), *Krosno. Studia z dziejów miasta i regionu*, Vol. I (up to 1918) (P.W.N.), Kraków 1972, 349 pp.

97 *Akta grodzkie i ziemskie, op. cit.*, Vol. III, No. 28, 'in omnibus iuribus et libertatibus, videlicet, depositione omnium mercanciarum, omnium et quorumlibet mercatorum ac transitu seu via ... volumus ... conservare' (dated 1379). In 1380 a merchant en route to or from Tartary must stay 14 days in Lwów, pay customs duty, and offer goods for sale. If they were not sold after a fortnight they could then be taken on further, either east or west. *Akta grodzkie i ziemskie, op. cit.*, Vol.

III, No. 58. See also M. Małowist, 'Les routes du commerce et les marchandises du Levant dans la vie de la Pologne au bas moyen âge et début de l'époque moderne', *Mediterraneo e Oceano Indiano*, Firenze, 1970, pp. 157–175; M. Dubiecki, 'Kaffa osada genueńska, i jej stosunek do Polski w XV w.', *Przegląd Powszechny*, No. 12, Kraków, 1886, pp. 12–22.

98 *Akta grodzkie i ziemskie, op. cit.*, Vol. III, No. 76; *Kodeks dyplomatyczny miasta Krakowa, op. cit.*, Vol. I, No. 108.

99 S. Lweicki, *Lemberg's Stapelrecht*, Lwów, 1909, p. 34, Ibid., No. 220, 223.

100 H. Simonsfeld, *Der Fondaco dei Tedeschi in Venidig und die Deutsch-Venetianischen Handelsbeziehungen* (2 Vols.) Stuttgart, 1887, Vol.II, p. 98.

101 *Kodeks dyplomatyczny miasta Krakowa, op. cit.*, Vol. I, No. 34.

102 H. Simonsfeld, *op. cit.*, Vol. I, No. 368.

103 M. M. Postan, 'The Trade of Medieval Europe: The North', *The Cambridge Economic History of Europe*, Vol. II, Cambridge, 1962, p. 149. More specifically see R. Grodecki, 'Znaczenie handlowe Wisły w epoce piastowskiej', *Studia historyczne ku czci St. Kutrzeby*, Vol. II, Kraków, 1938, pp. 277–303; T. Hirsch, *Danzigs Handels und Gwerbesgeschichte unter der Herrschaft des Deutschen Ordens*, Leipzig, 1858, pp. 178–180; E. R. Raths, 'Die Entwicklung des Weichselhandels von der Mitte des 13. bis zum 15. Jahrhundert', *Deutscher Blätter im Polen*, Nos. 6, 7, Poznań, 1927; pp. 13–15, 17–19; S. Kutrzeba, 'Wisła w historii gospodarczej dawnej Rzeczypospolitej Polskiej', *Monografia-Geografija Wisły*, Vol. XI, Warszawa, 1920, pp. 1–15; D. Krannhals, 'Die Rolle der Weichsel in der Wirtschaftsgeschichte des Ostens', in *Die Weichsel, ihre Bedeutung als Strom und Schiffahrtsstrasse und ihre Kulturarbeiten*, R. Winkel (ed.), Leipzig, 1939, pp. 77–152.

104 S. Kutrzeba, Wisła w historii . . . *op. cit.*, p. 3.

105 Wojewódzkie Archiwum Państwowe w Krakowie, *Acta Castrensia Cracoviensis*, Vol. IX, No. 30, 31, 38. Wood sent via Dunajec to the Vistula for Toruń and Gdańsk sent through Cracow by merchant from Sącz.

106 *Kodeks dyplomatyczny miasta Krakowa, op. cit.*, Vol. I, No. 16; *Kodeks dyplomatyczny Małopolski, op. cit.*, Vol. I, No. 178.

107 In 1378 the previous treaty was cancelled by King Ludwik Węgierski, and Sącz merchants were now free from paying Cracow's customs duty and for use of the river route. However, two years later Ludwik rescinded the new agreement and forbad Sącz merchants from sending goods such as copper, iron etc. from Hungary by the river route. *Kodeks dyplomatyczny miasta Krakowa, op. cit.*, Vol. I, No. 57; *Kodeks dyplomatyczny Małopolski, op. cit.*, Vol. I, No. 340.

108 For example in 1390 it was allowed by King Władysław Jagiełło. *Kodeks dyplomatyczny Małopolski op. cit.*, Vol. IV, No. 1007. See also J. W. Gan, *Z dziejów żeglugi śródlądowej w Polsce* (Książka i Wiedza), Warszawa, 1978, pp. 40–41.

109 *Hanserecesse: Die Recesse und Andere Acten der Hansetage von 1256–1430*, K. Kippmann (ed.), 8 Vols., Leipzig, 1870–1897; Vol. III, No. 422, 9, p. 436 (1875), document dated 4/IV/1389.

110 For further discussion see W. Kowalenko, 'Polska żegluga na Wiśle i Bałtyku w XIV i XV w.', *Roczniki Historyczne*, Vol. XVII, Poznań, 1948, pp. 336–337.

111 *Hanserecesse, op. cit.*, Vol. V, No. 118, 5, p. 80 document dated 3/III/1403. 'Item haben die heren von Thorun vor unsern herer, den homeister, gebrocht, wy das dy Crakower und andere geste boben der Wyssle keyne nyderloge czu Thorun halden, sunder ire gut dorch dys land czur zeewart furen und brengen; dergelich dy geste dy czur zeewart her in dys land kommen, das sy keyne nyderloge in den obirsteten halden, sunder mit erern gute varen und keren im land, wor sy wellen; und das hat unser here homeistr befoln eynem itzlichen in syme rate doremme czu sprachin, czum nesta tage widr unczubrengen'.

112 T. Korzon, *Dzieje wewnętrzne Polski za Stanisława Augusta*, Vol. II, Warszawa, 1897, p. 1; S. Krzyżanowski, 'Morsztynowi w XV w.', *Rocznik Krakowski*, Vol. I, Kraków, 1898, p. 354.

113 For example between 1422 and 1424. M. Dogiel, *Codex Diplomaticus Regni Polonia et Magni Ducatus Lituaniae in quo pacta, foedera, tractus pacis . . .* ', Wilno, 1758–1764, 3 Vols. (I, IV, and V), Vol. IV, p. 112 (document dated Malmö 1422).

114 O. Balzar, *Corpus Iuris Polonici*, Kraków, 1906, Vol. III (ann. 1506–1522), pp. 566–567.

115 For further discussion on this theme see H. Samsonowicz, 'Przemiany osi drożnych w Polsce późnego średniowiecza', *Przegląd Historyczny*, Vol. LXIV, No. 4, Warszawa, 1973, pp. 696–716.

116 The best overall introduction is P. Dollinger, *The German Hansa* (Macmillan) London, 1970, 400 pp.; Ibid., *La Hanse (XIIe–XVIIe siècles)*, Paris, 1964, 405 pp.; Ibid., *Dzieje Hanzy XII–XVII wiek* (Wyd. Morskie), Gdańsk, 1975, 407 pp.

117 A. Brandt, *Geist und Politik in der Lübeckischen Geschichte*, Lübeck, 1954, 375 pp.

118 W. G. East, *An Historical Geography of Europe* (Methuen & Co.), London, 1956, p. 339.

119 A. H. de Oliveira Marques, *Hansa e Portugal na Idade Media* (Univ. thesis) Lisbon, 1959, 265 pp.

120 H. Witthöft, Mass- und Gewichtsnormen im hansischen Salzhandel', *Hansische Geschichts-blätter*, Vol. 95, Lübeck, 1978, pp. 38–65.

121 V. N. Vernardski, *Novgorod i novgorodskaia zemla w XV veke*, Moscow–Leningrad, 1961, 328 pp.; A. L. Choroskevic, *Torgovla Velikogo Novgoroda s Pribaltikoi i Zapadnoi Evropoi*, Moscow, 1963, 263 pp.

122 J. A. van Houtte, 'The Rise and Decline of the Market in Bruges', *Economic History Review*, Vol. 19, No. 1, London, 1966, p. 35.

123 E. R. Daenell, 'Polen und die Hansa um die Wende des XIV Jahrhundert', *Deutsche zeitschrift für Geschichtswissenschaft*, N.F. 2, Vierteljahrshefte, Freiburg, 1897–1898, pp. 10–26; M. P. Lesnikov, Beiträge zur Baltisch-Niederländischen Handelsgeschichte am Ausgang des 14. und zum Beginn des 15. Jahrhundert', *Wissenschaftliche Zeitschrift der Karl-Marx Universitat* Vol. 7, Berlin, 1957–58, pp. 19–30; C. Sattler, *Die Handelsrechungen des Deutschen Ordens*, Leipzig, 1887, 99 pp.

124 L. Gilliodts-van Severen (ed.), *Cartulaire de l'ancienne Estaple de Bruges*, 4 Vols. (Société d'Emulation, Bruges, 1904–1906; R. Häpkt, *Brugges Entwicklung zum mittelalterlichen Weltmarkt*, Berlin, 1908, 438 pp.; J. A. van Houtte, *op. cit.*

125 W. Stieda, 'Revaler Zollbücher und Quittungen des 14. Jahrhunderts', *Hansis-*

ches Geschichtsquellen, No. 1381, Halle, 1887, pp. 50–54; K. H. Sass, *Hansischer Einfuhrhandel in Reval um 1430*, Marburg/Lahn, 1955, 284 pp. M. Hallmann, *Das Lettenland im Mittelalter*, Münster-Köln, 1954, 264 pp.

126 G. Hollihn, 'Die Stapel- und Gästepolitik Rigas in der Ordenzeit', *Hansische Geschichtsblätter*, Vol. 60, Lübeck, 1935, pp. 91–207; E. Tibero, 'Moskau, Livland und die Hanse, 1487–1547', *Hansische Geschichtsblätter*, Vol. 93, Lübeck, 1975, pp. 13–70.

127 E. Keyser, *Die Bevölkerung Danzigs und ihre Herkunft im 13 und 14 Jahrhundert* (2nd ed.), Gdańsk, 1928, 360 pp.; H. Samsonowicz, 'Handel zagraniczny Gdańska w drugiej połowie XV w.', *Przegląd Historyczny*, Vol. 42, Warszawa, 1956, pp. 283–352; Ibid., 'Struktura handlu gdańskiego w pierwszej połowie XV wieku', *Przegląd Historyczny*, Vol. 53, Warszawa, 1962, pp. 695–715; Ibid., *Untersuchungen uber das Danziger Bürgerkapital in der zweiten Hälfte des 15. Jahrhunderts*, Weimar, 1969, 256 pp.; Ibid., 'Über Fragen des Landhandels Polens mit Westeuropa im 15/16 Jahrhundert', *Neue Hansische Studien*, Berlin 1970, 321 pp.; H. Weczerka, 'Bevölkerungszahlen der Hansestädte (insbesondere Danzigs) nach H. Samsonowicz', *Hansische Geschichtsquellen*, Vol. 18, Halle, 1964, pp. 69–80.

128 E. Assmann, 'Die Stettiner Zollrolle des 13 Jahrhunderts', *Hansische Geschichtsblätter*, Vol. 71, Lübeck, 1952, pp. 49–58.

129 K. Koppman, 'Über die Pest des Jahres 1565 und die Bevölkerungstatistik Rostocks im 14, 15, und 16 Jahrhundert', *Hansische Geschichtsblätter*, Vol. 10, Lübeck, 1901, pp. 45–67; K. F. Olechnowitz, *Rostock von der Stadtrechtsbestatigung im Jahr 1218 bis zur Bürgerlich-Demokratischen Revolution von 1848/49*, Rostock, 1968, 261 pp.

130 M. Plesnikov, *Lübeck als Handelsplatz für Osteuropawaren im 14 Jahrhundert*, Berlin, 1961, 348 pp.

131 K. Maleczyński, *Najstarsze targi w Polsce i ich stosunek do miast przed kolonizacją na prawie niemieckim*, Lwów, p. 9; H. Lesiński, 'Początki i rozwój stosunków Polsko-Hanzeatyckich w XIII wieku', *Przegląd Zachodni*, No. 5–8, Poznań, 1952, pp. 130–145.

132 L. Koczy, '*Związki handlowe wroclawia z Polską do końca XVI wieku*', Katowice, 1936, 341 pp.; M. Małowist, 'Polish–Flemish Trade in the Middle Ages', *Baltic and Scandinavian Countries*, Vol. IV, No. 1(8), Gdynia, 1938, pp. 1–9.

133 E. C. Semple, 'The Development of the Hanse Towns in Relation to Their Geographic Environment', *Journal of the American Geographical Society*, Vol. XXXI, New York, 1899, p. 241.

134 Ibid., p. 242; M. Małowist, 'Poland, Russia and Western Trade in the Fifteenth and Sixteenth Centuries', *Past and Present*, Vol. 13, London, 1958, pp. 26–41.

135 *Hanserecesse*, *op. cit.*, Series A, Vol. I, No. 75, dated between 1297–1302.

136 Ibid., No. 479.

137 Ibid., Vol. II, No. 115, sent direct to Cracow and complains about the poor production of pewter.

138 Ibid., Vol. II, No. 361§7, document dated 1387.

139 Ibid., Vol. V, No. 7.

140 Ibid., Vol. III, No. 683, document dated 1424.

141 Ibid., Seria B, Vol. II, No. 216, in a letter from the Prussian towns to the Polish king, states, 'das euwer Graden stete alse Crakow und andere mete (i.e. to the Hanseatic League), ingehoren'.

142 K. Spading, *Holland und die Hanse im 15. Jahrhundert. Zum Problematik des übergangs vom Feudalismus zum Kapitalismus* (Publ. H. Böhlau), Weimer, 1973, 189 pp.

143 I. Czarciński, 'Przedstawiciele Torunia na zjazdach hanzeatyckich w XIV i pierwszej połowie XV w.', *Acta Universitatis Nicolai Copernici. (Historia)*, Vol. II, Toruń, 1977, pp. 21–31.

144 A. Brandt, 'Die Stadt des späte Mittelalters im hansischen Raum', *Hansische Geschichtsblätter*, Vol. 96, Lübeck, 1978, pp. 1–14; H. Planitz, *Die deutsche Stadt im Mittelalter* (Publ. Böhlau–Verlag), Weimar 1965, 219 pp.

145 A. Mączak and H. Samsonowicz, 'La zone baltique, l'un des elements du marché européen', *Acta Poloniae Historica*, Vol. II, Warszawa, 1965, pp. 71–99.

146 H. Schwarzwalder, 'Bremen im Mittelalter', *Studium Generale* (Zeitschrift für die Einheit der Wissenschaften im Zusammerhang ihrer Begriffsbildungen und Forschungsmethod) Vol. 16, Heidelberg, 1963, pp. 391–421.

147 H. van der Wee, *The Growth of the Antwerp Market and the European Economy, 14th to 16th Centuries*, 3 Vols., Louvain, 1963.

148 L. von Winterfeld, 'Handel, Kapital und Patriziat in Kölnbis 1400', *Pfingstblätter des hansischen Geschichtsverein*, Vol. XVI, Lübeck, 1925, pp. 10–24.

149 *Hanserecesse, op. cit.*, Series A. Vol. VIII, No. 670.

150 Ibid., Series B, Vol. I, No. 43 (dated 1431); Ibid., Nos. 260, 269, 295 (all dated 1434); Ibid., No. 480 (dated 1435); Ibid., Vol. II, No. 469 (dated 1448); Ibid., Vol. V, No. 85 (dated 1461); Ibid., Series C, Vol. II, No. 140 (dated 1487).

151 Ibid., Series A, Vol. VIII, No. 712 §27 (dated 1430) in which the Lübeck meeting placed Cracow in a list of towns who 'ungerhorsam geworden' that they would send delegates to the meetings.

152 As for example *Hanserecesse, op. cit.*, Series B, Vol. I, No. 85 (1431); Ibid., No. 480 (1435); Ibid., Vol. V, No. 85 (1461).

153 As for example, Ibid., Series B, Vol. III, No. 288 (dated 1447); Ibid., Vol. V, No. 794 (dated 1466); Ibid., Vol. VII, No. 1 (dated 1473).

154 For example, Toruń in 1470, Ibid., Series B, Vol. VI, No. 342, or Gdańsk, Ibid., No. 288 (1470); Ibid., Vol. VII, No. 231 (1473); Ibid., No. 324 (1476) and Series C, Vol. II, No. 144 (1487).

155 Copies of minutes from general meetings were sent to Cracow, e.g. 1434 in Bruges, Ibid., Series B, Vol. I, No. 406, as well as particular sessions in a Prussian town, e.g. Toruń in 1434, Ibid., No. 416.

156 Ibid., Series C, Vol. V, No. 116 (1494).

157 Ibid., Series A, Vol. V, No. 392 (dated 1407); Ibid., Vol. VIII, No. 172§18 (dated 1430); Cracow was asked to help the League in their war against Eric, King of the United Kingdoms (1396–1442), of Scandinavia, in 1427. Ibid., Series A, Vol. VIII, No. 184.

158 For example in 1405 concerning a dispute between Cracow merchants and Poznań, Ibid., Series A, Vol. VIII, No. 1026, and 1424 when misunderstandings over toll collection, 'pfundzoll', imposed by Prussia in defence of the

League's treaties, led to a Prussian–Polish meeting at Weiluń, in which Cracow sent representatives. Ibid., Vol. VII, No. 683.

159 Letter sent from Cracow to the Lübeck meeting in 1461, stating that they had complained in vain to the Bruges merchants about poor cloth measures. Ibid., Series B, Vol. V, No. 85.

160 M. Małowist, 'The Problem of the Inequality of Economic Development in Europe in the Later Middle Ages', *Economic History Review*, Vol. XIX, No. 1, London, 1966, p. 25.

161 N. Ellinger-Bang, *Tabeller over Skibsfart og varetransport gennem Øresund*, Copenhagen, 1906–1953. The earliest returns of the toll-stations at the Sound are for 1495, and in that year the Dutch vessels formed the bulk of all shipping bound for the Baltic.

162 M. M. Postan, The Trade of Medieval Europe . . . ', *op. cit.*, p. 256.

163 Ibid., p. 151.

164 J. Dąbrowski, 'Kraków a Węgry w wiekach średnich', *Rocznik Krakowski*, Vol. XIII, Kraków, 1911, p. 211.

165 Archiv miesta Bardejov, *O.A.B.*, No. 2343, 2368, dated 18/VI and 17/XII/1484, record attempts to bypass the customs house at Żmigród travelling from Cracow, on route to Bardejov.

166 E. Janota, 'Zapiski o Bardyowie i najbliższej okolicy', *Rocznik Towarzystwa Nauk Krakowa*, Vol. IV, Kraków, 1862, p. 97.

167 M. Cromeri, *Polonia sive situ populis, moribus, magistratibus et republica Regni Polonici libri duo* (Biblioteka Pisarzów Polski), Kraków, 1901, p. 22, which states 'Polonia: montes Sarmatici, qui Poloniam et Russiam ab Hungariam dirimunt, ardui et silvestres'.

168 F. Kiryk, 'Stosunki handlowe Jasła i miast okolicznych z miastami słowackimi w XV wieku', *Studia z Dziejów Jasla i powiatu Jasielskiego*, J. Garbacik (ed.), Kraków, 1964, p. 147.

169 M. M. Postan, The Trade of Medieval Europe . . . ', *op. cit.*, p. 151.

170 J. Dąbrowski, *op. cit.*, p. 212. For information on money values etc. see J. A. Szwagrzyk, *Pieniądz na ziemiach polskich X–XX w.*, Wrocław–Warszawa–Kraków–Gdańsk, 1973, 360 pp.

171 L. Fejérpataky, *Magyarországi városok régi számadáskönyvei*, Budapest, 1885, pp. 174, 230, 344, 376.

172 *Najstarsze księgi*, *op. cit.* (Liber percepta et distributum), passim.

173 L. Fejérpataky, *op. cit.*, p. 345.

174 J. Kulischer, *Allgemeine Wirtschaftsgeschichte des Mittelalters und der Neuzeit*, 2 Vols., Munich–Berlin, 1928–29, Vol. I, p. 301.

175 *Monumenta Poloniae Historica: Pomniki dziejowe Polski*, A. Bielowski (ed.), Akademia Umiejętności w Krakowie, 6 Vols., Lwów, 1864–93 (Vol. VI, Kraków), Vol. I, No. 173.

176 Ibid., No. 128; B. Wyrozumska, *op. cit.*, pp. 30–31.

177 Ibid., Vol. III, No. 671, 341.

178 S. Kutrzeba, 'Finanse Krakowa w wiekach średnich', *Rocznik Krakowski*, Vol. III, p. 81; see also, Ibid., 'Taryfy celne i polityka celna w Polsce od XIII–XIV w.', *Ekonomista*, Vol. 2, Warszawa, 1902, pp. 33–48, 187–211.

179 Wojewódzkie Archiwum Państwowe w Krakowie, *Kodeks dyplomatyczny*

miasta Krakowa, op. cit., Vol. II, No. 310 which states 'alzo sal man den Krokeshen czol ynnemen von den gesten, dy ire gutter her brengen, sunder der landman gibt halb alzo vyl'. The basic reasons for collecting customs duty are given in *Najstarsze księgi, op. cit.*, Vol. II, p. 85.

180 D. Csánski, *Hazánk kereskedelmi viszonyai I. Lajos Korában*, Budapest, 1880, p. 21.

181 Ibid., p. 17. Furthermore in 1368, Ludwik guaranteed that Polish merchants in Hungary would not be burdened with new customs duties recently imposed. *Kodeks dyplomatyczny miasta Krakowa, op. cit.*, Vol. I, No. 38.

182 *Monumenta Poloniae, op. cit.*, Vol. I, No. 148; *Kodeks dyplomatyczny Polski*, L. Rzyszczewski, A. Muczkowski and J. Bartoszewicz (eds.), 3 Vols., Warszawa, 1847–1858, Vol. III, No. 75.

183 Wojewódzkie Archiwum Państwowe w Krakowie, *Acta Scabinalia Sandec*, for 1488 and 1505, pp. 414, 519, 421.

184 Wojewódzkie Archiwum Państwowe w Krakowie, *Consularia Cracoviensia, op. cit.*, Vol. I, No. 429, p. 29.

185 Ibid., *Liber Testamentorum*, p. 40.

186 A. Pawiński, 'Notatki kupca krakowskiego w podróży do Flandri', *Biblioteka Warszawa*, Vol. 3, 1872, pp. 53–73, refers to a document dated 1402 which mentions copper being sent from Cracow to Gdańsk via Toruń. Transport costs for sending the load from Toruń to Gdańsk (1 łast and 107 cetnars, i.e. 3.5 łasts) came to 1 grzywna, 3 wity and 4 skojce, and 20 Prussian fenigs. For exact monetary values see J. A. Szwagrzyk, *op. cit.*

187 M. M. Postan, The Trade of Medieval Europe . . . , *op. cit.*, p. 152.

188 Ibid., p. 154.

189 W. Vogel, *Geschichte der deutschen Seeschiffahrt*, Vol. I, Berlin, 1915, p. 422; see also E. Cieslak 'Niektóre zagadnienia hanzeatyckiego handlu i transportu morskiego', *Przegląd Zachodni*, No. 2, Poznań, 1952, pp. 12–26; V. Lauffer, 'Danzigs Schiffs-und Waarenverkehr am Ende des XV, Jahrhunderts', *Zeitschrift des Westpreussischen Geschichtsvereins*, Vol. 33, Danzig, 1894, pp. 44–46; M. Biskup, 'Handel wiślany w latach 1454–1466', *Roczniki dziejów społecznych i gospodarczych*, Vol. XIV, Poznań, 1952, pp. 155–199; H. Samsonowicz, 'Handel zagraniczny . . . *op. cit.*; Ibid., Struktura handlu . . . *op. cit.*; J. Schildhauer, 'Zur Verlagerung des See-und Handelsverkehrs im nordeuropäischen Raum während des 15. und 16. Jahrhunderts', *Jahrbuch für Wirtschaftsgeschichte, 1968*, Vol. IV, Berlin, 1968, pp. 187–195, which deals with the maritime traffic of Gdańsk in 1460, 1476 and 1530.

190 *Hansisches Urkundenbuch, op. cit.*, Vol. IV, No. 1, pp. 407, 431, 433, 437, 439, 444, 445, which refer to the years 1439–1440.

191 For example, a case is quoted in which the cost of transporting salt totalled 48% of the selling price. J. Billioud, 'Le sel du Rhône. La ferme du tirage de l'Empire au XVIe siècle', *Bulletin philologique et d'histoire (jusqu'à 1715), du Comité des travaux historiques et scientifiques*, 1958, Paris, 1959, pp. 211–226.

192 M. M. Postan, The Trade of Medieval Europe . . . , *op. cit.*, p. 151.

193 F. Baudel, *Capitalism and Material Life 1400–1800* (Fontana/Collins), London, 1977, p. 318.

194 Ibid., p. 309.

195 F. W. Carter, *Dubrovnik (Ragusa): A Classic City-State* (Seminar Press), London/New York, 1972, p. 159.
196 D. Hay, *op. cit.*, p. 363, – in 1394 a journey from Paris to Avignon was covered in 4 days (*circa* 650 kms); Barcelona to Valencia in 2 days and 2 hours; Rome to Venice in a day and a half (576 kms); in 1482 a relay of riders between Newcastle and London took 2 days (320 kms).
197 Ibid., p. 364.
198 H. Heaton, *Economic History of Europe* (Harper & Brothers), New York, 1948, p. 158.
199 Ibid., p. 233.
200 D. Hay, *op. cit.*, p. 364.
201 M. M. Postan, The Trade of Medieval Europe . . . , *op. cit.*, p. 144.
202 D. Hay, *op. cit.*, p. 364.
203 Y. Renouard, in *L'histoire et ses méthodes*, I. Samarand (ed.), Paris, 1961, pp. 122–114; P. Sardella, *Nouvelles et spéculations à Venise*, Paris, 1948, p. 210; C. A. J. Armstrong, 'Distribution and Speed of News at the Time of the Wars of the Roses', *Studies in Medieval History Presented to F. M. Powicke*, Oxford, 1948, pp. 129–154.
204 F. Braudel, *op. cit.*, p. 316.
205 A. Pawiński, *op. cit.*, p. 60.
206 M. Biskup, Handel wiślany . . . , *op. cit.*, p. 174.
207 P. Čtvrtník, *op. cit.*, p. 29.
208 H. Samsonowicz, Struktura handlu gdańskiego . . . , *op. cit.*, p. 711.
209 N. J. G. Pounds, An economic history of medieval Europe, *op. cit.*, p. 320.
210 J. Ptaśnik, 'Studia nad patrycjatem krakowskim wieków średnich', *Rocznik Krakowski*, Vol. XV, Kraków, 1913, pp. 23–95, and Vol. XVI, 1914, pp. 1–90; K. Pieradzka, 'Przedsiębiorstwa kopalniane mieszczan krakowskich w Olkuszu od XV do początków XVII wieku', *Zeszyty Naukowe Uniwersytetu Jagiellońskiego Historia*, No. 3, Kraków, 1958, pp. 45–46.
211 J. Bradbeer, 'The Geography of Mining: Why Still a Neglected Field?', *Research Seminar Series*, Dept. of Geography, Portsmouth Polytechnic, Portsmouth 1978, p. 1; A. M. Hay, 'A Simple Location Theory for Mining Activity', *Geography*, Vol. 61, Pt.2, Sheffield 1976, p. 65; see also R. E. Murphy, The Geography of Mineral Production', in *American Geography: Inventory and Prospect*, P. E. James & C. F. Jones (eds.), Syracuse, New York, 1954, p. 279.
212 J. Bradbeer, *op. cit.*, p. 2.
213 H. Madurowicz-Urbańska, 'L'Influence de l'industrie minière-métallurgique sur la campagne et l'exploitation paysanne', *Studia Historiae Oeconomicae* (U.A.M.), Vol. 10, Poznan, 1975, pp. 71–73.
214 M. Małowist, *Wschód a Zachód Europy w XIII–XVI wieku* (Konfrontacja struktur społeczno-gospodarczych) (P.W.N.), Warszawa, 1973, p. 181.
215 S. Kutrzeba, Handel Krakowa w wiekach średnich . . . ', *op. cit.*, p. 69.
216 M. M. Postan, The Trade of Medieval Europe . . . , *op. cit.*, p. 128.
217 T. Dziekoński, *Metalurgia miedzi, ołowiu i srebra w Europie środkowej od XV do końca XVIII w.* (Instytut Historii Kultury Materialnej, Polskiej Akademii Nauk), Wrocław–Warszawa–Kraków, 1963, pp. 106–108.

218 D. Molenda, *Górnictwo kruszcowe na terenie złóż śląsko-krakowskich do połowy XVI wieku*, Wrocław–Warszawa–Kraków (P.A.N.), 1963, p. 416.

219 J. Długossi, *Senioris Canonici Cracoviensis Opere Omnia*, A. Przezdziecki (ed.), Kraków, 1873, Vol. X, p. 1152, mentions people from Bytom sending lead and silver to Cracow in 1367. See also H. Łabęcki, *Górnictwo w Polsce*, 2 Vols., Warszawa, 1841, Vol. I (Opis kopalnictwa i hutnictwa polskiego), 538 pp.

220 Wojewódzkie Archiwum Państwowe, w Krakowie; '*Najstarsze księgi i rachunki miasta Krakowa . . .*' *op. cit.*, Vol. II, No. 427, fol. 99: dated 13/VI/1407, refers to a merchant Ulrich Kamerer, sending goods from Košice to Cracow, a large part of the contents being silver ore. See also F. Malyusz, *Zsigmondkori oklevéltár*, Vol. I, Budapest, 1956, p. 376, No. 3397. Nuremburg enterprises in the Carpathian Mountains encouraged economic measures in Hungary detrimental to Polish interests. Nuremburg's influence in the Hungarian government after 1400 and their involvement in its mining industry, led to an embargo on gold, silver and non-ferrous metals on Poland. Moreover, the Nuremburg firm of Flextorfer-Zenner acquired the lead-purchasing monopoly for Hungary; its indispensability for separating silver from Hungarian copper (seigern) and lack of alternative uses, meant the firm could dictate the lead price to Cracow's lead merchants. See W. von Stromer, 'Nuremburg in the International Economics of the Middle Ages', *Business History Review*, Vol. XLIV, No. 2 (Harvard Univ. Press), New York, 1972, pp. 217–218.

221 W. Mück, *Der mansfelder Kupferschieferbergbau*, Vol. I, Eisleben, 1910, p. 57; Silver straining was invented in 1451 by John Funcke in Germany. See also J. U. Nef, 'Silver Production in Central Europe, 1450–1618', *Journal of Political Economy*, Vol. LXIX, Chicago, 1941, pp. 575–591. Wojewódzkie Archiwum Państwowe, w Krakowie; *Consularia Cracoviensia*, Vol. I, No. 429, p. 219 refers to the purchase in 1458 by Magdalena Morsztyn of silver worth 107 grzywien from Mikołaj Goder and his company from Bardejov.

222 D. Molenda, Miasta górnicze Europy środkowo-wschodniej w epoce feudalnej', in *Miasta doby feudalnej w Europie środkowo-wschodniej* (Przemiany społeczne a układy przestrzenne) (P.W.N.), Warszawa–Poznań–Toruń, 1976, p. 191.

223 E. Reinhardt, *Johann Thurzo von Bethlemfalva, Bürger und Konsul von Krakau, in Goslar, 1478–1496*, Goslar, 1928, 251 pp.; M. Skladaný, 'Ján Thurzo v Mogile (1466–1496)', *Zbornik Filozofickieckej Fakulty Univerzity Komenského*, Vol. XXIV–XXV (Historica) 1973–1–47, Bratislava, pp. 203–222; G. Pölnitz, *Jacob Fugger II*, Tübingen, 1951, pp. 21–23, 53–55; M. J. Jansen, 'Jacob Fugger der Reiche', *Studien und Quellen*, Leipzig, 1910, pp. 134–135; E. Koch, 'Das Hütten- und Hammerwerck der Fugger zu Hohenkirchen bei Geogenthal in Thüringen 1495–1549', *Zeitschrift des Vereins für thüringische Geschichte und Altertumskunde*, Weimar, 1926, pp. 284–295; F. Dobel, 'Der Fugger Bergbau und Handel in Ungarn', *Zeitschrift des historischen Vereines für Schwaben und Neuburg*, Vol. VI, Neuburg, 1879, pp. 25–40.

224 P. Ratkoš, 'Vznik a začiatky banských miest na Slovensku', *Historické Štúdie*, Vol. XIX, Bratislava, 1974, p. 43.

225 Privilege dated 6/IV/1496. See L. Schick, *Un grand homme d'affaires au début du XVIe siècle Jacob Fugger*, Paris, 1957, p. 47.

226 M. Gumowski, *Dzieje mennicy krakowskiej*, Poznań, 1927, 263 pp. Wojewódzkie

Archiwum Państwowe, w Krakowie; *Consularia Cracoviensia*, Vol. I, No. 430, p. 218, dated 1489 refers to a treaty about coining money and mentions the use of silver from Hungary in Cracow's mint.

227 D. Molenda 'W sprawie badań huty miedzi w Mogile pod Krakowem w XV i XVI wieku', *Przegląd Historyczny*, Vol. 66, No. 3, Warszawa, 1975, p. 378.

228 H. van der Wee, *The Growth of the Antwerp Market and the European Economy* (fourteenth-sixteenth centuries), The Hague, 1963, 3 Vols; Vol. II, pp. 80–85; see also J. U. Nef, 'Mining and Metallurgy in Medieval Civilisation' in *The Cambridge Economic History of Europe*, Vol. II ('Trade and Industry in the Middle Ages'), Cambridge, 1952, pp. 458–473.

229 J. A. van Houtte, 'La genèse du grand marché international d'Anvers à la fin du moyen âge', *Revue belge de Philologie et d'Histoire*, Vol. XIX, Brussels, 1940, pp. 87–126.

230 D. Molenda, 'Dzieje Olkusza do 1795 r.', in *Dzieje Olkusza i regionu olkuskiego*, F. Firyk and R. Kołodziejczyk (P.W.N.), Vol. 1, Warszawa, 1978, 576 pp.

231 In 1374, Elżbieta Łokietkówna gave the right to Cracow to explore the Olkusz area for metals – 'Ordinatio montium Ilcussiensium'. H. Łabęcki, *op. cit.*, Vol. I, p. 190; privileges granted by Władysław Jagiełło (1426), Kazimierz Jagiellończyk (1485) (1491). Ibid.

232 *Zarys dziejów górnictwa na ziemiach polskich*, Vol. I, Katowice 1960, p. 123.

233 'postremo constitiumus et ordinamus de plumbi fodinis zupæ nostræ Slavkoviensis', H. Kownacki, *O starożytności kopalni kruszców wyrabiania metallów czyli robót górniczych w kluczu Sławkowskim* (Dufour), Warszawa, 1791, p. 72.

234 Both Chrzanów and Trzebinia received mining permission under Magdeburg Law from King Władysław Jagiełło in 1415. H. Łabęcki, *op. cit.*, Vol. I, p. 193.

235 J. Ptaśnik, 'Bonerowie', *Rocznik Krakowski*, Vol. VII, Kraków, 1905, p. 15.

236 M. Małowist, 'The Problem of the Inequality of Economic Development in Europe in the Later Middle Ages', *Economic History Review*, Vol. XIX, No. 1, London, 1966, p. 21; see also D. Molenda, *Kopalnie rud ołowiu na terenie złóż śląsko – krakowskich w XVI–XVIII wieku*', Wrocław–Warszawa–Kraków–Gdańsk, 1972, 213 pp.

237 Wojewódzkie Archiwum Państwowe, w Krakowie; *Consularia Cracoviensia*, No. 429, p. 25. He often sent more than 100 cetnars (i.e. 6½ tons) from Cracow to Košice at a time. *Liber Testamentorum*, pp. 40–41.

238 Morsztyn sent over 1,000 cetnars of lead (65 tons) from Cracow to Hungary at the end of the century. *Magyar Történelmi Tár* (Hungarian Historical Archive), for 1892, Budapest, 1893, p. 569.

239 N.J.G. Pounds, *An Economic History of Medieval Europe*, *op. cit.*, p. 453.

240 J. Pelc, *Ceny w Krakowie w latach 1369–1900*, Lwów, 1935, Table 54, pp. 63–64.

241 Wojewódzkie Archiwum Państwowe, w Krakowie, *Kodeks dyplomatyczny miasta Krakowa*, *op. cit.*, Vol. I, No. 7; refers to the export of lead from Cracow to Košice in 1310.

242 S. Kutrzeba, Handel Krakowa w wiekach średnich . . . '. *op. cit.*, p. 61.

243 Wojewódzkie Archiwum Państwowe, w Krakowie; *Consularia Cracoviensia*, Vol. I, No. 427, p. 262; with reference to one Mark from Nuremburg, Cracow merchants stated 'eyn eiginschaft gemacht hatte . . . off . . . bley, also 'das nymant

torfte in Ungirlant bley keuffin wen ag her'. See also *Corpus iuris Hungarici*, I. Markus (ed.), Budapest, 1899, pp. 220, 222, 226.

244 Wojewódzkie Archiwum Państwowe, w Krakowie; *Consularia Cracoviensia*, Vol. I, No. 427, p. 315.

245 Ibid., *Kodeks dyplomatyczny miasta Krakowa op. cit.*, Vol. II, No. 310, dated 1432; Najstarsze księgi i rachunki miasta Krakowa od r. 1300 do 1400, *op. cit.*, Vol. II, p. 85.

246 N. J. G. Pounds, *An Economic History of Medieval Europe*, *op. cit.*, p. 330.

247 T. Dziekoński, *op. cit.*, p. 114.

248 Ibid., p. 397; see also J. Vlachović 'Hutnícke spracůvanie medených a strieborno mededných růd v Banskej Bystricy v 2 polovicy 16 stor'. *Historické Studie*, Vol. V, Bratislava, 1959, pp. 110–148 for a more technical explanation of the process in its historical context.

249 Tajný archív m. Košice, *Telonium*, No. 549, letter dated 1483.

250 T. Dziekoński, *op. cit.*, p. 392.

251 *Cracovia artificum 1300–1500* (Zródła do historii sztuki i cywilizacji w Polsce) J. Ptaśnik (ed.), Vol. I, Kraków, 1917 No. 973 refers to Jan Thurzo buying a mine at Olkusz in partnership with Piotr and Imbram Salomon in 1490. See also P. Ratkoš, 'Das Kupferwesen in der Slowakei vor der Entsehung der Thurzo-Fuggerschen Handelsgesellschaft', in *Der Aussenhandel Ostmitteleuropas 1450–1650, Die ostmittel-europäischen Volkswirtschaften in ihren Beziehungen zu Mitteleuropa*, I. Bog (ed.), Wien, 1971, pp. 584–599, Ibid., 'Předpoklady vzniku turzovsko-fuggerovskej společností r. 1495', *Československý časopis historický*, No. 5, Prague, 1966, pp. 758–762; G. Wenzel, *A Fuggerek jelentösége magyarórszag törtenetében* (Akadémiai értekezések a történelemtudomány köréből), Budapest, 1882, 194 pp. Ibid., 'Okamánytár a Fuggerek magyarországi nagykereskedése és rézvállalatának történetérez 1494–1551', *Történélmi Tár*, Budapest, 1882, pp. 12–19.

252 For example see C. Wagner, *Analecta Scepusii sacri et profani*, Vol. I, Vienna, 1774, p. 56; Wojewódzkie Archiwum Państwowe, w Krakowie; *Advocatialia Cracoviensia*, No. 87, p. 307.

253 M. Biskup, 'Handel wiślany w latach 1454–1466, *op. cit.*, Table 7, p. 185.

254 For a critical history of Hungarian mining see G. Wenzel, *Magyarország bányászatának kritikai története*, Budapest, 1880, 253 pp.

255 G. von Probszt, *Die niederungarischen Bergstädte, Ihre Entwicklung und wirtschaftliche Bedeutung bis zur Ubergang an das Haus Habsburg (1546)*, München, 1966, pp. 23, 133; J. Vlachović, *Slovenska měd' v 16 a 17 storoči*, Bratislava, 1964, pp. 22–25. See also A. Péch, *Alsó Magyarország bányamívelésének története*, Vol. 1, Budapest, 1884, 204 pp.

256 *Kodeks dyplomatyczny miasta Krakowa*, *op. cit.*, Vol. I, No. 4, p. 8.

257 Ibid., No. 7.

258 Wojewódzkie Archiwum Państwowe, w Krakowie; *Consularia Cracoviensia*, Vol. I, No. 429, p. 521; *Najstarsze księgi i rachunki miasta Krakowa*, *op. cit.*, Vol. II, p. 129. See also J. Vozař, 'Banské mestá ako osobitný typ miest na slovensku', *Historický Časopis*, Vol. 21, No. 3, Bratislava, 1973, pp. 387–396; P. Ratkoš, 'Vznik a začiatky banských miest na Slovensku', *Historické Štúdie*, Vol. 19, Bratislava, 1974, pp. 33–58; D. Dorotjak, 'Kompozičné zásady historic-

kých jadier stredoslovenských banských miest', *Architectúra a Urbanismus*, Vol. IV, No. 3 (Slovenská adadémia vied), Bratislava, 1970, pp. 46–59.

259 *Kodeks dyplomatyczny miasta Krakowa, op. cit.*, Vol. I, No. 17, refers to a trade treaty signed between Cracow and Košice in 1324; see also, E. Fugedi 'Kaschau, eine osteuropäische Handels-stadt am Ende des 15 Jhs', *Studia Slavica*, Vol. II (Academiae Scientiarum Hungaricae), Budapest, 1956. pp. 185–213.

260 Wojewódzkie Archiwum Państwowe, w Krakowie; *Consularia Cracoviensia*, Vol. I, No. 428, p. 354; '*Najstarsze księgi i rachunki miasta Krakowa, op. cit.*, Vol. II, pp. 129, 165.

261 Kielce was in the Bishopric of Cracow in the Middle Ages, and small amounts of copper were found at Miedziana Góra and Karczówka. H. Łabęcki, *Górnictwo w Polsce, op. cit.*, p. 300.

262 O. Paulinyi, 'A közepkori magyar réztermelés gazdasági jelentösége Bányamüvelés és polgári vagyon Besztercebányán', in: *Károly Arpad-Emlékkönyv*, Budapest, 1933, pp. 402–439.

263 In 1399 the Prussians confiscated Hungarian copper which had come via Cracow and Toruń to Gdańsk. Nuremburg merchants wished to send it from Gdańsk to Flanders by sea. *Hansisches Urkundenbuch, op. cit.*, Vol. IV, No. 374.

264 *Lites ac res gestae inter Polonos Ordinemque Cruciferorum*, I. Zakrzewski and J. Karwasinska (eds.), 3 Vols., Poznán/Warszawa (1890–1935), Vol. I, Poznań, 1890, pp. 385–386. For full text see Ch.IV, Footnote 137.

265 K. Pieradzka, 'Trzy wieki stosunków handlowych pomiędzy Gdańskiem a Węgrami' *Rocznik Gdański*, Vol. IX, Gdańsk, 1935, pp. 190–208.

266 L. Koczy, 'Dzieje wewnętrzne Torunia do roku 1793', *Dzieje Torunia*, Toruń, 1933, pp. 99–209; J. Ptaśnik, '*Akta nurymberskie do dziejów handlu z Polską w wieku XV* (Nak, Akad. Umiejętności Kraków, 1912, 67 pp. See also A. Pleidell, *A nyugatra irányuló magyar külkereskedelem a középkorban*, Budapest, 1925, 324 pp.; V. Mencl, *Středověká města na Slovensku*, Bratislava, 1938, 230 pp.; M. Skladaný 'Zápas o banskobystrickú med' v polovici 15 storočia, *Zborník Filozofickej Fakulty Univerzity Komenského: Historica*, Vol. 27, 1976, Bratislava, 1978, pp. 175–210.

267 Wojewódzkie Archiwum Państwowe, w Gdańsku, *333 U Abt 6 No. 29*, dated 8/V/1443, Merchants from Nuremburg usually dealt with Gdańsk through the Hanseatic League. See *Hanserecess, op. cit.*, Vol. I, Abt. IV T no. 539, paragraph 7 and 8 and no. 540; see also W. Stromer, 'Nürnberger Unter-nehmer im Karpatenraum. Ein oberdeutches Buntmetall-Oligopol 1396–1412', *Kwartalnik Historii Kultury Materialnej*, Vol. XVI, Warszawa, 1968, pp. 641–662.

268 Wojewódzkie Archiwum Państwowe, w Gdańsku, *300 U Abt. 7 Nos. 41, 42, 43, 44*.

269 Ibid. *300 U Abt. 2 No. 184; Abt. 6, Nos. 94, 95*; *Abt. 7, Nos. 63, 66, 67, 83 and 110*; *Hansisches Urkundenbuch op. cit.*, Vol. 10, Nos. 602, 608, 643. All refer to cooperation between Gdańsk and Cracow merchants for mining enterprises in Olkusz and Upper Hungary.

270 *Libri iuris civilis Cracoviensis*, K. Kaczmarczyk (ed.), Kraków, 1913, p. 234, No. 6852; J. Ptaśnik, 'Turzonowie w Polsce', *Przewodnik naukowy i literacki*, Kraków, 1905, p. 15.

271 I. Ipolyi, *Geschichte der Stadt Neusohl* (Polska Akademia Umiejętności) Kraków, 1918, No. 192, p. 45.
272 R. Ehrenberg, *Das Zeitalter der Fugger*, Vols. 1–2, Jena, 1922.
273 Land for the furnace was purchased from a Cistercian Abbey. Archiwum Cystersów w Mogile sygn. 25, 71, No. 355; see also D. Molenda, 'W sprawie badań huty miedzi w Mogile pod Krakowem w XV i XVI wieku', *Przegląd Historyczny*, Vol. 66, No. 3, Warszawa, 1975, pp. 369–382.
274 G. Wenzel, A. *Fuggerek jelentösége* . . . *op. cit.*, No. 5; agreement dated May 1495.
275 S. Czarnocki, *Objaśnienie do mapy bogactw kopalnych Polski*, Warszawa, 1931, pp. 52–55; B. Zientara, *Dzieje małopolskiego hutnictwa żelaznego* XIV–XVII *wiek* (P.W.N.), Warszawa, 1954, pp. 31–32.
276 In a document dated 1222 the Bishop of Cracow includes in the church income 'viginti tres salis, quadraginta urnas melis, centum ligaturas ferri'. For further discussion see W. A. Maciejowski, *Pamiętniki o dziejach pismiennictwa i prawodawstwie Słowian*, Vol. I (Druk, P. Barycki) Warszawa, 1939, p. 274.
277 Z. A. Helcel, *op. cit.*, p. 220; J. Dąbrowski, 'Kraków a Węgry w wiekach średnich', *op. cit.*, p. 223.
278 Kodeks dyplomatyczny miasta Krakowa, *op. cit.*, Vol. I, No. 57.
279 J. Dąbrowski, 'Kraków a Węgry . . . *op. cit.*, p. 223.
280 Wojewódzkie Archiwum Państwowe, w Krakowie; *Consularia Cracoviensia*, Vol. I, p. 262; see also T. Hirsch, *op. cit.*, p. 257.
281 Kodeks dyplomatyczny miasta Krakowa, *op. cit.*, Vol. II, No. 310, pp. 419–422.
282 Wojewódzkie Archiwum Państwowe, w Krakowie; *Acta Scabinalia Sandecki*, 1488–1505, pp. 120, 688; C. Wagner, *Analecta Scepusii*, *op. cit.*, Vol. I, p. 56. *Matricularum Regni Poloniae Summaria, escussis codocibus, qui in Chartophylacio Macimo Varsoviensi asservantur*, T. Wierzbowski (ed.), Vol. I, Warszawa, 1905, No. 272, refers to 15 'stufuntow' pieces of iron sent from Bardejov via Żmigród to Cracow in 1482, and also mentioned in Archiv mesta Bardejov, *O.A.B.*, No. 2235.
283 T. Hirsch, *op. cit.*, p. 257; C. Sattler, *Handelsrechnungen des Deutschen Ordens*, Leipzig, 1887, p. xxv–xxx. In Prussia, Hungarian iron was called 'osemund' i.e. Swedish iron.
284 *Bilder von der Kupferkies-Lagerstätten bei Kützbünel und den Schwefel-Lagerstätten bei Swoszowice* . . . ', F. M. Ritter von Friese (ed.), Vienna, 1890, pp. 25–34; J. Pazdur, 'Production du soufre en Pologne du XVe au XVIIIe siècle', *Kwartalnik Historii Kultury Materialnej*, Vol. XVI, No. 4 (P.W.N.), Warszawa, 1968, p. 6667.
285 W. Oczko, *Przymiot i cieplice*, E. Klink (ed.), Warszawa, 1881, p. 517.
286 Z. A. Helcel, *op. cit.*, Vol. II, Kraków, 1870, No. 1684, p. 245, and No. 1685, p. 254.
287 Wojewódzkie Archiwum Państwowe, w Krakowie; *Consularia Cracoviensia*, 1412–1449, p. 8; a document dated 1412 refers to one P. Bikarini buying kermes for 25 grzywien in Prague groschen from J. Hutter and sent to Venice and Florence, 'Petro Pykaran von Wenedien umb. 25 mr. prag. grosch. von des czirwicz wegin'.
288 Ibid., pp. 257–8, document dated 1429 when 25 kamień (stones) of kermes worth 2,673 talents were sent from Cracow to Venice. See also, C. M. Cipolla, 'The

Trends in Italian Economic History in the Later Middle Ages', *Economic History Review*, 2nd series, Vol. II, London, 1949–1950, p. 183.

289 Kodeks dyplomatyczny miasta Krakowa, *op. cit.*, Vol. II, No. 310, p. 420.

290 K. Gierdziejewski, *Zarys dziejów odlewnictwa polskiego*, Katowice, 1959, pp. 75–77.

291 A. C. Crombie, *Augustine to Galileo: The History of Science A.D. 400–1650*, London, 1952, p. 190; see also D. Molenda, 'Początek eksploatacji galmanu na ziemiach polskich do połowy XVII w', *Kwartalnik Historii Kultury Materialnej*, Vol. XXI, No. 7 (P.W.N.), Warszawa, 1973, pp. 43–46.

292 C. M. Cipolla, *Money, Price and Civilization in the Mediterranean World*, Princeton Univ. Press, Princeton, 1956, pp. 56–57.

293 F. B. Pegolotti, *La Practica della Mercatura* (A. Evans, ed.) (The Medieval Academy of America), Cambridge, Mass. 1936, 201 pp.

294 E. Gibbon, *The History of the Decline and Fall of the Roman Empire* (G. Bury, ed.), London, 1909, Vol. I, p. 62.

295 L. F. Saltzman, *English Trade in the Middle Ages*, Oxford, 1931, p. 420.

296 W. G. East, *op. cit.*, p. 315; see also, W. Heyd, *Geschichte des Levanthandels*, Vol. II, Leipzig, 1885, p. 182.

297 M. Małowist, 'Bałtyk i Morze Czarne w handlu średniowiecznym', *Janitar*, Vol. I, No. 2, Gdynia, 1937, pp. 69–74; Ibid., 'The Baltic and the Black Sea in Medieval Trade', *Baltic and Scandinavian Countries*, Vol. III, No. 5, Gdynia, 1937, pp. 1937, pp. 36–42; Ł. Charewiczowa, *Handel średniowiecznego Lwowa*, *op. cit.*, pp. 53–60.

298 Disimoni e Belgrano, *Documenti ed estratti inediti e poco noti; riguardanti la storia del commercio e della marina ligura*, Published by the Societa Ligura di Storia patria, Vol. V, No. 1, Genoa, 1901, pp. 373–383, refers to the presence of Genoese merchants in Antwerp in 1315.

299 J. Ptaśnik, 'Włoski Kraków za Kazimierza Wielkiego i Władysława Jagiełły', *Rocznik Krakowski*, Vol. XIII, Kraków, 1911, p. 51.

300 Ibid.

301 *Hansisches Urkundenbuch*, *op. cit.*, Vol. III, Nos. 156, 159, 171.

302 M. Małowist, 'Polish–Flemish Trade in the Middle Ages', *Baltic and Scandinavian Countries*, Vol. IV, No. I (8), Gdynia, 1938, p. 6.

303 S. Kutrzeba, *Handel Polski ze Wschodem w wiekach średnich*, Kraków, 1902, pp. 28–29.

304 Z. P. Pach, 'Levantine Trade and Hungary in the Middle Ages', *Études Historiques*, Vol. I (Studia Historica Academiae Scientiarum Hungaricae No. 97), Budapest, 1975, p. 300.

305 R. S. Lopez (ed.), *Medieval Trade in the Mediterranean world*, New York, 1955, p. 145; see also, S. E. Howe, *In Quest of Spices*, London, 1946, p. 19.

306 J. Kieniewicz, 'Droga morska do Indii . . . ', *op. cit.*, p. 587. A similar trend in pepper prices was also witnessed in Venice.

307 Archiwum miasta Lwowa, *Consularia Leopoliensia*, Vol. I, pp. 225, 287, 316.

308 Wojewódzkie Archiwum Państwowe w Krakowie, *Consularia Cracoviensia*, Vol. I, No. 428, pp. 368, 391; *Kodeks dyplomatyczny miasta Krakowa*, *op. cit.*, Vol. II, No. 310; *Najstarsze księgi . . . op. cit.*, Vol. II, p. 85.

309 For example, saffron, nutmeg and cinnamon could not be sold in quantities of

less than half a stone (kamień), cloves and ginger less than one stone, pepper and caraway seeds for less than one sack (worek). S. Kutrzeba, 'Handel Krakowa . . .', *op. cit.*, pp. 181–183; S. Lewicki, 'Prawo składu', *op. cit.*, pp. 57–59

310 M. W. Laborge, 'The Spice Account', *History Today*, Vol. XV, No. 1, London, 1965, p. 34.

311 J. Pelc, *Ceny w Krakowie*, *op. cit.*, p. 67.

312 C. Sattler, *Handelsrechnungen der deutschen Ordens*, *op. cit.*, pp. 451–522.

313 Ibid.

314 G. Strauss, *Nuremburg in the Sixteenth Century. City Politics and Life between Middle Ages and Modern Times* (Indiana Univ. Press), Bloomington and London, 1976, p. 128.

315 L. Deák, 'Bardejovský obchod a Bardejovska obchodna cesta v prve polovici 15 storičia', *Historica: Sbornik Filozofickej Fakulty University Komenského*, Vol. XIV, Bratislava, 1963, pp. 120–121. Evidence of a Cracow merchant named Nikelklen transporting pepper and saffron totalling 28 cetnars (1.81 tons), to Bardejov is recorded in B. Iványi, *Bártfa szabad királyi város levéltára*, Vol. I (1319–1501) Budapest, 1886, Document No. 507.

316 M. Dubiecki, 'Kaffa osada genueńska i jej stosunek do Polski w XV wieku', *Przegląd Powszechny*, Vol. XII, Kraków, 1886, pp. 56–64, and 216–227; M. Małowist, *Kaffa – kolonia genueńska na Krymie i problem wschodni w latach 1453–1475*, Warszawa, 1947, 213 pp.; J. Wyrozumski, *op. cit.*, p. 64.

317 N. Iorga, *Studii istorice asupra Chiliei şi Cetătii Albe*, Bucharest, 1900, pp. 124–126; Ł Charewiczowa, *Handel średniowiecznego Lwowa*, *op. cit.*, pp. 109–120; P. Panaitescu, *op. cit.*, p. 22.

318 C. H. H. Wake, 'The Changing Pattern of Europe's Pepper and Spice Imports, ca. 1400–1700', *Journal of European Economic History*, Vol. 8, No. 2, Rome, 1979, p. 361.

319 M. M. Postan, The Trade of Medieval Europe . . .*op. cit.*, p. 200; A. Agats, *Der hansische Baienhandel*, Heidelberg, 1904, pp. 3–6, 16, 23.

320 K. Skoczylas-Ciszewska and J. Poborski, 'Tectogenesis of the Miocene Evaporite Deposits at the Margin of the Carpathians, East of Cracow', *Twenty Third International Geological Congress: Proceedings*, Vol. 3, Prague, 1968, pp. 10–19; K. Maślankiewicz, *Z dziejów górnictwa solnego w Polsce* (Wyd. Naukowo-Techniczne), Warszawa, 1965, pp. 1–10.

321 A document from Chełm dated 1230 refers to 300 bottles of wine brought to Poland in payment for salt – 'Hy bevor pflag man zen polen stuckelin zu machin, der hiz eines in krusch. Der buste man do cry hundirt'. Z. A. Helcel, *Starodawne prawa polskiego pomniki*, *op. cit.*, Vol. II, p. 25. See also H. Łabięcki, 'Najdawniejsze dzieje salin krakowskich aż do żupnictwa Jana Bonera, czyli do roku 1515', *Biblioteka Warszawska*, Vol. II, Warszawa, 1856, pp. 265–306; F. Skibiński, 'Handel solny we wczesnym średniowieczu Polskim', *Księga Pamiątkowa ku uczczeniu dwudziestopięcioletniej działalności naukowej Prof M. Handelsmana*, Warszawa, 1929, pp. 451–464.

322 F. Skibiński, *op. cit.*, p. 455.

323 Salt from Cracow was first mentioned in Silesia in a document dated 1293 with reference to a Prince Bolesław of Opole who gave a local cloister 'braciom omnimodam libertatem transeundi telonia nostra licet sal vel annonam . . . tam

in Cracoviam vel Sleziam aducere vel deducere decreverint', G. Stenzel, *Liber fundationis claustri S.M. Virg. in Heinrichow*, Wrocław, 1854, Urk. 36.

324 B. Iványi, 'Két középkori sóbánya statutum', *Századok*, Vol. XLV, Budapest, 1911, p. 18; A. Keckowa, 'Saliny ziemi krakowskiej do końca XIII wieku', *Studia i materiały z historii kultury materialnej*, Vol. XXXIII, Wrocław–Warszawa–Kraków, 1965, pp. 17–20, 88.

325 J. Krzyżanowski, 'Statut Kazimierza Wielkiego dla krakowskich żup solnych', *Rocznik Krakowski*, Vol. XXV, Kraków, 1933, pp. 96–128; R. Grodecki, 'Ordynacjaa Kazimierza Wielkiego dla Krakowskich żup solnych z 1368 roku', *Studia i materiały do dziejów żup solnych w Polsce*, Vol. III (Muzeum Żup Krakowskich Wieliczka), Kraków, 1974, pp. 7–12.

326 J. Krzyżanowski, Statut Kazimierza . . . *op. cit.*, p. 118; Z. A. Helcel, *Starodawne prawa . . . op. cit.*, Vol. I, p. 219.

327 K. Wuttke, *op. cit.*, p. 283.

328 C. Wagner, *Analecta Scepusii . . . op. cit.*, Vol. I, p. 28; S. Kutrzeba (ed.) 'Akta odnoszące się do stosunków handlowych Polski z Węgrami głównie z archiwum koszyckiego z lat 1354–1505', *Archiwum Komisji Historycznej*, Vol. IX, Kraków, 1902, p. 412, No. 1; B. Ivańyi, 'Két középkori . . . ', *op. cit.*, p. 188.

329 J. Wyrozumski, 'Państwowa gospodarka solna w Polsce do schyłku XIV wieku', *Zeszyty Naukowe Uniwersytetu Jagiellońskiego No. CLXXVIII; Prace Historyczne*, Vol. 21, Kraków, 1968, pp. 115, 119–128; see also W. Fellmann, 'Die Salzproduktion im Hanseraum', *Hansische Studien Heinrich Sproemberg zum 70 Geburtstag* (Akademie Verlag) Berlin, 1961, pp. 13–23; H. Witthöft, Mass- und Gewichtsnormen, *op. cit.*, pp. 38–65.

330 C. Wagner, *Analecta Scepusii, op. cit.*, Vol. I, p. 105.

331 *Corpus iuris Hungarici, op. cit.*, Vol. I, Budapest, 1899, pp. 226–227, Article 20; "cum quodammodo pars sit magnas dementiae, id, quod de suo quisque habere potest, ab aliis mutuare"; see also M. C. Kovachich, *Supplementum ad vestigia-comitiorum apud Hungaros*, Budapest, 1798, p. 225.

332 Ibid., p. 283, Article 11.

333 For example Polish salt from Cracow arrived in Bardejov in 1418 . . . '40 gr. dedimus iudici Close ex parte saliun regis Poloniae', quoted in L. Fejérpataky, *Magyarországi várasok régi számadáskönyvei*, Budapest, 1885/1886, p. 163; for 1432, 'dem Peter Winter pro 1 centenario salis equis fl.5', Ibid., p. 303; for 1435, 'Vor 1 banc salcz . . . ', Ibid., p. 352; for 1437, 'Solvimus de vectura salis de Bochna, quod dedit Jacobus super Zupparn de centenariis 25 . . . fl. auri 5; Ibid., p. 376a; 'de vectura salis de Bochna de 25 centenariis fl. auri 5', Ibid., p. 376b; 'Item de sale quod datum est in debitis, quae tenetur Jacubus frater Zupparn ut ante continetur, centner salis 49 facit fl . . . ' Ibid., p. 453; for 1439, 'Item Nelner adduxit 40 centner salis de Bochna . . . ' Ibid., p. 612.

334 Salt from Bochnia was sent to Bardejov on four separate occasions between 1437–1439; L. Fejérpataky, *op. cit.*, p. 376b and 491; see also, L. Deák, 'Bardejovský Obchod . . . ', *op. cit.*, p. 120; O. R. Halaga, 'Polská sûl na Slovensku v stredověku', *Tisic let česko-polské vzájemnosti*, Opava, 1966, pp. 232–271; Ibid., 'Polská a uhorská sůl na Slovensku v stredověku', *Studia z Dziejów Górnictwa i Hutnictwa*, Vol. XII(XXX), Warszawa, 1968, pp. 28–75.

335 Archív miesta Bardejova. Mention of salt from Cracow may be found on several

occasions in the toll returns, – for 1444 (No. 433), for 1448 (No. 522), for 1448–1452 (No. 523), for 1452–1454 (No. 687). For example in 1454 one Katarína Gladischová paid 10 florins for Polish salt from Cracow. See B. Iványi, *Bártfa szabad . . .* ', *op. cit.*, Vol. I, Document No. 725.

336 Archív miesta Bardejov; document dated 27/XII/1492 refers to salt from Cracow to Sabin. (No. S 3083); document dated 1482 records salt from Cracow to Levoča (No. S 2222); Archív miesta Prešov; document dated 1429 (No. mag. 153), 'Item solvimus Scherer 29 den pro 3 bancis salis'. For other places mentioned, see Archív miesta Bardejov, No. * 5292/II, 'Intelleximus esse nonnullos qui passim sales externos as divagos . . . vendere consuevissent'.

337 Transylvanian salt supplies were often free from customs duty or tolls. See G. Wenzel, *Magyarország bányászatának kritikai története*, Budapest, 1880, Okmányi függelék, pp. 436–439; see also J. Mihályi, *Máramorosi diplomák a XIV és XV századból*, Máramaros-Sziget, 1900, p. 1, No. 1; p. 23, No. 12.

338 Monumenta Hungaria. Historia: (Magyar történelmi emlékek), Diplomataria, Vol. I, Budapest, 1857, No. 7.

339 K. Wutke, 'Die Versorgung . . . ', *op. cit.*, p. 284, footnote 1.; A. Mosbach, *Przyczynki do dziejów polskich z arachiwum miasta Wrocławia zebrał August Mosbach*, Poznań, 1860, p. 96.

340 S. Kutrzeba, 'Handel Krakowa w wiekach średnich . . . ', *op. cit.*, p. 83.

341 H. Łabęcki, *Górnictwo w Polsce*, *op. cit.*, p. 141.

342 J. V. Bandtke, *Jus Polonicum*, Warszawa, 1831, p. 307; Serafin paid 1600 grzywnas/year to the king for the mine lease. See also J. R. Hrdina, *Geschichte der Wieliczker Saline*, Vienna, 1842, 274 pp.

343 K. Wutke, 'Die Salzerschliessungsversuche in Schlesien in Vorpreuss', *Zeitschrift des Vereins für Geschichte und Alterthum Schlesiens*, Vol. XXVIII B. Breslau, 1894, p. 242; A. Illies, 'Drumurile şi transportul sării in Ţara Românească secolele XV–XIX', *Studii şi Materiale de Istorie Medie*, Vol. 7, Bucharest, 1974, pp. 223–242.

344 Cracow's salt monopoly was confirmed in the Ordinance of Kazimierz Wielki in 1368, J. V. Bandtke, *op. cit.*, p. 177.

345 J. Rutkowski, *Historia Gospodarcza Polski*, Warszawa, 1953, p. 50.

346 Ibid.

347 S. Kutrzeba, 'Handel Krakowa . . . ' *op. cit.*, p. 141.

348 *Akta grodzkie i ziemskie z czasów Rzeczpospolitej Polskiej z archiwum tak zwanego Bernardyńskiego we Lwowie* (wyd. K. Liske), Vol. IX, Lwów, 1883, No. 58. Salt from Rus was still being exported abroad in the Later Middle Ages and even beyond; see A. D. Gorskii, *Ocherki ekonomicheskogo polozhenija krest'jan severo-vostochnoi Rusi XIV–XV vv.*, Moscow, 1960, pp. 96–97; N. V. Ustiugov, *Solevarennaja promyshlennost' soli Kamskoi v XVII veke*, Moscow, 1957, pp. 1–15.

349 M. M. Postan, 'The Trade of Medieval Europe . . .' *op. cit.*, p. 120.

350 J. A. van Houtte, 'The Rise and Decline of the Market in Bruges', *op. cit.*, p. 35.

351 A. Attman, *The Russian and Polish Markets in International Trade 1500–1650* (Institute of Economic History of Göteborg University Studies, No. 26), Göteborg, 1973, p. 5.

352 N. J. G. Pounds, *An Economic History of Medieval Europe*, *op. cit.*, p. 377; see

also J. A. van Houtte, 'Bruges et Anvers marchés nationaux ou internationaux du XIVe siècle, *Revue du Nord*, Vol. XXXIV, Paris, 1952, pp. 89–108.
353 M. M. Postan, 'The Trade of Medieval Europe . . .' *op. cit*., p. 35.
354 H. Lesinski, 'Handel na wybrzeżu Słowiańskim w XII w. w świetle ceł morskich' *Przegląd Zachodni*, Vol. VII, No. 1/2, Poznań, 1951, pp. 53–64; W. Łega, *Kultura Pomorza we wczesnym średniowieczu na podstawie wykopalisk*, Toruń, 1930, p. 74.
355 K. Ślaşki, 'Pomorskie szlaki handlowe w XII i XIII w.', *Przegląd Zachodni*, Vol. IV (1 połrocze) No. 3, Poznań, 1948, pp. 285–290.
356 J. Rutkowski, *Historia Gospodarcza Polski*, Warszawa, 1953, p. 127.
357 L. Eberle, 'Polski rzemiosło skórzane w średniowieczu. Stan badań i kierunki poszukiwań' (Wyroby Rzemieślnicze w Polsce w XIV–XVIII wieku), *Studia i Matreriały z Historii Kultury Materialnej*, Vol. XLV (P.A.N.), Warszawa–Wrocław–Kraków–Gdańsk, 1971, pp. 9–21.
358 L. Koczy, *Handel miasta Poznania do połowy XVI w*, Poznań, 1930, pp. 49–51; M. P. Lesnikov, 'Der hansische Pelzhandel zu Beginn des 15 Jahrhunderts', in *Hansische Studien*, Berlin, 1961, pp. 219–222; J. Baranowski, *Przemysł polski w XVI wieku*, Warszawa, 1919, pp. 96, 112–115; K. von Loewe, 'Commerce and Agriculture in Lithuania, 1400–1600' *Economic History Review*, Vol. 26, No. 1, 1973, p. 31.
359 J. Ptaśnik, 'Studia nad patrycjatem krakowskim wieków średnich, *Rocznik Krakowski*, Vol. XV, Kraków, 1913, p. 73; K. Myśliński, 'Lublin a handel Wrocławia z Rusią w XIV i XV wieku', *Rocznik Lubelski*, Vol. III, Lublin, 1960, pp. 24–28; L. Białkowski, 'Lublin na starych szlakach handlowych', *Pamiętnik Lubelski*, Vol. III, Lublin, 1938, p. 290; H. Samsonowicz, 'Handel Lublina na przełomie XV i XVI w.', *Przegląd Historyczny*, Vol. LIX, No. 4, Warszawa, 1968, p. 622.
360 M. Małowist, 'The Problem of the Inequality . . . ' *op. cit*., p. 24.
361 A. Pawiński, 'Notatki kupca . . . ' *op. cit*., p. 64; a thousand skins cost 13 grzywnas.
362 *Najstarsze księgi i rachunki . . . op. cit*., Vol. II, p. 85; *Kodeks dyplomatyczny miasta Krakowa . . . op. cit*., Vol. II, No. 310.
363 M. Biskup, 'Handel Wiślany . . . ', *op. cit*., Table 10, p. 187; Table 3, pp. 190–191.
364 Monumenta Hungarica, Historia; Diplomataria, *op. cit*., p. 8. See also F. Kiryk, 'Stosunki handlowe Jasła i miast okololicznych z miastami Słowackimi w XV wieku', *Studia z dziejów Jasła i powiatu jasielskiego*, Kraków, 1964, pp. 143–162.
365 S. Kutrzeba and J. Ptaśnik, 'Dzieje handlu i kupectwa krakowskiego', *Rocznik Krakowski*, Vol. XIV, Kraków, 1912, pp. 92–96, based on an analysis of the *Libris iuris civilis* records in Wojewódzkie Archiwum Państwowe w Krakowie.
366 K. Pieradzka, 'Garbary, przedmieście Krakowa (1363–1587)', *Biblioteka Krakowska*, No. 71, Kraków, 1931, 154 pp.
367 J. Ptaśnik, 'Towns in Medieval Poland', Ch.2 in *Polish Civilization: Essays and Studies*, M. Giergielewicz and L. Krzyżanowski (eds.), New York Univ. Press, New York 1979, p. 36.
368 J. Pelc, *Ceny w Krakowie . . . op. cit*., p. 74.
369 K. Śląski, *op. cit*., p. 290.

370 W. Stieda, 'Revaler Zollbücher und-Quittungen des 14 Jahrhunderts', *Hansische Geschichtsquellen*, Vol. V, Halle, 1887, p. CXIII.
371 H. Samsonowicz, 'Handel Lublina', *op. cit.*, p. 622; H. Białowski, *op. cit.*, p. 290.
372 P. Panaitescu, *op. cit.*, p. 9. Also from Moldavia came the wax-moth (*Galleria mellonella*) which was used as wax for religious purposes. See Ł. Charewiczowa, *Handel średniowiecznego Lwowa*, *op. cit.*, p. 70. Cracow merchants also bought Russian wax in Lwów, S. Kutrzeba and J. Ptaśnik, 'Dzieje handlu . . .' *op. cit.*, p. 19; Archiwum miasta Lwowa, *Consilium Leopoldiensis*, Vol. 11, p. 263.
373 Wax from Hungary was already mentioned at the beginning of the fourteenth century in Cracow documents. *Kodeks dyplomatyczny miasta Krakowa op. cit.*, Vol. I, No. 7. Other examples include Ibid., Vol. II, No. 310; *Najstarsze księgi . . . op. cit.*, Vol. II, No. 85.
374 Along with lead, salt, furs and skins, wax was mentioned as exported from Cracow to Silesia in the fourteenth century. S. Kutrzeba and J. Ptaśnik, 'Dzieje handlu . . .' *op. cit.*, p. 24. No evidence however exists of wax being sent further to Nuremburg from Cracow via Wrocław. Ibid.
375 Wax was sent in small quantities via Gdańsk to western Europe mainly by the land route. See J. Rutkowski, *op. cit.*, p. 130. Wax was also sent down the Vistula from Cracow to Gdańsk. See M. Biskup, 'Handel wiślany . . .' *op. cit.*, p. 181, with reference to 18 kamieńs of wax (225 kgs.) sent in 1464.
376 S. Kutrzeba and J. Ptaśnik, 'Dzieje handlu . . .' *op. cit.*, p. 26 maintain that Cracow traders throughout the whole of the fourteenth and into the fifteenth century continued to transport the main Polish articles viz. lead, wax and wood, to Flanders.
377 M. Małowist, *Wschód a Zachód . . . op. cit.*, p. 167.
378 L. Eberle, *op. cit.*, p. 21.
379 A. Attman, *op. cit.*, p. 96.
380 S. Kutrzeba, 'Handel Krakowa w wiekach . . . ' *op. cit.*, p. 71; J. Pęckowski, *Dzieje miasta Rzeszowa, do końca XVIII w.* Rzeszów, 1913, p. 270.
381 *Ksiegi ławnicze (Przemyśla)*, J. Smolka and Z. Tymińska (eds.), Przemyśl, 1935, Vol. I, no.1541, refers to oxen bought by Krosno traders in 1424 for sending to Silesia via Cracow; Akta grodzkie i ziemskie . . . *op. cit.*, Vol. XI, Lwów, 1879, no.1522, gives similar information for 1442.
382 F. Kiryk, 'Stosunki handlowe Jasła . . . *op. cit.*, p. 160.
383 L. Białkowski, 'Lublin na starych szlakach . . .' *op. cit.*, p. 290; in 1493 the Polish king Jan Olbrecht declared that oxen en route from the Wołyń region must follow the road from Bełz, Hrubieszów, Lublin, to Radom, and from Radom the usual route (per vias solitas) to Silesia. The decree was renewed by King Alexander in 1503.
384 S. Kutrzeba, 'Handel Krakowa w wiekach . . . ', *op. cit.*, p. 170.
385 A. Mosbach, *Przyczynki . . . op. cit.*, p. 98 refers to a document from 1441 on Cracow's oxen trade; a similar one for 1499 is also mentioned, p. 120; for earlier evidence see Wojewódzkie Archiwum Państwowe w Krakowie. *Consularia Cracoviensia* Vol. II, p. 23 (1412), p. 46 (1413).
386 F. Kavka, 'Český obchod s textilními výrobky v rumunských zemích', *Sborník historický* Vol. V, Prague, 1957, p. 134.

387 N. Iorga, *Geschichte des rumänischen Volkes im Rahmen seiner Staatsbildungen* (2 Vols), Gotha 1905, Vol. I, pp. 148–149 discusses the importance of cattle rearing in Moldavia in the fifteenth century; Archiwum miasta Lwowa, *Consilium Leopoldiensis*, p. 482 stated, 'obligat se solvere, si boves vendiderit in Jaroslaw, si vero non vendiderit, cum eisdem versus Briga alias Brzeg ire contingerit'; Lwów merchants collected cattle from fairs at Halicz, Kołomyja, Drohobycz and Tysmienicz, and were continually in touch with Cracow merchants who were on contracts to buy them. (Archiwum miasta Lwowa, *Consilium Leopoldiensis*, Vol. XIV, no. 2596).

388 S. Kutrzeba, 'Handel Krakowa w wiekach . . . ' *op. cit.*, p. 255; L. Deák, 'Bardejovský obchod . . . ' *op. cit.*, p. 121.

389 Archív město Bardejov, *O.A.B.* No. 2583.

390 Ibid., Nos. 2332, 2214, 2027, 2028, 2296.

391 Ibid., No. 2322 dated 1485 records horses sent from Bardejov and Sabinov to Dębno, 'Accepimus certa relacione, quia sunt equi boni in Sobnow . . . '. Other evidence of horses sent from Bardejov to Tarnów and Zagórze, ibid., Nos. 2323, 2296, 2497, 2446 and 2625. They were also sent to Krosno, ibid., Nos. 2501, 2365.

392 *Akta grodzkie i ziemskie . . . op. cit.*, Vol. III, No. II.

393 The route through Ropczyca was preferred to the normal cattle road via Rzeszów to Cracow and Kleparz; Kodeks dyplomatyczny miasta Krakowa, *op. cit.*, Vol. I, No. 150.

394 S. Kutrzeba, 'Handel Krakowa w wiekach . . . ', *op. cit.*, p. 225.

395 L. Déak, 'Bardejovský obchod . . .' *op. cit.*, quotes two examples; 100 goats sent to Poland priced at 48 florens; same number sent for 28 florens, p. 122, footnote, 140.

396 J. Dąbrowski, *Ziemia Sanocka w XV stuleciu*, Lwów, 1931, p. 19.

397 J. N. Ball, *Merchants and Merchandise . . . op. cit.*, p. 135.

398 M. M. Postan, 'The Trade of Medieval Europe . . .' *op. cit.*, p. 122.

399 M. Małowist, 'Podstawy gospodarcze przywrócenia jedności państwowej Pomorza gdańskiego z Polską w XVw.', *Przegląd Historyczny*, Vol. 45, Warszawa, 1954, p. 164.

400 W. G. East, *An Historical Geography of Europe . . . op. cit.*, p. 328.

401 R. W. Unger, 'Dutch Herring, Technology and International Trade in the Seventeenth Century', *The Journal of Economic History*, Vol. XL, No. 2, Atlanta, Georgia, 1980, p. 254.

402 W. G. East, *op. cit.*, pp. 345–346.

403 H. Lesiński, 'Handel na wybrzeżu Słowiańskim . . . ' *op. cit.*, p. 54.

404 R. W. Unger, *op. cit.*, p. 263. See also W. Böhnke 'Der Binnenhandel des Deutschen Ordens in Preussen und seine Beziehung zum Aussenhandel um 1400', *Hansische Geschichsblätter*, Vol. 80, Leipzig, 1962, p. 27; V. Lauffer, 'Danzigs Schiffs-und Waarenverk am Ende des XV Jahrhunderts', *Zeitschrift des West preussischen Geschichtsvereins*, Vol. 38, Breslau, 1894, p. 44.

405 A. Jelicz, *W średniowiecznym Krakowie . . . op. cit.*, p. 107.

406 Ibid.

407 On occasions incorrect or false signs on barrels of herrings led to officials disposing of them in the river 'submergi debent de forma juris', e.g. Wojewódz-

kie Archiwum Państwowe w Krakowie; *Scabinalia Sandec 1488–1505*, p. 598; for a long time it was known that the supply of herrings was 'non valencia alias warunek, sed falsificata ... et ad usum et vescendum bonis hominibus non valencia'. Evidence of herrings sent from Gdańsk is referred to in M. Biskup, 'Z problematyki handlu Polsko–Gdańskiego drugiej połowy XV w.' *Przegląd Historyczny*, No. 2–3, Warszawa, 1954, p. 396; see also Ibid. 'Handel wiślany ... ' *op. cit.*, pp. 188–190; H. Samsonowicz, 'Struktura handlu gdańskiego ... ' *op. cit.*, pp. 707–708; Ibid., 'Handel zagraniczny Gdańska ... ' *op. cit.*, pp. 350–351.

408 *Najstarsze księgi ... op. cit.*, Vol. Ii, p. 85 dated 1393 in the Sącz tariff customs records 'de tunna piscum, de tunna alecum, de tunna agwillarum' (i.e. fish, caviar, eels).

409 Already quantities of herrings were sent from Cracow to Hungary; e.g. 1432 *Kodeks dyplomatyczny miasta Krakowa, op. cit.*, Vol. II, no. 310; see also *Advocatus Cracoviensis* No. 89, p. 347; No. 92, p. 25; L.. Déak, 'Bardejovský obchod ...' *op. cit.*, p. 115.

410 A. Klose, *Von Breslau ... op. cit.*, Vol. II, Pt.II, p. 358; K. Wüttke, 'Die Versorgung Schlesiens ... ' *op. cit.*, p. 289.

411 Wojewódzkie Archiwum Państwowe w Krakowie. *Advocatus Cracoviensis* No. 86, p. 398 for example refers to Jan Kugler, a Toruń merchant, who sold a trader from Levoča half a barrel of eels which were said to be falsely marked and badly preserved, 'Falsificata et sine cremato'. A similar case is recorded in S. Kutrzeba, 'Handel Krakowa w wiekach ... ' *op. cit.*, p. 190.

412 For example the *Accounts of the Jagiellonian court*, p. 569 refer to special varieties of fish being purchased. Quoted in J. Dąbrowski, 'Kraków a Węgry ... ' *op. cit.*, p. 224.

413 Beluga, a large Russian sturgeon (*Acipenser huso*) is mentioned in the City Council Book of 1395; *Najstarsze księgi ... op. cit.*, Vol. II, p. 128

414 Ł Charewiczowa, *Handel średniowiecznego Lwowa ... op. cit.*, p. 92; Ibid. 'Handel Lwowa z Mołdawią ... ' *op. cit.*, p. 51; S. Kutrzeba, 'Handel Polski ze Wschodem ... ' *op. cit.*, p. 79 refers to this route as the 'Fish Road'.

415 J. Pelc, *Ceny w Krakowie ... op. cit.*, pp. 161–162.

416 M. M. Postan, The Trade of Medieval Europe ... *op. cit.*, p. 125.

417 Ibid., p. 154.

418 Ibid., p. 125. For further discussion see T. Hirsch, *Danzigs Handels-und Gewerbgeschichte ... op. cit.*, pp. 253–255.

419 J. N. Ball, *Merchants and Merchandise ... op. cit.*, p. 22.

420 T. Korzon, *Wewnętrzne dzieje Polski za Stanisława Augusta (1764–1794)*, 4 Vols., Kraków, 1882–1885; Vol. II, pp. 69–70.

421 R. G. Albion, *Forests and Sea Power*, Cambridge, Mass., 1926, Vol. I, pp. 43–44; H. Kempas, *Seeverkehr und Pfundzoll im Herzogtum Preussen*, Bonn, 1964, pp. 292–293; M. Biskup, 'Handel ...' *op. cit.*, pp. 180–181.

422 Archiwum Państwowe w Gdańsku, *300,59* No. 7, p. 79–a; document dated 1476; S. Kutrzeba, 'Wisła w historji gospodarczej dawnej Rzeczpospolitej Polskiej', *Geografija Wisły*, Vol. XI, Warszawa, 1920, p. 10.

423 Ibid., p. 10.

424 Archiwum Państwowe w Gdańsku, *300 D.19.120*, gives reports of trade with

Holland in timber. See also H. A. Poelman, *Bronnen tot de Geschiedenis van den Oostzeehandel*, Vol. I, Pt.2, The Hague, 1917, No. 1968; T. Hirsch, *Danzigs Handels . . . op. cit.*, pp. 116, 120, 122.

425 O. R. Halaga, 'Polská sůl na Slovensku . . .' *op. cit.*, p. 244.

426 W. Naudé, *Die Getreidehandels politik der europäischen Staaten vom 13 bis zum 18 Jahrhundert*, Berlin, 1896, 501 pp.; C. Sattler, *op. cit.*, pp. 132–133, 248–251, 243. According to Attman 'Modern Polish research has provided a very detailed picture of the development of Polish agriculture from 1450 to 1650, of the division into regions, and of the social conditions within the rural economy'. A. Attman, *op. cit.*, p. 16; see also H. Samsonowicz, 'Przemiany gospodarcze w Polsce w I. połowie XV wieku', in S. Gawęda (ed.), *Dlugossiana studia historyczne w pięćsetlecie Śmierci Jana Długosza, Prace Historyczne*, Vol. 65 (Uniw. Jagiell.), P.W.N., Warszawa, 1980, pp. 11–24.

427 S. Kutrzeba, 'Handel Krakowa w wiekach . . . ' *op. cit.*, p. 57.

428 Ibid., p. 116.

429 M. Małowist, 'The Problem of the Inequality . . . ' *op. cit.*, p. 26.

430 Archiwum Państwowe w Gdańsku, *300 D, 3*, 248, document dated 1482 refers to Z., A. and S. Tęczynski, sending 3 rafts of wood and cereals from the Cracow and Sandomierz voivodships to Gdańsk, along the Vistula River; see also M. Biskup, 'Z problematyki handlu' *op. cit.*, p. 399.

431 Archiwum Państwowe w Toruniu, No. 2376, document dated 23/IV/1482, refers to 5 raft loads of wood and grain sent from Tarnów to Gdańsk along the Vistula.

432 D. Hay, *Europe in the Fourteenth and Fifteenth Centuries . . . op. cit.*, p. 382.

433 Document dated 12/X/1342 – a municipal ordinance states, 'Quicumque hospitum adducunt pannos, illos ulterius vendere non debent aliis hospitibus nisi in die forensi et non debent pauciores simul vendere quam sex pannos Flandrenses. Et istos pannos non debent alias vendere quam in cameris pannorum', *Monumenta medii aevi historica res gestas Poloniae illustrantia* (J. Szujski, ed.), Vol. VII, Kraków, 1882, p. 374.

434 Document dated 24/X/1364, *Monumenta medii aevi . . . op. cit.*, p. 375.

435 P. Vaczy, 'La transformation de la technique et l'organisation de l'industrie textile en Flandre aux XIe–XIIIe siècles' (Hungarian Academy of Sciences) *Études Historiques*, Vol. I, Budapest, 1960, pp. 291–316; H. Pirenne, 'Draps de Frise ou draps de Flandre' *Vierteljahrschrift für Sozial- und Wirtschaftsgeschichte*, Vol. VII, Nuremburg, 1909, pp. 308–315; G. de Poerck, *La Draperie médiévale en Flandre en Artois: technique et terminologie*, 3 Vols., Bruges (Rijksun universiteit te Gent), 1951, 304 pp.; G. Espinas, *La Draperie dans la Flandre française au moyen âge* (2 Vols.), Paris, 1923, Vol. I, pp. 31–35; J. A. van Houtte, 'The Rise and Decline . . . *op. cit.*, p. 29.

436 M. M. Postan, 'The Trade of Medieval Europe . . .' *op. cit.*, p. 126.

437 Ibid., No. 286; in 1390 the Cracow city council referred to Brussels cloth 'pro stamine Broslensi' as expensive; *Monumenta medii aevi . . . op. cit.*, Vol. IV, p. 227 (Libri antiquissimi civitatis Cracoviae saeculi decimi quarti); in 1391 it was mentioned as 'stamine de Brusil data regi' Ibid., p. 298.

438 *Kodeks dyplomatyczny miasta Krakowa, op. cit.*, Vol. II, Nos. 2 and 63; F. W. Carter, 'Cracow's Transit Textile Trade 1390–1795: A Geographical Assessment' *Textile History* Vol. 19 No. 1, 1988, pp. 23–60.

439 *Kodeks dyplomatyczny miasta Krakowa, op. cit.*, Vol. II, No. 286; *Monumenta medii aevi* (Libri antiquissimi ... *op. cit.*) pp. 88, 116, 117, 119. See also C. Verlinden, 'Brabantsche en Vlaamsche laken te Krakau op het einde der XIVe eeuw.' *Mededelingen van de Koniklijke Vlaamsche Academie* (Klasse der Leterren), Vol. V, No. 2, Brussels, 1943, pp. 10–25; Ibid., 'Deuz pôles de l'expansion de la draperie flamande et brabançonne au XIVe siècle; la Pologne et la Péninsule Ibérique', *Kwartalnik Historii Kultury Materialnej*, Vol. XVI, No. 4, Warszawa, 1968, pp. 679–689.

440 *Kodeks dyplomatyczny miasta Krakowa, op. cit.*, Vol. II, No. 286; Libri antiquissimi ... *op. cit.*, pp. 116, 117, 119. See also R. Sprandel, 'Zur Tuchproduction in der gegend von Ypern' *Vierteljahrschrift für Sozial- und Wirtschaftsgeschichte*, Vol. LIV, Nuremburg, 1967, pp. 336–340.

441 *Kodeks dyplomatyczny miasta Krakowa*, Vol. II, *op. cit.*, No. 286. See also H. Joosen, 'Recueil de documents relatifs à l'histoire de l'industrie drapière à Malines (des origines à 1384)', *Bulletin de la Commission Royale d'Histoire*, Vol. XCIX, Brussels, 1936, pp. 432–441.

442 *Kodeks dyplomatyczny miasta Krakowa*, Vol. II, p. 286.

443 Ibid., Nos. 286 and 310 §7.

444 Ibid., No. 286.

445 Ibid.

446 Ibid., No. 286 §5.

447 Ibid., No. 286.

448 Wojewódzkie Archiwum Państwowe w Krakowie, *Consularia Cracoviensia*, Vol. II, p. 64.

449 Ibid., p. 64 mentions 'pecia albi de Anglia', 'pecia albi de Anglicani', and pp. 198, 354 refer to 'stamen Lundencense'; Ibid., Vol. III, p. 47; English cloth is also mentioned in a document dated 24/X/1364 viz. 1364 die 24 mensis Octobris, 'Englesch czw VI hellern vnd vom landtuch czw fir hellern von ider elen vnd nicht anders' *Behem* fol. 219 from the original document printed in *Kodeks dyplomatyczny miasta Krakowa, op. cit.*, Vol. II, No. 286; Ibid., No. 310 §7.

450 *Kodeks dyplomatyczny miasta Krakowa, op. cit.*, Vol. II, No. 286; Ibid. No. 310 §7.

451 Ibid., No. 286 §5 refers to rough cloth produced in the county of Kent; Wojewódzkie Archiwum Państwowe w Gdańsku *A.M.G. Missiva 300*, 27/5, fol. llv–12v – refers to a merchant from Ruienmunde sending Colchester cloth to Cracow in 1447 in exchange for copper.

452 J. Ptaśnik, 'Włoski Kraków za Kazimierza Wielkiego i Władysława Jagiełły' *Rocznik Krakowski*, Vol. XIII, Kraków, 1911, pp. 49–110, especially pp. 65–73 which deals with high Italian officials in Cracow during the reign of Władysław Jagiełło (1386–1434).

453 Ibid., 'Z dziejów kultury włoskiego Krakowa' *Rocznik Krakowski*, Vol. IX, Kraków, 1907, 148 pp. especially pp. 6–10 which deal with the first evidence of Italians in Cracow.

454 *Najstarsze księgi i rachunki miasta Krakowa ... op. cit.*, Vol. II, p. 102. No. 885, refers to an Italian merchant, Jakub Pexa who traded in Italian cloth and silks and Eastern goods in Cracow and Kołomyja (Ukraine). See also *Kodeks*

dyplomatyczny katedry krakowskiej św. Wacława, F. Piekosiński (ed.) 2 Vols., Kraków, 1874–1883; Vol. II, pp. 182–185, No. 396.

455 M. Małowist, Z problematyki wzrostu gospodarczego . . . *op. cit.*, p. 666.

456 E. Perroy, 'Le commerce anglo-flamand au XIII siècle: La Hanse Flamande de Londres', *Revue Historique*, Vol. 98, No. 511, 1974, pp. 3–18.

457 A warp had different meanings and was a very uneven measure e.g. Gdańsk warp = 30 łokiec (17.86 metres); and an English warp had 44 łokiec (36.20 metres); half an English warp = 13.10 metres. The figures in Table 3 only serve to give some guide as to the amount of cloth imported by foreign merchants; the amount brought in by Cracow merchants is unknown because they did not pay customs duty.

458 Monumenta Hungarica Historia: (Magyar történelmi emlékek) . . . *op. cit.*, Vol. XX, No. 350. In a document dated 1265 mention is made of Polish cloth (salt and lead) being sent to Liptov – 'Item cum in Tordousina de panno integro, de sale et de plumbo et non de aliis rebus debuissent solvere, quodve tributum eis duximus misericorditer indulgendum, ut de sarcina unius equi solvere debeant 3 novos denarios vel 6/vet/eres et no de aliis rebus'; G. Wanzel, *Codex Diplomaticus Arpadianus*, Vol. XI, Budapest, 180, p. 499. See also M. Jeršová – Opočenská, *Slovenský diplomatár*, Bratislava, 1909, p. 5; V. Chaloupecký, *Staré Slovensko*, Bratislava, 1923, p. 90; A. Kavulják, *Hrad Orava*, Turčianski Teplice/Sv. Martin, 1927, p. 68.

459 *Kodeks dyplomatyczny miasta Krakowa*, *op. cit.*, Vol. II, No. 7.

460 L. Fejérpataky, Magyarországi várasok . . . *op. cit.*, p. 224.

461 Wojewódzkie Archiwum Państwowe w Krakowie, *Consularia Cracoviensia*, Vol. I, No. 429, pp. 14, 330; Ibid., Vol. III, p. 14 (dated 1450); *Kodeks dyplomatyczny miasta Krakowa*, *op. cit.*, Vol. II, No. 310.

462 Wojewódzkie Archiwum Państwowe w Krakowie, *Consularia Cracoviensia*, Vol. I, No.429, p. 330.

463 *Kodeks dyplomatyczny miasta Krakowa*, *op. cit.*, Vol. II, No. 330. S. Kutrzeba, 'Handel Krakowa w wiekach . . .' *op. cit.*, p. 62.

464 Wojewódzkie Archiwum Państwowe w Krakowie, *Consularia Cracoviensia*, Vol. I, No. 430, p. 262; Ibid., No. 428, p. 354.

465 S. Kutrzeba, 'Handel Krakowa w wiekach . . .' *op. cit.*, p. 62.

466 Wojewódzkie Archiwum Państwowe w Krakowie, *Consularia Cracoviensia*, Vol. I, No. 427, p. 231 (dated 1406). See also S. Kutrzeba, 'Akta odnoszące się do stosunków handlowych Polski z Wegrami głównie z archiwum koszyckiego z lat 1354–1505', *Archiwum Komisji Historycznej*, Vol. IX, Kraków, 1902, pp. 407–485, Nos. 4, 9; Archív m. Košíc, No. 43 (document dated 1405).

467 F. Fejérpataky, *Magyarországi várasok* . . . *op. cit.*, pp. 303, 433, 464.

468 Monumenta Hungarica Historia (Magyar történelmi emlékek) . . . *op. cit.*, Vol. XX, No. 350.

469 F. Kováts, *Nyugatmagyarország árúforgalma a XV században*, Budapest, 1902, p. 21; L. Fejérkpataky, *op. cit.*, pp. 224, 229, 230, 303, 308, 312, 326, 344, 346, 357, 432. One of the most famous Cracow wholesalers was Jan Kletner circa 1450. He supplied Bardejov with large amounts of cloth as evidenced by the amount of money received: Archív m. Bardejov – letters to town council; 21/XII/1454 = 718 Hungarian florens in 1453 and 798 Hung. florens in 1454,

20/VI/1455; 346 Hung. florens in 1458 and in 1460 = 4 bales of cloth worth 179 Hung. florens, 6/II/1460; Documents: *O.A.B.* Nos. 760, 800, 999, 1235, and 425. Other Cracow merchants involved in the cloth trade with Bardejov were Jan Teschner and Jan Dobczyk, a letter dated 10/I/1455 and two others undated, instruct payment for cloth to be made to Dobczyk's furrier in Cracow, *O.A.B.* Nos. 766, 899, 1011.

470 Wojewódzkie Archiwum Państwowe w Krakowie, *Consularia Cracoviensia*, Vol. I, p. 176 – document dated 1403, quantity of cloth sent to Košice from Cracow, to the firm of S. Talent & A. Czarnisch; document dated 1405, Košice firm bought 300 warps of Strigau (Strzygawa) cloth and 20 long cloths from Tirlemont for 2,410 gulden 'in good gold', Archív m. Košice, No. 43; document dated 1406, 110 warps of Görlitz (Zgorzelec) cloth sent from Cracow to Košice, *Consularia Cracoviensia*, Vol. I, No. 427, p. 231; document dated 1436, 400 warps of Görlitz (Zgorzelec) cloth sent from Cracow to Košice, *Consularia Cracoviensia*, Vol. II, p. 368; document dated 1450, Malines cloth sent from Cracow to Košice, Ibid., Vol. III, p. 14. See also J. Szücz, *A várasok és kézmüvesség a XV századi Magyarországon*, Budapest, 1955, pp. 53–222 (especially pp. 217–220); G. Wenzel, *Kassa város parkettké-szitése a XV sz. kezdetén*, Budapest, 1871, 435 pp.

471 Wojewódzkie Archiwum Państwowe w Krakowie, *Consularia Cracoviensia*, Vol. I, No. 428, pp. 344–345.

472 Other merchants included Andrzej Czarnisch, Marcin Scholtis, Henryk Smyt, Mikołaj and Piotr Glezer, J. Neysser, F. Koza, P. Crenmark: Wojewódzkie Archiwum Państwowe w Krakowie, *Consularia Cracoviensia*, Vol. I, No. 427, pp. 176, 231; No. 428, pp. 14, 354, 368; No. 429, p. 14; No. 430, p. 262.

473 Archiwum miasta Lwowa, *Consularia Leopoldensia*, Vol. I, 283, 291, pp. 558–559, 621, 858–8590; *Percepta et exposita civitatis Leopoldiensis od 1460–1518 r.* Lwów, 1884, p. 331. S. Kutrzeba, Handel ze Wschodem ... *op. cit.*, p. 127; Wojewódzkie Archiwum Państwowe w Krakowie, *Consularia Cracoviensia*, Vol. IV, p. 204, dated *circa* 1485 during the rule of Sultan Bayezid II (1481–1512).

474 For example in 1457 a Nuremburg merchant concluded an agreement with 4 traders viz. Gregorz from Branice (governor of Radom castle), Andrze from Tęczyn, Jan from Melsztyn, and Jan from Tarnów, for the supply of material valued at 6,000 grzywnas (particularly damask) and velvet valued at 2,000 grzywnas. L. Lepszy, 'Przemysł artystyczny i handel', *Rocznik Krakowski*, Vol. VI, Kraków, 1904, p. 302. See also H. Samsonowicz, 'Struktura handlu gdańskiego ...' *op. cit.*, pp. 707–708; Ibid., 'Handel zagraniczny Gdańska ...' *op. cit.*, p. 348.

475 Some of Cracow's cloth merchants had their own houses and shops in Lwów e.g. Sewer Betman, Jan Boner, Stano Zygmuntowicz, Jan Romer, Jan Schutcz, Jan Beck, Pawel and Hieronim Morsztyn, Archiwum miasta Lwowa, *Consularia Leopoliensis*, Vol. I, p. 549, 764.

476 In 1444 evidence of Cracow merchants at Jarosław selling western textiles at the August fair; A. Czołoski and F. Jawowski (eds.), *Pomniki dziejowe Lwowa z archiwum miasta*, Vol. IV (Księgi ławnicze miejskie 144–1448), Lwów, 1921, No. 1262, p. 167.

477 P. Panaitescu, 'La route commercial . . .' *op. cit.*, pp. 9–19; *Percepta et exposita . . . op. cit.*, pp. 5, 7, 12, 15, refers to a document dated 1460 in which two Cracow merchants Klemens and Michał Szwarc sent Prince Stephan the Great (Stefan cel Mare), the ruler of Moldavia, various types of cloth from Florence, London, Bruges, etc. See also Ł. Charewiczowa, 'Handel Lwowa z Mołdawią . . .' *op. cit.*, p. 48.

478 *Akta grodzkie i ziemskie*, E. Katużniacki (ed.), Vol. VII, No. 8, Lwów, 1878, p. 237.

479 *Codex Diplomaticus Silesiae*, K. Wüttke (ed.), Vol. VIII, Breslau, 1867, No. 4. The Act of 1303 regulated the retail cloth. Mention is made of cloth from Ghent and Ypres; Ibid., No. 39; Act of 1360 mentions cloth from Bruges, Poperinge, Ypres, Ghent, Malines, Courtrai, Brussels, Louvain etc.

480 L. Koczy, *Związki handlowe Wrocławia z Polską do końca XVI wieku*, Katowice, 1936, p. 8.

481 Wojewódzkie Archiwum Państwowe w Krakowie, *Consularia Cracoviensia*, Vol. I, p. 196, document dated 1404 and refers to 190 warps (3,393.40 metres) of Görlitz cloth being sent from Wrocław to Cracow; Ibid., Vol. I, p. 239, document dated 1406 refers to 120 warps (2,143.20 metres) sent from Wrocław to Cracow; Ibid., Vol. II, p. 7, document dated 1412 – 43 warps (767.98 metres) sent from Wrocław to Cracow; Ibid., Vol. III, p. 516 dated 1474 refers to a quantity (unspecified) of Görlitz cloth being sent from Wrocław to Cracow; *Kodeks dyplomatyczny miasta Krakowa*, *op. cit.*, Vol. II, No. 286; dated 1390 refers to 110 warps (1,964.6 metres) of Görlitz cloth sent from Wrocław to Cracow.

482 Ibid., Vol. II, No. 286; *Consularia Cracoviensia*, Vol. II, p. 247, document dated 1429.

483 In 1437, 100 warps (1,786 metres) of Opava cloth sent to Cracow together with 28 warps (500 metres) of ordinary 'langtuch' (long warp) cloth which arrived in the city from an Opava merchant. *Księgi ławnicze krakowskie*, ... *op. cit.*, p. 16, for the year 1437. Cloth from Wrocław workshops is mentioned for 1390, when 180 warps (3,214.8 metres) arrived in the city, *Consularia Cracoviensia*, Vol. I, p. 231. In 1456 a Cracow merchant received 'eynen ballen' of Wrocław cloth, *Consularia Cracoviensia*, Vol. III, p. 140.

484 S. Kutrzeba, 'Handel Krakowa w wiekach . . .' *op. cit.*, p. 85.

485 L. Koczy, *Związki handlowe Wrocławia . . . op. cit.*, p. 9.

486 M. Małowist, 'Evolution industrielle en Pologne du XIVe au XVIIe siècle', in *Studi in Onore di Armando Sapori*, Vol. I, Milan, 1957, p. 577.

487 A. Mączak, 'Sukiennictwo wielkopolskie XIV–XVII wieku', *Badania z dziejów rzemiosła i handlu w epoce feudalizmu*, Vol. III, Warszawa, 1955, 250 pp.

488 S. Herbst, *Toruńskie cechy rzemieślnicze*, Toruń, 1933, 234 pp.

489 J. Wyrozumski, *Tkactwo małopolskie w późnym średniowieczu*, Warszawa-Kraków, 1972, pp. 134–140.

490 *Kodeks dyplomatyczny miasta Krakowa*, *op. cit.*, Vol. II, No. 286.

491 Wojewódzkie Archiwum Państwowe w Krakowie, *Księga wójtowska*, for the years 1476–77.

492 *Kodeks dyplomatyczny miasta Krakowa*, *op. cit.*, Vol. II, No. 310 – cloth called

'sądecki' was available on the Cracow market at the beginning of the fifteenth century, and according to Dąbrowski played an important role in trade with Hungary, J. Dąbrowski, 'Kraków a Węgry . . .' *op. cit.*, p. 219.

493 *Kodeks dyplomatyczny miasta Krakowa, op. cit.*, Vol. II, No. 310, which was known in Cracow as 'krosnesche leymet' and dated 1432.

494 Ibid., Vol. I, No. 112 dated 1419 in which the king allowed cloth to be sold by the ell, and confirmed again in 1431, Ibid., Vol. I, No. 130. Towns like Lublin, Sandomierz, Wiślica and Lelów were against this ruling on cloth measurement for sale.

495 Ibid., Vol. I, No. 149.

496 Archív m. Košic, *H. pur 1*, fol. 52 (ed.c.č. 2583) dated Nov. 1398. See also O. R. Halaga, 'Textíle v Pol'sko-Pruskom obchode východoslovenských miest (13–15 stor)', *Historický Časopis*, Vol. 22, No. 1, Bratislava, 1974, p. 8.

497 Archív m. Bardejov, *O.A.B.*, No. 2481, document dated 1486.

498 O. R. Halaga, 'Textíle v Pol'sko-Pruskom obchode . . .' *op. cit.*, p. 6.

499 Archív m. Bardejov, *O.A.B.*, No. 1280/b, gives the following names of Cracow merchants: G. Schon, J. Weber, J. Kletner, P. Kaufman, A. Kustlor and M. Godfer.

500 Archív m. Bardejov, *O.A.B.*, Nos. 1701, 1591, 2365, 2728, 1662, 1579; also even small places like Osobnica are mentioned; *O.A.B.*, No. 1505, document dated 27/X/1463. See also F. Bujak, *Materiały do historii miasta Biecza (1361–1632)*, Kraków, 1914, p. 23, No. 70; S. Kutrzeba, 'Handel Krakowa w wiekach . . .' *op. cit.*, p. 63.

501 J. Wyrozumski, 'Rozwój życia miejskiego do połowy XVI w.', in *Krosno: Studia z dziejów miasta i regionu*, Vol. I (do roku 1918), Kraków, 1972, p. 102.

502 M. Biskup, 'Handel wiślany w latach . . .' *op. cit.*, Table 9, p. 186, with reference to years 1463–1465.

503 M. M. Postan, 'The Trade of Medieval Europe . . .' *op. cit.*, p. 200.

504 G. Strauss, *Nuremburg in the Sixteenth Century* . . . *op. cit.*, p. 201.

505 Y. Renourd, 'Le grand commerce du vin au moyen âge' *Etudes d'Histoire Médiévale* (SEVPEN) Paris, 1968, pp. 235–248.

506 J. Craeybeckx, *Un grand commerce d'importation – les vins de France aux anciens Pays-Bas (XIII–XVI siècle)*, Paris, 1958, pp. 15–17.

507 R. Schultze, *Geschichte des Weines und der Trinkgelage*, Berlin, 1867, p. 120.

508 B. Gigalski, *Der Weinbau im Lande des Deutschen Ordens während des Mittelalters; Ein Vortrag*, Brunsberg, 1908, pp. 8–9. See also M. Strzemski, 'Przemiany środowiska geograficznego Polski, jako tła przyrodniczego rozwoju rolnictwa na ziemiach polskich (od połowy trzeciego tysiąclecia p.n.e. do naszych czasów), *Kwartalnik Historii Kultury Materialnej*, Vol. IX, No. 3, Warszawa, 1961, pp. 331–357.

509 *Słownik Geograficzny Królestwa Polskiego i innych krajów słowiańskich*, B. Chlebowski, F. Sulimierski and W. Walewski (eds.), Vol. XIII, Warszawa 1893, with reference to villages named 'Winnica'.

510 K. Moldenhawer, 'Szczątki roślinne z wykopalisk z X w. na Ostrowiu Tumskim', *Przegląd Archeologiczny*, Vol. VI, Warszawa, 1938–39, pp. 226–227; J. Kwapieniowa, 'Początki uprawy winnej latorośli w Polsce' *Materiały Archeologiczne*, Vol. I, Kraków, 1959, p. 361.

511 Z. Morawski, 'Rozwój i upadek winiarstwa w Polsce (XII–XVI wiek)', *Kwartalnik Historii Kultury Materialnej*, Vol. XXVI, No. 1, Warszawa, 1978, p. 67.
512 Ibid., p. 71.
513 I. Acsády, *A Pozsonyi és Szepesi kamarák, 1565–1604*, Vol. I, of *Két Pénzügytörténelmi Tanulmány*, Budapest, 1894, p. 74, 75.
514 A. Ambrózy, *Tokay-Hegyala*, Budapest, 1933, p. 32. See also K. Pieradzka, *Handel Krakowa z Węgrami w XVI w.* (Biblioteka Krakowska, No. 87), Kraków, 1935, pp. 96–141 on the wine trade.
515 S. Kutrzeba 'Handel Krakowa w wiekach ...' *op. cit.*, p. 68.
516 J. Dąbrowski, 'Kraków a Węgry w wiekach ...' *op. cit.*, p. 222.
517 *Kodeks dyplomatyczny miasta Krakowa, op. cit.*, Vol. I, No. 7. See also A. Pleidell, *A nyugatra irányuló ... op. cit.*, p. 25; G. Komoróczy, *Borkivitelünk észak felé. Fejezet a Magyar kereskedelom történeteböl*, Kassa (Košice), 1944, 405 pp.
518 The four main production centres of Tokay wine were Tállya, Tarcal, Tolcsva and Tokaj; secondary centres included Ptak (Sárospatak), Sátoraljaújhely and Mád; J. Dąbrowski, *op. cit.*, p. 40.
519 *Rachunki dworu króla Władysława Jagiełły i królowej Jadwigi*, F. Piekosiński (ed.) Kraków, 1896, pp. 210, 213–214.
520 J. Dąbrowski, ... *op. cit.*, p. 225; F. W. Carter 'Cracow's Wine Trade (Fourteenth to Eighteenth Centuries)' *Slavonic and East European Review*, Vol. 65, No. 4, 1987, pp. 337–378.
521 S. Kutrzeba (ed.), 'Akta odnoszące ...' *op. cit.*, No. 50 dated 1482 in a dispute with Emeryk Zapolya, Count of Spiš. A similar dispute occurred with Košice, Ibid., No. 78, 79; Archív město Košic, *Suppl. H ad 1482* (maxime a Polonis et Ruthenis aliisque nationibus regnorum et regionum superiorum ... comparare soliti fuissent), also refers to competition between wine from Zemplín and Sremska Mitrovica.
522 Bardejov (Bartfa), did not produce wine itself and was mainly a collecting centre; it played a middleman role between Poland and the foothill region of Hegyalya. Examples of wine for sale in the town came from Sremska Mitrovica, Somogy, Baranya and Szikszó. Archív město Bardejov, documents for 1457 *No. 944/988* pp. 1–2; for 1458, *No. 1057/1053*, pp. 1–2; L. Fejérpataky, *Magyarországi városok régi ... op. cit.*, pp. 257, 360a. Evidence of wine being sent from Bardejov to Cracow found in: Wojewódzkie Archiwum Państwowe w Krakowie, *Consularia Cracoviensia*, Vol. III, pp. 154–156 dated 1451; also 1482 Sremska (Seremiensis) wine sent to Cracow, Archív miesto Bardejov, *Fasc. II, No. 621*; Archív miesto Košic, *No. 527*. L. Fejérpataky, *op. cit.*, p. 371; Sremska wine also mentioned in *Najstarsze księgi ... op. cit.*, p. 269 and other Hungarian wines in *Księgi ławnicze krakowskie ... op. cit.*, p. 117, for 1451.
523 Košice was the main market centre for wines from the Zemplín region especially the town of Abaújszantó. Quantities of wine known as 'Vinum terrestre Cassovianum', were sent to Cracow. (*Rachunki królewskie z lat 1471–1472 i 1476–1478*, S. Gawęda, Z. Perzanowska, A. Strzelecka (eds.), Wrocław-Kraków, 1960, No. 88, fol. 82); they were described as light table wines. Evidence of wine from Košice being sent to Cracow: Wojewódzkie Archiwum Państwowe w Krakowie, *Consularia Cracoviensia*, Vol. III, No. 249, pp. 154, 167, 334; No.

430, p. 358. Other wines came through Košice to Cracow from Szegszard, Ruszt and Tokay; A. Divéky, *Felsö – Magyarország kereskedelmi . . . op. cit.*, p. 63. See also E. Fügedi, 'Kaschau, Eine osteuropäische . . .' *op. cit.*, p. 201.

524 For example Jan Crenmark, who supplied wine to Cracow from his own vineyards at the beginning of the fifteenth century; Wojewódzkie Archiwum Państwowe w Krakowie, *Consularia Cracoviensia*, Vol. II, pp. 354–355; Jakob Swob sent wine through Bardejov; L. Fejérpataky, *op. cit.*, p. 433.

525 Jan Sweidniczer sent wine to Cracow from his vineyards near Košice, Wojewódzkie Archiwum Państwowe w Krakowie, *Consularia Cracoviensia*, Vol. II, pp. 429, 154–156, 167; *Liber Testamentorum*, p. 41 and had his own agent, Birhemar, there. Frantiszek Czotmer from Košice sent wine to Mikołaj Knodow through an agent, Wojewódzkie Archiwum Państwowe w Krakowie, *Consularia Cracoviensia*, Vol. II, No. 429, pp. 334, 342; towards the end of the fifteenth century wine was sent to Cracow from Hungary by Mikołaj Karl through his agents, *Consularia Cracoviensia*, Vol. II, No. 430, pp. 348, 359; the wife of Jan Turzon also sent wine to Cracow, Ibid., No. 429, p. 397.

526 For further discussion on these problems see A. Gardonyi, 'A Felsö-Magyarország kereskedelmi útjai a középkorban', *Közgazdasagi szemle*, Budapest, 1908, pp. 5–6.

527 For comparison, wine in France could be carried down the Garonne until about 11th Nov. See W. G. East, *A Historical Geography of Europe . . . op. cit.*, p. 107.

528 Barrels of wine sent from Bardejov by Stanisław Roschek were filled with one third 'feces alias lagyer'. Wojewódzkie Archiwum Państwowe a Krakowie, *Scabinalia Sandec. 1488–1505*, p. 638.

529 For example, wine from Bardejov was sent to Biercz on 14/IV/460; Archív město Bardejov *O.A.B.* No. 1208; in 1486, 11 barrels of wine sold by Bardejov merchant in Biecz, Ibid., No. 2451; wine from Bardejov in addition sent to Jasło, Ibid., No. 1681; Krosno likewise bought wine from Bardejov. Ibid., No. 1852; similarly to Dębno on 17/II/1482, 'Item dedimus 21. flor. pro vino, quod nobis missistis et duos cum medio flor. pro vectigale et ductura', Ibid., No. 2217; a letter dated 9/XII/1458 was sent from Dębno to Bardejov for wine ' . . . petimus . . . velitis . . . nobis transmittere unum medium vas (vini), quod esset bonum pro nostra parata peccunia seu florensis. I absencia nostra velitis transmittere nostra magnifice domine in Bycz', Ibid., No. 2437; even small settlements received Bardejov wine such as Brzozów; Ibid No. 2379; and Klimontów, Ibid., No. 1666.

530 *Percepta et exposita civitatis Leopoldiensis od 1460–1518r.* (Archiwum miasta Lwowa) Lwów, 1884, p. 620; *Consularia Leopoliensis*, Vol. I, p. 864.

531 *Monumenta Leopoliensia historica* (Pomniki dziejowe Lwowa), A. Czołowski (ed.), Vol. IV, Inventrum, p. 143, Lwów, 1921 refers to an Italian merchant who bought a vat of Malmsey wine from Moldavia in 1442. *Circa* 1400 a garniec (gallon) of Malmsey wine cost between 2.5 and 12 grosz, J. Pelc, *Ceny w Krakowie . . . op. cit.*, Table 24, p. 40.

532 Archiwum miasta Lwowa, *Consularia Leopoliensis*, Vol. I, pp. 147, 150–151.

533 *Codex diplomaticus ecclesiasticus ac civilis regni Hungariae . . . op. cit.*, Vol. X, Pt. 2, p. 259, No. 147, and Vol. IX, Pt. 5, p. 264, No. 129, refer to Košice's commercial contacts with Transylvania and Moldavia in 1378 and 1394.

534 See for example, Y. Renouard, 'La consommation des grands vins de Bourbonnais et de Bourbobne à la cour pontificale d'Avignon', *Annales Bourgogne*, Vol. XXIV, Dijon, 1952, pp. 221–244.
535 *Najstarsze księgi ... op. cit.*, Vol. II contains many references to German wines.
536 *Kodeks dyplomatyczny miasta Krakowa, op. cit.*, Vol. I, No. 72–75.
537 T. Hirsch, *Danzigs Handells-und Gewerbsgeschichte ... op. cit.*, pp. 261–262.
538 For example, in 1390; *Hansisches Urkundenbuch ... op. cit.*, Vol. IV, No. 1017, 1018, 1034.
539 W. Lauffer, 'Danzigs Schiff-und Waarenverkehr ... ' *op. cit.*, pp. 40–44.
540 J. Ptaśnik, 'Z dziejów kultury włoskiego Krakowa ... ' *op. cit.*, pp. 37–39.
541 M. M. Postan, 'The Trade of Medieval Europe ... ' *op. cit.*, p. 200.
542 For example, in Germany, in order to rise a big thirst, people ate salted bread or, if they had them 'Pfefferkuchen', highly peppered and spiced little cakes. See G. Strauss, *Nuremburg in the Sixteenth Century ... op. cit.*, p. 201.
543 J. Craeybeckx, *op. cit.*, pp. 15–17.
544 D. Hay, *Europe in the Fourteenth ... op. cit.*, p. 368; M. Biskup, 'Handel wiślany ...' *op. cit.*, p. 192; H. Samsonowicz, 'Struktura handlu gdańskiego ...' *op. cit.*, p. 707.
545 S. Kutrzeba, 'Piwo w średniowiecznym Krakowie', *Rocznik Krakowski*, Vol. I, Kraków, 1898, pp. 37–52; Ibid., 'Finanse Krakowa w wiekach ...' *op. cit.*, p. 85.
546 *Kodeks dyplomatyczny miasta Krakowa, op. cit.*, Vol. I. No. 146 refers to 40 carts of beer annually (1 cart carried *circa* 500 gallons: i.e. 20,000 gallons). Beer also came to Cracow from Opava (Troppau).
547 Evidence of Bardejov beer exported to Poland in 1453 when a new barrel was sent to Wielopole (near Muszyna); Archív m. Bardejov, *O.A.B.*, No. 712; this was again repeated in 1455, Ibid., No. 788. It is difficult to assess the amount consumed locally, and that sent further to Cracow for sale.
548 *Kodeks dyplomatyczny miasta Krakowa, op. cit.*, Vol. I, No. 161; in order to prevent competition from Kazimierz suburb, certain goods e.g. cattle, foreign beer, wine and salt had to be sold within Cracow's walls. S. Kutrzeba, Handel Krakowa w wiekach średnich ... *op. cit.*, pp. 149–150.
549 *Kodeks dyplomatyczny miasta Krakowa, op. cit.*, Vol. II, No. 209.
550 B. Behem, *Codex Picturatus*, Kraków, 1505 (Biblioteka Jagiellońska).
551 M. Małowist, 'W sprawie badań nad historią rzemiosła miejskiego w średniowiecnej Polsce', *Rocznik Dziejów Społeczno Gospodarczych*, Vol. III Warszawa, 1951, pp. 16–19; H. Samsonowicz, *Rzemioslo wiejskie w Polsce XIV–XVII w Badania z dziejów rzemiosła i handlu w epoce feudalizmu*, Vol. II Warszawa, 1954, 154 pp. Similar events were also taking place in Hungary at this time – see L. Szádeczky, '*Iparfejlödés és czéhek története Magyarországon*, 2 Vols. (Okirattárral 1307–1848), Budapest, 1913.
552 For example see B. Bucher (ed.), *Die alten Zunf- und Verkehrs-Ordnungen der Stadt Krakow*, Vienna 1889, 194 pp.
553 W. Łoziński, 'Złotnictwo lwowskie', *Bibljoteka Lwowska*, Vol. XV–XVIII Lwów, 1912, pp. 115–117, which refers to competition between Lwów and Kraków goldsmith products; J. Ptaśnik, 'Towns in Medieval Poland ...' *op. cit.*, p. 46.

554 Z. Bocheński, Krakowski Cech Mieczników', *Biblioteka Krakowska*, No. 92 Kraków, 1937, 90pp; J. Ptaśnik, *Cracovia artificum 1300–1500 ... op. cit.*, Nos. 124, 127, 130, 139, 147.
555 M. Kwapieniowa, 'Organizacja produkcji i zbytu wyrobów garncarskich w Krakowie w XIV–XVIII w.' *Studia nad Produkcją Rzemieślniczą w Polsce (XIV–XVIIIw.)* (P.A.N.), Wrocław–Warszawa–Kraków, 1976, pp. 7–87.
556 W. von Stromer, 'Nuremburg in the International Economics of the Middle Ages', *The Business History Review*, Vol. 44, Harvard, 1970, p. 212.
557 G. Strauss, *Nuremburg in the Sixteenth Century ... op. cit.*, pp. 135–136.
558 J. F. Roth, *Geschichte des Nuernbergischen Handels*, Vol. I, Leipzig, 1800, p. 41.
559 Wojewódzkie Archiwum Państwowe w Krakowie, *Consularia Cracoviensia*, Vol. I, p. 245 (dated 1406); Ibid., p. 319 (1408); Vol. II, p. 210 (1424); Ibid., p. 227 (1426); Ibid., p. 274 (1430); Ibid., p. 327 (1434); Ibid., p. 304 (1432); Ibid., p. 320 (1433); Ibid., p. 423 (1440); Ibid., p. 452 (1442); Ibid., p. 487 (1446); Ibid., p. 497 (1447); Vol. III p. 159 (1457); Ibid., p. 165 (1457); Ibid., p. 669 (1480); Vol. IV, p. 11 (1483); Ibid., p. 241 (1489); *Księgi ławnicze krakowskie ... op. cit.*, for 1409, p. 10; for 1420, p. 30; for 1444, p. 311; for 1446, p. 363; for 1454, p. 205; for 1457 p. 283; for 1456, p. 267.
560 Mention of knives, scythes and sickles for example. *Najstarsze księgi ... op. cit.*, Vol. II, p. 85; *Kodeks dyplomatyczny miasta Krakowa, ... op. cit.*, Vol. II, No. 310; also mention of caps, buttons and articles known as 'cromerey' *Consularia Cracoviensia*, Vol. I, No. 429, p. 749; *Advocatus Cracoviensia*, No. 87, p. 260; See also E. Fügedi, 'Kaschau, eine ...' *op. cit.*, p. 200.
561 M. Biskup, 'Handel wiślany ...' *op. cit.*, p. 183 for the years 1463–65.
562 W. A. Wagner, 'Handel dawnego Jarosławia do połowy XVII wieku', *Prace Historyczne* (Wydane ku uczczeniu 50-lecia akademickiego koła historyków universytetu Jana Kazimierza we Lwowie 1878–1928), Lwów, 1929, p. 137.
563 Ibid.
564 L. Lepszy, 'Przemysł artystyczny i handel' *Rocznik Krakowski*, Vol. VI, Kraków, 1904, p. 274.
565 J. Kalić, 'Stosunki handlowe krajów bałkańskich z Europą Centralną w XV w.' *Przegląd Historyczny*, Vol. 67, No. 4. Warszawa, 1976, p. 571.
566 Eastern carpets were very popular in Poland particularly for covering walls and was, according to Łoziński, 'without doubt sold in Cracow'. W. Łoziński, *Życie polskie w dawnych wiekach*, Lwów, 1912, p. 129.
567 L. Deák, 'Bardejovský obchod a ...' *op. cit.*, p. 119; See also A. Wyrobisz, *Szkło w Polsce od XV do XVII w.*, Wrocław, 1968, 340 pp.
568 M. M. Postan, 'The Trade of Medieval Europe ...' *op. cit.*, p. 205. See also W. Abel, *Agricultural Fluctuations in Europe from the Thirteenth to the Twentieth century* (Methuen & Co.), London, 1980, 363 pp.
569 F. Braudel and F. C. Spooner, 'Prices in Europe from 1450 to 1750', in Vol. IV, *Cambridge Economic History of Europe*, Cambridge, 1967, Ch.VII, p. 381. See also C. M. Cipolla, 'Currency Depreciation in Medieaval Europe', *Economic History Review*, Vol. XV, London, 1963–64, pp. 413–422.
570 B. Ulanowski, 'Kilka zabytków ustawodawstwa królewskiego i wojewodzińskiego w przedmiocie handlu i ustanawiania cen', *Archiwum Komisji Prawnicznej*, Vol. I (Akad. Krak.) Kraków, 1895, pp. 36–42.

571 *Kodeks dyplomatyczny miasta Krakowa,* ... *op. cit.*, Vol. II, No. 286, dated 1413.
572 Ibid., No. 340 (dated 1481), Nos. 451, 454.
573 Ibid., No. 287 (dated 1396), Nos. 298, 366.
574 G. Adler, 'Das grosspolnische Fleischergewerb vor 300 Jahren', *Zeitschrift der historische geschichte für der Provinz Posen*, Vol. IX, 1892, p. 282.
575 J. Pelc, *Ceny w Krakowie,* ... *op. cit.*; The credit for having undertaken during the inter-war period a systematic research on the history of prices in Poland goes to the Lwów School of Economic History directed by Prof. F. Bujak. The eleven volumes so far published contain the prices and earnings of the period from the fourteenth to the beginning of the twentieth century; five cities are involved: Gdańsk, Warsaw, Lublin, Cracow and Lwów. For further discussion see S. Hoszowski, 'The Revolution of Prices in Poland in the Sixteenth and Seventeenth Centuries', *Acta Poloniae Historica*, Vol. II, Warszawa, 1959, pp. 7–16; Ibid., 'L'Europe centrale devant la révolution des prix aux XVIe et XVIIe siècles', *Annales: Economies Sociétiés, Civilisations*, Vol. 16, No. 1, Paris (Armand Colin), 1961, pp. 441–456.
576 M. M. Postan, 'The Trade of Medieval Europe ...' *op. cit.*, p. 212.
577 A. Jelicz, *W średniowiecznym Krakowie* ... *op. cit.*, p. 109.
578 N. J. G. Pounds, *An Economic History of Medieval Europe* ... *op. cit.*, p. 481.
579 J. Pelc, *Ceny w Krakowie* ... *op. cit.*, p. 55.
580 J. N. Ball, *Merchants and Merchandise* ... *op. cit.*, p. 34; The English had their shops and warehouses not only in the Polish Baltic ports but also inland at Kowno, Troki, Kazimierz and Cracow. *The Hull Polish Record*, Vol. I, I. Wilson (ed.), London, 1932–33, pp. 80–81. See also Z. Taźbierski 'Uwagi o kontaktach handlowych Anglii z Polską od XV do XVIII w. w świetle literatury brytyskiej', *Rocznik Gdański*, Vol. XXXVIII, No. 1, Gdańsk, 1978, pp. 77–95.
581 G. Granasztói, 'La Ville de Kassa dans le Commerce Hungaro-Polonais au XVI siècle', in *La Pologne et la Hongrie aux XVIe–XVIIIe siècles*, V. Zimányi (edt.) (Akadémiai Kiadó), Budapest, 1981, p. 57.
582 *Kodeks dyplomatyczny miasta Krakowa,* ... *op. cit.*, Vol. I, No. 17; it was dated St. Matthew the Apostle's day according to S. Zarzecki, *Wiadomości o handlu miasta Krakowa na dochód Towarzystwa Dobroczynności*, No. 1, Kraków, 1812, p. 13.
583 *Kodeks dyplomatyczny miasta Krakowa,* ... *op. cit.*, Vol. I, No. 64.
584 Tajný archív miesto Košíc, *Depositorium*, No. 5; Ibid., *F. Telonium*, No. 9 'quod mercatoribus inter eos (Cassovienses) existentibus et versus regnum Polonie in Cracoviam vel ultra eandem cum eorum rebus et mercibus, cuiuscunque generis vel materiei existant, iugiter transeuntibus et victus eorum acquirentibus mediam partem veri thelonii seu tributi rerum eorum quarumlibet in loco dicti thelonii dare et persolvere consueti relaxasset', dated 16/IX/1378.
585 O. R. Halaga, *Košice – Balt. Výroba a obchod v styku vyslanectvo miest s Pruskom*, Košice, 1974, Dokumenty, No. 37–38. It is probable that the agreement had already been signed in 1390, and reconfirmed in 1394.
586 W. von Stromer, 'Nuremburg in the International ...' *op. cit.*, p. 210.
587 H. Ammann, 'Wirtschaftsbeziehungen zwischen Oberdeutschland und Poland im Mittelalter', *Vierteljahreschrift für Sozial-und Wirtschaftsgeschichte*, Vol. 48,

Wiesbaden, 1961, pp. 433–34; Ibid., 'Nürnberger unternehmer im Karpatenraum ...' *op. cit.*, p. 645; G. Strauss, *Nuremburg in the Sixteenth Century ... op. cit.*, p. 128, which states 'Nuremburgers were active in Łódź, Cracow, Warsaw, and Lublin'.

588 *Kodeks dyplomatyczny miasta Krakowa, ... op. cit.*, Vol. I, No. 148, dated 1449; in fact the practice was so abused by foreigners that in 1449 the king, Kazimierz Jagiellończyk, forbad all of them from obtaining the right of Polish town citizenship.

589 Ibid., Vol. II, No. 300§2.

590 Ibid., Vol. II, p. 487 'iuramentum civium in mercanciis suspectorum' and p. 489, 'iuramentum famulorum in mercanciis suspectorum'.

591 J. Ptaśnik, 'Z dziejów kultury włoskiego Krakowa', *Rocznik Krakowski*, Vol. IX, Kraków, 1907, pp. 11–14, which discusses in detail the *Libris Iuris Civilis.*

6: The Political situation of Cracow, 1500–1795

1 J. N. Ball, *Merchants and Merchandise: The Expansion of Trade in Europe, 1500–1630*, Croom Helm, London, 1977, p. 11, who states 'This emphasis on continuity is certainly preferable in the social and economic sectors to an overdramatisation of the revolutionary character of the early sixteenth century, yet it is by no means improper to call attention ... to the importance of short-term factors affecting economic development.'

2 J. M. Roberts, *The Pelican History of the World*, Penguin Books, Harmondsworth, 1981, p. 521.

3 For example see B. Chudoba, *Spain and the Empire 1519–1643* (Octagon), New York, 1977, 299 pp. (Reprint of the 1952 edn); an important study which uncovered entirely new material about the role of Spain in European history, especially during the Thirty Years War.

4 W. H. McNeill, *The Shape of European History* (Oxford University Press), New York, 1974, p. 121.

5 *Volumina legum. Prawa, konstytucye y przywileie Królestwa Polskiego y Wielkiego Xięstwa Litewskiego y wszystkich Prowincyi należących ... uchwalone*, Vol. I–VIII, Warszawa 1733–1782; Vol. IX, Kraków 1889; Vol. X, Poznań 1952; reconfirmed in the Union of Lublin, 1569.

6 Z. Wojciechowski, *Zygmunt Stary (1506–1548)*, Warszawa, 1946, 342 pp.; K. Kolankowski, *Zygmunt August-Wielki Książe Litwy do roku 1548*, Lwów, 1913, 254 pp.

7 J. von Hammer, *Geschichte der Chane der Krim unter osmanischer Herrschaft*, Vienna 1856, 413 pp.

8 A. Kaminski, *Lithuania and the Polish-Lithuanian Commonwealth 1000–1795*, Vol. IV, in A History of E.C. Europe Series (Univ. of California Press), Seattle and London, 1974, 404 pp.; M. Hrushevsky, *A History of the Ukraine*, O. J. Fredrikson (ed.), Yale Univ. Press, 1914, 392 pp.

9 For example see K. Lepszy (ed.), *Polska w okresie drugiej wojny północnej 1655–1660*, Vol. I–III, Warszawa, 1957.

10 F. Nowak, 'The Interregna and Stephen Batory, 1572–1586', Ch. XVIII in *The*

Cambridge History of Poland, Vol. I From the Origins to Sobieski (to 1696), W. F. Reddaway et al. (eds.), Octagon Books, New York, 1978, pp. 369–391.

11 O. Forst de Battaglia, 'Jan Sobieski, 1674–1699', Ch. XXIV in *The Cambridge History of Poland, ... op. cit.*, pp. 532–556; J. Wimmer, 'Jan Sobieski's Art of War, *Historia militaris polonica*, Warszawa, 1977, pp. 10–34.

12 T. Korzon, *Wewnętrzne dzieje Polski za Stanisława Augusta, 1764–1794* (wyd. drugue), Warszawa, 6 Vols., 1898; R. N. Bain, *The Last King of Poland and his Contemporaries* (Methuen) London, 1909, 212 pp. B. Dembiński, 'The Age of Stanislas Augustus and the National Revival', Ch. VI, in *The Cambridge History of Poland ... Vol. II, from Augustus II to Pilsudski (1697–1935)*, Octagon Books ... *op. cit.*, pp. 112–136.

13 For further discussion see M. Michnik & L. Mosler, *Historia Polski do roku 1795*, Warszawa, 1965, 472 pp.; T. Manteuffel (ed.), *Historia Polski*, Warszawa, 1965, Vol. I (do roku 1764) 835 pp.

14 J. von Hammer, *Geschichte der Osmanischen Reiches*, 10 vols., Pest, 1827–1835; J. W. Zinkeisen, *Geschichte des Osmanischen Reiches in Europa*, Gotha 1859, 7 vols.; P. Coles, *The Ottoman Impact on Europe*, London, 1968, 243 pp.; S. Shaw, *History of the Ottoman Empire and Modern Turkey*, Vol. I, Cambridge, 1976, 435 pp.

15 B. Nolde, *La Formation de l'Empire russe, Etudes, Notes et Documents*, Paris 1963, 356 pp.; H. von Staden, *The Land and Government of Muscovy: A Sixteenth-Century Account*, T. Esper (ed.) Standford Univ. Press, California, 1967, 405 pp.; S. von Herberstein, *Description of Moscow and Muscovy 1557*, B. Picard (ed.), London, 1969, especially pp. 61–62, which gives some idea of the formal and stage-managed procedure to which foreign ambassadors were subjected in the sixteenth century.

16 The other three were: first, hostility of the Persians; secondly, the alliance with France; and thirdly, trade and war with Venice. For further discussion see F. W. Carter, *Dubrovnik (Ragusa) – A Classic City-State*, Seminar Press, London and New York, 1972, p. 327.

17 H. İnalcik, 'The Rise of the Ottoman Empire', Ch. 2 in *A History of the Ottoman Empire*, M. A. Cook (ed.) (Cambridge Univ. Press) Cambridge, 1976, p. 50.

18 A. Gieysztor *et al.*, *History of Poland* (P.W.N.), Warszawa, 1968, p. 174.

19 F. W. Carter, *op. cit.*, p. 329.

20 The Poles did not wish to see the Danubian Principalities turned into Ottoman provinces. See C. M. Kortepeter, *Ottoman Imperialism during the Reformation: Europe and Caucasus* (New York Univ. Press), 1972, New York/London, pp. 143–146; Ibid., 'Ottoman Imperial Policy and the Economy of the Black Sea Region in the Sixteenth Century' *Journal of the American Oriental Society*, Vol. 86, No. 2, New York, 1966, p. 86–112.

21 S. M. Solov'yev, *Istoriya Rossii s drevneyshikh vremyon*, Vol. VII, Moscow, 1960, pp. 261–262; M. Kazimierski (ed.), 'Précis de l'Histoire des Khans de Crimée ... ' *Journal Asiatique* (2nd Ser.), Vol. XII, Paris, 1833, p. 428.

22 V. J. Parry, 'The Period of Murad IV, 1617–48', Ch. 5 in *A History of the Ottoman Empire ...*, *op. cit.*, p. 149.

23 O. Forst de Battaglia, 'Jan Sobieski ... ' *op. cit.*, p. 532.

24 A. N. Kurat and J. S. Bromley, 'The Retreat of the Turks 1683–1730', Ch. 7 in *A History of the Ottoman Empire ...*, *op. cit.*, pp. 178–180.

25 H. İnalcik, 'Osmali-Rus rekabetinin menşei ve Don-Volga kanali tesebbüsü (1569), *Turk Tarih Kurumu Basimevi*, No. 46, Ankara, 1948, pp. 54–55.
26 W. Pociecha, 'Zygmyunt (Sigismund), I, 1506–48', Ch. XV in *The Cambridge History of Poland . . ., op. cit.*, Vol. I, pp. 300–321.
27 A. Gieysztor et al., *History of Poland . . ., op. cit.*, p. 172.
28 Ibid., p. 182.
29 J. Pajewski, 'Zygmunt August and the Union of Lublin, 1548–72', Ch. XVII in *The Cambridge History of Poland . . ., op. cit.*, Vol. I, pp. 348–350.
30 R. M. Seltzer, *Jewish People, Jewish Thought: The Jewish Experience in History* (McMillan) London/New York, 1980, p. 476.
31 Ibid.
32 V. G. Kiernan, *State and Society in Europe, 1550–1650* (B. Blackwell), Oxford, 1980, p. 203.
33 F. Nowak, 'The Interregna and Stephen Batory . . .', *op. cit.*, p. 391.
34 A. Gieysztor et al., *History of Poland . . ., op. cit.*, p. 246.
35 W. Konopczyński, 'Early Saxon Period 1697–1733', Ch. I, in Vol. II, *The Cambridge History of Poland, from Augustus II to Piłsudski (1697–1935) . . ., op. cit.*, p. 7.
36 Ibid., p. 12.
37 S. M. Solov'yev, *op. cit.*, Vol. XIII, p. 361; see also I. de Madariaga, *Russia in the Age of Catherine the Great* (Weidenfeld & Nicolson), London, 1981, p. 191.
38 R. M. Seltzer, *op. cit.*, p. 482.
39 S. M. Solov'yev, *op. cit.*, Vol. XIV p. 251; I. de Madariaga, *op. cit.*, p. 204.
40 K. Lutosłanski, *Recueil des actes diplomatiques, traités de documents concernant la Pologne*, Vol. I, Lausanne-Paris, 1918, p. 36, Nos. 30, 34. Russia acquired 93,000 km^2 from Poland.
41 R. Dyboski, *Outlines of Polish History* (G. Allen & Unwin), London, 1931, p. 96.
42 Ibid., p. 97.
43 O. Halecki, *A History of Poland* (J. M. Dent & Sons), London, 1961, p. 137.
44 F. Nowak, 'Sigismund III, 1587–1632', Ch. XXI, in *The Cambridge History of Poland . . . op. cit.*, Vol. I, p. 452.
45 M. Frančić, *Kalendarz dziejów Krakowa* (Wyd. Lit.), Kraków, 1964, p. 72.
46 K. Bąkowski, *Historya Krakowa w zarysie* (Druk, 'Czasu'), Kraków, 1898, p. 76.
47 F. Nowak, 'Sigismund III . . . ' *op. cit.*, p. 454; J. A. Gierowski, 'The International Position of Poland in the Seventeenth and Eighteenth Centuries', Ch. 11 in *A Republic of Nobles: Studies in Polish History to 1864*, J. K. Fedorowicz (ed.) (Cambridge Univ. Press), Cambridge, London, New York, New Rochelle, Melbourne, Sydney, 1982, p. 218.
48 R. A. Kann, *A History of the Habsburg Empire 1526–1918* (Univ. of California Press), Berkeley, Los Angeles, London, 1974, p. 41.
49 F. Nowak, 'Sigismund III . . . ' *op. cit.*, p. 463.
50 R. Dyboski, *op. cit.*, p. 98.
51 The terms of the 6 years' truce signed in Altmark (1629) placed all the Prussian ports under Swedish control, with the exception of Puck, Gdańsk and Königsberg, as well as Livonian territory up to the Dvina river. See A. Gieysztor et al., *History of Poland op. cit.*, p. 221.
52 F. Nowak, 'Sigismund III . . . ' *op. cit.*, p. 474.

53 O. Halecki, *op. cit.*, p. 158; A. Gieysztor, et al., *op. cit.*, p. 247.
54 W. Tomkiewicz, 'The Reign of John Casimir: Part II, 1654–1668', Ch. XXIII (B), in *The Cambridge History of Poland ... op. cit.*, Vol. I, p. 520; J. Wisłocki 'Walka z najazdem szwedzkim w Wielkopolsce w latach 1655–56' *Studia i Materiały do dziejów Wielkopolski i Pomorza*, Vol. II, No. 2, Poznań, 1957, pp. 61–64. R. Frost, *After the Deluge: Poland, Lithuania and the Second Northern War, 1655–60* (C.U.P.), Cambridge 1993, 211 pp.
55 A. Gieysztor et al., *op. cit.*, p. 247.
56 S. Ranotowicz, *Opisanie inkursji Szwedów do Polski i do Krakowa (1655–1657)* (Warszawa dawniej i dziś, No. 11), Warszawa 1958, p. 14; M. Frančić, *op. cit.*, p. 86.
57 L. Sikora, *Szwedzi i Siedmiogrodianie w Krakowie od 1655 do 1657 roku* (Biblioteka Krakowska), No. 39, Warszawa, 1908, p. 5.
58 M. Frančić, *op. cit.*, p. 84; K. Bąkowski, *op. cit.*, p. 99.
59 L. Sikora, *op. cit.*, p. 5.
60 Ibid., p. 41; H. Landberg, 'Finansowanie wojny i zaopatrywanie garnizonów. Szwedzki zarząd okupacyjny w Krakowie podczas wojny polskiej Karola X Gustawa', *Studia i Materiały do Historii Wojskowości*, Vol. 19, No. 2, Warszawa, 1973, pp. 171–216.
61 K. Bąkowski, *op. cit.*, p. 104; Charles X Gustav negotiated the compact of Radnoth (6/XII/1656) with Rákóczi, which announced a partition of Poland among Sweden, Brandenburg, Transylvania, the Cossack Ukraine and the Lithuanian magnate Radziwiłł. W. Tomkiewicz, *op. cit.*, p. 523.
62 S. Ranatowicz, *op. cit.*, p. 21.
63 Biblioteka Ossolińskich Wrocław, *Rękopis* No. 189, p. 350 (Kronika Szymona Starowolskiego). See also S. Świszczowski, 'Mury miejskie Krakowa', *Ochrona Zabytków*, Vol. VIII, Warszawa, 1955, pp. 162, 163; E. Ligęza, *Wodociągi dawnego Krakowa do połowy XVII wieku*' (Kraków dawniej i dziś, No. 18), Kraków, 1971, p. 30.
64 Wojewódzkie Archiwum Państwowe w Krakowie, *Acta iuris supremi arcis Cracoviensis*, Vol. 24, p. 450 in which the king freed damaged houses and other buildings from all taxes and rates.
65 W. Konopczyński, 'Early Saxon Period, 1697–1733', Ch. I in *The Cambridge History of Poland ... op. cit.*, Vol. II, pp. 4–5.
67 O. Halecki, *op. cit.*, p. 181; W. Konopczyński, *op. cit.*, p. 6.
68 K. Bąkowski, *op. cit.*, p. 122; M. Frančić, *op. cit.*, p. 90.
69 A. Górny and K. Piwarski, *Kraków w czasie drugiego najazdu Szwedów na Polskę, 1702–1709*, Kraków, 1932, pp. 25–35.
70 J. Garlicki, J. Kossowski and L. Ludwikowski, *A Guide to Cracow* ('Sport i Turystyka' Publ.), Warszawa, 1969, p. 31; The Polish Diet of 1710 decreed that a part of the contributions exorted from the city, to the amount of half a million Polish florins, should be repaid by the realm, but Cracow received nothing. See L. Lepszy, *Cracow, The Royal Capital of Ancient Poland: Its History and Antiquities*, London, 1912, p. 123.
71 M. Frančić, *op. cit.*, p. 91.
72 Ibid.
73 Ibid.

74 Russian troops under General Lamoth occupied the castle until 26/IX/1716; M. Frančić, *op. cit.*, p. 91.
75 A. Gieysztor et al., *op. cit.*, p. 277.
76 M. S. Anderson, *Europe in the Eighteenth Century 1713–1783* (Longmans), London, 1963, p. 185.
77 L. W. Cowie, *Eighteenth-Century Europe* (G. Bell & Sons Ltd.), London, 1963, p. 163.
78 W. Konopczyński, 'Later Saxon Period, 1733–1763', Ch. II in *The Cambridge History of Poland ... op. cit.*, Vol. II, p. 29.
79 G. Slocombe, *op. cit.*, p. 194.
80 Ibid. p. 195.
81 A. Gieysztor, et al., *op. cit.*, pp. 285–286.
82 M. Frančić, *op. cit.*, p. 92; K. Bąkowski, *op. cit.*, p. 126.
83 R. Dyboski, *op. cit.*, p. 118.
84 K. Bąkowski, *op. cit.*, p. 126.
85 M. Frančić, *op. cit.*, p. 92.
86 J. Garlicki, J. Kossowski and L. Ludwikowski, *op. cit.*, p. 31; G. Slocombe, *op. cit.*, p. 205.
87 The Seven Years War (1756–1763) was a European and colonial and naval war between Great Britain and Prussia on the one side, and France, Austria, Russia, Sweden and Saxony on the other. See E. N. Williams, *The Penguin Dictionary of English and European History, 1485–1789* (Penguin Books), Harmondsworth, 1980, pp. 403–407.
88 D. B. Horn, *Frederick the Great and the Rise of Prussia* (The English Universities Press Ltd.), London, 1964, p. 84.
89 M. Frančić; *op. cit.*, p. 93; K. Bąkowski, *op. cit.*, p. 126; D. Rederowa, 'Lata upadku (od połowy wieku XVII do r. 1775), in *Kraków: Studia nad rozwojem miasta*, J. Dąbrowski (ed.), Wyd. Lit., Kraków, 1957, p. 238.
90 Ibid.
91 Signed 15/II/1763 between Prussia, Austria and Saxony which restored the pre-war situation in which Prussia kept Silesia. Russia made a separate treaty with Prussia in late April 1762. See E. N. Williams *op. cit.*, p. 407.
92 R. E. Dickinson, *The West European City: A Geographical Interpretation* (Routledge & Kegan Paul Ltd.), London, 1951, p. 216.
93 R. H. Osborne, *East-Central Europe: A Geographical Introduction to Seven Socialist States* (Chatto & Windus) London, 1966, p. 270.
94 R. E. H. Mellor, *Eastern Europe: A Geography of the Comecon Countries* (MacMillan Press), London, 1975, p. 153.
95 A. Davies, 'A Study of City Morphology and Historical Geography', *Geography* Vol. 18, Sheffield, 1933, p. 31.
96 *Grand Larousse encyclopédique en dix volumes*, Vol. X, Paris, 1964, p. 690.
97 *Enciclopedia italiana di scienze, lettre ed art*, Vol. 34, Roma, 1937, p. 1019.
98 *Enciclopedia universal ilustrada Europeo-Americana*, Vol. 15, Madrid (no publ. date), p. 1468.
99 Ibid. Vol. 67, p. 116.
100 *The Encyclopedia Americana*, Vol. 16, New York, 1961, p. 537. The 1972 edn gives 1609.

101 Ibid. Vol. 28, p. 716. The 1972 edn. for Warsaw gives 1596, p. 362.
102 *Der grosse Brockhaus, 16 Aufl.* Vol. 6, Wiesbaden 1955, p. 605.
103 Ibid. Vol. 12, Wiesbaden 1957, p. 338. *The Brockhaus Enzyklopädie 20 Aufl.* Wiesbaden 1970, gives 1596 for Cracow (Vol. 10, p. 576) but 1611 for Warsaw, not 1609 (Vol. 20, p. 29).
104 *Svensk uppslagsbok*, Vol. 16, Malmö, 1959, column 1010.
105 Ibid. Vol. 30, Malmö, 1961, column 1088. According to Małecki it fails to mention however, that one of the principle reasons for the move was Sigismund III's desire to have his capital nearer to Sweden; J. M. Małecki, 'Kiedy i dlaczego Kraków przestał być stolicą Polski' *Rocznik Krakowski*, Vol. XLIV, Kraków, 1973, pp. 21–36.
106 *Encyclopaedia Britannica*, Vol. 23, London-Chicago-Geneva-Sydney-Toronto, 1964, p. 373. For Cracow the date of the capital's movement is given as 1609. Ibid., Vol. 6, p. 690.
107 *Bolshaja Sovetskaja Enciklopedija*, Vol. 7, Moskva, 1951, p. 15.
108 Ibid. Moskva, 1953, p. 202.
109 *Słownik Geograficzny Królestwa Polskiego i innych krajów Słowiańskich*, F. Sulimierski, B. Chlebowski and W. Walewski (eds), Vol. X, Warszawa, 1893, p. 36 gives 1596 as the date Warsaw became capital, but Vol. IV, Warszawa, 1883, p. 603 states Cracow was capital until 1610; *Mały słownik historii Polski*, wyd. 4, Warszawa 1967, p. 347, refers to Sigismund III moving his capital to Warsaw in 1596, where the Parliament (Sejm) had been located since 1569. On p. 394 it states that Sigismund III Waza brought the remainder of his capital to Warsaw in 1596; W. Kurkiewicz, A. Tatomir and W. Żurawski, *Tysiąc lat dziejów Polski. Przegląd ważniejszych wydarzeń z historii i kultury*, wyd. 2, Warszawa, 1962, p. 437 gives the exact date when the capital moved to Warsaw, namely 18th March 1596; *Miasta polskie w tysiącleciu*, Vol. 1, Wrocław, 1965, pp. 203, 623 states only that the king moved his residence to Warsaw, but does not mention his capital; J. Garlicki, J. Kossowski and L. Ludwikowski, *op. cit.*, p. 30, quote 'Sigismund III Vasa (1587–1632) after the 1595 fire at Wawel, transferred the court to Warsaw. Although Cracow preserved the title of capital and coronation city . . . '; M. Frančić *op. cit.*, p. 78 gives a precise date, 25 May 1609, when the capital was moved to Warsaw; *Encyklopedia Powszechna P W N* (P.W.N.), Vol. 2, Warszawa, 1974, p. 593, under the entry for Cracow gives 1596 as date of transference. However, Vol. 4, Warszawa, 1976, p. 585 under the section of Warsaw gives the impression of a gradual transfer between 1596 and 1611.
110 *Ukrains'ka Radjans'ka Entsiklopedija*, Vol. 7, Kiev, 1962, p. 324.
111 Ibid., Vol. 2, Kiev, 1960, p. 228.
112 *A Pallas Nagy Lexikona*, Vol. XI, Budapest, 1895, p. 32.
113 Ibid., Vol. XVI, Budapest, 1897, p. 667.
114 *Enciclopedia Română*, C. Diaconovich (ed.), Vol. II, Sibiu, 1900, p. 32.
115 Ibid. Vol. III, Sibiu, 1904, p. 1195.
116 J. U. Niemcewicz, *Dzieje panowania Zygmunta III*, Vol. 1–3, Warszawa, 1819; M. Baliński and T. Lipiński, *Starożytna Polska pod względem historycznym, jeograficznym i statystycznym*, Vol. 1, Warszawa, 1843, p. 419; Ibid., Vol. 2, Warszawa, 1844, p. 45, footnote 1.

117 I. Zagórski, *Daty pobytu królow polskich w Warszawie między rokiem 1526 a 1596*, (Biblioteka Warszawska, Vol. 4), Warszawa, 1847, p. 207; F. M. Sobieszczański, *Rys historyczno-statystyczny wrostu i stanu miasta Warszawy od najdawniejszych czasów aż do 1847 roku*, Warszawa, 1848, p. 25.
118 M. Frančić, *op. cit.*, p. 75.
119 Ł. Gołębiowski, *Opisania historyczno-statystyczne miasta Warszawy*, 2 edn. Warszawa, 1827, pp. 13–14.
120 A. Wejnert, *Starożytności warszawskie*, Vol. 2, Warszawa, 1848, pp. 90–122.
121 K. Bąkowski, *op. cit.*, p. 91; Ibid., *Dzieje Krakowa*, Kraków, 1911, p. 418.
122 F. Nowak, 'Sigismund III ...' *op. cit.*, p. 465.
123 Jan Wielewicki noted 'Discessit Cracovia in Lituaniam Serenissimus Rex 28 Maii nec plus usque ad mortem Cracoviam repetit' quoted in S. Windakiewicz, *Dzieje Wawelu*, Kraków, 1925, pp. 123–4.
124 F. Nowak, 'Sigismund III ...' *op. cit.*, p. 466.
125 J. Lileyko, *A Companion Guide to the Royal Castle in Warsaw* (Interpress Publ.) Warsaw, 1980, pp. 33–35; W. Tomkiewicz, 'Warszawa stolicą Rzeczypospolitej', *Rocznik Warszawski*, Vol. 7, Warszawa, 1966, p. 103.
126 K. Konarski, *Warszawa w pierwszym jej stołecznym okresie*, Warszawa, 1970, p. 11; J. M. Małecki, 'Kiedy i dlaczego Kraków przestał być stolicą Polski', *Rocznik Krakowski*, Vol. XLIV, Kraków, 1973, p. 26.
127 *Volumina Legum* (Wyd. Ohryzki), Petersburg, 1859–1860, Vol. 5, p. 227 (for 1674), p. 672 (for 1683); Vol. 6, p. 192 (for 1710), p. 589 (for 1763).
128 Ibid. Vol. 9, p. 133.
129 M. Tomasini, 'Kiedy Warszawa stała się stolicą Polski ze stanowiska prawnego', *Palestra*, Warszawa, 1931, pp. 70–74; *Volumina Legum, op. cit.*, Vol. 9, Vol. 10.
130 *Volumina Legum, op. cit.*, Vol. 10, p. 172; K. Konarski, *op. cit.*, p. 11.
131 *Volumina Legum, op. cit.*, Vol. 10, pp. 198, 209, 224.
132 Ibid., Vol. 3, p. 945.
133 Ibid., Vol. 5, p. 50.
134 Ibid., Vol. 7, pp. 102–103.
135 *Chambers Twentieth Century Dictionary*, A. M. Macdonald (edt.), Edinburgh, 1974, p. 193.
136 J. Pajewski, 'Zygmunt August and the Union of Lublin, 1548–72', Ch. XVII in *The Cambridge History of Poland ..., op. cit.*, Vol. 1, p. 358; F. Bujak, 'Stolice Polski (Gniezno-Kraków-Warszawa)', in Ibid., *Studia geograficzno-historyczne*, Warszawa, 1925, p. 287.
137 J. Gierowski, 'Rzeczpospolita szlachecka wobec absolutystycznej Europy', *Pamiętnik X Powszechnego Zjazdu Historyków Polskich w Lublinie*, Vol. 3, Warszawa, 1971, p. 116.
138 R. N. Bain, *The Last King of Poland ... op. cit.*; T. Korzon, *Wewnętrzne dzieje Polski za Stanisława Augusta: ... op. cit.*
139 I. de Madariaga, *Russia in the Age of Catherine the Great* (Weidenfeld & Nicolson) London, 1981, p. 197.
140 D. B. Horn, *op. cit.*, pp. 114–115; see also J. Dygdała, 'Opozycja Prus Królewskich wobec cła generalnego w latach 1764–1766', *Zapiski Historyczne*, Vol. 42, No. 2, Warszawa, 1977, pp. 25–44.

141 H. M. Scott, 'France and the Polish Throne, 1763–1764', *Slavonic and East European Review*, Vol. 53, No. 132, London, 1975, p. 371.
142 R. Dyboski, *op. cit.*, p. 123.
143 L. W. Cowie, *op. cit.*, p. 283; J. Krasicka, *Kraków i ziemia krakowska wobec konfederacji barskiej*, Kraków, 1929, pp. 98–99, 147.
144 K. Bąkowski, *op. cit.*, pp. 131–132; J. Garlicki, J. Kossowski and L. Ludwikowski, *op. cit.*, pp. 31–32.
145 W. F. Reddaway, 'The First Partition', Ch. V in *The Cambridge History of Poland ... op. cit.*, Vol. II, p. 88–111.
146 D. B. Horn, *op. cit.*, p. 123.
147 A. Gieysztor et al., *History of Poland, op. cit.*, pp. 329–330.
148 For example see D. B. Horn, *British Public Opinion and the First Partition of Poland* (Oliver & Boyd Publ.) London and Edinburgh, 1945, 356 pp.
149 I. de Madariaga, *op. cit.*, p. 615, footnote 56.
150 P. S. Wandycz, *The Lands of Partitioned Poland, 1795–1918* (A History of East Central Europe Vol. VII), Univ. of Washington Press, Seattle and London, 1974, p. 11.
151 A. Cygielman, 'Cracow' contribution in *Encyclopaedia Judaica*, Vol. 5 (MacMillan), Jerusalem, 1971, column 1032. Until infilling of the former bed of the Old Vistula (Stara Wisła) in the nineteenth century, Kazimierz was physically separated from Cracow. Kazimierz was therefore located on an island surrounded by two arms of the river which joined together again at the foot of Wawel hill.
152 *Miasta polskie w tysiącleciu*, Vol. I, Warszawa, 1965, p. 627.
153 F. Bardel, *Miasto Podgórze, jego powstanie i pierwszych 50 lat istnienia*, Kraków, 1901, p. 17.
154 K. Bąkowski, *op. cit.*, p. 140.
155 J. Louis, *Kupcy krakowscy w epoce przejściowej, 1773–1846* (Kalendarz Krakowski Jozefa Czecha 1883), Kraków, 1883, p. 10.
156 D. B. Horn, *Frederick the Great ..., op. cit.*, p. 124.
157 J. Garlicki, J. Kossowski and L. Ludwikowski, *op. cit.*, p. 32.
158 L. Lepszy, *Cracow, The Royal Capital of Poland: Its History and Antiquities* (T. Fisher Unwin) London, 1912, p. 123.
159 S. Gierszewski, 'Wpływ wojny szwedzko-rosyjskiej 1788–1790 na sytuację w portach pomorskich', *Ars historica Warszawa*, 1976, p. 677–689.
160 W. F. Reddaway, 'The Second Partition', Ch. VII in *The Cambridge History of Poland, op. cit.*, Vol. II, p. 137.
161 R. Dyboski, *op. cit.*, p. 284–285.
162 H. Frankel, *Poland: The Struggle for Power, 1772–1939* (Lindsay Drummond Ltd. Publ.), London, 1946, pp. 28–29.
163 R. H. Lord, *The Second Partition of Poland* (Harvard Historical Studies, Vol. XXIII), Cambridge, Mass., 1916, 231 pp.; Lord Eversley, *The Partitions of Poland* (T. Fisher Unwin, London), 1915, 354 pp.
164 W. F. Reddaway, 'The Second Partition', *op. cit.*, p. 153; I. de Madariaga, *op. cit.*, p. 427–440.
165 A. Whitridge, 'Kościuszko: Polish Champion of American Independence', *History Today*, No. 7, London, 1975, pp. 453–461; M. M. Gardner, *Kościuszko:*

A Biography of the Polish Patriot (Allen & Unwin Publ.), London, 1920, 253 pp.; E. Krakowski, *Histoire de la Pologne. La Nation polonaise devant l'Europe* (Denoël & Steele Publ.), Paris, 1934, pp. 279–304; D. Guerrini, *La Russia, La Polonia e la Scandinavia* (Vallardi Publ.), Milano, 1915, pp. 206–207; R. Grodecki, K. Lepszy and J. Feldman, *Kraków i ziemia krakowska* (Państwowe Wydania Książek Szkolnych), Lwów, 1934, p. 187–196; W. Bortnowski, *O powstaniu kościuszkowskim*, Warszawa, 1951, p. 39–42.

166 A. Gieysztor et al., *History of Poland, op. cit.*, p. 383–390.

167 R. H. Lord, 'The Third Partition of Poland', *The Slavonic and East European Review*, Vol. III, No. 9, London, 1925, p. 483–498; M. Kukiel, 'Kościuszko and the Third Partition', Ch. VIII in *The Cambridge History of Poland, op. cit.*, Vol. II, p. 154–176; I. de Madariaga, *op. cit.*, p. 441–451.

168 M. Frančić, *op. cit.*, p. 105; K. Bąkowski, *op. cit.*, p. 161–163.

169 A. Podraza, 'Początki rządów austriackich w krakowskiem (1772–1795)' *Nauka dla Wszystkich*, No. 55 (P.A.N.), Kraków, 1967, p. 19–20; D. Rederowa, 'Kraków porozbiorowy (1796–1815)', in *Kraków jego dzieje i sztuka*, J. Dąbrowski (ed.) (Wyd. Arkady) Warszawa, 1965, pp. 397–406.

170 K. Lepszy, *op. cit.*, p. 123.

171 Ibid.

172 K. Brodziński, *Wspomnienia mojej młodości*, 1833 (unpubl.), opr. A. Łucki 1910. See entry for K. Brodziński by A. Łucki and I. Chrzanowski, *Bibliotece Narodowej*, ser. I, No. 113, Warszawa, 1935, pp. 17–18.

7 The commerce of Cracow, 1500–1795

1 W. Pociecha, 'Zygmunt (Sigismud) I, 1506–1548', Ch. XV in *The Cambridge History of Poland from the Origins to Sobieski (to 1696)*, W. F. Reddaway et al. (eds.) Octagon Books, New York, 1978, pp. 300–321.

2 J. Rutkowski, 'The Social and Economic Structure in the Fifteenth and Sixteenth Centuries', Ch. XX(B) in *The Cambridge History of Poland ... op. cit.*, pp. 441–450.

3 For a wider discussion see H. R. Trevor-Roper, 'The General Crisis of the Seventeenth Century', *Past and Present*, No. 16, London, 1959, pp. 31–64.

4 J. M. Małecki, 'Kiedy i dlaczego Kraków przestał być stolicą Polski', *Rocznik Krakowski*, Vol. XLIV, Kraków, pp. 21–36.

5 W. Rusiński, 'The Role of Polish Territories in the European Trade in the Seventeenth and Eighteenth Centuries', *Studia Historiae Oeconomica*, No. 3, Poznań, 1969, pp. 115–134; Z. P. Pach, 'The Shifting of International Trade Routes in the 15th–17th Centuries', *Acta Historica Scientiarum Hungarica*, Vol. 14, Budapest, 1968, pp. 287–319.

6 More general discussion may be found in J. Topolski, 'Economic Decline in Poland from the Sixteenth to Eighteenth Centuries', in *Essays in European Economic History 1500–1800*, P. Earle (ed.), Clarendon Press, Oxford 1974, pp. 127–142.

7 C. H. H. Wake, 'The Changing Pattern of Europe's Pepper and Spice Imports, ca. 1400–1700', *Journal of European Economic History*, Vol. 8, No. 2, Rome, 1970, pp. 361–404; J. L. Azevedo, *Épocas de Portugal económico; esboças de*

historia, 2nd edn, Lisbon, 1947, pp. 90–110; D. Lach, *Asia in the Making of Europe*, Vol. 1 (The Century of Discovery), Chicago 1965, pp. 140–160; V. Magalhaes-Godinho, 'Crises et changements géographiques et structuraux au XVIe siècle', *Studi in Onore di Armando Sapori*, Vol. II, Milan 1957, pp. 981–998; Ibid., *L'Economie de l'empire portugais aux XVe et XVIe siècles*, Paris, 1969, pp. 72–735.

8 For further discussion see F. W. Carter, *Dubrovnik (Ragusa) – A Classic City-State* (Academic Press), London-New York, 1972, pp. 349–354.

9 P. Jeannin, 'Anvers et la Baltique au XVIe siècle', *Revue du Nord*, Vol. 37, No. 146, Lille 1955, pp. 93–114; Ibid., 'Les relations économiques des villes de la Baltique avec Anvers au XVIe siècle', *Virteljahrschrift für Sozial-und Wirtschaftsgeschichte*, Vol. XLIII, 1956, Leipzig, pp. 193–217.

10 M. Bogucka, 'Amsterdam and the Baltic in the First Half of the Seventeenth Century', *Economic History Review*, Vol. 26, London, 1972, pp. 433–447.

11 S. Kutrzeba, *Handel Polski ze Wschodem w wiekach średnich*, Kraków 1903, pp. 125–126; M. Berindei, 'Contribution a l'étude du commerce Ottoman des fourres Moscovites. La route moldavo-polonaise 1453–1700', *Cahiers du Monde russe et soviétique*, Vol. XII, No. 4, Paris-The Hague-New York, 1971, p. 398.

12 W. Łozinski, *Patrycyat i mieszczaństwo lwowskie w XVI i XVII wieku*, Lwów, 1892, pp. 36–37.

13 J. Rutkowski, 'The Social and Economic Structure in the Fifteenth and Sixteenth Centuries', Ch. XX(B), *The Cambridge History of Poland* ... *op. cit.*, p. 447.

14 M. Małowist, 'Poland, Russia and Western Trade in the Fifteenth and Sixteenth Centuries', *Past and Present*, Vol. 13, London 1958, p. 30.

15 Archiwum Główne Akt Dawnych w Warszawie, Archiwum Koronne, *Dzial Turecki*, Vol. 36, No. 80, contains numerous letters concerning the trade activities of A. Halkokondil, a Greek merchant dealing in spices etc. at the end of the sixteenth century; see also A. Dziubiński, 'Drogi handlowe polsko-tureckie w XVI stuleciu', *Przegląd Historyczny*, Vol. LVI, No. 2, Warszawa 1965, p. 235; E. Nadel-Golobić, 'Armenians and Jews in Medieval Lvov: Their Role in Oriental Trade', *Cahiers de Monde russe et soviétique*, Vol. XX, Nos. 3–4, Paris-The Hague-New York, 1979, pp. 345–388.

16 M. Dan and S. Goldenburg, 'Le commerce balkano-levantin de la Transylvanie au course de la seconde moitié du XVIe siècle et début de XVIIe siècle', *Revue des études sud-est européenes*, Vol. 5, No. 1–2, Bucureşti 1967, p. 114; N. Iorga 'Acte privitoare la comerţul românesc cu Lembergul', *Studii şi documente*, Vol. XXIII, Bucureşti 1913, pp. 328–399, 410–454.

17 The voyage from Goa to Lisbon took five months. See G. Luzzato *Storia economica dell'eta moderna e contemporanea*, Padova 1950, p. 160.

18 L. Koczy, *Handel Poznania do polowy XVI wieku*, Poznań 1930, p. 48.

19 For example in 1511 King Zygmunt I forbad the import of spices to Poland from Silesia. As a result Armenian merchants brought them in from Turkey. M. Bielski, *Kronika Polska*, Vol. II, Sanok, 1856, p. 961; in 1534 Turkish merchants came to Poland with merchandise, *Biblioteka Czartoryskich w Krakowie*, Vol. IV, No. 611, pp. 107–108; in 1536 a Turkish merchant named Said visited Cracow with merchandise including spices, *Biblioteka Czartoryskich w Krakowie*, Vol. IV, No. 611, p. 125. See also C. Georgieva 'Les rapports de commerce entre

l'empire Ottoman et la Pologne et les terres Bulgares en XVIe siècle', *Bulgarian Historical Review*, Vol. 6, No. 3, Sofia 1978, pp. 38–51.

20 E. Nadel-Golobić, *op. cit.*, p. 355.

21 D. Quirini-Popławska, 'Die italienische Einwandere in Krakau und ihr Einfluss auf die polnischen Wirtschaftbeziehungen zu österreichischen und deutschen Städten im 16 Jahrhundert', *Wissenschaftliche Zeitschrift de Friedrich Schiller Universität*, Iena, Gesellschaft-und Sprachwissen-schaftliche Reihe 26, 1977, pp. 337–354.

22 H. Samsonowicz, 'Handel Lublina na przełomie XV i XVI wieku', *Przegląd Historyczny*, Vol. 59, No. 4, Warszawa 1969, p. 622.

23 J. Małecki, 'Handel zewnętrzny Krakowa w XVI wieku', *Zeszyty Naukowe Wyższej Szkoły Ekonomicznej w Krakowie*, No. 11 (Prace z zakresu historii gospodarczej), Kraków, 1960, pp. 104–105.

24 Wojewódzkie Archiwum Państwowe w Krakowie, *Księgi celne*, rkp. 2116 (1593), fol. 25, 82.

25 R. Rybarski, *Handel i polityka handlowa Polski w XVI wieku*, Vol. 2, Poznań, 1929, pp. 185–187; 202–203.

26 H. Kellenbenz, 'Der Pfeffermarkt und die Hansestädte', *Hansische Geschichtsblätter*, Vol. 74, Lübeck, 1956.

27 S. Kutrzeba 'Wisła w historji gospodarczej dawnej Rzeczpospolitej Polskiej', *Geografia Wisły*, Vol. XI, Warszawa, 1921, pp. 43–44 in which analysis is made of the customs books (*Księgi celne/Register Thelonei*) for the port of Włocławek in the lower Vistula, in the sixteenth century.

28 *Danziger Inventar*, P. Simons (ed.), Munich-Leipzig, 1913, Doc. no. 1070.

29 Wojewódzkie Archiwum Państwowe w Gdańsku', *Missiva* 300, 27/18, K.181 v–182.

30 J. M. Małecki, *Związki handlowe miast polskich z Gdańskiem w XVI pierwszej połowie XVII wieku* P.A.N. Odd. w Krakowie (Prace Komisji Nauk Historycznych No. 20) Wrocław-Warszawa-Kraków, 1968, pp. 150–151. See also H. Samsonowicz, 'Relations commerciales entre la Baltique et la Mediterranée aux XVIe et XVIIe siècles: Gdańsk et l'Italie', *Mélanges en L'Honneur de Ferdinand Braudel (Histoire Économique de Monde Mediterranéen 1450–1650)*, Vol. I, Toulouse 1973, pp. 537–545.

31 Sugar worth 40 grzywnas sent from Gdańsk to Cracow in 1538/1539. R. Rybarski, *op. cit.*, p. 185.

32 J. M. Małecki, *Związki handlowe* ... *op. cit.*, p. 152.

33 P. Jeannin, 'The Sea Borne and the Overland Trade Routes to Northern Europe in the XVIth and XVIIth Centuries', *Journal of European Economic History*, Vol. 11, No. 1, Rome 1982, pp. 49–50.

34 J. M. Małecki, *Związki handlowe* ... *op. cit.*, Table XVIII c.d. and A (Pt. 3), c.d., pp. 239–240.

35 H. Kellenbenz, 'Autour de 1600: Le commerce du poivre des Fugger et le marché international du poivre', *Annales Économies-Sociétés-Civilisations*, Vol. XI, Paris, 1956, p. 7.

36 J. N. Ball, *Merchants and Merchandise: The Expansion of Trade in Europe 1500–1630*, Croom Helm, London, 1977, pp. 146–147; M. Bogucka, 'Handel Gdańska z Półwyspem Iberyjskim w pierwszej połowie XVII wieku', *Przegląd Histo-*

ryczny, Vol. 60, No. 1, Warszawa, 1969, pp. 8–9; V. Barbour, 'Dutch and English Merchant Shipping in the Seventeenth Century', *Economic History Review*, Vol. II, London 1930, pp. 261–290.

37 J. M. Małecki, *Związki handlowe* ... *op. cit.*, Table XVIII c.d. and A (Pt. 3), c.s. pp. 239–240; see also A. Manikowski, 'Zmiany czy stagnacja? Z problematyki handlu polskiego w drugiej połowie XVII wieku', *Przegląd Historyczny*, Vol. 64, No. 4, Warszawa, 1973, pp. 771–791 which gives some comparisons for the second half of the seventeenth century.

38 J. Michalewicz, 'Z badań nad konsumpcją w Polsce. Kuchnia królewska Zygmunta III', *Kwartalnik Historii Kultury Materialnej*, Vol. XIII, No. 4, Warszawa, 1965, p. 707.

39 J. M. Małecki, *Związki handlowe* ... *op. cit.*, Table XVIII c.d. and A (Pt. 3), c.d., pp. 239–240.

40 H. Obuchowska-Pysiowa, 'Trade between Cracow and Italy from the Customshouse Registers of 1604', *Journal of European Economic History*, Vol. 9, No. 3, Rome 1980, p. 648.

41 J. Tazbir, 'Konsumpcja cytrusów w Polsce XV–XVIII wieku', in *Pożywienie w dawnej Polsce* (Studia i materiały z historii kultury materialnej), Warszawa, 1967, pp. 105, 115.

42 H. Obuchowska-Pysiowa, *op. cit.*, p. 648; J. Louis *Kupcy Krakowscy w epoce przejściowej 1773–1846* (Kalendarz Krakowski Jozefa Czecha 1883), Kraków 1883, p. 10 refers to Jews from Italy who first came to Cracow to sell spices, *circa* 1612.

43 H. Obuchowska-Pysiowa, 'Struktura handlu Krakowa z krajami południowowschodnimi i z państwem moskiewskim w 1604 r.', *Roczniki Dziejów Społecznych i Gospodarczych*, Vol. XXXVIII, Warszawa-Poznań, 1977, p. 97.

44 Arhivele Statului Sibiu. *Zwanzig-und Dreisig Rechnungen*, cutia XXVII, No. 1–16.

45 Ibid. No. 5 II (1673); Ibid., No. 10 II (1682); Ibid. (1683–85); Ibid. (1687).

46 Ibid. No. 1–16.

47 G. Netta, *Curs de istoria comerţului*, Bucureşti 1937, p. 469.

48 Ibid. See also F. Marinescu 'The Trade of Wallachia with the Ottoman Empire 1791 and 1821', *Balkan Studies*, Vol. 22, No. 2, Thessaloniki, 1981, p. 287.

49 M. Kulczykowski, *Kraków jako ośrodek towarowy Małopolski zachodniej w drugiej połowie XVIII wieku*, Warszawa (P.W.N.), 1963, 173 pp.; J. M. Małecki, 'Le Role de Cracovie dans l'Économie Polonaise aux XVIe, XVIIe, et XVIIIe siècles', *Acta Poloniae Historica*, Vol. 21, Wrocław-Warszawa-Kraków, 1970, pp. 108–122.

50 Wojewódzkie Archiwum Państwowe w Krakowie, *Księgi kupieckie*, nr. dawny 1596, dated 24/IV/1732, refers to 'Inwentarz korzennych towarów różnych po nieboszczce Jey Mci pani Franciszce Wilhejmowey ex post Dziergninowey radczyny Krakowskiey tak w sukienniach iako też w kamienicznym sklepie y po inszych mieyscach dla konserwacyi bedących. Przezemnie nizey podpisanego eo nomine a nobili ac spectabili magistratu Cracoviensi d. 24 Aprilis anno dni 1732 deputowanego spisany', with reference to spices bought in Gdańsk; see also E. Cieślak, 'Sea-borne trade between France and Poland in the XVIIIth Century', *Journal of European Economic History*, Vol. 6, No. 1, Rome 1977, pp. 55–56;

W. S. Unger, 'Trade through the Sound in the Seventeenth and Eighteenth Centuries', *Economic History Review* (2nd Series), Vol. 12, 1959, p. 212; A. R. Disney, *Twilight of the Pepper Empire. Portuguese Trade in South-west India in the Early Seventeenth Century* (Harvard Historical Studies 95), Cambridge, Mass. and London, 1978, 220 pp.

51 K. Kuklińska, 'Commercial Expansion in XVIIth Century Poland: The Case of Poznań', *Journal of European Economic History*, Vol. 6, No. 2, Rome 1977, pp. 443–460.

52 Wojewódzkie Archiwum Państwowe w Krakowie, *Księgi kupieckie* nr. dawny 29 49, dated 1725 states that an unknown Cracow merchant had obtained from Gottfryd Oberman from Wrocław, ginger, violet, spices, anchovies, liquorice cake, herbal tea, etc. Similarly Hilzer Samuel received from Wrocław, olives from Seville, pepper from Holland, sugar, cinnamon, saffron, ginger, caraway seeds, capers, almonds, cloves etc.

53 Z. P. Pach, 'Le commerce du Levant et la Hongrie au XVIe siècle', in *La Pologne et la Hongrie aux XVIe–XVIIIe siècles*, V. Zimanyi (ed.), Akadémiai Kiadó, Budapest, 1981, p. 51; Ibid., 'The Shifting of International Trade Routes . . .' *op. cit.*, pp. 287–319.

54 G. Granasztói, 'La ville de Kassa dans le commerce Hungaro-Polonais au XVIe siècle', in *La Pologne et la Hongrie ... op. cit.*, p. 61; see also P. Horváth, 'Príspevok k obchodńym stykom Slovenska so Sliezskom a Moravou v prvnéj polovici 16 stor', *Historické Studie*, Vol. 11, Bratislava, 1966, p. 191.

55 K. Pieradzka, 'Handel Krakowa z Węgrami w XVI w.', *Biblioteka Krakowska*, No. 87, Kraków, 1935, pp. 244–245.

56 Wojewódzkie Archiwum Państwowe w Krakowie, *Księgi celne*, rkp. No. 2142 (dated 1624), and No. 2143 (dated 1629). See also R. Fišer, 'Orientace a skladba zahraničního obchodu Levoče v první pol. 17 století', *Historický Časopis*, Vol. XXI, No. 3, Bratislava 1973, p. 436; F. Hejl, 'Východoslovenská města a jejich místo ve struktuře středoevropských obchodních vztahů v období rozvitého a pozdního feudalismu', *Historický Časopis*, Vol. XXI, No. 3, Bratislava, 1973, p. 404.

57 F. Hejl, 'Český obchod na Krakovském trhu po Bílé Hoře', *Sborník Prace Filosofické Fakulty Brneňské University*, Vol. 10 (řada historická No. 8), Brno, 1961, p. 242.

58 Ibid.

59 Ibid.

60 M. P. Dan, 'Le commerce de la Transylvanie avec la Pologne au XVI siècle', *Revue Roumaine d'Histoire*, Vol. 8, No. 3, Bucureşti, 1969, pp. 621; 634; Ibid. 'Din relaţiile comerciale ale Transilvaniei cu Slovacia în sec XVI. relaţiile cu Levoča', *Studia Universitatis Babeš-Bolyai* (Ser. Historia) Vol. 14, No. 1, Cluj, 1969, pp. 17–28; Ibid. 'Negustorii clujeni la Cracovia in ultimul deceniu al secolului al XVI-lea', *Acta Musei Napocensis*, Vol. 8, 1971, Cluj, pp. 205–217; Ibid. and S., Goldenburg, 'Le commerce balkano-levantin de la Transylvanie au cours de la seconde moitié du XVIe siècle et au début du XVIIe siècle', *Revue des études sud-est européenes*, Vol. 5, No. 1–2, Bucureşti, 1967, p. 28; Ibid., 'Marchands Balkaniques et Levantins dans le commerce de Transylvanie aux XVIe et XVIIe siècles', in *Actes du Premier Congrès International des Études Balkaniques*

et Sud-Est Européenes, Vol. III, Sofia 1969, p. 647; S. Panova, Targovski vrazki meždu balgarskite zemi i Transylvanija prez XVII v.', *Istoričeski Pregled*, Vol. 32, No. 2, Sofia, 1976, p. 53.

61 L. Demeny 'Comerţul de tranzit spre Polonia prin Ţara Românească si Transilvania (ultimul sfert al secolului al XVII-lea), *Studii Revista de Istorie*, Vol. 22, No. 3, Bucureşti, 1969, p. 474; F. Pap, 'Rute comerciale şi localităţi Transilvane în comerţul clujului cu produse agricole şi vite (prima jumătate a sec. XVII), *Acta Musei Napocensis*, Vol. XV, Cluj, 1978, pp. 352–354.

62 M. P. Dan, 'Relaţiile comerciale dintre Oradea şi Cracovia la sfîrşitul secolului al XVI-lea', *Crisia*, No. 3, Bucureşti, 1973, p. 176.

63 H. G. Koenigsberger and G. L. Mosse, *Europe in the Sixteenth Century* (General Europe Series), Longmans, London 1979, pp. 22–23.

64 F. Braudel and F. C. Spooner, 'Prices in Europe from 1450–1750', Ch. VII in *The Cambridge History of Europe*, Vol. IV, Cambridge 1967, pp. 374–486.

65 J. Dąbrowski, 'Consequenze economiche delle scoperte geografiche nel territorio dal Baltico al Mar Mero', in *La Pologne au VIIe Congrès International des Sciences Historiques, Varsovie, 1933*, Vol. 3, Warszawa 1933, pp. 9–14.

66 C. H. H. Wake, *op. cit.*, p. 390.

67 S. Hoszowski, 'L'Europe centrale devant la révolution des prix aux XVIe et XVIIe siècles', *Annales: Économies, Sociétés Civilisations*, Vol. 16, No. 1, Paris 1961, p. 451; Ibid., 'The Revolution of Prices in Poland in the 16th and 17th Centuries' *Acta Poloniae Historica* Vol. II (P.A.N.: Instytut Historii) Warszawa 1959, pp. 7–16.

68 E. Tomaszewski, *Ceny w Krakowie w latach 1601–1795*, Lwów 1934, pp. 66–7.

69 M. Hroch, 'Die Rolle des Zentraleuropäischen Handels im Ausgleich der Handelsbilanz zwischen Ost- und Westeuropa 1550–1650', in *Der Aussenhandel Ostmitteleuropas 1450–1650*, I, Bog (ed.), Köln-Wien, 1971, pp. 12–14.

70 D. H. Pennington, *Seventeenth Century Europe* (General History of Europe Series), Longmans, 1980, p. 77; see also M. Bogucka, 'The Monetary Crisis of the XVIIth Century and Its Social and Psychological Consequences in Poland', *Journal of European Economic History*, Vol. 4, No. 1, Rome 1975, pp. 137–52.

71 G. D. Hundert, 'Jews, Money and Society in the Seventeenth-Century Polish Commonwealth: The Case of Kraków', *Jewish Social Studies*, Vol. 43, New York, 1981, p. 269.

72 A. Keckowa, 'Polish Salt-Mines as a State Enterprise XIIIth–XVIIIth Centuries', *Journal of European Economic History*, Vol. 10, No. 3, Rome 1981, p. 619.

73 G. Otruba, 'Quantitative Aspekte der Salzproduktion in der österreichischen Reichshälfte unter besonderer Berücksichtigung der alpinen Salinen im 19 Jahrhundert', in *Österreichisches Montanwesen: Produktion, Verteilung, Sozialformen*, M. Mitterauer (ed.), Vienna 1974, p. 29.

74 A. Keckowa, *op. cit.*, p. 626; see also the important study of R. Rybarski, *Wielickie żupy solne w latach 1497–1594*, Warszawa 1932, Table 1, p. 175.

75 I. Pajdak, 'Udział Wieliczan w handlu solą (do 1772 r.), *Studia i materiały do dziejów żup solnych w Polsce*, Vol. VIII, Kraków/Wieliczka, 1979, pp. 109–24.

76 H. Łabęcki, *Górnictwo w Polsce*, Vol. I, Warszawa 1841, p. 141.

77 J. Wyrozumski, *Państwowa gospodarka solna w Polsce do schyłku XIV wieku, Kraków 1968*, pp. 83–4, 93–4.

78 A. Keckowa, *op. cit.*, p. 623.

79 J. Olkiewićz, *Opowieści o Włochach i Polakach* (Lud. Społ. Wyd.), Warszawa, 1979, p. 45.

80 J. Siemienski, 'Constitutional Conditions in the Fifteenth and Sixteenth Centuries, Ch. XX(A) in *The Cambridge History of Poland ... op. cit.*, p. 428.

81 M. Bogucka, 'Polish Towns Between the Sixteenth and Seventeenth Centuries', Ch. 7, in *A Republic of Nobles: Studies in Polish History to 1864*, J. K. Fedorowicz (ed.), Cambridge Univ. Press, Cambridge, London, New-York, Rochelle, Melbourne-Sydney, 1982, p. 146.

82 D. Quirini-Popławska, 'Żupnicy krakowscy Lorenzo Justimonti i Giovanni Battista Cecci', *Prace Historyczne*, Vol. 56, Warszawa-Kraków 1977, pp. 150, 152.

83 D. Quirini-Popławska, 'Z działalności, Włochów w Polsce w I połowie XVI wieku. Nowe szczegóły o Guccich i Thedaldich', *Studia Historyczne*, Vol. XII, No. 2, Kraków 1969, p. 180.

84 *Acta Tomiciana, (Cracov et Plocen, Canonicum, ejusdem Petri Tomicii, post, Serenissime bone sfforcie, regine Poloniae, Secretarium, collecte, et in tomos XXVI digeste)*, S. Górski (ed.), Vol. III Poznań, 1853, pp. 22, 108, 150, 275.

85 For example see, J. J. Bouquet, 'Le problème du sel au pays de Vaud jusqu'au début du XVIIe siècle', *Schweizerische Zeitschrift für Geschichte*, Vol. 7, Zurich 1957, pp. 289–344; R. Racour, 'Traffic du sel en Valais et rivalité franco-espagnole', *Cahiers d'Histoire*, Vol. 12, Grenoble 1967, pp. 283–288.

86 Wojewódzkie Archiwum Państwowe w Poznaniu, *Biblioteka Raczyńskich*, No. 36, document dated 1527, Salt, 410 bałwans (i.e. 44 tons) sent from Wieliczka to Warsaw. Transport cost 48 grosz/bałwan (1,075 kg). When various other costs had been added e.g. loaders/unloaders, person in charge of salt movement ('hutman'), etc., total transport costs to Vistula and by boat to Warsaw was 882 złotys 14 grosz, or 2 złotys 15 grosz/bałwan. In Wieliczka a bałwan cost 4 złotys 4 grosz. Therefore, salt movement to Warsaw equalled 49 per cent of original price/bałwan.

87 Archiwum Glówne Akt Dawnych w Warszawie, *Akta Metryki Koronnej*, v. 77; f. 91; dated 1549, with reference to one Cracow merchant, Valentinus (Foltyn) Konrad who received permission from king to send salt from Cracow to Warsaw and Płock; similar example for 1540 by same merchant. See Archiwum Główne Akt Dawnych w Warszawie, *Salinarum Vieliciensium*, No. 39; for general discussion see A. Keckowa, 'Królewskie składy soli w Warszawie w XVI i pierwszej połowie XVII wieku', *Kwartalnik Historii Kultury Materialnej*, Vol. XXV, No. 4, Warszawa 1977, pp. 507–27.

88 A. Keckowa, 'Frochtarze żup krakowskich w XVI-XVIII wieku', *Studia Historyczne* (Księga jubileuszowa z okazji 70. rocznicy urodzin prof. dra. Stanisława Arnolda), Warszawa 1965, pp. 166–77.

89 R. Rybarski, *op. cit.*, p. 97.

90 Ibid., In 1587/8, 2,639 bałwans of salt sent from Wieliczka to Kazimierz nad Wisłą and only 540 (i.e. 17 per cent) to Stryjów on the land route. The average cost of sending a bałwan to Kazimierz nad Wisłą was 17 grosz cheaper than to Stryjów.

91 B. Jewsiewicki, 'Tradycyjny transport włóczny w Karpatach polskich w sezonie

bezśnieżnym', *Kwartalnik Historii Kultury Materialnej*, Vol. XIII, No. 2, Warszawa 1965, pp. 337–50.

92 H. Samsonowicz, 'Struktura handlu gdańskiego w pierwszej połowie XV wieku', *Przegląd Historyczny* Vol. LIII, No. 4, Warszawa 1962, pp. 695–713; ibid., 'Handel zagraniczny Gdańska w drugiej połowie XV wieku', *Przegląd Historyczny*, Vol. XLVII, No. 2, Warszawa 1956, pp. 283–352; ibid., 'Le commerce maritime de Gdańsk dans la première moitié du XVI siècle', *Studia Historiczne Oeconomicae*, Vol. 9, Poznań 1974, pp. 147–65; M. Bogucka, 'Handel Gdańska z Półwyspem Iberyjskim...' *op. cit.*, pp. 1–23; Ibid., *Handel zagraniczny Gdańska w pierwszej połowie XVII wieku*, Wrocław 1970, pp. 45–7; ibid., 'Sól w handlu Bałtyckim w pierwszej połowie XVII wieku', *Zapiski Historyczne*, Vol. XXXVI, No. 1, Warszawa 1971, pp. 101–10.

93 *Volumina legum*, J. Ochryzko (ed.), St Petersburg 1859–60, Vol. I, pp. 265, 368, 395, 540; Vol. II, pp. 947, 1276; Vol. IV, p. 502.

94 A. Keckowa, 'Polish Salt-Mines...' *op. cit.*, p. 628.

95 Ibid.

96 L. Finkel, 'Opis Rzeczypospolitej Polskiej z roku 1574-go, według relacyji Ludwika Gonzagi, ks. de Nevers i Rhétel', *Biblioteka Warszawska*, Vol. IV, Warszawa, 1887, pp. 380, 381.

97 J. Morawski, 'La Pologne vue par deux voyageurs français du XVIe siècle', *Le Pologne. Revue mensuelle*, Vol. XIII, No. 4, Paris 1932, pp. 706–7. See also, M. Komaszyński 'Krakowskie żupy solne w relacjach Francuzów XVI i SVII wieku', *Studia Historyczne*, Vol. XVI, No. 3 (62), Kraków 1973, pp. 323–42.

98 Biblioteka Czartoryskich w Krakowie rkp. 1021; J. Wyrozumski, 'Warzelnie soli krakowskiej na pograniczu śląsko-polskim w drugiej połowie XVI i pierwszej połowie XVII wieku', *Prace Historyczny*, Vol. 4 (Historia Vo. 4), Kraków, 1960, pp. 33–62.

99 Wojwódzkie Archiwum Państwowe w Krakowie, *Księgi celne*, rkp. 2116 dated (1592) as a typical example.

100 This included places like Strzelce Opolskie, Ostrożnica, Olesńo, Głogówek, and Lubniewice.

101 Wojewódzkie Archiwum Państwowe w Krakowie, *Księgi celne*, rkp. 2116, 18 v.; P. Horváth, 'Príspevok k obchodným stykom...' *op. cit.*, p. 186; F. Matějek, 'K dejinám českého obchodū s oblastí Haličsko-Karpatskou koncem 16 století. Krakov jako transitní středisko českého zboží', *Slovanské Historické Studie*, Vol. 3, Praha 1960 , p. 204.

102 Státní Ústřední Archiv v Praze, *rkp. R.G. 60*, pp. 105–6.

103 B. Iványi, 'Két középkori sóbánya-statutum', *Századok*, Vol. XLV, Budapest 1911, p. 187.

104 Ibid., p. 189.

105 L. Lasztókay, *Eperjes szabad királyi város levéltarában található nevezetesebb okiratok ismertelése*, Eperjes (Prešov), 1881, pp. 11, 50.

106 B. Iványi, *op. cit.*, p. 188.

107 L. Lasztókay, *op. cit.*, p. 50.

108 M. Horváth, *Az ipar és kereskedés története Magyarország a három u tolsó század alatt*, Vol. III, Budapest 1868, p. 89. Three Slovakian towns (Orava, Liptov and Turiec) had already received permission to import Polish salt in

1548; *Codex Iuris Hungarici,* Vol. I, Budapest 1822, art. 29 (dated 1548), p. 406. See also V. Černý, 'Polská sůl na Oravě', *Rocznik Dziejów Społecznych i Gospodarczych,* Vol. III, Lwów 1934, p. 150.

109 K. Pieradzka, 'Handel Krakowa z Węgrami ...' *op. cit.,* p. 228.

110 A. Ilieş, 'Drumurile şi transportul sării în Ţara Românească secolele XV–XIX, *Studii şi Materiale de Istorie Medie,* Vol. 7, Bucareşti 1974, pp. 223–42.

111 J. M. Małecki, 'Handel zewnętrzny Krakowa ...' *op. cit.,* p. 100.

112 K. Pieradzka, 'Handel Krakowa z Węgrami ...' *op. cit.,* p. 229; see also R. Rybarski, *Wielickie żupy solne ... op. cit.,* p. 154.

113 K. Baran, *Prawa i przywileje królewskiego wolnego miasta Nowego Targu,* Kraków 1980, No. 19.

114 In 1582 a total of 87 cetnars (i.e. 5½ tons) of salt sent to Levoča; in 1595 this reached 142 cetnars (i.e. 9¼ tons). P. Horváth, 'Obchodné styky Levoče ...' *op. cit.,* p. 131.

115 Státní Ústřední Archiv v Levoče, *Arch. Komitety,* arch. 13 miast I No. 24.

116 J. Le Laboureur, *Relation du voyage de la Royaume de Pologne et du retour de Madame la Mareschalle de Guébriant, ambassadrice extraordinaire ...,* Paris 1647, Pt. III, pp. 44–6; see also 'Wypis ostatni z podróży Pani de Guébriant posłowey nadzwyczayney do Polski za Władysława IV-go', *Nowy Pamiętnik Warszawski,* Sept. 1801, pp. 285–7; M. Komaszyński, *op. cit.,* p. 337.

117 In 1606 Bohemia imported a total of 43 bałwans (i.e. 4½ tons) of salt of which 28 bałwans (*circa* 3 tons) bought from Poles in Jablonec nad Nisou. Similar cases for years 1617, 1618, 1628. V. Černý, 'Polská sůl na Oravě ...' *op. cit.,* p. 156. See also M. Marečková, *Dálkový obchod Prešova v prvých třech deseitletích 17 století a Krakovsky trh,* Brno 1971, 180 pp.

118 V. Černý, 'Polska sůl ...' *op. cit.,* pp. 156, 161.

119 Wojewódzkie Archiwum Państwowe w Krakowie, *Księgi celne,* rkp. 2142 (dated 1624). During 1624 salt is mentioned as imported from Wieliczka, 18 times by carters from Opava, 50 times for Jičín, and 5 times for Ostrava. See also F. Hejl, 'Český obchod na Krakovskem trhu ...' *op. cit.,* p. 241.

120 For example 1659 reference to annually buying large quantities of Polish salt 'Salis lapidei maioris polonici', sent to the border town of Jablonec nad Nisou, and from there sent further on to the whole of Hungary. V. Černý, *op. cit.,* p. 164; further example in 1681 on the import of Polish salt. *Corpus Iuris Hungarici,* Vol. II, Budapest 1899, art. 15, p. 146. Another example dated 11/II/1660. Archiv v Oravského Podzámku, *sig. 86*/1,2,3.

121 'Northern Poland used salt imported from France, Germany and Sweden', J. Rutkowski, 'The Social and Economic Structure ...', *op. cit.,* p. 448; see also W. S. Unger, 'Trade through the Sound in the 17th and 18th Centuries', *Economic History Review* (2nd Series), Vol. 12, London, 1959, pp. 207–8; M. Hroch, 'Obchod mezi východní a zapadní evropou v období počátků kapitalizmu', *Československý Časopis Historický,* Vol. 11, No. 4, Praha 1963, pp. 488–9; M. Bogucka, 'Handel Gdańska z półwyspem Iberyjskim ...', *op. cit.,* pp. 8–9; ibid., 'Sól w handlu Bałtyckim ...', *op. cit.,* pp. 101–10; ibid., 'Le sel sur le marché de Gdańsk au cours de la première moitié du XVIIe siècle', *Studia Historiae Oeconomicae,* Vol. 11, Poznań 1976, pp. 57–69.

122 A. Keckowa, 'Polish Salt-Mines ...', *op. cit.,* p. 626; this is also reflected in the

small amount of research undertaken on the salt trade in the eighteenth century. See A. Keckowa, 'Przegląd literatury dotyczący dziejów górnictwa soli w Polsce', *Studia z dziejów górnictwa i hutnictwa w Polsce*, Vol. I, Warszawa 1957, pp. 77–89.

123 A. Keckowa, 'Polish Salt-Mines ...', *op. cit.*, p. 629.

124 M. Kulczykowski, *op. cit.*, pp. 74–5.

125 Ibid. See also E. Cieślak, 'Memoriał z r. 1772 o handlu Polski z Francją przez Gdańsk', *Zapiski Historyczne*, Vol. XXVII, No. 1, Warszawa 1962, pp. 79–86, which suggests the increasing importance of west European salt exports particularly from France to northern Poland, during the second half of the eighteenth century.

126 Österreiches Hof- und Kammerarchiv, Wien, *Hoffinanz Ungarn*, document dated 21/II/1721; see also V. Černý, *op. cit.*, 168.

127 I. Dordea and V. Wollmann, 'Exploatarea sării în Transilvania şi Maramureş in veacul al XVIII-lea', *Anuarlu Institului de Istoire şi Arheologie Cluj-Napoca*, Vol. 20, Cluj 1977, pp. 163–203.

128 H. Hlam, 'Österreich und Neurussland', *Jahrbücher für Geschichte Osteuropas*, Vol. VI, No. 2–4, München 1941, pp. 31–42; J. Reychman, 'Le commerce polonais en mer Noire au XVIIIe siècle par le port de Kherson', *Cahiers du monde russe et soviétique*, Vol. VII, No. 2, Paris-The Hague-New York, 1966, p. 246.

129 R. Rybarski, *Wielickie żupy solne ...*, *op. cit.*, p. 138, gives the following data: 1499, one bałwan cost 4 złotys, 3 grosz; 1532 = 4 złotys, 4 grosz; ditto 1560; 1571 = 4 złotys, 6 grosz; 1587/88 = 4 złotys, 4 grosz.

130 H. Łabęcki, '*Górnictwo w Polsce* ...', *op. cit.*, 150: R. Rybarski, *Wielickie ... op. cit.*, p. 140. During the sixteenth century the nobility bought salt at the privileged price of 6 grosz/cetnar (65 kg) providing they did not use it for resale. The usual price paid by Polish merchants was 8 grosz/cetnar, but to the nobility for resale, 12 grosz, and to foreign merchants 18 grosz. According to A. Keckowa, 'Polish Salt-Mines ...' *op. cit.*, p. 631, the Polish nobility totalled ten percent of the population.

131 L. Zeiszner, *Wieliczka, krótki opis historyczny, geologiczny i geograficzny*, Berlin 1843, p. 26, states that the actual weight of a barrel was always in some doubt, but between 1750 and 1762 it contained 6 cetnar (i.e. 390 kg).

132 Ibid, p. 20.

133 V. Černý, 'Polská sůl ...' *op. cit.*, p. 178, during the second half of the eighteenth century, Polish supplies were replaced by Austrian salt.

134 E. Schremmer, 'Saltmining and the Salt-trade: a State-Monopoly in the XVIth–XVIIth Centuries. A Case-Study in Public Enterprise and Development in Austria and the South German States', *Journal of European Economic History*, Vol. 8, No. 2, Rome 1979, pp. 291–312.

135 M. Bogucka, 'Sól w handlu Bałtyckim ...' *op. cit.*, pp. 106–7.

136 K. Gierdziejewski, *Zarys dziejów odlewnictwa polskiego*, Katowice 1954, p. 99.

137 M. Małowist, 'Poland, Russia and Western Trade ...' *op. cit.*, p. 30.

138 S. Hoszowski, 'L'Europe centrale devant la révolution des prix ...' *op. cit.*, p. 450.

139 F. Piekosiński and S. Krzyżanowski (eds.), *Prawa, przywileje i statuta miasta Krakowa, 1507–1795*, Vol. II, Kraków 1909, No. 1065.

140 J. Topolski, 'Economic Decline in Poland from the Sixteenth to Eighteenth Centuries', Ch. 6 in *Essays in European Economic History 1500–1800*, ed. for Economic History Society by P. Earle (Clarendon Press) Oxford 1974, p. 134.

141 J. Nax, *Wykład początkowych prawideł ekonomiki politycznej*, Warszawa 1790, pp. 54–5, 204, on which W. Rusiński ('The Role of Polish Territories...', *op. cit.*, p. 127) based his comments.

142 H. Łabęcki, *op. cit.*, pp. 189–292, 294–303; T. Dziekoński, *Metalurgia miedzi, ołowiu i srebra w Europie środkowiej od XV do końca XVIII wieku*, Wrocław-Warszawa-Kraków, 1963, p. 106–114; D. Molenda, *Górnictwo kruszcowe na terenie złóż śląsko-krakowskich do połowy XVI wieku* (P.A.N.), Wrocław-Warszawa-Kraków, 1963, 425 pp; Ibid., *Kopalnie rud ołowiu na terenie złóż śląsko-krakowskich w XVI-XVIII wieku*, Wrocław 1972, 140 pp.

143 H. Łabęcki, *op. cit.*, p. 193, reference to documents dated 1502, 1504 and 1555. See also D. Molenda 'Inwestycje produkcyjne i kulturalne w miastach górniczych Europy środkowo-wschodniej w XIII+XVII wieku', *Kwartalnik Historii Kultury Materialnej*, Vol. XXVI, No. 1, Warszawa 1978, pp. 15–27; K. Pieradzka, 'Przedsiębiorstwa kopalniane mieszczan krakowskich w Olkuszu od XV do początków XVII wieku', *Zeszyty Naukowe Uniwersytetu Jagiellońskiego*, No. 16 (Historia No. 3), Kraków 1958, p. 35–80.

144 D. Molenda, *Górnictwo kruszcowe ... op. cit.*, p. 420.

145 *Zarys dziejów górnictwa na ziemiach polskich*, Vol. I, Katowice 1960, p. 129.

146 Wojewódzkie Archiwum Państwowe w Krakowie, *Consularia Cracoviensis*, No. 434, p. 243 dated 1530, refers to first arrival of Tratkop family in Cracow; of the Haller family, Krystof and Stefan were most closely connected with Olkusz mining. In 1536 Stefan Haller received citizenship of Cracow. Wojewódzkie Archiwum Państwowe w Krakowie, *Libri Iuris Civilis* (Księgi przyjęć prawa miejskiego), rkp. No. 1422, p. 307.

147 Wojewódzkie Archiwum Państwowe w Krakowie, *Consularia Cracoviensis* (plenipotentiae), rkp. No. 756, p. 369 dated 1589.

148 *Sprawozdania z poszukiwań na Węgrzech*, Kraków 1919, Nos. 332 and 341, quoted in K. Pieradzka, 'Handel Krakowa ...' *op. cit.*, p. 195.

149 Arhiva Statului Braşov, *col. Schnell*, Vol. II, p. 68, lead sent from Cracow to Košice on 20/XI/1522, whence sent to Braşov; Lead sent to Vilna, J. M. Małecki, 'Handel zewnętrzny ...' *op. cit.*, p. 100.

150 A. Divéky, *Felsö Magyarország kereskedelmi összeköttetése Lengyelországgal föleg a XVI–XVII században*, Budapest 1905, p. 59; Biblioteka Czartoryskich w Krakowie, *rkp No. 1025*, p. 32, dated 28/VII/1552 – letter to Polish king 'Non parva summa plumbi S.R.M. apud me existente quod quanto maiori potuissem precio ... vendidisse cupiebam. Et quia sub id tempur rumor increbuerat plumbum Polonum Ungaros et praesertim Bistricienses et Schemniczenses non emptoros quod plumbi e Villaco Carinthiae haberent copiam'.

151 For example Státní Archiv v Banské Štiavnicé, *Copirbuch 1564*, on 13–14/IX/1564, 1,500 cetnars (150 tons) of lead sent from Cracow to Banská Štiavnica; Ibid., *Protocoll 1570–1577 (LX–590)*, 1,700 cetnars (170 tons) of lead sent from Cracow to Banská Štiavnica-document dated 22/II/1570; Ibid., *Copirbuch 1591–1593 (IX-6)* lead sent from Cracow to Banská Štiavnica documents dated 23/III/1592 and 18/VI/1592.

152 J. Vozár, 'Banské mestá ako osobitný typ miest na Slovensku', *Historický Časopis*, Vol. 21, No. 3, Bratislava 1973, pp. 387–396. Already in 1519, 271 horses (i.e. 135,500 cetnars/135½ tons) of lead sent to Hungary from Cracow. R. Rybarski, *Handel i polityka* ... *op. cit.*, Vol. II, p. 178.

153 M. Dan, 'Din relaţiile comerciale ale Transilvaniei cu Slovacia ... ' *op. cit.*, p. 25; P. Horváth, 'Obchodné styky Levoče ... ' *op. cit.*, p. 131.

154 For example 1588, 1,240 cetnars (124 tons) of lead sent from Cracow, via Włocławek to Gdańsk, R. Rybarski, *Handel i polityka* ... *op. cit.*, Vol. I, p. 40; 1588, 90 pieces of lead sent from Cracow via Warsaw and Gdańsk to Spain together with a quantity of litharge (lead monoxide), *Danziger Inventar 1531–1591*, Bearbeitet von P. Simson, München-Leipzig, 1913 (Inventare Hansischer Archive des Sechzehnten Jahrhunderts, Bd.3), No. 9785, 9791; in 1590, 2,000 cetnars (200 tons) of lead sent from Cracow via Gdańsk to Hamburg. Ibid., No. 10041; in 1593–94 a total of 16 carts (8,000 cetnars = 80 tons), 5 firkins (faska) and 2 stones (kamień) of lead and litharge sent to Gdańsk from Cracow, Wojewódzkie Archiwum Państwowe w Krakowie, *Księgi celne*, rkp 2117; Ibid., *Księgi celne*, rkp 2118 k, fols 39v, 56v, dated 1594 refer to lead sent from Cracow to Gdańsk.

155 J. M. Małecki, *Związki handlowe miast polskich* ... *op. cit.*, p. 140. Lead from Cracow sent to Jičín (170 cetnars = 17 tons) and to Olomouc (46 cetnars = 5 tons) in 1591, F. Matějek, 'K dějinám českého obchodu ...' *op. cit.*, p. 205. Possible that lead sent from Cracow to the mining towns of Slovakia, on the road via Jablonec nad Nisou to Žilina, may have sold some en route in Silesia. P. Horváth, 'Príspevok k obchodným stykom ... ' *op. cit.*, p. 186.

156 Wojewódzkie Archiwum Państwowe w Krakowie, *Księgi celne*, rkp. 2116 k, fols. 1v, 3v, 19v, 28, 44v etc.

157 V. Černý, 'Polská sůl ...' *op. cit.*, p. 154.

158 Wojewódzkie Archiwum Państwowe w Krakowie, *Księgi celne*, rkp. 2121, fol. 253, document dated 16/I/1600.

159 M. Bogucka, 'Handel Gdańska z Półwyspem Iberyjskim ...' *op. cit.*, Table C, p. 18.

160 Wojewódzkie Archiwum Państwowe w Krakowie, *Księgi celne*, rkp. 2126, pp. 203–245, documents dated 1604; Ibid., rkp. 2138, p. 87 (1616); Ibid., *Księgi wójtowskie*, rkp. 237, p. 115, document dated 1620.

161 Ibid., *Księgi celne*, rkp. 2216, document dated 1750, refers to 39 cetnars (4 tons) of lead and 96 cetnars (10 tons) of litharge sent from Olkusz to Cracow; Ibid., rkp. 2231, document dated 1760 refers to 26 pieces, 7 cetnars and 11 kopas of lead sent from Olkusz to Cracow (1 kop = 60 pieces).

162 M. Kulczykowski, *Kraków jako ośrodek* ... *op. cit.*, p. 75.

163 *Dziennik Handlowy*, Warszawa 1788, p. 752.

164 S. Hoszowski, 'L'Europe centrale devant la révolution des prix ...' *op. cit.*, p. 450.

165 H. Łabęcki, *op. cit.*, p. 189–200; T. Dziekoński, *Wydobywanie i metalurgia kruszców na dolnym Śląsku od XIII do połowy XX wieku* (P.A.N.), Instytut Historii Kultury Materialnej, Wrocław-Warszawa-Kraków-Gdańsk, 1972, pp. 358–362.

166 J. U. Nef, 'Silver Production in Central Europe, 1450–1618', *Journal of Political Economy*, Vol. XLIX, Chicago 1941, p. 578.

167 Z. Ameisenowa, *Kodeks Baltazara Behema* (Auriga: Oficyna Wydawnicza w Warszawie), Warszawa 1961, p. 36.
168 For example see J. Jaňaček, 'Die Fugger und Joachimstal', *Historia*, Vol. 6, Prague 1963, p. 109–144.
169 R. Rybarski, *Handel i polityka ... op. cit.*, Vol. II, p. 235, with reference to precious metals sent to Kamieniec Podolski in 1551/1552. See also A. Attman, *The Russian and Polish Markets in International Trade 1500–1650* (Institute of Economic History, Göteborg Univ.), Göteborg 1973, p. 102; K. Pieradzka, 'Handel Krakowa z Węgrami ...' *op. cit.*, p. 2.11.
170 S. Kutrzeba and J. Ptaśnik, 'Dzieje handlu i kupiectwa krakowskiego: Handel od XVI do XVIII wieku', *Rocznik Krakowski*, Vol. XIV, Kraków 1912, p. 34.
171 B. Geremek, 'Relacja Jakuba Esprincharda z podróży przez Śląsk i Małopolskę', *Kwartalnik Historii Kultury Materialnej*, Vol. VII, Warszawa, 1959, pp. 449–450.
172 J. Pazdur, 'Production de soufre en Pologne du XVe au XVIIIe siècle', *Kwartalnik Historii Kultury Materialnej*, Vol. XVI, No. 4, Warszawa 1968, p. 676.
173 Wojewódzkie Archiwum Państwowe w Krakowie, *Consularia Cracoviensis*, No. 449, p. 687 dated 1586 refers to 24 tons of sulphur sent by Erazm Strusz, Cracow merchant, via Vistula to Gdańsk. Evidence from the mid-seventeenth century also exists: Ibid. *Księgi wielkiej wagi*, rkp. 2279, pp. 27, 31 refers to 180 cetnars of sulphur sent from Cracow to Gdańsk in 1645; Ibid., 1648, 50 firkins (faska) and 1651, 3 firkins – rkp. 2280, p. 57; rkp. 2282, p. 48.
174 M. P. Dan, 'Relaţiile comerciale dintre Oradea ...' *op. cit.*, p. 177.
175 Státní archiv v Banské Bystrice, *Gemeiner Stadt Neu-Sol Waag Reitung 1679–1702.*
176 J. Pazdur, *op. cit.*, p. 676.
177 M. Wolański, *Związki handlowe Śląska z Rzeczpospolitą w XVII wieku ze szczególnym uwzględnieniem Wrocławia*, Wrocław 1961, p. 252.
178 M. Kulczykowski, *op. cit.*, p. 75.
179 Wojewódzkie Archiwum Państwowe w Krakowie, *Teki Schneidera* No. 93, dated 1792 which gives the annual value of sulphur that year as 11,800 złotys.
180 One of the leading entrepreneurs was Pawel Kaufman who founded a wire and tin plate works in 1525; in 1540 two Nuremburg merchants established a brass foundry in Cracow. See D. Molenda, 'Początek eksploatacji galmanu na ziemiach polskich (do połowy XVII wieku)', *Kwartalnik Historii Kultury Materialnej*, Vol. XXI, No. 1, Warszawa 1973, pp. 47, 59.
181 D. Krannhals, *Danzig und der Weichselhandel in seiner Blützeit vom 16. bis zum 17. Jahrhundert*, Leipzig 1942, p. 33–34.
182 Wojewódzkie Archiwum Państwowe w Krakowie, *Księgi wielkiej wagi*, rkp. 2279, 2280 for the years 1645, 1548. See also D. Molenda 'Początek eksploatacji ... ', *op. cit.*, Table I, p. 67.
183 W. Kula, 'Kopalnia ołowiu i galmanu w hrabstwie tęczyńskim, 1721–1781', in *Szkice o manufakturach w Polsce XVIII wieku*, Warszawa 1956, Pt. 1–2 (1720–1780), pp. 118–135.
184 J. Rutkowski, 'The Social and Economic Structure ... ' *op. cit.*, p. 446.
185 Archiwum Główne Akt Dawnych w Warszawie, *Akta Metryki Koronnej*, R.R. 26, p. 39, which refers to 'donatio mineræ feræ rudy'.

186 H. Łabęcki, *op. cit.*, p. 45.

187 P. Horváth, 'Príspevok k obchodným stykom ...' *op. cit.*, p. 184; usually only about 30–40 cetnars were exported from places like Levoča to Cracow, but the largest consignment from there came in 1587 (136 cetnars = 14 tons). See P. Horváth, 'Obchodné styky Levoče ... ' *op. cit.*, p. 127; it was also exported to Cracow from Moravia 'ferrum Moravicum' as mentioned in the city statute of 1533, whilst in 1584, 120 cetnars (12 tons) of it was sent to Cracow. See B. Zientara, *Dzieje Małopolskiego hutnictwa żelaznego XIV–XVII wieku*, Warszawa, 1954, pp. 162, 163.

188 A. Dunin-Wąsowicz, *Kapitał mieszczański Nowego Sącza na przełomie XVI/XVII wieku*, Warszawa, 1967, pp. 123–124.

189 F. Piekosiński and S. Krzyżanowski, '*Prawa, przywileje i statuta* ...' *op. cit.*, Vol. I, No. 370, p. 462 ('Statut kowali' dated 1533); see also B. Zientara, *op. cit.*, p. 160.

190 In 1572, Gdańsk merchant, Hans Lange, sent 700 waggons (i.e. 35,000 cetnars/ 350 tons) of iron down the Vistula from Cracow. S. Kutrzeba, 'Wisła w historji ... ', *op. cit.*, p. 29.

191 Z. Guldon, 'Eksport żelaza świętokrzyskiego na Litwę i Białoruś na przełomie XVI i XVII wieku', *Rocznik Świętokrzyski*, Vol. 6, Kielce, 1977 (publ. 1978), pp. 11–20.

192 S. Cynarski, 'Krosno w XVII w XVIII wieku', in *Krosno: Studia z dziejów miasta i regionu*, Vol. I (do roku 1918), Kraków, 1972, p. 223.

193 W. S. Unger, 'Trade through the Sound ...', *op. cit.*, p. 216; M. Hroch, 'Obchod mezi východní a zapadní evropou ... ' pp. 485–486.

194 J. Osiński, *Opisanie polskich żelaza fabryk*, Warszawa, 1782, 92 pp.; see also J. Pazdur, 'Materiały do dziejów hutnictwa żelaza w Polsce w XVIII wieku', *Studia z Dziejów Górnictwa i Hutnictwa*, Vol. I, Wrocław, 1957, pp. 319–34459; Z. Guldon and L. Stępkowski, *Z dziejów handlu Rzeczpospolitej w XVI–XVIII wieku: studia i materiały* (Wyższa Szkoła Pedagogiczna im. J. Kochanowskiego), Kielce, 1980, Tables 21–26, pp. 257–267.

195 M. Kulczykowski, 'Handel Krakowa w latach 1750–1772', *Prace Historyczne*, No. 4 (Historia No. 4), Kraków, 1960, p. 89.

196 M. P. Dan, 'Le commerce de la Transylvanie ... ' *op. cit.*, p. 626.

197 For example, in 1589 a total of 152 stones (kamień) or 2 tons sent from Cracow to Gdańsk, Wojewódzkie Archiwum Państwowe w Krakowie, *Księgi celne*, rkp. 2115 (after 12/VI/1589).

198 F. W. Carter, *Dubrovnik (Ragusa)* ..., *op. cit.*, p. 138.

199 J. Reychman, 'Le commerce polonais en mer Noire ... ', *op. cit.*, p. 241.

200 T. Korzon, *Wewnętrzne dzieje Polski za Stanisława Augusta, 1764–1794*, Warszawa, 1897, Vol. II, p. 131.

201 For example, 400 stones (kamień) or 7½ tons sent to Wrocław in 1755. M. Kulczykowski, *Kraków jako ośrodek* ..., *op. cit.*, p. 81.

202 N. J. G. Pounds, *An Historical Geography of Europe 1500–1840* (Cambridge Univ. Press), Cambridge, London, New York, New Rochelle, Melbourne, Sydney, 1979, p. 60 and Fig. 1.10, p. 51.

203 G. Wenzel, 'Okmánytár a Fuggerek Magyarországi nagykereskedése és rézvállalatának történetérez 1494–1551', *Történelmi Tár*, Budapest, 1882, pp. 25–54;

M. Jansen, *Jakob Fugger der Reiche Studien und Quellen* Leipzig, 1910, 240 pp.; M. Baruch, 'Stosunki Fuggerów z Polską', *Sprawozdanie z posiedzenia Towarzystwa Naukowego w Warszawie* (Wydz. II), Vol. 9, Warszawa, 1916, pp. 69–87; L. Schick, *Un grand homme d'affaires au début du XVIe siècle – Jacob Fugger*, Paris, 1957, 344 pp.; G. von Pöllnitz, *Jakob Fugger*, Tübingen, 1961, 295 pp.

204 G. Wenzel, *A Fuggerek jelentösége Magyarország törtenetében* (Akadémiai értekezések a történelemtudomány köréböl), Budapest, 1882, 197 pp. gives the family background of the Thurzo family in Slovakia. He died in 1508; see D. Molenda, 'W sprawie badań huty miedzi w Mogile pod Krakowem w XV i XVI wieku', *Przegląd Historyczny*, Vol. 66, No. 3, Warszawa, 1975, p. 374.

205 T. Ipolyi, *Geschichte der Stadt Neusohl* (Polska Akademia Umiejętności, Kraków, 1918, No. 192, p. 11; see also P. Ratkoš, 'Das Kupferwesen in der Slowakei vor der Entstehung der Thurzo-Fuggerschen-Handelsgesellschaft', in *Der Aussenhandel Ostmitteleuropas 1450–1650*, I. Bog. (ed.), Wien, 1971, pp. 583–596.

206 G. Wenzel, *Magyarország bányászatának kritikai története*, Budapest, 1880, p. 75, 77, 98.

207 Ibid., A. Péch, *Alsó Magyarország bányamívelésének története*, Vol. I, Budapest, 1884, 205 pp; S. Szilágyi, 'Bethlen Gábor és a Bányavárosok', *Történelmi Tár*, Budapest, 1893, document XII.

208 E. Fink, 'Die Bergwerkunternehmungen der Fugger in Schlesien', *Zeitschrift für Geschichte und Kulturgeschichte Österreichisch-Schlesiens*, Vol. 28, Troppau, 1913, p. 297.

209 K. Pieradzka, 'Handel Krakowa z Węgrami ...' *op. cit.*, p. 174.

210 For example this firm paid taxes of 500 grosz annually to the city budget for use of Cracow's transit facilities between 1562–1569; F. Piekosiński and S. Krzyżanowski, *Prawa, przywileje ... op. cit.*, pp. 1053, 1057, 1065, 1069, 1073, 1078, 1081.

211 *Codex Diplomaticus Silesiae*, K. Wutke (ed.), Vol. XVII, Wrocław, 1900, pp. 46–47.

212 K. Pieradzka, 'Handel Krakowa z Węgrami ...' *op. cit.*, p. 180.

213 In 1538/1539 a total of 927 cetnars (92½ tons) was recorded as passing through the Cracow Chamber of Commerce; R. Rybarski, *Handel i polityka ... op. cit.*, Vol. II, p. 183; 1584 this was 3,776 cetnars (377½ tons), Ibid. p. 201; in 1593 it was 1,571 cetnars (157 tons). Wojewódzkie Archiwum Państwowe w Krakowie, *Księgi celne*, rkp. 2117.

214 J. Vlachovič, 'Produktion und Handel mit ungarischen Kupfer im 16 und im ersten viertel des 17 Jahrhunderts' in *Der Aussenhandel Ostmitteleuropas ... op. cit.*, pp. 600–627; Ibid., *Slovenska měd v 16 a 17 storiča*, Bratislava 1964, p. 100–104; G. Ember, 'Zur Geschichte des Aussenhandels Ungarns im 16ten Jahrhundert', *Studia Historica*, Vol. 44, Budapest, 1960, pp. 33–35; J. Hartung, 'Aus dem Geheimbuche etnes deutschen Handelshauses im 16 Jahrhundert', *Zeitschrift für Sozial-und Wirtschaftsgeschichte*, Vol. 8, Hamburg, 1898, p. 40; F. Dobel, 'Der Fugger Bergbau und Handel in Ungarn', *Zeitschrift des Historischen Vereines für Schwaben und Neuburg*, Vol. VI, Stuttgart, 1879, pp. 30–36.

215 Wojewódzkie Archiwum Państwowe w Krakowie, *Księgi celne*, rkp. 2142, document dated 10/I/1624 refers to Wolfgang Paler, owner of a mine in Banská

Bystrica, who paid a customs duty of 118 złotys 18 grosz for sending 2,291 cetnars (229 tons) of copper through Cracow since 23/II/1623; see also J. Vlachović, 'Slovak, Copper Boom in World Markets of the Sixteenth and in the First Quarter of the 17th Centuries', *Studia Historica Slovaca*, Vol. I, Bratislava 1963, p. 88.

216 M. Kulczykowski, 'Handel Krakowa w latach . . .' *op. cit.*, p. 91 and Table (no number), p. 89.

217 Ibid., p. 91.

218 J. A. Goris, *Étude sur les colonies marchandes méridionales (Portugais, Espagnols, Italiens) à Anvers de 1488 à 1567*, Louvain, 1925, p. 275; H. van der Wee, *The Growth of the Antwerp Market and the European Economy*, 2 vols., Louvain 1963, especially Vol. I, pp. 130–207; E. Coornaert, 'Note sur le commerce d'Anvers au XVIe siècle', *Revue de Nord*, Vol. 13, Paris, 1927, pp. 123–7; its preeminence lasted until destruction by the Spaniards in 1576; thereafter Amsterdam became the leading copper market. N. J. G. Pounds, *An Historical Geography . . .*, *op. cit.*, p. 270.

219 M. Jansen, *op. cit.*, p. 143.

220 L. Bechtel, *Die Fugger in Danzig und im nordosteuropäischen Raum*, Dissertation, Universitätsbibliothek, München, 1944, p. 15.

221 Wojewódzkie Archiwum Państwowe w Krakowie, *Acta Consularia Cracoviensis*, 1500–1512, p. 171 refers to a consignment of copper being held by one M. Keiper, in a warehouse 4 kilometres from Cracow, in 1502. It belonged to Jan Thurzo and was bound for Gdańsk.

222 In 1512 Toruń merchants bought 2,220 cetnars (220 tons) of copper in Cracow and in 1514, 9,337 cetnars (940 tons). J. M. Małecki, 'Handel zewnętrzny Krakowa . . .', *op. cit.*, p. 105.

223 E. Fink, *op. cit.*, p. 329, A. Divéky, *op. cit.*, p. 54; in 1519 copper sent from Cracow via Gdańsk to Portugal. *Hanserecesse: Die Recess und Andere Acten der Hansetage von 1477–1530* (D. Schäfer, ed.) Abt III, Vol. 7, Leipzig, 1888, No. 197. Personal connections between Gdańsk and Portugal existed, and Hungarian copper was sent to King Emmanuel. See E. Kestner, 'Danzigs Handel mit Portugal im 16 Jahrhundert', *Zeitschrift der Westprussiens Geschichtesverein*, Vol. I, 1880, Breslau, pp. 98, 99.

224 Wojewódzkie Archiwum Państwowe w Gdańsku 300 U Abt 12 No. 2, *rąchunki miejskie*, pp. 408, 409, document dated 1531 and refers to a Gdańsk agent Hans Brede who ordered from a Cracow agent Jerzy Hegel, transport for copper and gunpowder; payment was in the form of money from Spain.

225 In 1525, Jakob Fugger obtained, through his agent, the monopoly rights for supplying Denmark with copper. G. Wenzel, *A Fuggerek jelentösége . . . op. cit.* No. 27.

226 In 1511 the Fugger company sent copper from Cracow to Holland. The boat left Gdańsk and was attacked by merchants from Lübeck. J. Ptaśnik (ed.), *Obrazki z przeszłości Krakowa*, Seria 1 (Biblioteka Krakowska No. 21), Kraków, 1902, pp. 79–80. For further discussion see H. Kellenbenz (ed.), *Schwerpunkte der Kupferproduktion und Kupferhandels in Europa: 1500–1650* (Böhlau), Köln, 1977, 412 pp.; P. Jeannin, 'Le cuivre, les Fugger et la Hanse', *Annales: Economies, Sociétés Civilisations*, Vol. 10, Paris, 1955, pp. 229–36; K. Pieradzka, 'Trzy wieki stosunków handlowych pomiędzy Gdańskiem a

Węgrami', *Rocznik Gdański*, Vols. IX–X, Gdańsk, 1935–36 (publ. 1937), pp. 189–208.

227 P. Jeannin, 'Le cuivre, les Fugger ...' *op. cit.*, p. 233.

228 Large annual amounts became less frequent in the documentary evidence. Notable ones in the second half of the sixteenth century were: 1568, 6,282 (630 tons) and 1574, 9,000 cetnars (900 tons), recorded as coming from Cracow through the Włocławek customs post (Wojewódzkie Archiwum Państwowe w Warszawie, *Księgi komory włocławskiej*, for respective years en route to Gdańsk.

229 For example, 1523, 300 cetnars (30 tons) of copper sent from Cracow to Lithuania. A. Wawrzyńczyk, *Studia z dziejów handlu Polski z Wielkim Księstwem Litewskim i Rosją w XVI wieku*, Warszawa, 1956, p. 65.

230 In 1538–1539, 27 cetnars (2½ tons) of copper sent from Cracow to Troki. R., Rybarski, *Handel i polityka ... op. cit.*, Vol. II, p. 183.

231 In 1564, copper from N. Hungary sent via Cracow to Vilna, *Biblioteka Czartoryskich* (Muzeum Narodowe w Krakowie), rkp. No. 1041 (rachunki Decjusza), p. 24.

232 In 1530–1531, 4 carts of copper (2,000 cetnars or 200 tons) sent from Cracow to Lublin. R. Rybarski, *Handel i polityka ... op. cit.*, Vol. II, p. 250.

233 J. Vlachovič, *Slovenska měd ... op. cit.*, p. 119.

234 M. Bogucka, 'Handel Gdańska z Półwyspem Iberyjskim ...' *op. cit.*, Table C, p. 18.

235 J. Vlachovič, *Slovenska měd ... op. cit.*, p. 226.

236 H. Obuchowska-Pysiowa, 'Handel wiślany ...' *op. cit.*, p. 163.

237 Ibid.; in 1650, 735 cetnars (73½ tons) and 1651, 411 cetnars (41 tons) of copper left Cracow but only recorded as being sent to Warsaw.

238 Usually referred to in documents as 'stannum elaboratum'. P. Horváth, 'Príspevok k obchodným stykom ...' *op. cit.*, p. 186.

239 For example, in 1538/1539, 166.5 cetnars (16½ tons) and 2 firkins (fas) of tin sent from Wrocław to Cracow; R. Rybarski, *Handel i polityka ... op. cit.*, Vol. II, p. 184; in 1584 18 firkins were sent from Wrocław to Cracow; Ibid., p. 141; in 1591/1592, 10 firkins of tin sent from Wrocław to Cracow and 3 firkins from an unknown destination. Wojewódzkie Archiwum Państwowe w Krakowie, *Księgi celne*, rkp. 2116.

240 In 1538/1539, 30 cetnars (3 tons) of tin sent to Cracow from Nuremburg. R. Rybarski, *Handel i polityka ... op. cit.*, p. 184.

241 Ibid., In 1538/1539, 40 cetnars (4 tons) of tin sent to Cracow from Prostějov, in Moravia. See also J. Mayer, *Težba cínu ve Slavkovském Lese v 16 století*, Prague, 1970, 224 pp.

242 Wojewódzkie Archiwum Państwowe w Krakowie, *Księgi celne*, rkp. 2116; 16.5 cetnars (1½ tons) of tin sent to Hungary in 1591/1592; Ibid., 11 cetnars (1 ton) of tin sent to Gdańsk.

243 P. Horváth, 'Príspevok k obchodným stykom ...' *op. cit.*, p. 183; K. Pieradzka, 'Handel Krakowa z Węgrami ...' *op. cit.*, p. 181.

244 In 1537, 2 barrels (beczki) of vitriol were dispatched to Gdańsk and recorded in the customs books at Włocławek; in 1558, 35 barrels were recorded for the same journey. S. Kutrzeba and F. Duda, *Regestra thelonei aquatici, Wladislaviensis saeculi XVI*, Kraków, 1915, pp. XLII–XLIV.

245 In 1589, one firkin was recorded in the Cracow customs as being sent to Gdańsk, and in 1597 1.5 firkins were noted for the same journey. Wojewódzkie Archiwum Państwowe w Krakowie, *Księgi celne*, rkp. 2115, 2120.
246 J. Vlachovič, 'Slovak Copper Boom...', *op. cit.*, p. 88.
247 F. Braudel and F. C. Spooner, 'Prices in Europe from 1450 to 1750', Ch. VII in *Cambridge Economic History of Europe*, Vol. IV, Cambridge, 1967, p. 425.
248 S. Hoszowski, 'L'Europe centrale devant la révolution des prix ...', *op. cit.*, p. 445.
249 K. Pieradzka, 'Trzy wieki stosunków ...', *op. cit.*, p. 298.
250 F. Braudel and F. C. Spooner, *op. cit.*, p. 424.
251 Ibid., 'Lwów showed a very slow rise followed by a short decline and a level stretch to 1661, then finally after 1718 a marked rise which apparently brought its prices closer to the international market level.'
252 Ibid.
253 D. Molenda, 'Początek eksploatacji galmanu ...' *op. cit.*, p. 61; in 1653 a cetnar of lead in Cracow cost 10–16 złotys, whilst a cetnar of calamine was only 3 złoty; see also E. Abt, 'Von der älteren Geschichte der Galmay Gräberey in Oberschlesien', *Schlesische Provinzalblätter*, Breslau, 1790, p. 146.
254 W. Kula, 'Historie et économie: la longue durée', *Annales: Economies, Sociétés Civilisations*, Vol. 15, Paris, 1960, p. 360.
255 F. Braudel and F. C. Spooner, *op. cit.*, p. 422.
256 N. J. G. Pounds, *An Historical Geography of Europe* ... *op. cit.*, pp. 2–3.
257 K. Kuklińska, *op. cit.*, p. 447.
258 R. Rybarski, *Handel i polityka* ..., *op. cit.*, Vol. I, p. 83; J. Topolski, 'Rola Gniezna w handlu europejskim od XV do XVII wieku', *Studia i Materiały do Dziejów Wielkopolski i Pomorza*, Vol. VII, No. 2, Poznań, 1962, p. 60; M. Grycz, *Handel Poznania, 1550–1665*, Poznań, 1964, 340 pp.
259 'Thus in 1580, Poznań had 75 tanners and Cracow, probably 160.' M. Małowist, 'Evolution industrielle en Pologne du XIV au XVII siècle', in *Studi in Onore di Armando Sapori*, Vol. I, Milan, 1957, p. 594; see also I. Turnau, 'Życie i praca w Polskich warsztatach skórniczych w XVIII wieku', *Roczniki Dziejów Społecznych i Gospodarczych*, Vol. XL, Warszawa-Poznań, 1979, pp. 71–85; I. Baranowski, *Przemysl polski w XVI wieku*, Warszawa, 1919, pp. 97–100; K. Pieradzka, 'Garbary: przedmieście Krakowa (1363–1587)', *Biblioteka Krakowska*, No. 71, Kraków, 1931, pp. 11–12.
260 K. Pieradzka, 'Handel Krakowa z Węgrami ...' *op. cit.*, p. 243.
261 H. Samsonowicz, 'Handel Lublina na przełomie XV i XVI wieku', *Przegląd Historyczny*, Vol. 59, No. 4, Warszawa, 1968, p. 622.
262 E. Nadel-Golobić, *op. cit.*, pp. 356–7; 370–1.
263 J. Małecki, *Studia nad rynkiem regionalnym Krakowa w XVI wieku* (P.W.N.), Warszawa, 1963, p. 161.
264 J. Rutkowski, *Historia Gospodarcza Polski (do 1864 r.)* (Książka i Wiedza), Warszawa, 1953, p.125.
265 According to Rybarski, in 1584, 25,491 raw cattle hides were declared to Cracow's customs officers. R. Rybarski, *Handel i polityka* ... *op. cit.*, Vol. II, p. 194.
266 Státní Archiv v Praze, *rkp. 2054*, zápis 29 I; 11 I; 13, III and 18 XI, with reference

to 1597, when 925 cattle skins and 220 cordovan (goat) skins sent from Cracow to Prague; see also J. Heřman, 'Obchod pražských židů s Polskem do roku 1541', in *Tisíc Let Česko-Polské Vzájemnosti*, Opava, 1966, p. 202.

267 Wojewódzkie Archiwum Państwowe w Krakowie, *Księgi celne*, rkp. 2115, 2116, gives examples for 1591–1592; 973 cattle skins sent from Cracow to Wrocław rkp. 2116, k.23.

268 J. Małecki, 'Przczynek do dziejów handlu Gdańska w drugiej połowie XVI wieku (Związki handlowe z Krakowem)', *Studia Gdańsko-Pomorskie*, Gdańsk, 1964, p. 40.

269 P. Jeannin, 'The Sea-borne and Overland Trade . . .', *op. cit.*, p. 27.

270 M. Dan, 'Negustorii clujeni la Cracovia in ultimul deceniu al secolului . . .', *op. cit.*, p. 210; Wojewódzkie Archiwum Państwowe w Krakowie, *Księgi celne*, rkp. 2119, p. 335 (dated 1595), Ibid., p. 193; Ibid., 237.

271 J. Ptaśnik, 'Z dziejów kultury włoskiego Krakowa', *Rocznik Krakowski*, Vol. IX, Kraków, 1907, p. 38; M. Korzeniowski, *Analecta Roman, Pauli Giovannini: Relazione di Polonia, 1561*, Warszawa, 1913, p. 195.

272 A. Attman, *op. cit.*, p. 15; J. Topolski, 'Faktoren der Entstehung einer Internationalen Jahrmarktnetzes in Polen im 16 und 17 Jh.', *Studia Historiae Oeconomicae*, Vol. 5, Poznań, 1970, p. 104; M. Małowist, 'Poland, Russia and Western Trade . . .', *op. cit.*, p. 29.

273 H. Samsonowicz, 'Handel Lublina . . .' *op. cit.*, p. 622.

274 L. Koczy, 'Handel Litwy przed połową XVII wieku', *Pamiętnik VI Powszechnego Zjazdu Historyków Polskich w Wilnie*, Vol. I, Lwów, 1935, p. 276; H. Łowmiański, 'Handel Mohilewa w XVI w', in *Studia ku czci. S. Kutrzeby*, Vol. II, Kraków, 1938, pp. 20–34.

275 A. Wawrzyńczyk, *op. cit.*, p. 46.

276 J. M. Małecki, 'Handel zewnętrzny Krakowa . . .' *op. cit.*, p. 130.

277 Wojewódzkie Archiwum Państwowe w Krakowie, *Księgi celne*, rkp. 2116.

278 R. Rybarski, *Handel i polityka . . . op. cit.*, Vol. II, p. 103.

279 Reference is made to a Cracow fur trader in Transylvania in 1578, one Georges 'pellifex'. Arhivele Statului Cluj, *Archive Bistritza*, No. 140 (22/XII/1578). See also *Sprawozdanie z poszukiwań . . . op. cit.*, p. 128, No. 380.

280 E. Nadel-Golobić, *op. cit.*, p. 356.

281 Wojewódzkie Archiwum Państwowe w Krakowie, *Księgi celne*, rkp. 1213, p. 81, document dated 1577. Cracow furriers also had their own agents in Gdańsk. Ibid., rkp. K.L.29, pp. 691–2.

282 J. Heřman, *op. cit.*, pp. 202–203.

283 Wojewódzkie Archiwum Państwowe w Krakowie, *Księgi celne*, rkp. 2119, p. 335 refers to one Andris Baychel from Cluj who bought 1500 grey squirrel furs in 1595; Ibid., pp. 193 mentions Gaspar Diak from Debrecen, who also bought skins in Cracow; Ibid., p. 237, one Helias from Tokaj who bought skins along with other goods.

284 A. Divéky, *Felsö Magyarország . . . op. cit.*, p. 60, No. 5.

285 M. Berindei, 'Contribution a l'étude du commerce Ottoman des fourrures Moscovites. La route Moldavo-Polonaise 1453–1700', *Cahiers du Monde Russe et Soviétique*, Vol. XII, No. 4, Paris-The Hague-New York, 1971, p. 400.

286 Archiwum Główne Akt Dawnych w Warszawie, Archiwum Koronne, *Dzialy*

Tureckie, Vol. 18, No. 39; see also M. V. Fehner, 'Torgovlja russkogo gosudarstva so stranami Vostoka v XVI veke', *Trud Gosudarstvennogo Istoricheskogo Muzeja*, Vol. XXI, Moskva, 1952, p. 104; C. Georgieva, *op. cit.*, p. 43.

287 H. Obuchowska-Pysiowa, 'Struktura handlu Krakowa ...', *op. cit.*, p. 83; E. Podgradskaja, *Torgovye svjazi Moldavii so L'vovom v XVI–XVII vekach*, Kishinev, 1968, pp. 123–4, 179, which gives details on large quantities of skins, and fewer furs, arriving on the Lwów market.

288 H. Obuchowska-Pysiowa, 'Handel wiślany ...' *op. cit.*, pp. 151, 289.

289 M. Horn, 'Rzemiosło skórzane w miastach ziemi przemyskiej i sanockiej w latach 1550–1650', *Kwartalnik Historii Kultury Materialnej*, Vol. XX, No. 1, Warszawa, 1972, p. 79196; Ibid., 'Handel wołami na Rusi Czerwonej w pierwszej połowie XVII wieku', *Rocznik Dziejów Społecznych i Gospodarczych*, Vol. XXIV, Warszawa, 1963, pp. 74–6.

290 Z. Guldon and L. Stępkowski, 'Handel Torunia z Wielkim Księstwem Litewskim w początkach XVII wieku', *Zapiski Historyczne*, Vol. XLIV, No. 2, Warszawa-Poznań-Toruń, 1979, pp. 100–2.

291 H. Obuchowska-Pysiowa, 'Struktura handlu Krakowa ...' *op. cit.*, p. 105.

292 Wojewódzkie Archiwum Państwowe w Krakowie, *Księgi celne*, rkp, 2121, p. 20; 2122, p. 91, refers to one Kaspar Szulc from Cracow importing goods from Elbląg on 14/VII/1599 and 27/VI/1600; also Jakub Lewis received a large consignment including rabbit skins; Ibid., 2127, p. 155, document dated 14/XI/1605.

293 Turku City Archives, syg. G.I.1; Riksarkivet Helsinki syg. 235 c., refer to various goods including skins and furs sent from Turku to Gdańsk *circa* 1606; see also M. Bogucka, 'Z badań na handlem Gdańsk-Turku w XVI i pierwszej połowie XVII wieku', *Zapiski Historyczne*, Vol. XLV, No. 3, Warszawa-Poznań-Toruń, 1980, p. 29.

294 L. Demény, *op. cit.*, p. 473.

295 S. Panova, *op. cit.*, p. 51.

296 Wojewódzkie Archiwum Państwowe w Krakowie, *Księgi celne*, rkp. 2142; other towns included: Bilovec, Frýdek, Krnov, Opava, Prostějov and Příbor.

297 R. Fišer, 'Orientace a skladba ... ', *op. cit.*, p. 435; Ibid. 'Les contacts commerciaux ... ', *op. cit.*, p. 849.

298 Wojewódzkie Archiwum Państwowe w Krakowie, *Księgi celne*, rkp. 2176 (1660) the first year after the Swedish wars; Ibid., rkp. 2202 (1688) the last customs register preserved from the seventeenth century. See also J. M. Małecki, 'Rola Krakowa w handlu ... ', *op. cit.*, p. 184.

299 Wojewódzkie Archiwum Państwowe w Gdańsku, *Missiva*, 300, No. 70, k. 203–204 (1645); Ibid., 27, 71, k. 647–647v (1647).

300 H. Obuchowska-Pysiowa, 'Trade between Cracow and Italy ... ', *op. cit.*, p. 649.

301 Wojewódzkie Archiwum Państwowe w Krakowie, *Księgi celne*, rkp. 2232, document dated 1763; 57 horses needed to transport cart loads of ram skins from Nowy Targ and Hungary.

302 Ibid. rkp. 2231–2232.

303 M. Kulczykowski, *Kraków jako ośrodek ... op. cit.*, p. 131.

304 A. Attman, 'The Russian Market in World Trade 1500–1860', *Scandinavian Economic History Review*, Vol. XXIX, No. 3, Jyväskyla, 1981, p. 202.

305 W. S. Unger, 'Trade through the Sound ... ', *op. cit.*, p. 211.
306 J. Reychman, 'Le commerce polonais ... ', *op. cit.*, p. 242.
307 H. Obuchowska-Pysiowa, 'Trade between Cracow and Italy ... ', *op. cit.*, p. 650, footnote 38.
308 R. Rybarski, *Handel i polityka ... op. cit.*, Vol. I, p. 103–104.
309 W. Tomkiewicz, 'Kultura artystyczna' in *Polska XVII wieku; Państwo-społeczeństwo-kultura*, J. Tazbir (ed.), Warszawa, 1969, p. 268.
310 J. M. Małecki, *Studia nad rynkiem ... op. cit.*, p. 183 and Map 10.
311 E. Nadel-Golobić, *op. cit.*, p. 356.
312 P. Jeannin, 'The Sea-borne and Overland Trade ...' *op. cit.*, p. 39.
313 Ibid.
314 Ibid., L. Koczy, 'Handel Litwy ...' *op. cit.*, p. 273.
315 J. Riabianin, *Materiały do historii miasta Lublina, 1317–1792*, Lublin, 1938, Nos. 110a; 136.
316 M. Dan, 'Din relaţiile comerciale ale Transilvaniei cu Polonia la sfîrşitul secolului al XVI-lea. Produse textile importate de clujeni de la Cracovia', in *Omagiu lui P. Constantinescu-Iaşi cu prilejul împlinirii a 70 de ani* (Ed. Acad. R.P.R.), Bucureşti, 1965, p. 293.
317 K. Pieradzka, 'Handel Krakowa z Węgrami ...' *op. cit.*, p. 241; even so in 1595 one Jurek Kewal from Kežmarok succeeded in buying 25 kamieńs of tallow in Cracow together with 10 kamieńs of turpentine, sheep and wether skins; Wojewódzkie Archiwum Państwowe w Krakowie, *Księgi celne*, rkp. 2119, p. 3.
318 F. Engel-Janosi, 'Zur Geschichte der Wiener Kaufmannschaft von der Mitte des 15 bis zur Mitte des 16 Jahrhunderts', *Mitteilungen des Vereins für Geschichte der Stadt Wien*, Vol. 6, Wien, 1926, p. 130.
319 R. Rybarski, 'Handel i polityka ...' *op. cit.*, Vol. 2, p. 198; a total of 17,030 kamieńs went to Vienna from Cracow.
320 P. Horváth, 'Obchodné styky Levoče ... ', *op. cit.*, p. 126; F. Matějek, 'K dějinám českého ... ', *op. cit.*, p,. 204.
321 J. M. Małecki, 'Handel zewnętrzny Krakowa ... ', *op. cit.*, p. 136.
322 In 1511 Jan Boner in one transportation alone sent 5 cart loads of wax from Cracow to Germany; J. Ptaśnik, 'Bonerowie' *Rocznik Krakowski*, Vol. 7, Kraków, 1905, p. 15.
323 Wojewódzkie Archiwum Państwowe w Krakowie, *Księgi celne*, rkp. 2116; in 1591/1592, of the total wax (18,812 kamieńs) exported abroad, one Cracow merchant, A. Rotermund sent 13,646 kamieńs i.e. 72.53%.
324 R. Rybarski, *Handel i polityka ... op. cit.*, Vol. II, p. 176, for the year 1519.
325 J. M. Małecki, *Studia nad rynkiem ... op. cit.*, p. 181, and Map 10; see also A. Żabko-Potopowicz, 'Dzieje bartnictwa w Polsce', *Roczniki Dziejów Społecznych i Gospodarczych*, Vol. XIV, Warszawa, 1953, pp. 10–21.
326 M. Dan, 'Mărfuri exportate de negustorii clujeni la Cracovia, la sfîrşitul secolului al XVI-lea', *Lucarari Ştiinţifile Oradea Istorica*, Oradea, 1973, pp. 37–43; Ibid. 'Din relaţiile comerciale ale Transilvaniei ... *op. cit.*, p. 292.
327 J. Rutkowski, *Historia gospodarcza Polski ... op. cit.*, p. 174.
328 R. Rybarski, *Handel i polityka ... op. cit.*, Vol. I, p. 187; K. Pieradzka, 'Handel Krakowa z Węgrami ... ', *op. cit.*, p. 241.
329 R. Rybarski, *Handel i polityka ... op. cit.*, Vol. I, p. 62.

330 M. Gromyko, 'Russko-niderlandskaja trgovlja na murmanskom beregu v XVI', *Srednie veka*, Moscow, 1960, p. 246; V. Doroshenko, 'Eksport sel'skochozjajstvenno produkscii vostochnoj pribaltiki v 1562–1620 gg', *Ezhegodnik po agrarnoj istorii vostochnoj Evropy*, Vilnius, 1963, p. 187.
331 H. Obuchowska-Pysiowa, 'Struktura handlu Krakowa . . .' *op. cit.*, p. 109.
332 R. Fisěr, 'Les contacts commerciaux . . . ', *op. cit.*, p. 849.
333 M. Wolański, *Związki handlowe Śląska z Rzeczpospolitą w XVII wieku*, Wrocław, 1961, p. 232.
334 M. Bogucka, 'Handel Gdańska z Półwyspem Iberyjskim . . . ', p. 18, Table B. gives wax exports from Gdańsk to the Iberian peninsula, for the years 1607–1618.
335 H. Obuchowska-Pysiowa, 'Trade between Cracow and Italy . . .', *op. cit.*, p. 109, Table 16 which gives a total of 358½ quintals of wax sent to Italy from Cracow in 1604.
336 M. Marečková 'Prešov v Uhersko-Polských obchodních vztazích počátkem 17 století', *Historický Časopis*, Vol. XXI, No. 3, Bratislava, 1973, p. 432.
337 R. Fišer, 'Les contacts commerciaux . . . ', *op. cit.*, p. 849; see also Z. Kirilly et al., 'Production et productivité agricoles en Hongrie a l'époque de féodalisme tardif (1550–1850), *Nouvelles Études Historiques*, Vol. I, Budapest, 1965, pp. 11–15.
338 K. Kuklińska, Handel Poznania . . . *op. cit.*, p. 80.
339 J. Louis, *op. cit.*, p. 22.
340 M. Kulczykowski, *Kraków jako ośrodek . . . op. cit.*, p. 63.
341 J. Louis, *op. cit.*, p. 22.
342 M. Kulczykowski, 'Handel Krakowa w latach 1750–1672 . . .', *op. cit.*, p. 96.
343 M. Wolański, *op. cit.*, p. 283–284.
344 M. Kulczykowski, 'Handel Krakowa . . . ', *op. cit.*, pp. 95.
345 M. Kulczykowski, *Kraków jako ośrodek . . . op. cit.*, Table 13, p. 55 and Map 2, p. 56.
346 For example in 1750, Cracow received 230 barrels (*circa* 63,000 litres) of honey from Jewish merchants in Tarnogród, which was 17% of the whole of Cracow's honey imports that year. Similarly, a quarter of all honey traded in Cracow was in the hands of Jewish merchants. M. Kulczykowski, *Kraków jako ośrodek . . . op. cit.*, p. 53.
347 Honey imports from Galicia to Cracow in 1755 totalled 1,263 barrels of which only 35 went for export (i.e. 3%); 1780, 1437 barrels, 86 for export (6%); 1785, 829 barrels, 31 for export (4%). M. Kulczykowski, *Kraków jako ośrodek . . . op. cit.*, p. 53.
348 M. Kulczykowski, 'Handel Krakowa . . .' *op. cit.*, p. 96.
349 P. Jeannin, 'The Sea-borne and the Overland Trade . . .' *op. cit.*, p. 7.
350 For example E. Westermann (ed.), *Internationaler Ochsenhandel in der frühen Neuzeit, 1450–1750* (Klett-Cotta), Stuttgart, 1979, 300 pp; Ibid., 'Zum Handel mit Ochsen aus osteuropa im 16 Jahrhundert. Materialen und Geschichtspunkte', *Zeitschrift für Ostforschung*, Vol. 22, No. 2, Marburg, 1973, pp. 234–276; R. Klier, 'Der schlesische un polnische Transithandel durch Böhmen nach Nürnberg in den Jahren 1540 bis 1576', *Mitteilungen des vereins für Geschichte der Stadt Nürnberg*, Vol. 53, Nuremburg, 1965, pp. 195–228;

J. Baszanowski, *Z dziejów handlu polskiego w XVI–XVIII wieku: handel wołami* (Gdańsk Tow. Nauk), Ossolineum, Gdańsk 1977, 240 pp.

351 H. Samsonowicz, 'Przemiany osi drożnych . . . ', *op. cit.*, p. 708, Table 1.

352 Z. P. Pach, 'The Shifting of International Trade Routes . . .' *op. cit.*, pp. 310–311.

353 J. Rutkowski, 'The Social and Economic Structure . . . ', *op. cit.*, p. 447.

354 J. Baszanowski, *op. cit.*, Maps 2 and 3.

355 M. Dan and S. Goldenburg, 'Le commerce balkano-levantin de la Transylvanie au cours de la seconde moitié du XVI siècle et au début du XVIIIe siècle', *Revue des études sud-est européennes*, Vol. 5, No. 1, Bucureşti, 1967, p. 97.

356 E. Nadel-Golobić, *op. cit.*, p. 356.

357 J. M. Małecki, *Studia na rynkiem . . . op. cit.*, p. 151 and Map 6, p. 152.

358 J. Wyrozumski, 'Rozwój życia miejskiego do połowy XVI wieku', in *Krosno: Studia dziejów miasta i regionu*, Vol. I (do roku 1918), Kraków, 1972, 93.

359 A. Wagner, 'Handel dawnego Jarosławia do połowy XVII wieku', *Prace Hist. wyd. ku uczczeniu 50-lecia akademickiego koła historyków univ. Jana Kaz. we Lwowie*, Lwów, 1929, p. 145; see also H. Wiese and J. Böltz, *Rinderhandel und Rinderhaltung im nordwesteuropäischen Küstengebiet vom 15 bis zum 19 Jahrhundert*, Stuttgart, 1966, p. 15, who discuss some of the outlets for the cattle trade through Hamburg and Lübeck.

360 For example Wojewódzkie Archiwum Państwowe w Lublinie, *Consularia*, rkp. 144, p. 80, refers to cattle sent direct from Lublin to Silesia (Brzeg) by Cracow merchants in 1536.

361 E. Westermann, 'Zum Handel mit Ochsen . . . ', *op. cit.*, pp. 255–256.

362 Ibid., pp. 264–266.

363 P. Jeannin, 'The Sea-borne and the Overland Trade . . .' *op. cit.*, p. 8.

364 F. W. Henning, 'Der Ochsenhandel aus den Gebieten nördlich der Karpaten im 16 Jahrhundert', *Scripta Mercaturae*, Brandenburg, 1973, pp. 31–32.

365 R. Klier, *op. cit.*, p. 203.

366 Ibid., p. 206; see also J. M. Małecki, 'Die Wandlungen im Krakauer und Polnischen Handel zur zeit der Türkenkriege des 16 und 17 Jahrhunderts', in *Die Wirtschaftliche auswirkungen der Türkenkriege*, Graz, 1971, p. 185.

367 F. Matějek, 'K dějinám českého obchodů . . . ', *op. cit.*, p. 204.

368 Arhivele Statului Cluj, *Protocole de Bistritza*, Vol. III, No. f. 14–15 (dated 30/V/1571) refers to one Jan Luchiczki who came to Transylvania on behalf of Samuel Sboroczi (Zborowski) to buy horses for Cracow.

369 P. Horváth, 'Obchodné styky Levoče . . . ', *op. cit.*, p. 124; see also E. Fugedi, 'A bártfai XVI század eleji bor-és lókivitel kérdése', *Agrártörteneti Szemle*, Vol. 14, No. 1–2, Budapest, 1972, pp. 41–89.

370 In 1584 a total of 684 horses were noted in the customs books of Cracow; R. Rybarski, *Handel i polityka . . . op. cit.*, Vol. 2, p. 194; in 1591/1592 these totalled just over 400, Wojewódzkie Archiwum Państwowe w Krakowie, *Księgi celne*, rkp. 2116.

371 H. Obuchowska-Pysiowa, 'Handel wołami w świetle rejestru celnego komory krakowskiej z 1604 roku', *Kwartalnik Historii Kultury Materialnej*, Vol. XXI, No. 3, Warszawa, 1973, p. 512.

372 M. Horn, 'Handel wołami na Rusi Czerwonej w pierwszej połowie XVII

wieku', *Roczniki Dziejów Społecznych i Gospodarczych*, Vol. XXIV, Warszawa, 1962, p. 86.

373 H. Obuchowska-Pysiowa, 'Struktura handlu Krakowa . . . ', *op. cit.*, p. 508.

374 W. Rusiński, 'The Role of Polish Territories . . . ', *op. cit.*, p. 127.

375 Archiv v Oravského Podzámku, No. 217/27 (dated 1621); sheep were als purchased in Orava in 1619, Ibid. No. 218/38.

376 V. Zimany, 'Les problèmes principaux du commerce extérieur de la Hongrie a partir du milieu du XVIe jusqu'au milieu du XVIIe siècle', in *La Pologne et la Hongrie aux XVIe–XVIIIe siècles*, V. Zimányi (ed.) (Akadémiai Kiadó) Budapest 1981, p. 39; Z. P. Pach, The Role of East-Central Europe . . . *op. cit.*, p. 258. V. Zimányi and H. Prickler, 'Konjuktúra es depresszió a XVI–XVII századi Magyarországon az ártörténet és harmincadbevételek tanúsagai alapján: kitekintés a XVIII századra', *Agrártörteneti Szemle*, Vol. 16, No. 1–2, Budapest, 1974, p. 79–201.

377 A. Attman, *The Russian and Polish markets . . . op. cit.*, p. 164.

378 J. Topolski, 'Faktoren der Entstehung . . . ', *op. cit.*, p. 116; H. Rachel, 'Polnische Handels-und Zollverhältnisse im 16 bis 18 Jahrhundert', *Jahrbuch für Gesetzbegung, Verwaltung und Volkswirtschaft im Deutschen Reich*, Vol. 33, No. 2, Leipzig, 1909, p. 59; H. Samsonowicz, 'Les foires en Pologne au XVe et XVIe siècle sur la toile de fond de la situation économique en Europe', in *Der Aussenhandel Ostmitteleuropas 1450–1650*, I. Bog (ed.), Köln-Wien, 1971, p. 246.

379 M. Kulczykowski, *Kraków jako ośrodek . . . op. cit.*, p. 64.

380 K. Kuklińska, 'Commercial Expansion . . .' *op. cit.*, p. 453.

381 W. Pruski, *Hodowla zwierząt gospodarskich w Galicji w latach 1772–1918* (P.A.N.) Wrocław-Kraków-Gdańsk, 1975, Vol. I, p. 99; M. Kulczykowski, 'Handel Krakowa w latach 1750–1772 . . . ', *op. cit.*, p. 98.

382 Wojewódzkie Archiwum Państwowe w Krakowie, *Księgi celne*, rkp. 2232 (dated 1762), p. 22, 23, 113, 116, 132, 140. Ibid. rkp. 2231 (1759), p. 84.

383 K. Kuklińska, 'Commercial Expansion . . .' *op. cit.*, *op. cit.*, p. 453.

384 Wojewódzkie Archiwum Państwowe w Krakowie, *Regestra pensae minoris civitatis cracoviensis* (mała waga), rkp. 2313 (dated 1771).

385 Ibid. *Protocalla, causarium criminalium officii Consularis cracoviensis*, rkp. 887, pp. 457–8 (dated 1760); Ibid. *Księgi celne*, rkp. 2230 (1757), p. 102; Ibid. rkp. 2232, p. 92.

386 M. Kulczykowski, *Kraków jako ośrodek . . . op. cit.*, p. 66.

387 F. Braudel and F. C. Spooner, *op. cit.*, p. 414.

388 J. Rutkowski, *Historia Gospodarcza Polski . . . op. cit.*, p. 87.

389 S. Kutrzeba, Wisła w historji . . . *op. cit.*, p. 41.

390 J. Rutkowski, *Historia Gospodarcza Polski . . . op. cit.*, p. 88.

391 Biblioteka Czartoryskich w Krakowie, rkp. 1732 (*rachunki Decjusza*), p. 17, dated 1533 and referring to 'emporio piscium'.

392 H. Samsonowicz, 'Le commerce maritime de Gdańsk . . .' *op. cit.*, p. 51 Table I.

393 K. Pieradzka, 'Handel Krakowa z Węgrami . . .' *op. cit.*, p. 233; R. Rybarski, *Handel i polityka . . . op. cit.*, Vol. II, pp. 192–3.

394 Wojewódzkie Archiwum Państwowe w Krakowie, *Księgi celne*, rkp. 2117–2118.

395 T. Chudoba, 'Z zagadnień handlu wiślanego Warszawy w XVI wieku', *Przegląd Historyczny*, Vol. 50, No. 2, Warszawa, 1959, p. 306.
396 M. Rawicz-Witanowski, *Monografia Łęczycy*, Kraków, 1898, p. 185, for example refers to a consignment of eels and herring sent from Łęczyca to Cracow in 1564.
397 Archiwum Główne Akt Dawnych w Warszawie, *Archiwum Skarbu Koronnego*, Oddział XIII, No. 3 dated 1533–1534 refers to 58 vats of Lwów fish, and about 2,000 barrels of other fish from Lublin and Krasnystaw arrived in Cracow en route to Hungary. Quoted in K. Pieradzka, 'Handel Krakowa z Węgrami ...' *op. cit.*, pp. 231–2.
398 Ibid., 233
399 Wojewódzkie Archiwum Państwowe w Krakowie, *Księgi celne*, rkp. 2117, 2118, for 1594; see also J. M. Małecki, *Studia nad rynkiem* ... *op. cit.*, p. 164.
400 K. Pieradzka, 'Handel Krakowa z Węgrami ...' *op. cit.*, p. 231.
401 M. Dan, Negustorii clujeni la Cracovia ... *op. cit.*, p. 210.
402 Wojewódzkie Archiwum Państwowe w Krakowie, *Księgi celne*, rkp. 2117 (1593), and 2119 (1595); R. Rybarski, *Handel i polityka* ... *op. cit.*, Vol. I, pp. 75–86; M. Dan, 'Relaţiile comerciale dintre Oradea ...' *op. cit.*, p. 180.
403 K. Pieradzka, 'Handel Krakowa z Węgrami ...' *op. cit.*, pp. 231–2.
404 Ibid., p. 232.
405 P. Horváth, 'Obchodné styky Levoče ...' *op. cit.*, pp. 129–30; Ibid., 'Príspevok k obchodným stykom ...' *op. cit.*, pp. 185–6; F. Matějek, K dějinám českého obchodu ...' *op. cit.*, p. 204.
406 A. E. Christensen, *Dutch Trade to the Baltic abnout 1600 – Studies in the Sound Toll Registers and Dutch Shipping Records*, Copenhagen/The Hague, 1941, p. 17.
407 R. W. Unger, 'Dutch Herring, Technology, and International Trade in the Seventeenth Century', *Journal of Economic History*, Vol. XL, No. 2, Atlanta, Georgia, 1980, p. 263; S. Haak, 'Brielle als vrije en bloeinde Handelsstad in de 15de eeuw', *Bijdragen voor Vaderlandsche Geschiedenis en Oudheidkunde* (4th series), Antwerp, 1907, pp. 40–3; N. Gottschal, *Fischereigewerbe und Fischandel der Nierdländischen Gebiete im mittelalter*, Bod Wörishofen, 1927, pp. 39–42.
408 R. W. Unger, *op. cit.*, p. 263; see also W. S. Unger, 'De Sonttabellan', *Tijdschrift voor Geschis*, Vol. 41, Antwerp, 1926, p. 144; N. E. Bang and K. Korst (eds.), *Tabeller over Skifsfart og Varentransport gennem Øresund, 1498–1783*, Copenhagen/Leipzig, 1906–1953, and compare this with H. Obuchowska-Pysiowa, 'Handel wiślany w pierwszej ...' *op. cit.*, pp. 16–17, Tables 44–5. Data is missing for certain years.
409 For further discussion see M. Bogucka, 'Handel Bałtycki Amsterdamu w pierwszej połowie XVII wieku w świetle kontraktów frachtowych', *Zapiski Historyczne*, Vol. XXXIV, No. 2, Toruń, 1969, p. 185; Ibid., 'Z zagadnień obrotów wewnętrznych regionu Bałtyckiego; Handel Gdańsk-Sztokholm w 1643 roku, *Zapiski Historyczne*, Vol. XLIII, No. 4, Toruń, 1978, p. 48; A. Groth, 'Wybrane problemy handlu Elbląga w latach 1698–1711', *Rocznik Gdański*, Vol. XXXIX, No. 1, Gdańsk, 1979, pp. 133–44; M. Hroch, 'Obchod mezi východni a zapadní Evropou ...' *op. cit.*, p. 483; N. Kostomarov, 'Ocherk torgovli Moskovskogo gosudarstva v XVI i XVII stol.', *Sobranie sochinenii*, kn. 8, T.20, St Petersburg, 1905, pp. 278, 391, 406.

410 H. Obuchowska-Pysiowa, 'Handel wiślany w pierwszej ...' *op. cit.*, p. 149.

411 J. Sygański, *Analekta sandeckie do XVI i XVII wieku*, Lwów, 1905, pp. 20–1, refers to herrings sent from Gdańsk via Lublin to Cracow and exported to Bardejov totalling 4 barrels and costing 14½ talers (i.e. 580 grosz) in 1609.

412 M. Marečková, 'Prešov v uhersko-polských obchodních ...' *op. cit.*, p. 432.

413 Wojewódzkie Archiwum Państwowe w Krakowie, *Księgi celne*, rkp. 2142 (dated 1624) refers to 23 barrels of herrings and one barrel of salted fish sent to Jičín from Cracow by merchants from Międzyrzecz and Šenov. Ibid., rkp. 2202 (dated 1688) refers to a small quantity of herrings sent from Cracow to northern Hungary.

414 For example Arhivele Statului Cluj, *Socotelile Oraşului Cluj*, 14a, XXII, p. 5 (dated 17/1/1617).

415 W. S. Unger, 'Trade through the Sound in the Seventeenth and Eighteenth Centuries', *Economic History Review* (2nd Series), Vol. 12, London, 1959, p. 208.

416 Wojewódzkie Archiwum Państwowe w Krakowie, *Księgi celne*, rkp. 2232 (dated 1763) in which a total of 364 barrels of herrings were transported to Cracow.

417 K. Kuklińska, Handel Poznania w drugiej ... *op. cit.*, pp. 101–2.

418 Wojewódzkie Archiwum Państwowe w Krakowie, *Archiwum Depozytów*, rkp. 161, p. 27. In 1755, salted fish including sturgeon were sent to Cracow weighing 84 cetnars (*circa* 8 tons); Ibid. p. 68. In 1763, 11 barrels of sturgeon sent to Cracow from Lwów, Jarosław and Stanisławów.

419 M. Kulczykowski, *Kraków jako ośrodek ... op. cit.*, p. 64.

420 Wojewódzkie Archiwum Państwowe w Krakowie, *Archiwum Depozytów*, rkp. 161, pp. 27, 30–1, 45, 48, 62.

421 M. Kulczykowski, *Kraków jako ośrodek ... op. cit.*, p. 67.

422 A. E. Christensen, *op. cit.*, p. 465.

423 For comparison, see herring prices in Gdańsk 1550–1780 in R. W. Unger, *op. cit.*, p. 271, Fig. 2.

424 W. Naudé, *Die Getreidehandelspolitik der europaischen Staaten vom XIII bis zum XVIII Jahrhundert*, Berlin, 1896, especially Ch. 4.

425 A. E. Christensen, *op. cit.*, pp. 465–7; M. Hroch, 'Obchod mezi východni ...' *op. cit.*, p. 490.

426 J. Topolski, 'Commerce des denrées agricoles et croissance économique de la zone baltique aux XVIe et XVIIe siècles', *Annales: Économies, Sociétés, Civilisations*, Vol. 29, No. 2, Paris, 1974, pp. 425–35; A. Wyczański, 'La base intérieure de l'exportation polonaise des céréales dans la seconde moitié de XVI e siècle', in *Der Aussenhandel Ostmitteleuropas ... op. cit.*, p. 261.

427 *Regestra thelonei aquatici Wladislaviensisis saeculi XVI*, S. Kutrzeba and F. Duda (eds.), Cracow, 1915, pp. 105, 130, 131, 135. In 1556, 40 łasts (i.e. *circa* 90 tons) of cereals sent down Vistula to Gdańsk by Cracow merchants; 1557, 12 łasts of rye (64 tons), and 70 łasts (160 tons) of cereals and in 1575 a further 20 łasts (46 tons) of cereals recorded on this route. See also J. M. Małecki, 'Przczynek do dziejów ...' *op. cit.*, p. 144.

428 A. Olejarczuk, 'Towarowość Małopolskiej producji zobżowej w XVI wieku. Próba rejonizacji', *Przegląd Historyczny*, Vol. LXIV, No. 4, Warszawa, 1971, Table 2, p. 731; R. Rybarski, *Handel i polityka ... op. cit.*, Vol. II, Table 4.

429 In 1584, 42 horses (i.e. 21 tons) of cereals and 25 horses (12½ tons) of millet and

buckwheat left Cracow for Bielsko; R. Rybarski, *Handel i polityka ... op. cit.*, Vol. II, p. 191; in 1591/92 a total of 205 ćwiertinas and 67 horses (i.e. 36½ tons) of cereals, 54 ćwiertinas (¾ ton) of rye, 6 horses (3 tons) of barley, and 9 horses (4½ tons) of buckwheat and peas left Cracow for Silesia; Wojewódzkie Archiwum Państwowe w Krakowie, *Księgi celne*, rkp. 2116, k. 75 v; 76 v; 79. See also J. M. Małecki, *Studia nad rynkiem ... op. cit.*, p. 147.

430 J. M. Małecki, *Studia nad rynkiem ... op. cit.*, p. 147.

431 Ibid.

432 H. Obuchowska-Pysiowa, 'Handel wiślany w pierwszej ...', *op. cit.*, Table 24, p. 150 in which it was calculated that Cracow held twelfth place in order of grain exported (in łasts).

433 L. Lepszy, 'Przemysł artystyczny i handel ...', *op. cit.*, p. 300.

434 P. Jeannin, 'The Sea borne and the Overland ...', *op. cit.*, p. 52.

435 H. Madurowicz and A. Podraza, *Regiony gospodarcze Małopolski zachodniej w drugiej połowie XVIII wieku*, Wrocław, 1958, pp. 153–5.

436 M. Kulczykowski, *Kraków jako ośrodek ... op. cit.*, p. 39.

437 Ibid., pp. 40–1.

438 J. L. Louis, *op. cit.*, p. 12.

439 For example, Wojewódzkie Archiwum Państwowe w Krakowie, *Libri clavigerorum civitatis Cracoviae*, rkp. 1507 (dated 23/II-1/IX/1769); Ibid. rkp. 1508 (dated 1769–1791) note about 40 places sending cereals to Cracow from the local region; cereals from Żywiec – for example 1755: *Księgi celne*, rkp. 2223, pp. 17, 21, 22, 33, 35, 91, 109, 145, 163, 165. Sometimes wood was brought to Cracow in exchange for cereals (wheat and rye), e.g. 1750 from Jordanów, *Księgi celne*, 2216, p. 117, from Maków, *Księgi celne*, rkp. 2216, p. 117, and *Varia Civitatis*, Inwentarz Tymczasowy, rkp. 279 (dated 1764).

440 J. Reychman, 'Le commerce polonais ...', *op. cit.*, p. 247.

441 *Edicta et mandata universalia in Regnis Galiciae et Lodomeriae*, Kraków, 1785, p. 202, refers to law dated 22/XII/1785 and also mentions a lower customs duty on imported cereals.

442 *Dziennik Handlowy*, Kraków, 1787, p. 574. One of the more active Cracow merchants in this trade was Piotr Nowicki who sent consignments of cereals down the R. Vistula to Gdańsk, e.g. Wojewódzkie Archiwum Państwowe w Krakowie, *Korespondencja handlowa i prywatna, bilanse, i akta kupca krakowskiego, W. Piotra Nowickiego*, rkp. 3218, letters of 10/V/1769, 1/V/1770, 8/V/1770; the letter of 1/VI/1770 refers to the purchase of 1,000 korzecs (¾ ton) of barley. For other examples see also Ibid. rkp. 3157 (dated 1757).

443 M. Frančić, 'Kraków produkujący i konsumujący', *Studia z Historii Społeczno-Gospodarczej Małopolski*, No. 6 (Instytut Historii (P.A.N.), Warszawa, 1963, pp. 157–8.

444 H. Madurowicz and A. Podraza, *op. cit.*, p. 137, refers to 140 carts (109 tons) of flour sent to these towns in 1780, and 150 carts (116 tons), in 1785.

445 J. Pelc, *op. cit.*, pp. 60–2; E. Tomaszewski, *op. cit.*, pp. 54–5; F. Braudel and F. C. Spooner, *op. cit.*, pp. 392–8.

446 S. Hoszowski, 'The Revolution of Prices in Poland in the 16th and 17th Centuries', *Acta Poloniae Historica*, Vol. II (P.A.N. Instytut Historii), Warszawa, 1959, p. 10.

447 Ibid., 'L'Europe Centrale devant ...' *op. cit.*, p. 449.
448 Ibid., 'The Revolution of Prices ...', *op. cit.*, p. 11.
449 Ibid., p. 12.
450 Ibid.,
451 K. Kuklińska, 'Les rôles joués par les marchés intérieurs et extérieurs dans le développement du commerce polonais au XVIIIe siècle', *Studia Historiae Oeconomicae*, Vol. 11 (U.A.M.) Poznań, 1976, p. 97; Ibid. Commercial Expansion in XVIIIth ..., *op. cit.*, p. 459.
452 W. Kula, *Teoria ekonomiczna ustroju feudalnego*, Warszawa, 1962, pp. 104–6; see also H. Madurowicz-Urbańska, *Ceny zboża w zachodniej Małopolsce w drugiej połowie XVIII wieku*, Warszawa, 1963.
453 H. Obuchowska-Pysiowa, 'O handlu drewnem w Polsce w XVI wieku', *Sylvan: A 108*, No. 4, Warszawa, 1964, pp.55–64.
454 J. M. Małecki, *Studia nad rynkiem ...*, *op. cit.*, p. 166, Table 30.
455 Wojewódzkie Archiwum Państwowe w Krakowie, *Księgi dochodów i wydatków 1570–1588*, rkp. 1632–1653.
456 M. Kromer, *Polania sive de origine et rebus gestis*, Basle, 1589, p. 808, states 'habet opportunitatem Vistulae fluvii, quo pisces, ligna et omnis aedificatiorum materia et alia queadam e vicina Silesia importantur'. For further discussion see H. Barycz, *Szlakami Dziejopisarstwa Staropolskiego: Studia nad Historiografią w XVI–XVIII wieku* (P.A.N.), Wrocław-Warszawa-Kraków-Gdańsk, 1981, pp. 83–123.
457 Ibid., *Księgi celne*, rkp. 2115, 2117, 2118, for the years 1589, 1593, 1594.
458 Ibid.
459 Ibid., rkp. 2115, 2117.
460 For example, in 1579, Biblioteka Czartoryskich w Krakowie, rkps. 1043, pp. 83–4; in 1586, Wojewódzkie Archiwum Państwowe w Krakowie, *Księgi dochodów i wydatków, rkp. 1649, p. 27.*
461 R. G. Albion, *Forests and Sea Power*, Cambridge, Mass., 1926, pp. 43–5; H. S. K. Kent, *War and Trade in the Northern Seas*, Cambridge, 1973, pp. 39–54; H. Kempas, *Seeverkehr und Pfundzoll im Herzogtum Preussen*, Bonn, 1964, pp. 292–3; A. Soom, 'Der Ostbaltische Holzhandel und die Holzindustrie im 17 Jahrhundert', *Hansische Geschichtsblätter*, Vol. 79, Lübeck, 1961, pp. 80–100; R. Rybarski, *Handel i polityka ...*, *op. cit.*, Vol. I, p. 50; see also M. Bogucka, Z zagadnień obrotów ..., *op. cit.*, p. 48, Table 4.
462 W. Rusinski, 'The Role of Polish Territories ..., s*op. cit.*, p. 121; E. Cieślak, 'Sea-borne Trade between France and Poland in the XVIIIth Century'. *Journal of European Economic History*, Vol. 6, No. 1, Rome 1977, p. 55.
463 J. Broda, 'Cech włóczków krakowskich', *Roczniki Dziejów Społecznych i Gospodarczych*, Vol. 16, Warszawa-Poznań, 1954, pp. 228, 293.
464 M. Kulczykowski, *Kraków jako ośrodek ...*, *op. cit.*, p. 70.
465 Ibid.
466 J. Pelc, *op. cit.*, p. 70; E. Tomaszewski, *op. cit.*, p. 71.
467 F. Braudel and F. C. Spooner, *op. cit.*, p. 418, and Fig. 28.
468 E. Tomaszewski, *op. cit.*, p. 71.
469 N. J. G. Pounds, *An Historical Geography of Europe 1500–1840 ... op. cit.*, pp. 217–18.

470 I. Bog, 'Wachstumsprobleme der oberdeutschen Wirtschaft 1540–1618', *Jahrbücher für National-ökonomie und Statistik*, Vol. 179, Stuttgart, 1966, pp. 493–537.

471 S. Inglot, 'Economic Relations of Silesia with Poland from the XVI Century to the Beginning of the XVIII Century', *Annales Silesiae*, Vol. 1, No. 2, Wrocław, 1960, pp. 153–70.

472 N. J. G. Pounds, *An Historical Geography ... op. cit.*, p. 218.

473 F. J. Fisher, 'Commercial Trends and Policy in Sixteenth Century England', *Economic History Review*, Vol. 10, November 1940, London, pp. 95–117; A. Friis, *Alderman Cockayne's project and the cloth trade in the reign of James I: the commercial policy of England in the main aspects, 1603–25*, Copenhagen-London 1927, p. 230; about 1500 some 50,000 cloths: by 1528, 75,000 cloths were exported from London. In 1540 the figure already exceeded 100,000 cloths and ten years later 130,000. In 1606, the Merchant Adventurers Company of London alone exported 100,000 ('white' and dressed) cloths.

474 W. S. Unger, 'Trade through the Sound ...' *op. cit.*, pp. 209–10; H. Zins, *England and the Baltic in the Elizabethan era*, Manchester Univ. Press/Rowman and Littlefield, Totowa, New Jersey, 1972, p. 182; M. Hroch, 'Die Rolle des Zentraleuropäischen Handels ... *op. cit.*, pp. 18–19; J. Rutkowski, *Historia Gospodarcza ... op. cit.*, p. 130, refers to over 8,000 postaws (warps) of cloth coming via the Sound to Gdańsk in the second half of the sixteenth century, most important and expensive of which were Dutch and English cloths; see also A. Mączak, *Między Gdańskiem a Sundem: Studia nad handlem bałtyckim od połowy XVI do połowy XVII wieku* (P.W.N.) Warszawa, 1972, pp. 118, 126.

475 A. Klíma, *Manufakturní období v Čechách*, Praha 1955, pp. 127–53; A. Mączak, *Sukiennictwo wielkopolskie w XIV–XVII wieku*, Warszawa, 1955, pp. 244–68.

476 M. Małowist, Poland, Russia and Western Trade ... *op. cit.*, p.30; E. Nadel-Golobić, *op. cit.*, p. 356.

477 For example see R. Rybarski, *Handel i polityka ... op. cit.*, Vol. I, p. 204, which gives 'A list of Fairs to Attend' based on a document dated 1584 produced for an attorney or customs official in Greater Poland, and gives the more important fairs to attend in Silesia and N. Poland.

478 J. M. Małecki, *Studia nad rynkiem ... op. cit.*, p. 187.

479 J. Ptaśnik, 'Bonerowie' ... *op. cit.*, p. 17; L. Petry, *Die Popplau. Eine schlesische Kaufmannsfamilie des 15 und 16 Jahrhunderts*, Breslau, 1935, 204 pp.

480 H. Zins, ... *op. cit.*, p.182; it was also known as 'Lundisch' together with some descriptive adjective e.g. 'gemein Lundisch' (common Luński), 'breite Lacken' (broadcloth), 'fein Lundisch' (falendisch) etc. Wojewódzkie Archiwum Państwowe w Gdańsku, *Libri Portorii Elbingenses*, for the years 1586, 1587, 1596, 1599.

481 R. Rybarski, *Handel i polityka ... op. cit.*, Vol. 1, pp. 163–4; A. Mączak, *Sukiennictwo ... op. cit.*, p. 230.

482 H. Zins, *op. cit.*, p. 182.

483 J. Pelc, *Ceny w Krakowie ... op. cit.*, p. 69.

484 A. Wagner, Handel dawnego Jarosławia ... *op. cit.*, p. 134.

485 R. Rybarski, *Handel i polityka ... op. cit.*, Vol. II, pp. 186–7.

486 For example see S. Schneebalg-Perelman, 'Les tapisseries flamandes au château de Wawel à Cracovie', in *La tapisserie flamande et le grand témoignage du Wawel* (Fonds Mercator), Antwerp, 1972, pp. 377–437.
487 M. Wolański, *Związki handlowe Śląska ... op. cit.*, pp. 24–5; R. Rybarski, *Handel i polityka ... op. cit.*, Vol. 2, pp. 203–5 refers to 128 bales (51,440 metres) of 'gierlicz' (Zgorzelec:Gorlitz) cloth sent from Wrocław to Cracow in 1584 together with smaller quantities of other cloth.
488 M. Wolański, *op. cit.*, p. 20.
489 Cloth from Moravia was recorded in documents according to place of origin e.g. 'iczinskie' (Jičín), 'Prosczionskie' (Přerov), 'przeborskie' (Příbor), 'ołomunieckie' (Olomouc)' etc. See I. Krypiakevych, 'Materialy do istorii torhovli Lvova', *Zapysky naukovoho Tovarystva imeny Shevchenka*, Vol. XIV, No. 65, Lwów, 1905, pp. 4, 27, 34, 37.
490 P. Horváth, 'Príspevok k obchodným stykom ...' *op. cit.*, p. 174.
491 J. Chylík, *Přehled dějin moravského průmyslu*, Vol. I, Brno, 1948, p. 23.
492 J. Janáček, *Dějiny obchodu v předbělohorské Praze*, Praha, 1955, p. 147; see also Ibid., *Řemeslná výroba v českých městach v 16 století*, Praha, 1961, 156 pp.
493 A. Mączak, *Sukiennictwo wielkopolskie ... op. cit.*, pp. 244–5.
494 T. Ślawski, *Produkcja i wymiana towarowa Biecza w XVI i XVII wieku*, Rzeszów, 1968, p. 92–113.
495 R. Rybarski, *Handel i polityka ... op. cit.*, Vol. II, p. 186 refers to 6 horses of cloth (i.e. 3,600 metres) sent from Biecz together with 10 horses (6,000 metres) from Krosno and 11 horses (6,600 metres) from Stary Sącz. See also A. Lewicka, *Krosno w wiekach średnich*, Krosno, 1933, pp. 93–4, which refers to Krosno cloth sales in Cracow during the fifteenth century.
496 W. Łuszczkiewicz, *Sukiennice Krakowskie*, Kraków 1899, p. 30; Italian merchants in Cracow at this time included Giulio and Luca del Pace, Paolo Cellari, Antonio de Stesi, Filippo Falducci, Geronimo Oremus, Sebastiano Monteluppi and Giovanni Baptista Cechi; see also D. Quirini-Popławska, 'Die italienische ...' *op. cit.*, p. 330; Ibid. *Działność Wlochów w Polsce w połowie XVI wieku na dworze królewskim w dypolomacji i hierarchii kościelnej*, Wrocław, 1973 139 pp; J. Olkiewicz, *Opowieśći O Wlochach ... op. cit.*, pp. 35–47.
497 J. Ptaśnik, Z dziejów kultury włoskiego Krakowa, *Rocznik Krakowski*, Vol. IX, Kraków, 1907.
498 A. Mączak, *Sukiennictwo wielkopolskie ... op. cit.*, p. 230.
499 Wojewódzkie Archiwum Państwowe w Krakowie, *Consularia Cracoviensia*, rkp. 448, pp. 223–7; 342; 392–8; 641; 654; 657; 659; 825; 829; 842; 844; Ibid., rkp. 449, pp. 11; 12–15.
500 M. Hrushevsky, *Istoriia Ukrainy-Rusy*, Vol. VI, Kiev, 1907, p. 90; I. Krypiakevych, *op. cit.*, p. 45.
501 Z. Wojciechowski, *Zygmunt Stary, 1506–1548*, Warszawa, 1946, pp. 309–48.
502 J. Tazbir, 'Poland's "Golden Age" (1492–1586)', Ch. VII, in *History of Poland*, A. Gieysztor et al. (eds.) (P.W.N.), Warszawa, 1968, p. 207.
503 Ibid.
504 S. Kutrzeba, 'Dzieje handlu i kupiectwa krakowskiego', *Rocznik Krakowski*, Vol. XIV, Kraków, 1910, p. 8; I. Krypiakevych, *op. cit.*, p. 45.
505 E. Nadel-Golobić, *op. cit.*, p. 356; J. M. Małecki, *Studia nad rynkiem ... op. cit.*, pp. 47–8.

506 A. Mańkowski, 'Kupcy ormiańscy w Gdańsku', *Zapiski Towarzystwa Naukowego w Toruniu*, Vol. 4, Toruń, 1917–1919, pp. 182–3.
507 K. Pieradzka, 'Handel Krakowa z Węgrami . . .' *op. cit.*, p. 213.
508 R. Rybarski, *Handel i polityka . . . op. cit.*, Vol. II, pp. 186–7, 203–5.
509 Ibid.
510 I. T. Baranowski, *Przemysł polski w XVI wieku*, Warszawa, 1919, pp. 148–53.
511 For example in 1552, Biecz cloth cost 3.2 złotys/warp, R. Rybarski, *Handel i polityka . . . op. cit.*, Vol. II, p. 304; in 1565, Zgorzelec cloth cost 18 florins/warp, Ibid., p. 302; Lwówek cloth throughout much of the sixteenth century cost between 4.66 and 7 zołtys/warp; Ibid., pp. 300–1; Kersey cloth from 1555–1590 on average cost 17–20 grosz/ell, Ibid., p. 303; London cloth varied in price between 20 and 40 złotys/warp; Ibid., pp. 300–1.
512 K. Pieradzka, 'Handel Krakowa z Węgrami . . .' *op. cit.*, p. 213; see also P. Horváth, Obchodné styky Levoče . . .' *op. cit.*, p. 134; Ibid., 'Príspevok k obchodným stykom Slovenska . . .' *op. cit.*, p. 187; F. Matějek, 'K dějinám českého obchodu . . .', *op. cit.*, p. 199; V. Zimány 'Les problèmes principaux du commerce extérieur . . .', *op. cit.*, p. 41; G. Granasztói, 'La ville de Kassa . . .', *op. cit.*, p. 61.
513 Wojewódzkie Archiwum Państwowe w Krakowie, *Księgi celne*, rkp. 2118, k. 14, p. 56 refers to linen cloth sent from Cracow to Debrecin and Koloszvar (Cluj), dated 1594; see also R. Rybarski, *Handel i polityka . . . op. cit.*, Vol. II, p. 187.
514 In 1528, Hungary tried to monopolize the linen trade concentrating it in the hands of Bardejov merchants; Bardejov specialized in calico-making and bleaching. E. Janota, *Bardyów, historyczno-topograficzny opis miasta i okolic*, Kraków, 1862, p. 25; failure to exclude Polish linen imports led to further attempts to restrict its sale through the Spiš Chamber of Commerce in 1590. Magyar Országos Levéltar (Budapest Archive), *Camera Scepusensis Libri Expeditionum ad Maiestatem*, Vol. 9, folio 141.
515 K. Pieradzka, 'Handel Krakowa z Węgrami . . .' *op. cit.*, p. 227.
516 D. C. Giurescu, *Illustrated History of the Romanian People* (Editura Sport-Turism), Bucureşti, 1981, p. 216.
517 A. Mączak, *Sukiennictwo wielkopolskie . . . op. cit.*, p. 232.
518 Arhivă de State de Braşov, *Gewandhandelregiser*, 1579, Vol. III, A/15; I. Bogdan, *Documente privitoare la relaţiile Ţarii Românesti cu Braşovul si cu Ţara Ungurească în sec. XV şi XVI*, Bucureşti, 1905, pp. 233–5.
519 L. Demény, 'Comerţul de tranzit . . .', *op. cit.*, p. 465.
520 G. Berchet, *La Republica de Venezia e la Persia*, Torino, 1865, No. 28, p. 191; M. Małowist, 'Poland, Russia . . .', *op. cit.*, p. 30.
521 M. Hrushevsky, *op. cit.*, Vol. VI, p. 67; S. Kutrzeba, Dzieje handlu i kupiectwa . . . *op. cit.*, p. 6; E. Nadel-Golobič, *op. cit.*, p. 380; see also E. C. G. Brunner, 'De ontwikkeling van het Handelsverkeer van Holland met Oosteuropa tot einde der 16 eeuw', *Tijdschrift voor Geschiedenis*, Vol. XLI, Groningen, 1926, 14–18.
522 I. Krypiakevych, *op. cit.*, pp. 2–6, 11. Between 1600–1604 the Lwów firm of M. Scholtz and P. Boim bought 16,877 ells of cloth (8,438½ metres) from Cracow merchants out of their total purchases of 59,525 ells (29, 762½ metres) i.e. 28 per cent.

523 W. Łoziński, *Patrycyat i mieszczaństwo lwowskie w XVI i XVII wieku*, Lwów 1892, p. 118.
524 H. Samsonowicz, *Handel Lublina ... op. cit.*, p. 622.
525 For example in 1552. R. Rybarski, *Handel i polityka ... op. cit.*, Vol. II, p. 238.
526 Wojewódzie Archiwum Państwowe w Krakowie, *Advocatus Cracoviensia*, No. 207, p. 858 (dated 2/II/1586).
527 Wojewódzkie Archiwum Państwowe w Krakowie, *Księgi celne*, rkp. 2119, p. 138 (dated 1595); Ibid., rkp. 2121, p. 254 (dated 1600) and refers to a Vilna merchant who transported 2 horses (i.e. 1,200 metres) of cloth from Cracow to his home town.
528 A. Leszczyński, 'Handel Żydów ziemi bielskiej od XVI wieku do 1795 roku', *Biuletyn Żydowskiego Instytutu Historycznego*, Vol. 119, No. 2, Warszawa, 1979, pp. 33–51.
529 K. Pieradzka, 'Handel Krakowa z Węgrami ...' *op. cit.*, p. 221.
530 C. M. Cipolla, 'The Decline of Italy: The Case of a Fully Matured Economy', *Economic History Review*, Vol. V (2nd series), London, 1952, pp. 178–87.
531 J. K. Fedorowicz, *England's Baltic Trade in the Early Seventeenth Century: A Study in Anglo-Polish Commercial Relations* (Cambridge Studies in Economic History) Cambridge-London, 1980, 347 pp.; H. Fiedler, 'Danzig und England: die Handelsbestrebungen der Engländer vom Ende des 14 bis zum Ausgang des 17 Jahrhunderts', *Zeitschrift des westprussischen Geschichtsvereins*, Vol. LXVIII, Leipzig, 1928, pp. 70–2; H. Zins, *England and the Baltic ... op. cit.*, pp. 153–215.
532 Wojewódzkie Archiwum Państwowe w Gdańsku, *Missiva* 300, 27/44, k. 414–15; Ibid., 27/44 k. 102 v = 103; Ibid., 27–46 k 8 v, concerning a dispute lasting from 1596–1598 between Cracow and Gdańsk merchants over 7 warps of cloth (i.e. 140 metres) from London, in which blame was attached to a trader for using different cloth measures than those normally used.
533 R. Rybarski, *Handel i polityka ... op. cit.*, Vol. II, pp. 204–6; Wojewódzkie Archiwum Państwowe w Gdańsku, *Księgi handlowe* 300, R/F no. 19a, k. 78 v – 79, 218 v – 219; Ibid., no. 19b, k. 70 – 72, 76 v – 77. One Gdańsk merchant alone between 1623–1644 sent cloth valued at several thousand złoty to Cracow; Wojewódzkie Archiwum Państwowe w Krakowie, *Księgi celne*, rkp. 2133, p. 257, dated 12/VIII/1611, refers to Franciszek Hoffman from Elbląg, sending 100 warps (2,000 metres) of Kersey cloth and 7 warps (140 metres) of fine cloth ('Falendysz') to Cracow. See also J. M. Małecki, 'Związki handlowe miast polskich z Elblągiem w XVI i pierwszej połowie XVII wieku', *Rocznik, Elbląski* Vol. 5, Warszawa-Poznań, 1972, pp. 129–38. J. K. Fedorowicz, 'The Struggle for the Elbing Staple: An Episode in the History of Commercial Monopolies', *Jahrbücher für Geschichte Osteuropas*, Vol. 27, No. 2 (F. Steiner Verlag GmbH), Wiesbaden 1979, pp. 224, 226.
534 Wojewódzkie Archiwum Państwowe w Krakowie, *Księgi wójtowskie*, rkp. 239, pp. 1338–9, 1387 (dated 1625) refers to a Cracow merchant, H. Konrad, whose home was searched by customs officials; it revealed textiles bought from a Gdańsk merchant including 65 półsztuk (1,040 metres) of varied cloth and several tens of pieces of other cloth, including English in different colours; a similar case of cloth confiscation by customs officials from the home of a Cracow

merchant occurred in 1632. Wojewódzkie Archiwum Państwowe w Krakowie, *Księgi wójtowskie*, rkp. 240, p. 1720.

535 D. Sella, *Commercio e industria a Venezia nel secolo XVII*, Venezia, 1961, p. 127, Table II; see also A. Manikowski, 'Les soieries italiennes et l'activité des commerçants italiens de soieries en Pologne au XVII siècle', *Mélanges de l'Ecole française de Rome: Moyen Age – Temps Modernes*, Roma, Vol. 2, 1976, pp. 823–43.

536 H. Obuchowska-Pysiowa, 'Trade between Cracow and Italy ...' *op. cit.*, p. 647.

537 H. Barycz, *Spojrzenie w przeszłość polsko-włoską* (Ossolineum), Warszawa, 1965, p. 50, refers to 13 representatives of Italian families in Cracow, of which, according to A. Manikowski ('Zmiany czy stagnacja? Z problematyki handlu polskiego w drugiej połowie XVII wieku', *Przegląd Historyczny*, Vol. LXIV, No. 4, Warszawa, 1973, p. 787) they accounted for almost a third of the councillors in Cracow; see also J. Rutkowski, 'Studia nad położeniem włościan w Polsce w XVII wieku', *Ekonomista*, Vol. 14, Warszawa, 1914, pp. 25–37.

538 H. Obuchowska-Pysiowa, 'Trade between Cracow and Italy ...' *op. cit.*, p. 643.

539 Wojewódzkie Archiwum Państwowe w Krakowie, *Księgi kupieckie*, rkp. 3231, p. 38 (dated 1636); Ibid., rkp. 819, pp. 14, 15 (Lorenzo Tucci, dated 1647), refer to 1,114 ells (557 metres) of Lucca taffeta, together with 1,039 ells (519½ metres) of Florentine and 65 ells (32½ metres) of Neapolitan taffetas.

540 Wojewódzkie Archiwum Państwowe w Krakowie, *Księgi kupieckie*, rkp. 3228, p. 87 (Lorenzo Tucci and Giovanni Bernado San Pietro, dated 1631).

541 Wojewódzkie Archiwum Państwowe w Krakowie, *Consularia Cracoviensia*, rkp. 458, p. 1082 (Antonio Viviani, dated 1617) refers to 183 ells (91½ metres) of patterned and 382 ells (191 metres) of smooth Florentine satin.

542 Wojewódzkie Archiwum Państwowe w Krakowie, *Księgi kupieckie*, rkp. 819, p. 13 (dated 1647) refers to 836 ells (418 metres) of smooth and 71 ells (35½ metres) patterned satin from Lucca.

543 Wojewódzkie Archiwum Państwowe w Krakowie, *Księgi kupieckie*, rkp. 3228, pp. 185; 240 (Genoa); Ibid. *Advocatus Cracoviensia*, rkp. 240, p. 1584 (Naples); Ibid., rkp. 237, p. 1011 (Milan); Ibid., rkp. 241, p. 2137, rkp. 3226, p. 88, rkp. 3254, pp. 157, 158 (Bologna).

544 Wojewódzkie Archiwum Państwowe w Krakowie, *Księgi kupieckie*, rkp. 3230, pp. 4, 8–10, 13–14 (L. Tucci, dated July 1636) refers to 3,588 ells (1,794 metres) of Genoese, 1,072 ells (536 metres) of Luccese and 851 ells (425½ metres) of Venetian silk damask.

545 M. Taszycka, *Wloskie jedwabne tkaniny odzieżowe w Polsce w pierwszej polowie XVII wieku* (Ossolineum), Wrocław-Warszawa-Kraków-Gdańsk, 1971, p. 40.

546 Wojewódzkie Archiwum Państwowe w Krakowie, *Advocatus Cracoviensia*, rkp. 237, p. 1011; Ibid., rkp. 239, p. 1453; *Consularia Cracoviensia*, rkp. 460, pp. 171–172, 176, 178 (Genoa); Ibid., *Księgi kupieckie*, rkp. 3228, pp. 79, 83, 240, 355; rkp. 3230, p. 7 ('velluti alla fiorentina'); Ibid., *Advocatus Cracoviensia*, rkp. 239, p. 1453; *Księgi kupieckie*, rkp. 3215, p. 52 (Venice); Ibid., *Advocatus Cracoviensia*, rkp. 237, p. 1011, rkp. 239, p. 1453; rkp. 240, p. 38; rkp. 241, p. 2189; *Księgi kupieckie*, rkp. 3226, p. 36; rkp. 3254, pp. 130–131 (Lucca). Ibid.

Advocatus Cracoviensia, rkp. 240, p. 1580; *Consularia Cracoviensia*, rkp. 460, p. 182; *Księgi kupieckie*, rkp. 3228, pp. 84, 239, 261; rkp. 3230, p. 15 (Naples).

547 Wojewódzkie Archiwum Państwowe w Krakowie, *Księgi kupieckie*, rkp. 3228, p. 545; rkp. 3251, p. 11 (knotted velvet from Milan); Ibid. (smooth sheared velvet from Reggio).

548 Wojewódzkie Archiwum Państwowe w Krakowie, *Advocatus Cracoviensia*, rkp. 237, p. 1008 (A. Viviani, dated 1617) refers to 59 ells (29½ metres) of cloth-of-gold; Ibid., *Księgi kupieckie*, rkp. 3230, p. 3 (L. Tucci, dated 1636) refers to 124¾ ells (62 metres) of Florentine cloth-of-gold for the royal court.

549 Wojewódzkie Archiwum Państwowe w Krakowie, *Księgi kupieckie*, rkp. 3228, p. 262 (L. Tucci, dated 1633) refers to brocade manufactured in Milan; Ibid., rkp. 819, pp. 2, 15 (L. Tucci, dated 1647) refers to Florentine and Milanese brocade in his shop inventory.

550 P. Thornton, *Baroque and Rokoko Silks*, London, 1965, p. 47–48.

551 M. Taszycka, 'Włoskie tkaniny jedwabne w Krakowie w drugiej połowie XVII wieku', *Kwartalnik Historii Kultury Materialnej*, Vol. XXVI, No. 2, Warszawa, 1978, p. 136.

552 Wojewódzkie Archiwum Państwowe w Krakowie, *Advocatus Cracoviensia*, rkp. 256, p. 386–387 (Guglielmo Orsetti, January 1655) received 198½ ells (99¼ metres) of taffeta from Lucca; Ibid., p. 1027 (Bartolomo Constanzi, February 1669) received 289 ells (144½ metres) of Luccese taffeta. Ibid., p. 387 (Orsetti, 1655) 376½ ells (188¼ metres) of Florentine taffeta received from Italy; Ibid., p. 1028 (Constanzi, 1669) 237 ells (118½ metres) of taffeta from Florence mentioned in shop inventory; *Księgi kupieckie*, rkp. 3205, p. 148, 187, 296, 406; Ibid., rkp. 3207, p. 186; Ibid., rkp. 3209, p. 78; Ibid., rkp. 3255, p. 526, 528, 616, all refer to taffeta 'alla mantovana i alla pisana'.

553 Wojewódzkie Archiwum Państwowe w Krakowie, *Advocatus Cracoviensia*, rkp. 256, p. 380, 386–387, 396, 405, 1029; *Księgi kupieckie*, rkp. 3209, p. 59, refer to imports of Florentine satin; Ibid., *Advocatus Cracoviensia*, rkp. 256, p. 379, 387, 396, 399, 403–404, 1029; *Consularia Cracoviensia*, rkp. 446, p. 238 all refer to satin from Venice; Ibid., rkp. 256, p. 379–380, 385–388, 396–397, 404–406, 1029; *Księgi kupieckie*, rkp. 3204, p. 78; Ibid., 3207, p. 131; Ibid., rkp. 3209, p. 11, 293, 301, refer to satin from Lucca; Ibid., *Advocatus Consularia*, rkp. 256, p. 380, 405, 1029 – satin from Bologna.

554 Wojewódzkie Archiwum Państwowe w Krakowie, *Advocatus Cracoviensia*, rkp. 256, p. 380, 386, 397, 405–406, 543, 1028–1029; Ibid., *Księgi kupieckie*, rkp. 3207, p. 105, 245, 265; Ibid., 3209, p. 180, 270, 278, 283, 292, 297, 301; Ibid., rkp. 3255, p. 100 all refer to damask cloth from Lucca; Ibid., rkp. 3209, p. 262, damask from Florence; Ibid., *Advocatus Cracoviensia*, rkp. 256, p. 406, 1031, damask from Venice.

555 Wojewódzkie Archiwum Państwowe w Krakowie, *Advocatus Cracoviensia*, rkp. 256, p. 378–379, 388, 395, 407, 1032; *Księgi kupieckie*, rkp. 3222, p. 63, 79, 83, refer to Venetian velvet imports.

556 Wojewódzkie Archiwum Państwowe w Krakowie, *Advocatus Cracoviensia*, rkp. 256, p. 407, 1032; Ibid., *Księgi kupieckie*, rkp. 3205, p. 402, mentions velvet from Genoa; Ibid., rkp. 256, p. 388; Ibid., *Księgi kupieckie*, 3207, p. 153, 275; Ibid., rkp. 3222, p. 23, 31, 48, 85, velvet from Lucca; Ibid., rkp. 3222, p. 25, 48, velvet from Florence.

557 Wojewódzkie Archiwum Państwowe w Krakowie, *Księgi kupieckie*, rkp. 3255, p. 211 (dated 1676) is the only reference made to 'velluto persechino di Cremona.

558 Wojewódzkie Archiwum Państwowe w Krakowie, *Księgi kupieckie*, rkp. 3209, p. 259, 278 (dated 1692) mentions '2 resti di drappi d'oro'; Ibid., rkp. 3204, p. 63; rkp. 3205, p. 64, 99, 129 (dated 1680), 194, 296, 333, 487; rkp. 3207, p. 118 (dated 1686); rkp. 3209, p. 97, 103, 121, 292; rkp. 3255, p. 326, 362, 494, 530, all refer to Italian brocade.

559 Wojewódzkie Archiwum Państwowe w Krakowie, *Księgi kupieckie*, rkp. 3205, p. 173, 291 (M. A. Frederici, July 1679; March 1682) lamé sent from Venice, 143 ells (71½ metres) and 222 ells (111 metres).

560 Wojewódzkie Archiwum Państwowe w Krakowie, *Księgi kupieckie*, rkp. 3204, p. 69 (M. A. Frederici, March 1676), 5 bolts (155 metres) of samite ordered from Venice; Ibid., rkp. 3209, p. 259 (dated April 1677) Frederici obtained a further 5 bolts of samite from Venice; Ibid., rkp. 3255, p. 305 (dated 11 July 1692) refers to 3 bolts (93 metres) of samite from Venice sold in Cracow.

561 Wojewódzkie Archiwum Państwowe w Krakowie, *Księgi kupieckie*, rkp. 3205, p. 402, 427, 436; Ibid., rkp. 3207, p. 9, 49, 72, 148, 229, 277; Ibid., rkp. 3209, p. 76, 77, 147, 301, all refer to purchase and selling of Venetian silk gauze in Cracow.

562 Wojewódzkie Archiwum Państwowe w Krakowie, *Księgi kupieckie*, rkp. 3204, p. 80, 84; Ibid., rkp. 3205, p. 333, all refer to a shop inventory of 1682 containing 111 ells (55½ metres) of silk poplin.

563 Wojewódzkie Archiwum Państwowe w Krakowie, *Advocatus Cracoviensia*, rkp. 256, p. 408 (Milan); Ibid., *Księgi kupieckie*, rkp. 3207, p. 126, 146, 169 (Florence).

564 H. Obuchowska-Pysiowa, 'Trade between Cracow and Italy ... ', *op. cit.*, p. 647.

565 M. Taszycka, 'Włoskie tkaniny jedwabne ... ', *op. cit.*, p. 152.

566 L. Brentano, 'Uber den grundherrlichen Charakter des hausindustriellen Leinengewerbs in Schlesian', *Zeitschrift für Sozial-und Wirtschaftsgeschichte*, Vol. I, Leipzig, 1983, p. 320–323; J. Horner, *The Linen Trade of Europe during the Spinning Wheel Period* (McCaw, Stevenson & Orr Ltd.), Belfast, 1920, p. 389–390.

567 H. Kisch, 'The Textile Industries in Silesia and the Rhineland: A Comparative Study in Industrialisation', *Journal of Economic History*, Vol. 19, No. 4, New York, 1959, p. 543.

568 P. Jeannin, 'The Sea-borne and the Overland Trade ...' *op. cit.*, p. 14.

569 S. Inglot, *op. cit.*, p. 159.

570 A. Mączak, *Sukiennictwo wielkopolskie ... op. cit.*, 246, 263: A. Wyrobisz, 'Functional Types of Polish Towns in the XVIth–XVIIth Centuries', *Journal of European Economic History*, Vol. 12, No. 1, Rome 1983, p. 81.

571 S. Inglot, *op. cit.*, p. 153.

572 M. Marečková, 'Prešov v Uhersko-Polských obchodních ... ', *op. cit.*, p. 433.

573 J. Bieniarzówna, Mieszczaństwo krakowskie w 17 wieku ... *op. cit.*, p. 43–46.

574 Wojewódzkie Archiwum Państwowe w Krakowie, *Księgi celne*, rkp. 2142 dated 1631, refers to 2350½ warps (47,010 metres) of different linen, 2295¾

warps (45,915 metres) of silk textiles, 143 warps of mohair (2,850 metres) and 66 warps (1,320 metres) of Milan cloth sent from Cracow to northern Hungary.

575 Stefan Bocskay, Prince of Transylvania led an uprising of the peasantry in north and eastern Hungary against Hapsburg rule in 1604. The conflict lasted for two years. See W. Felczak, *Historia Węgier* ... *op. cit.*, p. 137–139.

576 M. Suchý, *Dějiny Levoče*, Vol. I, cap. III, pt. 7, Košice 1974, p. 191.

577 For example in 1624, 1,861 ells (930½ metres) of velvet sent from Cracow to Levoča, Wojewódzkie Archiwum Państwowe w Krakowie, *Księgi celne*, rkp. 2142.

578 The number of merchants mentioned in Cracow's customs registers for 1624 and 1629 were 34 and 14 respectively compared with 88 and 61 for Levoča and 50 and 32 for Prešov; Wojewódzkie Archiwum Państwowe w Krakowie, *Księgi celne*, rkp. 2142, 2143.

579 Z. P. Pach, 'The Shifting of International ... ', *op. cit.*, p. 313.

580 Ibid.

581 Ibid., p. 314; English cloths were classified according to quality i.e. ordinary was 'londis', and finer cloth, 'fájlondis' (in Hungarian) 'falendysz' (in Polish).

582 J. Topolski, 'Rola Gniezna ...' *op. cit.*, p. 34–41.

583 F. Kavka, 'Obchod s českými a slovenskými textiliemi v rumunských zemích (do pol. 17 stol.), *Historický sborník*, Vol. V, Praha, 1957, p. 146–147. See also A. V. Florovski, *České sukno na východoevropském trhu v XVI–XVIII věku*', Praha, 1947, 324pp.

584 Arhivele Statului Cluj, fondul Arhiva oraşului, *Socotelile Oraşului Cluj*, – 13a, XVI, p. 6 (dated 1613). Besides kersey (5 veg = 87½ metres) and linen (6 veg = 105 metres) also mention half a veg (8¾ metres) of fine London cloth 'Fajnlondisz'. See also S. Panova, 'Targovski vrazki ...' *op. cit.*, p. 37.

585 Wojewódzkie Archiwum Państwowe w Krakowie, *Księgi celne*, rkp. 2142, for example refer to 586 bolts (18,166 metres) and 18 ells (9 metres) were sent to Transylvania and Hungary (excluding Slovakia) in 1624.

586 A. Hlavová, *K otázkam struktury krakovského obchodu v roce 1617*, diplomová práce, Brnenské University, Brno, 1967, p. 119.

587 M. Marečková, 'Prešov v Uhersko-Polských ... ' *op. cit.*, p. 430.

588 M. Zakrzewska-Dubasowa, *Ormianie zamojscy i ich rola w wymianie handlowej i kulturalnej między Polską a Wschodem*, Lublin, 1966, p. 104.

589 L. A. Demény, 'Le commerce de la Transylvanie avec les régions du sud du Danube effectué par la douane de Turnu Roşu en 1685', *Revue Roumaine d'Histoire*, Vol. VII, No. 5, Bucureşti, 1968, p. 765; S. Panova, Targovski vrazki ... *op. cit.*, p. 37.

590 Wojewódzkie Archiwum Państwowe w Krakowie, *Księgi kupieckie*, rkp. 3227 (dated 1629); Ibid., rkp. 3228, p. 92–93 (dated 1631) refers to R. Bandinelli, J. Bonafede, and A. Ubaldini as having branch shops in Lwów.

591 Wojewódzkie Archiwum Państwowe w Krakowie, *Consularia Cracoviensis*, rkp. 466, p. 237–239 (dated 1666) with reference to G. Franchi.

592 P. Cernovodeanu, *England's Trade Policy in the Levant and Her Exchange of Goods with the Romanian Countries under the Latter Stuarts (1660–1714)*, Bibliotheca Historica Romaniae, Economic History Section Studies 41(2), Bucharest, 1972, p. 64; E. M. Podgradskaja, *Torgovye svjazi Mołdavii so L'vovom v XVI–XVII vekach*, Kishinev, 1968, p. 158–160.

593 H. Obuchowska-Pysiowa, 'Trade between Cracow and Italy ...' *op. cit.*, p. 643.
594 P. Jeannin, 'The Sea-borne and the Overland ...' *op. cit.*, p. 57; D. Quirini-Popławska, 'Die italienische Einwanderer in Krakau ... *op. cit.*, p. 344.
595 Wojewódzkie Archiwum Państwowe w Krakowie, *Księgi kupieckie*, rkp. 3227, p. 478 (dated 1629) refers to L. Tucci from Cracow having a branch shop in Lublin.
596 Ibid.
597 Wojewódzkie Archiwum Państwowe w Krakowie, *Księgi celne*, rkp. 2142 refers to Lublin merchants having bought 248 warps (4,960 metres) and 20 bolts (620 metres) of Moravian cloth.
598 H. Łowmiański, 'Handel Mohylowa ...' *op. cit.*, p. 528 mentions that on occasions transport costs were a quarter or more of the value of a load.
599 H. Obuchowska-Pysiowa, Struktura handlu Krakowa ... *op. cit.*, p. 137.
600 M. Taszycka, 'Włoskie tkaniny jedwabne ... ', *op. cit.*, p. 137.
601 M. Kulczykowski, 'Handel Krakowa w latach ...' *op. cit.*, p. 86.
602 H. Kisch, *op. cit.*, p. 545.
603 These Balkan merchants were referred to as 'turco-meritzi'. See P. Cernovodeanu, *op. cit.*, p. 108–109; also O. Cicanci, *Companiile Grecceşti din Transylvania şi comerţul european în anii 1636–1746* (Inst. Studii Europene; Biblioteca Istorica LIV), Bucureşti, 1981, 207 pp.
604 V. Coibanu, *Relaţiile politice româno-polone între 1699 şi 1848* (Academiei), Bucureşti, 1980, p. 31–55.
605 S. Inglot, *op. cit.*, p. 166.
606 H. Kisch, *op. cit.*, p. 545.
607 W. S. Unger, 'Trade through the Sound ...' *op. cit.*, p. 211.
608 M. Kulczykowski, *Kraków jako ośrodek ... op. cit.*, p. 81.
609 Wojewódzkie Archiwum Państwowe w Krakowie, *Księgi kupieckie*, rkp. 3190 (Inventarz handlu bławatnego spadkobierców Jana Bajera w Krakowie (1752–1754)).
610 These included fine expensive English fabrics 'sajeta', falendysz, Dutch 'perpetulle', English beaver cloth, wincey, flannel etc.
611 W. Kula, *Szkice o manufakturach w Polsce XVIII wieku*, Vol. I, Pt. 2, Warszawa, 1956, p. 587–761.
612 J. Louis, *Kupcy Krakowscy ... op. cit.*, p. 50.
613 W. Kula, *op. cit.*, p. 595.
614 Wojewódzkie Archiwum Państwowe w Krakowie, *Księgi celne*, rkp. 2224, p. 1, dated 1755, refers to 30 warps of cloth produced in Staszów, dyed in Silesia and returned to Cracow for sale.
615 M. Bałaban, *Dzieje Żydów w Galicji i Rezeczpospolitej Krakowskiej (1772–1868)*, Lwów, 1914, p. 113.
616 J. Louis, *op. cit.*, p. 51.
617 Muzeum Narodowe w Krakowie. *Zbiory Czartoryskich*, rkp. 1174, p. 513.
618 Wojewódzkie Archiwum Państwowe w Krakowie, *Księgi celne*, rkp. 2224, 2228, 2230, 2231, 2232.
619 H. Madurowicz and A. Podraza, *Regiony gospodarcze ... op. cit.*, p. 111–114; F. Kotuli, 'Łańcucki ośrodek tkacki w XVII i XVIII wieku', *Kwartalnik Historii Kultury Materialnej*, Vol. II, No. 4, Warszawa, 1954, p. 644–675.
620 M. Kulczykowski, *Kraków jako ośrodek ... op. cit.*, p. 109.

621 Wojewódzkie Archiwum Państwowe w Krakowie, *Inwentarz dóbr ruchomych po Stanisławce Fachinettini pozostałych (1762)*, rkp. 815.
622 Wojewódzkie Archiwum Państwowe w Krakowie, *Inwentarz dóbr ruchomych po F. Kramarzu (1772)*, rkp. 826.
623 Wojewódzkie Archiwum Państwowe w Krakowie, *Teki Schneidera*, No. 93 (dated 1792); H. Grosman, *Österreichs Handelspolitik mit Bezung auf Galizien in der Periode 1772–1790*, Vienna, 1914, p. 432–433.
624 Wojewódzkie Archiwum Państwowe w Krakowie, *Księgi celne*, rkp. 2230, p. 64, 140, 142, 144, 160, 162, 164, 172, 176, 178; rkp. 2231, p. 179; rkp. 2232, p. 107, 138, 140, 164, 215, 242, 252, 254, 286.
625 Wojewódzkie Archiwum Państwowe w Krakowie, *Teki Schneidera*, No. 94 (1792).
626 F. Braudel and F. C. Spooner, 'Prices in Europe . . . ', *op. cit.*, p. 419–422.
627 Ibid., Table 31, p. 481.
628 Ibid., p. 422.
629 F. Braudel, *Capitalism and Material Life 1400–1800* (Fontana/Collins), 1977, p. 158.
630 Z. Kuchowicz, 'Uwagi o konsumpcji produktów destylacji alkoholowej w Polsce w XVI wieku', *Kwartalnik Historii Kultury Materialnej*, Vol. XIX, No. 4, Warszawa, 1971, p. 675.
631 Z. Morawski, 'Rozwój i upadek winiarstwa w Polsce (XII–XVI Wiek)', *Kwartalnik Historii Kultury Materialnej*, Vol. XXVI, No. 1, Warszawa, 1978, p. 67.
632 J. Rutkowski, 'The Social and Economic Structure . . . ', *op. cit.*, p. 447–448.
633 K. Pieradzka, 'Handel Krakowa z Węgrami . . .', *op. cit.*, p. 94–141; G. Komoróczy, *Borkivitélünk észak felé. Fejezet a magyar kereskedelem történetéből*, Kassa, 1944, 194 pp.
634 R. Rybarski, *Handel i polityka . . . op. cit.*, Vol. II, p. 178, 182–183, 198–199.
635 S. Lewicki, *Prawa składu w Polsce*, Lwów, 1910, p. 138–139; S. Kutrzeba and J. Ptaśnik, 'Dzieje Handlu i Kupiectwa . . . ', *op. cit.*, p. 29.
636 L. Makkai, 'Agrarian Landscapes of Historical Hungary in Feudal Times', in *Études Historiques Hongroises 1980*. D. Nemes et al. (eds.) (Akadémiai Kiadó), Budapest, 1980, p. 203.
637 L. Boros, 'A természetföldrajzi tenyézok szerepe a Tokaji-hegy és környékenek földhasznositásában', *Földrajzi Értesito*, Vol. 31, No. 1, Budapest, 1982, p. 41–65.
638 I. Balassa, 'A szölömüveles és borkezelés változása a XVI–XVII században Tokajhegyalján', *Agrártörténeti Szemle*, Vol. 15, Budapest, 1973, p. 31–52.
639 Arhiv miesto Košic, *No. Akt. D 59* (dated 19/II/1522), 'nullus omnino Vina . . . venditionis causa de hoc regno ad Poloniam . . . ultra Cassoviam audeat et permittat ferre et deferri facere, sed cum huiusmodi vinis . . . educendis Cassoviam tamquam locum depositionis ingredi et intrare debeant!'. See also G. Granasztoi, 'La Ville de Kassa . . . ', *op. cit.*, p. 58–59.
640 E. Janota, *Bardyów. Historyczno-topograficzny opis Miasta i okolic*, Kraków, 1862, p. 19; E. Fügedi, 'A bártfai XVI század eleji bor-és lókivitrl kérdése', *Agrártörténeti Szemle*, Vol. 14, No. 1–2, Budapest, p. 41–89; J. Gecsényi, 'Bártfa város hegyaljai szöllögazdálkodása 1485–1563', *Agrártörténeti Szemle*, Vol. 8, Budapest, 1966, p. 485–497. Other places exporting wine to Cracow

included Prešov and Levoča. See P. Horváth, 'Obchodné styky Levoče . . . ', *op. cit.*, p. 125. For a more general discussion on this topic see E. Fügedi, 'Der Aussenhandel Ungarns am Anfang des 16 Jahrhunderts', in *Der Aussenhandel Ostmitteleuropas 1450–1650*, I. Bog (ed.) . . . *op. cit.*, p. 69–77.

641 Magyar Országos Levéltár, Budapest, *Camera Scepusensis Benigna Mandata*, Nr. 90 (dated 11/X/1596).

642 Muzeum Narodowe w Krakowie, *Biblioteka Czartoryskich*, rkp. No. 1033, p. 143–144 (dated 1544) refers to the amount for transporting wine from Sopron to Cracow. A purchase of 220¼ litres of local 'edenburski' wine was made in Sopron for 395 florens (11,850 grosz). Transport and carriage cost 201 florens (6,030 grosz) whilst other formalities added a further 93 florens (2,790 grosz). A total expense of 689 florens (20,670 grosz) was incurred of which transport and carriage cost 29.17%.

643 Wojewódzkie Archiwum Państwowe w Krakowie, *Libri iuris civilis*, rkp. No. 1422, p. 24.

644 Archiwum Główne Akt Dawnych w Warszawie, *No. 149a*, fol. 107 and verso, refers to Peter and Baptist Cellario.

645 Magyar Országos Levéltár, Budapest, *Camera Scepusensis Libri Expeditionum ad Maiestatem*, Vol. 6, fol. 99.

646 Ibid. fol. 160 (dated 1575); *Camera Scepusensis Benigna Mandata*, No. 21 (dated 1579) refers to wine from Tarcel in the Tokaj district sent to Cracow for the king's court; similarly wine from nearby Tallija sent to Cracow for the court, document dated 26/XI/1571, Arhiv miesto Košic, *No. 3078/7*.

647 S. Ingot, *Sprawy gospodarcze Lwa Sapiehy, 1588–1607*, Lwów, 1931, p. 30; G. Komoroczy, 'Uwagi na temat . . .', *op. cit.*, p. 91.

648 Magyar Országos Levéltár, Budapest, *Camera Scepusensis Benigna Mandata*, No. 171 (document dated 6/VIII/1571).

649 Calculated from information given in K. Pieradzka, 'Handel Krakowa z Węgrami . . .', *op. cit.*, p. 125; of the 1,133,510 litres of wine imported by the city that year, 448,584 litres consisted of Hungarian wine, i.e. 40%.

650 K. Pieradzka, *op. cit.*, p. 125.

651 Wojewódzkie Archiwum Państwowe w Krakowie, *Księgi celne*, rkp. 2119 (dated 29/IV/1591–21/IV/1592) which noted a total of 349,702 litres of Hungarian wine imported compared with 29,933 litres of Moravian vintage.

652 J. Małecki, *Studia nad rynkiem . . .*, *op. cit.*, p. 34.

653 E. Alberi, *Le Relazioni degli ambasciatori veneti*, Ser. III, Vol. II (relazioni Lorenzo Bernado, 1592), Firenze, 1846, p. 412; I. Nistor, *Die auswärtigen Handelsbeziehungen der Moldau im XIV, XV und XVI juhrhundert. Nach Quellen dargestellt*, Gotha, 1911, p. 74 (document referring to 1587); W. Łoziński, *Patrycjat i mieszczaństwo . . .*, *op. cit.*, p. 50 (document related to 1570).

654 N. Iorga, *Istoria comerţului românesc, Epoca veche*, Bucureşti, 1925, p. 203.

655 *Quellen zur Geschichte der stadt Kronstadt in Siebenbürgen*, Vol. III, Braşov, 1896, p. 347.

656 E. Veress, *Báthory István erdélyi fejedelem és Lengyel király levelezése*, Vol. I, Cluj, 1944, p. 151.

657 R. Rybarski, *Handel i polityka . . .*, *op. cit.*, Vol. 2, p. 199; Wojewódzkie Archiwum Państwowe w Krakowie, *Consularia Cracoviensia*, Vol. II, No. 446 (docu-

ment dated 1575) refers to Cracow wine merchant A. Baldi, sending his agent to Lwów to buy malmsey and muscatel wine.

658 A. Dziubiński, 'Drogi handlowe . . .', *op. cit.*, p. 240.

659 S. F. Tocmalaev, *Capitalul comercial şi profitul comercial*, Bucureşti, 1951, pp. 11–12, B. F. Porsnev, *Studii de economie politică a feudalismului*, Bucureşti, 1957, p. 109.

660 M. Dan, 'Le commerce de la Transylvanie . . .', *op. cit.*, p. 630; see also Ibid., 'Mărfuri exportate de negustorii . . .', *op. cit.*, pp. 37–43.

661 Wojewódzkie Archiwum Państwowe w Krakowie, *Księgi celne*, rkp. 2121, fol. 31 (document dated 28/I/1600).

662 J. Rutkowski, *Historia Gospodarcza* Polski . . ., *op. cit.*, p. 105.

663 Magyar Országos Levéltár, Budapest, *Camera Scepusensis Libri Expeditionum ad Maiestatem*, Vol. 7, fol. 334–334 verso; *Volumina legum* . . ., *op. cit.*, Vol. II, p. 1001.

664 G. F. Steckley, 'The Wine Economy of Tenerife in the Seventeenth Century: Anglo-Spanish Partnership in a Luxury Trade', *Economic History Review*, Vol. 33, No. 3, London, 1980, p. 337.

665 J. Ptaśnik, 'Z dziejów kultury włoskiego . . .', *op. cit.*, p. 35.

666 Ibid.

667 Ibid., p. 36; K. Pieradzka, 'Handel Krakowa z Węgrami . . .', *op. cit.*, pp. 129; 136.

668 J. Ptaśnik, Z dziejów kultury włoskiego . . ., *op. cit.*, p. 37.

669 G. Strauss, *Nuremburg in the Sixteenth Century* . . ., *op. cit.*, p. 128.

670 R. Rybarski, *Handel i polityka* . . ., *op. cit.*, Vol. II, p. 183.

671 Ibid., p. 199.

672 J. M. Małecki, *Związki handlowe miast polskich z Gdańskiem* . . ., *op. cit.*, p. 141.

673 R. Rybarski, *Handel i polityka* . . ., *op. cit.*, p. 199.

674 S. Kutrzeba and F. Duda, *Regestra thelonei aquatici Wladislaviensis saeculi XVI*, Kraków, 1915, p. 571.

675 Wojewódzkie Archiwum Państwowe w Krakowie, *Księgi celne*, rkp. 2112, which included 816 litres of Spanish dry wine.

676 Ibid., rkp. 2115–2120.

677 R. Rybarski, *Handel i polityka* . . ., *op. cit.*, Vol. II, p. 338 refers to wine being sent from Lublin to Brześć (1,087 litres, and 136 litres of muscatel); to Vilna (6,253 litres) and Pińsk (816 litres). Moreover in 1585 wine from the Carpathian foothill towns was sent to Brześć (13,593 litres) and Vilna (31,809 litres), *Księga celna komory brzeskiej z 1583 r.*, p. 297. See also A. Wawrzyńczyk, *Studia z dziejów handlu Polski* . . ., *op. cit.*, p. 62.

678 G. Komoroczy, 'Uwagi na temat . . .', *op. cit.*, p. 87.

679 Wojewódzkie Archiwum Państwowe w Krakowie, *Księgi celne*, rkp. 2116, p. 51.

680 K. Pieradzka, 'Handel Krakowa z Węgrami . . .', *op. cit.*, p. 13.

681 I. N. Kiss, 'Die Rolle der Magnaten-Gutswirtschaft im Grosshandel Ungarns im 17 Jahrhundert, in Der Aussenhandel Ostmitteleuropas 1450–1650 . . .', *op. cit.*, p. 480; S. Takáts, 'Borkivitelünk Lengyelországba 1610–ben es 1611 ben' *Magyar Gazdaságtörténeti Szemle*, Budapest, 1899, pp. 85–90.

682 P. Horváth, 'Príspevok k obchodným stykom východoslovenských miest s

Pol'skom a Sedmohradskom v 16–17 storočí, *Nové Obzory*, Vol. 7, Bratislava, 1965, p. 135.

683 G. Komoroczy, 'Uwagi na temat...', *op. cit.*, p. 100.

684 S. Tóth, *Sáros vármegye monografiája*, Vol. III, Budapest, 1912, p. 529.

685 M. Marečková 'Prešov v uhersko-polských...', *op. cit.*, p. 431.

686 R. Fišer, 'Obchodní styky Levoča...', *op. cit.*, p. 857.

687 M. Marečková, 'Majetková struktura Bardějovských obchodníků v prvé polovině 17 století', *Sborník Prací Filozofické Fakulty Brněnské University*, Vol. 27–28 (C 25–26) (Řada Hist.), Brno, 1978–1979, p. 134.

688 V. Zimányi and H. Prickler, 'Konjunktúra és depresszió a XVI–XVII szażadi...', *op. cit.*, pp. 79–201.

689 J. Perenyi, 'Wirtschaftliche und soziale umgestaltung in Ungarn unter der Türkenherrschaft im XVI und XVII Jahrhundert', in *Otázky dějin střední a východní Europy*, Brno, 1971, pp. 85–103.

690 Wojewódzkie Archiwum Państwowe w Krakowie, *Księgi celne*, rkp. 2142.

691 W. Felczak, *Historia Węgier...*, *op. cit.*, p. 147.

692 Wojewódzkie Archiwum Państwowe w Krakowie, *Akta giełdy kupieckiej*, rkp. 8, pp. 473–491 (dated 1671–1675).

693 Wojewódzkie Archiwum Państwowe w Krakowie, *Księgi celne*, rkp. 2202.

694 W. Rusiński, 'The Role of Polish territories...', *op. cit.*, p. 129.

695 These also included A. Lipski, J. Komornicki, Dembiński, Grodecki, S. Zaleski, K. Sulowski, Kormanicki and S. Rogożyński.

696 Wojewódzkie Archiwum Państwowe w Krakowie, *Księgi celne*, rkp. 2142 (dated 1624), gives a total of 13,018 litres equivalent of Moravian wine entering Cracow that year.

697 Wojewódzkie Archiwum Państwowe w Krakowie, *Księgi celne*, rkp. 2142; for example on 10/XII/1624, Ondřej Wiessner from Jičín sent 2,576 litres equivalent of Moravian wine and 23,828 litres equivalent of west Slovakian wine; Ibid., 16/XII/1624, Marek, a Jew from, Holešov in southern Moravia, delivered 4,186 litres equivalent of Moravian wine and 43,792 litres equivalent of 'Hungarian' wine.

698 Wojewódzkie Archiwum Państwowe w Krakowie, *Księgi celne*, rkp. 2202.

699 F. Pap, 'Rute comerciale şi localităţi...', *op. cit.*, p. 351.

700 M. Bogucka, 'Handel Gdańska z Półwyspem Iberyjskim...', *op. cit.*, Table B, p. 20.

701 Wojewódzkie Archiwum Państwowe w Krakowie, *Księgi celne*, rkp. 2122–2127; 2132. In 1601, one pipe (477⅓ litres) of Canary wine arrived in Cracow; 1602 – 10 pipes (4,773 litres); 1605 – 3 kufa (1,631 litres) and in 1611 – 2 pipes (954½ litres).

702 J. M. Małecki, *Związki handlowe miast Polskich...*, *op. cit.*, p. 245, refers to one pipe (477⅓ litres) of sweet Spanish wine 'sek', arriving in Cracow from Gdańsk.

703 Wojewódzkie Archiwum Państwowe w Krakowie, *Księgi celne*, rkp. 2126 (dated 1604) mentions one kufa (544 litres) of wine from Alicante reaching Cracow from Gdańsk; J. M. Małecki, *Związki handlowe miast polskich...*, *op. cit.*, p. 245, – one barrel (272 litres) in 1618 and 1619, one pipe (477⅓ litres) of the same wine.

704 H. Huteville, 'Relacja historyczna o Polsce', in *Ambroży Contarini, Podróż przez Polskę, 'Cudzoziemcy o Polsce. Relacje i opinie'*, J. Gintel (ed.), Vol. I, Warszawa, 1971, p. 328.
705 Wojewódzkie Archiwum Państwowe w Krakowie, *Księgi celne*, rkp 2126 (dated 1604).
706 H. Obuchowska-Pysiowa, 'Handel wiślany ...', *op. cit.*, Table 17, p. 143.
707 In 1744, Alexander Pulszky of Eperjes (Prešov) gave the town of Miskolc a loan of 40,000 florins (£4,000). H. Marczali, *Hungary in the Eighteenth Century* (Cambridge University Press) Cambridge, 1910, p. 25.
708 Arhivu v Oravského Podzámku, *sign. 86/20* (dated 1720). See also V. Černý, 'Polska sůl ...', *op. cit.*, p. 173; A. Fournier 'Handel und Verkyhr in Ungarn und Polen um die Mitte des XVIII Jahrhunderts', *Archiv für Österreichische Geschichte*, No. 69, Vienna, 1887, 165 pp.
709 Wojewódzkie Archiwum Państwowe w Krakowie, *Księgi celne*, rkp. 2216 (dated 1750). In that year for example, craftsmen alone imported 819 barrels (222,663 litres) of Hungarian wine, i.e. 27% of the year's total wine imports.
710 H. Marczali, *op. cit.*, p. 96.
711 H. Grosmann, *Österreichs Handel-Politik mit Bezung auf Galizien in der Periode 1772–1790*, Vienna, 1914, p. 195; G. Komoroczy, 'Borkivitelünk észak felé ...', *op. cit.*, pp. 245–254.
712 H. Marczali, *op. cit.*, p. 58.
713 *Dziennik Handlowy*, Warszawa, 1789, pp. 495–497; G. Duzinchevici, 'Contribuţii la istoria legaturilor comerciale româno-polone în secolul al XVIII-lea', *Revista Istorica*, Vol. 7, No. 9, Bucureşti, 1935, pp. 237–339.
714 E. Cieślak, 'Sea Borne Trade between France and Poland ...', *op. cit.*, p. 53; M. Kulczykowski, 'Handel Krakowa w latach ...'; *op. cit.*, p. 93.
715 M. Kulczykowski, 'Handel Krakowa w latach ...', *op. cit.*, p. 93.
716 Wojewódzkie Archiwum Państwowe w Krakowie, *Rachunki miasta Kazimierza*, rkp. 660 (dated 1775) records 42 barrels (11,419 litres) sent from Cracow to Galicia; Ibid. rkp. 665 (dated 1780) notes a total of 57 barrels (15,496 litres); Ibid. rkp. 670 (dated 1785) records only 13 barrels (3,534 litres) being sent there.
717 A. Wagner, 'Handel dawnego Jarosławia ...', *op. cit.*, p. 139.
718 F. Braudel and F. C. Spooner, 'Prices in Europe ...', *op. cit.*, pp. 407–408.
719 Ibid.
720 Ibid.
721 Ibid.
722 Ibid.
723 J. Pelc, *Ceny w Krakowie ...*, *op. cit.*, p. 66.
724 F. Braudel, *The Mediterranean and the Mediterranean World in the Ages of Philip II*, Vol. II (Fontana), London, 1973, Ch. IV, pp. 1087–1142.
725 N. J. G. Pounds, *An Historical Geography of Europe ...*, *op. cit.*, p. 191.
726 P. Horváth, 'Príspevok k obchodným stykom Slovenska ...', *op. cit.*, p. 184.
727 G. Lefebvre, *Les paysans du nord pendant la Révolution Française*, Lille, 1924, p. 195; W. Abel, *Geschichte der deutschen Landwirtschaft*, Stuttgart, 1962, p. 209.
728 R. J. Forbes, 'Food and Drink' in *A History of Technology*, C. Singer *et al.* (eds.), Vol. III, Oxford, 1957, p. 11.

729 F. Braudel and F. C. Spooner, 'Prices in Europe...', *op. cit.*, p. 410.
730 J. M. Małecki, *Studia nad rynkiem regionalnym...*, *op. cit.*, p. 35.
731 R. Rybarski, *Handel i polityka...*, *op. cit.*, Vol. I, p. 115.
732 F. Matějek, 'K dějinám českého obchodû...', *op. cit.*, p. 201.
733 S. Kutrzeba, 'Wisła w historji...', *op. cit.*, p. 40.
734 Wojewódzkie Archiwum Państwowe w Krakowie, *Księgi celne*, rkp. 2119–2121; R. Rybarski, *Handel i polityka...*, *op. cit.*, Vol. II, p. 198. In 1584, 4 barrels (1,087 litres) of Gdańsk beer arrived in Cracow; in 1594, one barrel (272 litres), 1595 and 1598, three barrels each (816 litres twice).
735 Wojewódzkie Archiwum Państwowe w Krakowie, *Księgi celne*, rkp. 2142.
736 Ibid. Bonkewiecki brought 3 sud (966 litres) of Opava beer to Cracow.
737 S. Cynarski, 'Krosno w XVII wieku...', *op. cit.*, p. 207.
738 K. Dziwik, 'Rozwój przestrzenny miasta Nowego Sącza od XIII do XIX wieku na tle stosunków gospodarczych', *Rocznik Sądecki*, Vol. V, Nowy Sącz, 1962, p. 194; A. Artymiak, *Z przeszłości Nowego Sącza*, Nowy Sącz, 1930, p. 46.
739 A. Falniowska-Gradowska and I. Rychlikowa (eds.), *Lustracja województwa krakowskiego 1789*, Pt. I, Wrocław-Warszawa-Kraków, 1972, p. 4.
740 M. Kulczykowski, 'Handel Krakowa w latach...', *op. cit.*, p. 94; Ibid., *Kraków jako ośrodek...*, *op. cit.*, p. 63.
741 Wojewódzkie Archiwum Państwowe w Krakowie, *Księgi celne*, rkp. 2216 (dated 1750) refers to 38 barrels (10,331 litres) of Żilina beer arriving in Cracow; Ibid. rkp. 2231 (dated 1763) mentions 327 barrels (88,902 litres) imported from the same brewery.
742 Z. Kuchowicz, 'Uwagi o konsumpcji produktów destylacji alkoholowej w Polsce w XVI wieku', *Kwartalnik Historii Kultury Materialnej*, Vol. XIX, No. 4, Warszawa, 1971, p. 669.
743 R. Rybarski, *Handel i polityka...*, *op. cit.*, Vol. I, p. 123.
744 Z. Kuchowicz, *op. cit.*, p. 670.
745 *Volumina Legum...*, *op. cit.*, Vol. II, pp. 39, 62, 178.
746 Z. Kuchowicz, *op. cit.*, p. 671.
747 F. Hejl, 'Český obchod na Krakovském trhu...', *op. cit.*, p. 241; F. Matějek, 'K dějinám českého obchodû...', *op. cit.*, p. 204.
748 Wojewódzkie Archiwum Państwowe w Krakowie, *Księgi celne*, rkp. 2142. For example Jakub, a Jew from Prostějov declared 150 garniec (566½ litres); Jurek Miloš from Opava, 100 (378 litres); Tomaš Piš from Přibor, 203 (766½ litres); Kryštof Kulmirz from Opava, 200 (755 litres); Jan Zelenka from Opava, 150 (566½ litres) and Matouš Pailer from Jičín, 114 (430½ litres).
749 A. Artymiak, *op. cit.*, p. 46.
750 N. J. G. Pounds, *An Historical Geography of Europe...*, *op. cit.*, p. 188.
751 J. A. Faber, 'Het problem van de dalende graansanveer uit de Oostzeelanden in de tweede heft van de zewentiende eeuw', *A A G Bijdragen IX Afdeling Agrarische Geschiedenis Landbouwhogeschool*, Wageningen, 1963, p. 7.
752 Wojewódzkie Archiwum Państwowe w Krakowie, *Rachunki miasta Kazimierza*, rkp. 660, 670.
753 M. Kulczykowski, *Kraków jako ośrodek...*, *op. cit.*, p. 59.
754 Ibid.

755 J. Reychman, 'Zatarg handlowy polsko-turecki o handel wódką XVIII w.', *Roczniki Dziejów Społecznych-Gospodarczych*, Vol. XIII, Poznań, 1951, p. 215.
756 T. Czacki, 'Raport do Komisji Skarbu Koronnego', *Dziennik Handlowy*, Warszawa, 1788, pp. 663–665; Ibid. 1786, p. 376; J. Reychman, 'Le commerce polonaise . . .', *op. cit.*, p. 242.
757 J. M. Małecki, *Studia nad rynkiem . . .*, *op. cit.*, p. 183.
758 F. Matějek, K dějinám českého obchodū . . ., *op. cit.*, p. 201.
759 Wojewódzkie Archiwum Państwowe w Krakowie, *Księgi celne*, rkp. 2231 (dated 1760). From Levoča, 115 barrels (31,265 litres) of mead sent to Cracow.
760 E. Tomaszewski, *Ceny w Krakowie w latach 1601–1795 . . .*, *op. cit.*, p. 64.
761 M. Bogucka, 'Handel Gdańska z Półwyspem Iberyjskim . . .', *op. cit.*, Table D. p. 21.
762 M. Berg, P. Hudson and M. Sonenscher, *Manufacture in Town and Country before the Factory* (Cambridge University Press), Cambridge, 1984, 250 pp.
763 G. Strauss, *Nuremburg in the Sixteenth Century . . .*, *op. cit.*, p. 135.
764 A. Mączak, H. Samsonowicz and B. Zientara, *Z dziejów rzemiosła w Polsce*, Warszawa, 1957, 195 pp.; J. Janaček, *Řemeslna výroba v českých městach v 16 stoleti*, Praha, 1961, 205 pp.; L. Szádeczky, *Iparfejlödés és a czéhek története Magyarországon* (Okirattárral 1307–1848), 2 vols., Budapest, 1913.
765 S. Herbst, *Toruńskie cechy rzemieślnicze*, Toruń, 1933, 183 pp., T. Jasiński, 'Rzemiosła kowalskie średniowiecznego Torunia', *Kwartalnik Historii Kultury Materialnej*, Vol. 23, No. 2, Warszawa, 1975, pp. 225–235.
766 J. Kwaśniewicz, 'Kleparz do 1795 roku', in *Z Dziejów Kleparza*, Kraków, 1968, p.
767 A. Jelicz, *W średniowiecznym Krakowie . . .*, *op. cit.*, p. 110.
768 Z. Ameisenowa, *Kodeks Baltazara Behema . . .*, *op. cit.*, pp. 34–44.
769 A. Franaszek, *Działalność wielkorządców krakowskich w XVI wieku* (Biblioteka Wawelska 5), Kraków, 1981, pp. 59–65; J. M. Małecki, *Studia nad rynkiem . . .*, *op. cit.*, pp. 109–129.
770 S. Kutrzeba and J. Ptaśnik, 'Dzieje handlu i kupiectwa krakowskiego . . .', *op. cit.*, p. 80.
771 A. Wyrobisz, 'Functional Types of Polish Towns in the XVIth-XVIIth Centuries', *Journal of European Economic History*, Vol. 12, No. 1, Rome, 1983, Table 1, p. 94.
772 B. Zientara, *Dzieje małopolskiego hutnictwa . . .*, *op. cit.*, pp. 162–4.
773 Ibid., p. 172; K. Sękowski, *Odlewnictwo krakowskie w pierwszej połowie XVI wieku* (Wyd. Inst. Odlewn.) Kraków, 1980, 59 pp.; J. M. Małecki, *Studia nad rynkiem . . .*, *op. cit.*, p. 183.
774 For example in 1584, steel imports from Silesia, Austria and Moravia totalled 81 cetnars (8 tons) and in 1591/2, 197 cetnars (20 tons). R. Rybarski, *Handel i polityka . . .*, *op. cit.*, Vol. II, p. 198; Wojewódzkie Archiwum Państwowe w Krakowie, *Księgi celne*, rkp. No. 2116.
775 Z. Bocheński, *Krakowski Cech Mieczników* (Biblioteka Krakowska No. 92), Kraków, 1937, pp. 12–20.
776 Ibid., pp. 81–84.
777 L. Lepszy, 'Cech złotników w Krakowie', *Rocznik Krakowski*, Vol. 1, Kraków, 1898, pp. 135–268.

778 J. Ptaśnik and M. Friedberg, *Cracovia artificum 1501–1550*, Kraków, 1948, p. XX.
779 J. M. Małecki, 'Rola Krakowa w handlu środkowej ...', *op. cit.*, p. 176.
780 Wojewódzkie Archiwum Państwowe w Krakowie, *Księgi celne*, rkp. 2117; for example on 4/I/1593, Ezechiel Gilar from Opava declared 7 vats of scythes to Cracow customs officials; on 10/I/1593, he declared a further 300 scythes in a box for Bartel Sambel in Cracow.
781 For example in 1538/39, 20,600 scythes arrived in Cracow from Wrocław, and a further 11,330 from Krems in Austria. R. Rybarski, *Handel i polityka ...*, *op. cit.*, Vol. II, pp. 178–9, 184.
782 Wojewódzkie Archiwum Państwowe w Krakowie, *Księgi celne*, rkp. 2117; R. Rybarski, *Handel i polityka ...*, *op. cit.*, Vol. 2, p. 209.
783 Ibid.
784 P. Horváth, 'Obchodné styky Levoče ...', *op. cit.*, p. 128.
785 For example in 1518, 60 metal hoes arrived in Wieliczka and Bochnia in exchange for 5 'Mensura' of salt (sexagena eorum pro mensuris 5, modernis tamen temporibus raro adducuntur); A. Keckowa and A. Wolff (eds.), 'Opis Żup Krakowski z roku 1518', *Kwartalnik Historii Kultury Materialnej*, Vol. 9, Warszawa, 1961, p. 96.
786 J. M. Małecki, *Studia nad rynkiem ...*, *op. cit.*, p. 184.
787 A. Chmiel, 'Dawne wyroby nożowników krakowskich i znaki na nich', *Rocznik Krakowski*, Vol. II, Kraków, 1899, pp. 89–108.
788 Wojewódzkie Archiwum Państwowe w Krakowie, *Księge celne*, rkp. 2117 (dated 1593); Ibid., rkp. 2118, k. 14; 56 (dated 1594); Archiwum Główne Akt Dawnych w Warszawie, *Archiwum Skarbu Koronnego*, Oddział XIII: 19 (dated 12/XII/1555).
789 M. P. Dan, 'Relaţiile comerciale dintre Oradea ...', *op. cit.*, p. 175.
790 Wojewódzkie Archiwum Państwowe w Krakowie, *Księgi celne*, rkp. 2118, k. 12 (dated 1594).
791 Ibid., rkp. 2116 (dated 1591/92), when 10 firkins (fas) (1 fas = about ½ cwt) of scythes, 4 cetnars (½ ton) of wire and 2 firkins (1 cwt) of sheet metal was sent from Cracow to Hungary.
792 W. Kierst, 'Wielkorządcy krakowscy w XIV–XVI st.', *Przegląd Historyczny*, Vol. 10, Kraków, 1910, pp. 291–2.
793 A. Dziubiński, 'Drogi handlowe polsko-tureckie ...', *op. cit.*, p. 250.
794 Wojewódzkie Archiwum Państwowe w Krakowie, *Księgi radzieckie*, rkp. 432, pp. 417–19 (dated 1519). Scythes sent from Cracow to Vilna; evidence from 1583 shows that 3,700 knives were sent from Cracow to Vilna, 'Księgi celne komory brzeskiej z 1583 (fragment)', *Archeograficheskij Sbornik dokumientov otnosiashchikhsja k istorii Sieviero-Zapadnoj Rossiji*, Vilna, 1897, p. 309.
795 Ibid., in 1583, 6,000 scythes went from Cracow via Brześć to Moscow, p. 307; also 35 firkins (17 cwt) of knives made in Cracow and 113 tacher (i.e. 1,356 knives) from Świdnica were sent to Moscow. R. Rybarski, *Handel i polityka ...*, *op. cit.*, Vol. II, pp. 337, 339; A. Wawrzyńczyk, *Studia z dziejów handlu Polski z Wielkim Księstwem ...*, *op. cit.*, p. 61.
796 S. V. Bachrushin, 'Ocherki po istorii remesla, torgovli i gorodov russkogo centralizovannogo gosudarstva XVI–XVII vv', *Nauchnye trudy*, Vol. I,

Moskva, 1952, pp. 47–8; Ibid., 'Ocherki po istorii kolonizacii Sibiri v XVI i XVII vv.', *Nauchnye trudy*, Vol. III, Moskva, 1955, p. 164; N. Kostomarov, *Ocherk torgovli Moskovskogo gosudarstva v XVI i XVII stoletijach*, St Petersburg, 1862, p. 355.

797 A. Chmiel, *op. cit.*, pp. 98–9; N. Gąsiorowska, *Przemysł metalowy polski w rozwoju dziejowym*, Warszawa, 1929, p. 19.

798 In 1530/31, 18 horses of steel (9 tons) sent from Cracow to Lublin. R. Rybarski, *Handel i polityka . . ., op. cit.*, Vol. II, pp. 180, 250.

799 F. Braudel, *Capitalism and Material Life . . ., op. cit.*, p. 298.

800 Ibid. pp. 295–300; C. Clair, *A History of European Printing*, London, 1976, p. 35.

801 J. M. Małecki, *Studia nad rynkiem . . ., op. cit.*, p. 116.

802 I. T. Baranowski, *Przemysł polski w XVI wieku* (ed. K. Tymieniecki), Warszawa, 1919, pp. 96–7; J. Ptaśnik, 'Papiernie w Polsce XVI wieku', in *Papiernie w Polsce XVI wieku (Prace F. Piekosińskiego J. Ptaśnika, K. Piekarskiego)*, W. Budka (ed.), Wrocław, 1971, pp. 6–40; W. Budka, 'Papiernia w Młodziejowicach', in *Studia nad książką poświęcone pamięci K. Piekarskiego*, Wrocław, 1951, pp. 199–210; Ibid., 'Papiernia w Mniszku', *Przegląd Papierniczy*, Vol. 7, Warszawa, 1951, pp. 121–5; Ibid., 'Papiernie w województwie krakowskim według rejestru poborowego z 1595 r.', *Silva Rerum*, Vol. 5, Warszawa, 1930, pp. 145–8.

803 J. Ptaśnik, 'Papiernie w Polsce . . .', *op. cit.*, p. 33.

804 Ibid. pp. 16–18; W. Budka, 'Papiernie poznańskie', *Przegląd Papierniczy*, Vol. 10, Warszawa, 1954, pp. 216–21; 251–3.

805 W. A. Chatto, *Facts and Speculations on the Origin and History of Playing Cards*, London, 1930, pp. 10–12; E. S. Taylor, *History of Playing Cards*, London, 1926, 14–18.

806 Wojewódzkie Archiwum Państwowe w Krakowie, *Księgi radzieckie*, rkp. 445, pp. 191–2, 405–6; 558–9 (dated 1562); A. Chmiel, 'Krakowskie karty do gry XVI-go wieku', *Rzeczy Piękne* Vol. 2, No. 2, pp. 17–20; No. 3, pp. 10–17, Kraków, 1919.

807 W. Szelińska, *Drukarstwo krakowskie 1474–1974*, Kraków, 1974, pp. 13–15; J. Bieniarzówa and J. M. Małecki, *Dzieje Krakowa (Kraków w wiekach XVI–XVIII)*, Vol. 2 (Wydawnictwo Literackie), Kraków, 1984, pp. 146–50; B. Klimaszewski, *An Outline History of Polish Culture* (Inter-Press), Warsaw, 1984, pp. 84–5.

808 W. Szelińska, *op. cit.*, p. 16; J. Ptaśnik, 'Cracovia impressorum XV et XVI saeculorum', in *Monumenta Poloniae typographica XV et XVI saeculorum*, Vol. 1, J. Ptaśnik (ed.), Lwów, 1922, pp. 37–41; M. Bałaban 'Drukarstwo żydowskie w Polsce XVI w.', in *Pamietnik Zjazdu Naukowego im. Jana Kochanowskiego*, Kraków, 1931, pp. 102–16; J. S. Bandtkie, *Historia drukarń krakowskich od zaprowadzenia druków do tego miasta aż do czasów naszych, wiadomością o wynalezieniu sztuki drukarskiej poprzedzona*, Kraków, 1815 (Warszawa, 1974), 504 pp.

809 W. Szelińska, *op. cit.*, p. 17.

810 I. Baranowski, *op. cit.*, p. 89.

811 F. Zuman, 'Knižka o papieru', *Roczniki Dziejów Społecznych i Gospodarczych*, Vol. 10, Warszawa, 1948, p. 370.

812 Wojewódzkie Archiwum Państwowe w Krakowie, *Księgi celne*, 2116, p. 5 (dated 1591).
813 In 1538/39 Paper sent from Cracow to Hungary, R. Rybarski, *Handel i polityka* ..., *op. cit.*, Vol. 2, pp. 188, 210; Ibid., 1584, 35 reams sent to Hungary; 1591/92, 52 reams sent to Hungary; K. Pieradzka, 'Handel Krakowa z Węgrami ...', *op. cit.*, p. 242, P. Horváth, Obchodné styky Levoče ..., *op. cit.*, p. 139.
814 I. Baranowski, *op. cit.*, p. 77; J. Ptaśnik, 'Papiernie w Polsce ...', *op. cit.*, p. 33; A. Wawrzyńczyk, *op. cit.*, p. 68; W. Budka, 'Papiernia w Balicach', *Archeion*, Vol. 13, Warszawa, 1935, p. 43; Ibid., 'Papiernia w Młodziejowicach ...', *op. cit.*, p. 206; A. G. Mankov, *Ceny i jich dvizhenije v russkom gosudarstve XVI veka*, Moskva-Leningrad, 1951, p. 241.
815 J. M. Małecki, *Związki handlowe miast* ..., *op. cit.*, p. 235, refers to books sent from Cracow to Gdańsk in 1594.
816 K. Pieradzka, *op. cit.*, p. 242.
817 A. Ballagi, *A magyar nyomdászat történélmi fejlödése (1472–1877)* (Franklin publ.), Budapest, 1878, p. 31, gives details of the first printing works established in Braşov in 1546.
818 M. Bałaban, *op. cit.*, p. 108; Wojewódzkie Archiwum Państwowe w Krakowie, *Księgi celne*, rkp. 2116, p. 10, 34v, 35v, 38v, 39, 42v, 43, 92, 92v, 93v (dated 1591).
819 M. Bałaban, *op. cit.*, pp. 103–8.
820 M. Kwapieniowa, 'Organizacja produkcji i zbytu wyrobów garncarskich w Krakowie w XIV–XVII w.', in *Studia nad produkcją rzemieślniczą w Polsce (XIV–XVIII w.)*, M. Kwapień, J. Maroszek and A. Wyrobisz (eds.), Wrocław-Warszawa-Kraków, 1976, p. 26.
821 J. M. Małecki, *Studia nad rynkiem* ..., *op. cit.*, pp. 190–3.
822 I. Bojarska, 'Krakowska ceramika ludowa', *Polska Sztuka Ludowa*, Vol. 3, Warszawa, 1949, pp. 140–8.
823 J. M. Małecki, *Studia nad rynkiem* ..., *op. cit.*, pp. 193–4.
824 Ibid., Przyczynek do dziejów handlu ..., *op. cit.*, p. 33; Ibid., *Związki handlowe miast polskich* ..., *op. cit.*, p. 234.
825 A. Wyrobisz, 'Warunki rozwoju przemysłu w Polsce w XVI i pierwszej połowie XVII w.', in *Studia nad produkcją rzemieślniczą w Polsce* ..., *op. cit.*, p. 241.
826 E. Baasch, 'Verkehr mit den Kriegsmaterialien aus und nach den Hansestadten von Ende des 16. bis Mitte des 17. Jahrhunderts', *Jahrbücher für Nationalökonomie und Statistik*, Vol. 137, No. III, F, 82, Berlin, 1932, pp. 538–41; V. A. Kordt, 'Ocherk snoshenie Moskovskogo gosudarstva s Respublikou Soedinennich Niderlandov po 1631 g.', *Sbornik Imperialisticheskii Istoricheskogo obshestva*, Vol. 116, St Petersburg, 1902, p. LXXX; W. S. Unger, 'Trade through the Sound ...', *op. cit.*, p. 216.
827 W. Rusiński, 'The Role of Polish Territories ...', *op. cit.*, p. 128.
828 Wojewódzkie Archiwum Państwowe w Krakowie, *Księgi kupieckie*, fragment, dated 1650 'Inventarium mercium Hybleynera'; S. Kutrzeba and J. Ptaśnik, 'Dzieje handlu i kupiectwa krakowskiego ...', *op. cit.*, p. 80.
829 S. Inglot, 'Economic Relations of Silesia ...', *op. cit.*, p. 160.
830 Z. Guldon and L. Stępkowski, 'Handel Torunia ...', *op. cit.*, p. 96. M. Wolański, *Związki handlowe Śląska* ..., *op. cit.*, pp. 191–2.

831 Wojewódzkie Archiwum Państwowe w Krakowie, *Księgi celne*, rkp. 2142 (dated 1624) and 2143 (dated 1629); R. Fišer, 'Obchodní styky Levoča ...', *op. cit.*, p. 856.
832 Wojewódzkie Archiwum Państwowe w Krakowie, *Księgi celne*, rkp. 2142; in 1624 a total of 14 barrels (7,700) and 400 scythes (8,100) were sent from Opava and Bruntál to Cracow.
833 F. Hejl, Český obchod na Krakovském trhu ..., *op. cit.*, p. 239.
834 M. Marečková, 'Prešov v uhersko-polských ...', *op. cit.*, p. 432.
835 Wojewódzkie Archiwum Państwowe w Krakowie, *Księgi celne*, rkp. 2142; in 1624, 630 scythes were sent from Cracow to towns in Slovakia (Hlohovec, Nové Mesto nad Váhom and the Orava valley).
836 H. Obuchowska-Pysiowa, 'Struktura handlu Krakowa z krajami ...', *op. cit.*, Table 1, p. 91.
837 S. Kutrzeba and J. Ptaśnik, 'Dzieje handlu i kupiectwa krakowskiego ...', *op. cit.*, p. 87.
838 B. Zientara, *Dzieje małopolskiego hutnictwa żelaznego* ..., *op. cit.*, p. 179.
839 M. Marečková, 'Majetková struktura Bardějovských obchodníků ...', *op. cit.*, p. 138.
840 Wojewódzkie Archiwum Państwowe w Krakowie, *Księgi celne*, rkp. 2142; in 1624, Jan Süssler from Krnov obtained 13 reams of paper in Cracow.
841 L. Demény 'Comerţul de tranzit spre Polonia ...', *op. cit.*, p. 474; F. Hejl, Český obchod na Krakovském trhu ..., *op. cit.*, p. 239.
842 W. Szelińska, *op. cit.*, p. 37.
843 B. Klimaszewski, *op. cit.*, p. 85. W. Budka, 'Z dziejów papiernictwa krakowskiego w XVII w.', in *Prace z dziejów Polski feudalnej ofiarowane Romanowi Grodeckiemu w 70 rocznicę urodzin*, Warszawa, 1960, pp. 484–8.
844 Wojewódzkie Archiwum Państwowe w Krakowie, *Księgi celne*, rkp. 2156 from Warsaw in 1642; rkp. 2158, from Lublin in 1644; rkp. 2158–2161, 2163, 2166–2169, from Gdańsk. See also, M. Pelczar, 'Förster Jerzy', *Polski Słownik Biograficzny*, Vol. 7, Kraków, 1958, p. 73; J. M. Małecki, *Związki handlowe miast polskich* ..., *op. cit.*, Table 25, p. 144.
845 D. Sella, *Commercio e industria e Venezia* ..., *op. cit.*, p. 3, 66.
846 H. Obuchowska-Pysiowa, 'Trade between Cracow and Italy ...', *op. cit.*, p. 649.
847 Wojewódzkie Archiwum Państwowe w Krakowie, *Księgi celne*, rkp. 2142; in 1624, Dominik Fišer from Lipník brought window-panes to Cracow for sale valued at 15 złoty.
848 W. Kula, *Szkice o manufakturach w Polsce XVIII wieku* ..., *op. cit.*, Vol. 2, pp. 854–5.
849 Ibid. Vol. 1, pp. 67–117.
850 Wojewódzkie Archiwum Państwowe w Krakowie, *Księgi celne*, rkp. 2216, p. 59, dated 1750; 4 cetnars (½ ton) of Swedish iron arrived in Cracow from Warsaw.
851 Ibid. pp. 125, 205; *Dziennik Handlowy*, Kraków, 1786, p. 46.
852 Wojewódzkie Archiwum Państwowe w Krakowie, *Księgi celne*, rkp. 2216 (dated 1750) p. 171.
853 Wojewódzkie Archiwum Państwowe w Krakowie, *Acta pupilaria et succesionalia in officio consularii cracoviensis*, sygn. 808, 815, 8;25, 826.

854 M. Kulczykowski, *Kraków jako ośrodek . . .*, *op. cit.*, p. 122.
855 Ibid., Handel Krakowa w latach 1750–1772 . . ., *op. cit.*, p. 100.
856 *Dziennik Handlowy*, Kraków, 1786, p. 45.
857 Z. Bocheński, *Krakowski cech mieczników . . .*, *op. cit.*, p. 35.
858 Wojewódzkie Archiwum Państwowe w Krakowie, *Księgi celne*, rkp. 2232, dated 1763, 20 firkins and 11 packets of nails sent from Biała to Cracow; Ibid. *Protocollo causarum criminalium officii Consularis Cracoviensis*, sygn. 890, consignment including nails sent from Sułkowice to Cracow.
859 M. Kulczykowski, 'Handel Krakowa w latach 1750–1772 . . .', *op. cit.*, p. 91.
860 Wojewódzkie Archiwum Państwowe w Krakowie, *Księgi celne*, rkp. 2232 (dated 1760).
861 K. Modelski, 'Rzemiosła metalowe w miastach wielkopolskich w drugiej połowie XVIII wieku', *Studia i Materiały do Dziejów Wielkopolski i Pomorza*, Vol. 14, No. 1, Poznań, 1980, pp. 77–92.
862 K. Kuklińska, Handel Poznania w drugiej połowie XVIII wieku . . ., *op. cit.*, pp. 131–3.
863 W. Budka, 'Z dziejów papiernictwa krakowskiego . . .', *op. cit.*, p. 485.
864 M. Opałek, *Obrazki z przeszłości Lwowa*, Lwów, 1923, p. 77; M. Kulczykowski, *Kraków jako ośrodek . . .*, *op. cit.*, p. 122.
865 J. Bieniarzówna and J. M. Małecki, *Dzieje Krakowa*, *op. cit.*, Vol. 2, pp. 144–50.
866 B. Klimaszewski, *op. cit.*, pp. 115–17.
867 W. Szelińska, *op. cit.*, pp. 44–4.
868 M. Kulczykowski, *Kraków jako ośrodek . . .*, *op. cit.*, p. 124.
869 K. Kuklińska, Handel Poznania . . ., *op. cit.*, Table 33, p. 134.
870 F. Braudel and F. C. Sponner, Prices in Europe . . ., *op. cit.*, p. 388; J. Pelc, *op. cit.*, p. 74.
871 K. Kuklińska, 'Les rôles joués par les marchés . . .', *op. cit.*, p. 99.
872 A. Wyrobisz, 'Functional Types of Polish Towns . . .', *op. cit.*, p. 84.
873 Ibid.

8 Cracow: a final appraisal

1 F. W. Carter, *Dubrovnik (Ragusa): A Classic City-State*, Seminar Press, London, and New York, 1972, 710 pp.
2 N. Davies, *Heart of Europe: A Short History of Poland*, Oxford University Press, Oxford and New York, 1986, pp. 291–296.
3 M. Keen, *The Pelican History of Medieval Europe*, Penguin Books, 1978, pp. 300–312.
4 L. Ładogórski, *Studia nad zaludnieniem Polski XIV wieku* (P.A.N., Instytut Historii), Wrocław, 1958, 229 pp.
5 M. Małowist, 'The Trade of Eastern Europe in the Later Middle Ages', Ch. VIII, *The Cambridge Economic History of Europe*, Vol. II, Trade and Industry in the Later Middle Ages, M. M. Postan and E. Miller (eds.) Cambridge University Press, Cambridge, London, New York, New Rochelle, Melbourne, Sydney, 198, p. 553.
6 J. Ptaśnik, *Italia Mercatoria apud Polonos saeculo XV ineunte*, Rome, 1910

(Lowscher & Co), 108, pp.; Ibid., 'Włoski Kraków za Kazimierza Wielkiego i Władysława Jagiełły' *Rocznik Krakowski*, Vol. XIII, Kraków, 1910, pp. 49–110; Ibid., *Kultura włoska wieków średnich w Polsce*, Warszawa, 1922, pp. 23–117.

7 K. Glamann, 'European Trade 1500–1750', Ch. 6 in C. Cipolla, *Fontana Economic History of Europe*, Vol. II, London, 1974, p. 477.

8 J. Topolski, 'A model of East-Central European continental commerce in the sixteenth and the first half of the seventeenth century', Ch. 9 in A. Mączak, H. Samsonowicz and P. Burke, *East-Central Europe in Transition: From the fourteenth to the seventeenth century*, Cambridge University Press, Cambridge, London, New York, New Rochelle, Melbourne, Sydney, 1985, p. 136.

9 N. J. G. Pounds, *An Economic History of Medieval Europe* (Longman), London/New York, 1974, p. 477.

10 P. W. Knoll, 'The Urban Development of Medieval Poland, with Particular Reference to Kraków', Ch. 2 in *The Urban Society of Eastern Europe in Premodern Times*, B. Krekić (ed.), University of California Press, Berkeley/Los Angeles/London, 1987, p. 95.

11 J. A. Szwagrzyk, *Pieniądz na ziemiach polskich X–XX w.*, Wrocław-Kraków-Warszawa-Gdańsk, 1973, p. 17; See also A. Szelągowski, *Pieniądz i przewrót cen w XVI i XVII w Polsce*, Lwów, 1902; Z. Sadowski, *Pieniądz a początek upadku Polski w XVII w.*, Warszawa, 1964; R. Kiersnowski, *Wstęp do numizmatyki polskiej wieków średnich*, Warszawa, 1964, pp. 128–131.

12 J. Pelc, *Ceny w Krakowie w latach 1369–1600* (Badania z Dziejów Społecznych i Godpodarczych, Nr. 14), Lwów, 1935, p. 51.

13 J. Topolski, 'Economic Decline in Poland from the Sixteenth to the Eighteenth Centuries', Ch. 6, in *Essays in European Economic History, 1500–1800* (P. Earle, ed.), Clarendon Press, Oxford, 1974, pp. 140–141.

14 M. Małowist, 'The Problems of the Inequality of Economic Development in Europe in the Later Middle Ages', *Economic History Review*, Vol. XIX, No. 2, London, 1966, pp. 255–256.

15 A. Wyrobisz, 'Functional Types of Polish Towns in the XVIth–XVIIIth Centuries', *Journal of European Economic History*, Vol. 12 (No. 1), Rome, 1983, p. 84.

16 P. Bushkovitch, 'Polish Urban History: Sixteenth and Seventeenth Centuries', *Polish Review*, Vol. 18, No. 3, New York, 1973, p. 89.

17 Ibid.

18 C. L. Salter, *The Cultural Landscape* (Duxbury Press), Belmont, California, 1975, p. 265.

19 *Time Magazine* (14/XI/1969), New York, p. 48.

Select bibliography

Attman, A. *The Russian and Polish Markets in International Trade, 1500–1650* (Institute of Economic History of Göteborg University) No. 21, Göteborg, 1973.

The Struggle for Baltic Markets. Powers in Conflict, 1558–1618, Göteborg, 1979.

Bieniarzówna, J. and J. M. Małecki, *Dzieje Krakowa (Kraków w wiekach XVI–XVIII)*, Vol. 2 (Wydawnictwo Literackie), Kraków, 1984.

Bogucka, M. 'The Role of Baltic Trade in European Development from the XVIth to the XVIIIth Centuries', *The Journal of European Economic History*, Vol. 9, No. 1, Rome, 1980, pp. 5–20.

Carter, F. W. 'Cracow's Early Development', *The Slavonic and East European Review*, Vol. 61, No. 2, London, 1982, pp. 177–225.

'Cracow as Trade Mediator in Polish–Balkan Commerce, 1590–1600' in *Trade and Transport in Russia and Eastern Europe* (eds M. McCauley and J. E. O. Screen (Occasional Papers, No. 2, S.S.E.E.S.) London, 1985, pp. 45–76.

'Cracow's Wine Trade (Fourteenth to Eighteenth Centuries), *The Slavonic and East European Review*, Vol. 65, No. 4, London, 1987, pp. 337–78.

'Cracow's Transit Textile Trade, 1390–1795 (A Geographical Assessment)', *Textile History*, Vol. 19, No. 1, Leicester, 1988, pp. 23–60.

Dąbrowski, J. 'Kraków a Węgry w wiekach średnich', *Rocznik Krakowski*, Vol. 13, Kraków, 1911, pp. 187–250.

Dan, M. 'Negustorii clujeni la Cracovia în ultimul decenju al secolui al XVI-lea', *Acta Musea Napocensis*, Vol. 8, Cluj, 1971, pp. 205–17.

Davies, N. *God's Playground. A History of Poland*, Vol. 1, 'The Origins to 1795', Oxford University Press, Oxford, 1982.

Heart of Europe. A Short History of Poland, Oxford University Press, Oxford, 1986.

East, W. G. *An Historical Geography of Europe* (Methuen), London, 1956.

Endrei, W. G. 'English Kersey in Eastern Europe with Special Reference to Hungary', *Textile History*, Vol. 5, Westbury, 1974, pp. 90–9.

Fedorowicz, J. K. (ed.), *A Republic of Nobles. Studies in Polish History to 1964*, Cambridge University Press, Cambridge, 1982.

Fišer, R. 'Orientace a skladba zahraničního obchodu Levoče v první polovice 17 století', *Historický Časopis*, Vol. XXI, No. 3, Bratislava, 1973, pp. 434–36.

Georgieva, C. 'Les rapports de commerce entre l'Empire Ottoman et la Pologne et les terres bulgares en XVIe siècle', *Bulgarian Historical Review*, Vol. 6, No. 3, Sofia, 1978, pp. 38–51.

Gieysztor, A. 'Le origini delle città nella Polonia medievale', in *Scritti in Onore di Armando Sappori*, Vol. 1, Milan, 1957, pp. 129–45.

Halaga, O. R. 'Krakovské pravo skladu a sloboda obchodu v Uhersko-Pol'sko-Pruskej politike pred Grunwaldom', *Slovanské Štúdie*, Vol. XVI, Bratislava, 1975, pp. 163–91.

Hejl, F. 'Městsky archiv v Krakové a jeho význam pro studium Česko-Polských a Slovensko (Uhersko) – Polských vztahů od konce XVI do konce XVIII století', *Časopis Matice Moravské*, Vol. 79, Brno, 1960, pp. 288–90.

'Český obchod na Krakovském trhu po Bílé Hoře', *Sborník Prace Filosofické Fakulty Brněnské University*, Vol. 10 (řada historická č.8), Brno, 1961, pp. 228–50.

Inglot, S. 'Economic Relations of Silesia with Poland from the XVI century to the beginning of the XVIII century', *Annales Silesiae*, Vol. 1, No. 2, Wrocław, 1960, pp. 153–70.

Innocenti, P. *La città di Cracovia. Origini, aspetti a funzioni dell' organismo urbano. Suo ruole nella struttura geografico-economico della Repubblica populare polacca.* Firenze (Memorie dell' instituto di geografia economica dell' Università degli studi di Firenze), 1973.

Jeannin, P. 'Le cuivre, Les Fuggers et la Hanse', *Annales: Economiques, Sociétés, Civilisations*, Vol. 10, Paris 1955, pp. 229–36.

Jelicz, A. 'Życie codzienne w średniowiecznym Krakowie' (wiek XIII–XV) (Państwowy Instytut Wydawniczy), Warszawa, 1965.

Kaczmarczyk, K. *Das Historische Archiv der Stadt Krakau, seine Geschichte, Bestände und Wissenschaftliche Erforschung, Mitteilungen des k.k. Archivrates*, Bd. I, Krakau, 1914, pp. 155–200.

Keckowa, A. *Żupy krakowskie w XVI–XVIII wieku (do 1772 roku)*, Zak. Nar.Im.Ossolinskich wyd.Pol.Akad. Nauk, Wrocław-Warszawa-Kraków (Instytut Historii Kultury Mat.) (Pol.Akad. Nauk).

Kieniewicz, I. 'Droga morska do Indii i handel korzenny w latach 1498–1522', *Przegląd Historyczny*, Vol. 55, No. 4, Warszawa, 1964, pp. 573–603.

Knoll, P. W. 'The Urban Development of Medieval Poland, with Particular Reference to Kraków', in *Urban Society of Eastern Europe in Premodern Times*, B. Krekić (ed.) (University of California Press), Berkeley-Los Angeles-London, 1987, pp. 63–136.

Krawczyk, A. 'Present-day Aims and Methods of Historical Geography' *Prace Geograficzne*, Vol. 61, Kraków, 1984, pp. 81–84.

Kuklińska, K. 'Central European Towns and the Factors of Economic Growth in the Transition from Stagnation to Expansion between the XVIth and XVIIIth Centuries', *Journal of European Economic History*, Vol. 11, No. 1, Rome, 1982, pp. 105–15.

Kulczykowski, M. *Kraków jako ośrodek towarowy Małopolski zachodniej w drugiej połowie XVIII wieku*, Warszawa (P.W.N.), 1963.

Kutrzeba, S. 'Handel Krakowa w wiekach średnich na tle stosunków handlowych Polski', *Rozprawy Akademii Umiejętności (Hist.Phil.)*, Vol. XLIV, Kraków, 1902.

Kutrzeba, S. and F. Duda (eds.), *Regestra Thelonei Aquatici Wladislaviensis saeculo XVI*, Cracoviae, 1915.

Kutrzeba, S. and J. Ptaśnik, 'Dzieje handlu i kupiectwa krakowskiego', *Rocznik Krakowski*, Vol. XIV, Kraków, 1912.

Lepszy, L. *Cracow, the Royal Capital of Ancient Poland*, London, 1912.

Louis, J. *Kupcy krakowscy w epoce przejściowej 1773–1846*, (Kalendarz Krakowski Jozefa Czecha, 1883), Kraków, 1883.

Mączak, A., H. Samsonowicz and P. Burke, *East-Central Europe in Transition. From the Fourteenth to the Seventeenth Century* (Cambridge University Press), Cambridge, 1985.

Małecki, J. M. 'Handel zewnętrzny Krakowa w XVI wieku', *Zeszyty Naukowe Wyższej Szkoły Ekonomicznej w Krakowie*, No. 11 (Prace z zakresu historii gospodarczej) Kraków, 1960, pp. 73–152.

Studia nad rynkiem regionalnym Krakowa w XVI wieku, Warszawa (P.W.N.), 1963.

'Rola Krakowa w handlu Europy Środkowej w XVI i XVII wieku', *Studia z historii gospodarczej i demografii historycznej*, Kraków, 1975, pp. 173–88.

Małowist, M. 'The Baltic and the Black Sea in Medieval Trade', *Baltic and Scandinavian Countries*, Vol. III, No. 1(5), Gdynia, 1937, pp. 36–42.

Wschód a Zachód Europy w XIII–XVI wieku, Warszawa, 1973.

Matějeck, F. 'K dějinám českého obchodu s oblastí Haličsko-Karpatskou 16 století. Kraków jako transitní středisko českého zboží', *Slovanské Historické Studie*, Vol. 3, Praha, 1960, pp. 185–214.

Mitkowski, J. 'Nationality Problems and Patterns in Medieval Polish Towns. The Example of Cracow', *Zeszyty Naukowe Uniwersytetu Jagiellońskiego (Prace Historyczne No. 59)*, Vol. DIII, Warszawa-Kraków, 1978, pp. 31–42.

Molenda, D. 'W sprawie badań huty miedzi w Mogile pod Krakowem w XV i XVI wieku', *Przeglad Historyczny*, Vol. 66(3), Warszawa, 1975, pp. 369–82.

Obuchowska-Pysiowa, H. 'Handel wołami w świetle rejestru celnego komory krakowskiej z 1604 roku', *Kwartalnik Historii Kultury Materialnej*, Vol. 21, Warszawa, 1973, pp. 507–12.

'Struktura handlu Krakowa z krajami południowo-wschodnimi i z państwem Moskiewskim w 1604 roku', *Roczniki Dziejów Społecznych i Gospodarczych*, Vol. 38, Warszawa-Poznań, 1977, pp. 81–114.

'Trade between Cracow and Italy from the Customs-house Registers of 1604', *Journal of European Economic History*, Vol. 9(3), Rome, 1980, pp. 633–53.

Pach, Z. P. 'The Shifting of International Trade Routes in the 15th–17th Centuries', *Acta Historica Scientiarum Hungarica*, Vol. 14, Budapest, 1968, pp. 287–319.

Panaitescu, P. 'La Route Commerciale de Pologne à la Mer Noire au Moyen Âge', *Revista Istorică Română*, Vol. 3(2–3) (Extras din), Bucureşti, 1934.

Pawiński, A. 'Notatki kupca krakowskiego w podróży do Flandrii', *Biblioteka Warszawska*, Vol. 3, Warszawa, 1872, pp. 53–73.

Pelc, J. *Ceny w Krakowie w latach 1369–1600*, Lwów, 1933.

Piekosiński, F. (ed.), *Kodeks dyplomatyczny miasta Krakowa 1257–1506*, Kraków, 1879–1882, Vols. 1–2 (4 parts).

Piekosiński, F. and S. Krzyżanowski, *Prawa, przywileje i statuta miasta Krakowa 1507–1795, (1696)*, Kraków, 1885–1909, Vols. 1–2.

Piekosiński, F. and J. Szujski, *Najstarsze księgi i rachunki miasta Krakowa*, Kraków, 1885–1909, Vols. 1–2.
Pieradzka, K. 'Handel Krakowa z Węgrami w XVI w.', *Biblioteka Krakowska*, No. 87, Kraków, 1935.
Postan, M. M. and E. Miller, *The Cambridge Economic History of Europe, Vol. II Trade and Industry in the Middle Ages* (Cambridge University Press), Cambridge, 1987.
Pounds, N. J. G. *An Economic History of Medieval Europe* (Longman), London/New York, 1974.
An Historical Geography of Europe, 1500–1840 (Cambridge University Press), Cambridge, 1979.
Ptaśnik, J. *Italia Mercatoria apud Polonos Saeculo XV ineunte*, Rome (Loescher & Co.), 1910.
'Towns in Medieval Poland', Ch. 2 in *Polish Civilization. Essays and Studies*, M. Giergielewicz (ed.) (New York University Press), New York, 1979, pp. 25–50.
Quirini-Popławska, D. 'Z działalności Włochów w Polsce w I połowie XVI w. Nowe szczegóły o Guccich i Thedaldich', *Studia Historyczne*, Vol. 12(2), Warszawa, 1969, pp. 171–99.
'Die italienische Einwanderer in Krakau und ihr Einfluss auf die polnischen Wirtschaftbeziehungen zu österreichischen und deutschen Städten im 16 Jahrhundert', *Wissenschaftliche Zeitschrift de Friedrich Schiller Universität*, Iena, (Gesellschaft-und Sprachwissensschaftliche Reihe 26), 1977, pp. 337–354.
Radwański, K. *Kraków przedlokacyny. Rozwój przestrzenny*, Kraków, (Pras.Zakl-.Graf.), 1975, 403 pp.
Reddaway, W. F. (et al.), eds. *The Cambridge History of Poland from its Origins to Sobieski (to 1696)*, (Octagon Books), New York, 1978, 579 pp.
Rutkowski, J. *Historia gospodarcza Polski*, Warszawa, 1953, 193 pp.
Rybarski, R. *Handel i polityka handlowa Polski w XVI stuleciu*, Warszawa, 1958, Vols. I–II.
Samsonowicz, H. 'Przemiany strefy bałtyckiej w XIII–XV wieku', *Roczniki Dziejów Społecznych i Gospodarczych*, Vol. XXXVII, Warszawa, 1976, p. 47–61.
'Przemiany osi drożnych w Polsce późnego średniowiecza', *Przegląd Historyczny*, Vol. LXIV(4), Warszawa, 1973, pp. 697–716.
Stachowska, K. 'Prawo składu w Polsce do 1565 r.', *Sprawozdania P.A.U.*, No. 9, Warszawa, 1950, p. 586.
Sujkowski, A. *Geografja ziem dawniej Polski*, Warszawa-Lwów-Lublin-Łódź-Wilno, (Wyd.M.Arcta), 1921, 195 pp.
Szwagrzyk, J. A. *Pieniądz na ziemiach Polskich X–XX w.*, Wrocław-Warszawa-Kraków-Gdańsk, 1973, 360 pp.
Taszycka, 'Włoskie tkaniny jedwabne w Krakowie w drugiej połowie XVII wieku', *Kwartalnik Historii Kultury Materialnej*, Vol. XXV(2), Warszawa, 1978, p. 133–153.
Tobiasz, M. *Pierwsze wieki Krakowa*, (Woj.Oś.Inf.Turyst.) Kraków, 1974, 151 pp.
Tomaszewski, E. *Ceny w Krakowie w latach 1601–1795*, Lwów, 1934, 101 pp.
Topolski, J. *Gospodarka polska a europejska w XVI–XVIII wieku*, (Wyd. Pozn.), Poznań, 1977, 287 pp.

Unger, R. W. 'Dutch Herring, Technology and International Trade in the Seventeenth Century', *The Journal of Economic History*, Vol. XL(2), Atlanta Georgia, 1980, pp. 253–80.

Verlinden, C. 'Brabantsche en Vlaamsche laken te Krakau op het einde der XIVe eeuw', *Mededelingen van de Koniklijke Vlaamse Academie (Klasse der Letteren)*, Vol. 17(2), Brussels, 1943, p. 16–33.

Vetulani, A. and L. Wyrostek, 'Bibliografia prac Prof. Dr. S. Kutrzeby', *Studia Historyczne ku czci Stanisława Kutrzeby*, Kraków, 1938, Vol. I, p. ix–xxxiii.

Vlachovič, J. *Slovenska Měd v 16 a 17 storoči*, Bratislava, 1964, 345 pp.

Wasowiczówna, T. 'Research on the Medieval Road System in Poland', *Archaeologia Polonia*, Vol. II, Warszawa, 1959, p. 125–140.

Wawrzyńczyk, A. *Studia z dziejów handlu Polski z Wielkim Księstwem Litewskim i Rosją w XVI w.*, Warszawa, 1956, 231 pp.

Wojtowicz, B. *Katalog dawnych map Wielkoskalowych Krakowa XVI–XIX wieku*, (P.W.N.). Warszawa-Kraków, 1981, 286 pp.

Wyrobisz, A. 'Functional Types of Polish Towns in the XVIth–XVIIth Centuries', *Journal of European Economic History*, Vol. 12(1), Rome, 1983, pp. 69–103.

Wyrozumska, B. *Drogi w ziemi krakowskiej do końca XVI wieku*, (P.A.N.), Prace Komisji Nauk Historycznych, No. 41, Kraków, 1977, 256 pp.

Wyrozumski, J. 'Handel Krakowa ze wschodem w średniowieczu', *Rocznik Krakowski*, Vol. 50, Kraków, 1980, pp. 57–64.

Wyrozumski, J., *Dzieje Krakowa (Kraków do schyłku wieków Średnich)* Vol. 1 (Wydawnictwo Literackie) Kraków 1992.

Zaki, A. *Początki Krakowa*, (Wyd. Literackie Kraków), Kraków, 1965, 153 pp.

Zientara, B. *Dzieje małopolskiego hutnictwa żelaznego XIV–XVII wieku*, Warszawa, 1954, 304 pp.

Zimányi, V. (ed.), *La Pologne et la Hongrie aux XVIe–XVIIIe siècles*, (Textes du Colloque Polono-Hongrois de Budapest), Budapest, (Akad. Kiadó), 1981, 148 pp.

Zins, H. *England and the Baltic in the Elizabethan Era*, (Manchester University Press), Manchester, 1972, 347 pp.

Index

Cambridge Studies in Historical Geography

---

1 Period and place: research methods in historical geography. *Edited by* A. R. H. BAKER *and* M. BILLINGE
2 The historical geography of Scotland since 1707: geographical aspects of modernisation. DAVID TURNOCK
3 Historical understanding in geography: an idealist approach. LEONARD GUELKE
4 English industrial cities of the nineteenth century: a social geography. R. J. DENNIS*
5 Explorations in historical geography: interpretative essays. *Edited by* A. R. H. BAKER *and* DEREK GREGORY
6 The tithe surveys of England and Wales. R. J. P. KAIN *and* H. C. PRINCE
7 Human territoriality: its theory and history. ROBERT DAVID SACK
8 The West Indies: patterns of development, culture and environmental change since 1492. DAVID WATTS*
9 The iconography of landscape: essays in the symbolic representation, design and use of past environments. *Edited by* DENIS COSGROVE *and* STEPHEN DANIELS*
10 Urban historical geography: recent progress in Britain and Germany. *Edited by* DIETRICH DENECKE *and* GARETH SHAW
11 An historical geography of modern Australia: the restive fringe. J. M. POWELL*
12 The sugar-cane industry: an historical geography from its origins to 1914. J. H. GALLOWAY
13 Poverty, ethnicity and the American city, 1840–1925: changing conceptions of the slum and ghetto. DAVID WARD*
14 Peasants, politicians and producers: the organisation of agriculture in France since 1918. M. C. CLEARY
15 The underdraining of farmland in England during the nineteenth century. A. D. M. PHILLIPS
16 Migration in Colonial Spanish America. *Edited by* DAVID ROBINSON
17 Urbanising Britain: essays on class and community in the nineteenth century. *Edited by* GERRY KEARNS *and* CHARLES W. J. WITHERS
18 Ideology and landscape in historical perspective: essays on the meanings of some places in the past. *Edited by* ALAN R. H. BAKER *and* GIDEON BIGER

19 Power and Pauperism: the workhouse system, 1834–1884. FELIX DRIVER
20 Trade and urban development in Poland: an economic geography of Cracow from its origins to 1795. F. W. CARTER
21 An historical geography of France. XAVIER DE PLANHOL

*Titles marked with an asterisk * are available in paperback*